ENERGY

Sources, Utilization, Legislation, Sustainability, Illinois as Model State

ENERGY

Sources, Utilization, Legislation, Sustainability, Illinois as Model State

G Ali Mansoori
University of Illinois at Chicago, USA

Nader Enayati

L Barnie Agyarko

World Scientific

NEW JERSEY • LONDON • SINGAPORE • BEIJING • SHANGHAI • HONG KONG • TAIPEI • CHENNAI • TOKYO

Published by

World Scientific Publishing Co. Pte. Ltd.
5 Toh Tuck Link, Singapore 596224
USA office: 27 Warren Street, Suite 401-402, Hackensack, NJ 07601
UK office: 57 Shelton Street, Covent Garden, London WC2H 9HE

British Library Cataloguing-in-Publication Data
A catalogue record for this book is available from the British Library.

ENERGY
Sources, Utilization, Legislation, Sustainability, Illinois as Model State

ISBN 978-981-4704-00-7

Typeset by Stallion Press
Email: enquiries@stallionpress.com

Printed in Singapore

Preface

The focus of this book is energy sources, utilizations, legislations and sustainability as it relates to a state, a province, a country, or a community within a state. We present various kinds of energy sources, ways to convert energy for end use, better use of energy towards conservation and energy and environmental sustainability. As a very proper model-state we chose the State of Illinois which has the largest overall fossil energy reserves, including the largest strippable bituminous coal reserves; the largest user of nuclear energy in the USA and investing in all kinds of renewable energies including wind energy, solar energy, biofuels, geothermal energy and various energy storage options.

This book consists of 11 chapters and an extensive glossary. Topics of the chapters are:

Chapter 1 Introduction
Chapter 2 Coal
Chapter 3 Petroleum
Chapter 4 Natural Gas
Chapter 5 CO_2: Emissions, Capture, Sequestration and Utilization
Chapter 6 Nuclear Energy
Chapter 7 Biofuels
Chapter 8 Wind Energy
Chapter 9 Solar Energy

In writing this book we used our personal funds and our own free time. We neither solicited nor received financial support or incentives from any individual or organization in writing this book. Our motivation to do this project was merely to educate the public (including students, energy engineers and planners, as well as state-wide policy makers) about all aspects of energy and to pay back and serve the State of Illinois, where we have lived most of our adult lives and have enjoyed the opportunities we have been given.

Many colleagues, friends, family members, students and energy experts have encouraged us in various ways to write this book and assisted us in reading the chapters, collaborating with us on related research, or helped us understand a topic from their lectures and writings, which they provided to us. We like to specifically thank the following individuals (in alphabetical order): P. Aich, M. Amin, E. Araújo, P. Araújo, A. Aslan, R. Bagherian, A. Bulvan, F. Civan, G. Crabtree, L.A. Curry, N. DeRose, H. Ebneyoussef, A. Esche, M. Fathizadeh, M. Heidari, H. Kendy, A.F. Koster van Gross, L. Masu, M.M. Moyeri, S. Pandya, A. Saber, V. Sadighian, M. Sarshar, N. Tadayyon, T. Wolf, and F. Zouras. The reader should bear in mind that we have made every effort to find and include proper references to study and report in this book. The authors apologize to those whose publications and energy-related works are not included.

We are grateful to the staff of World Scientific Publishing and specially Ms Chelsea Chin, for their patience and support in the preparation of this book.

About the Authors

G. Ali Mansoori

Dr. Mansoori is a professor of chemical engineering, bioengineering and physics at the University of Illinois at Chicago. He holds degrees of BSc, MSc and PhD and postdoctoral training in chemical engineering. He has been teaching various university level courses on energy and related topics which include (EnrE 100: Energy: Sources, Conversion, Conservation; ChE 201 & EnrE 201: Engineering Thermodynamics; Hon 201: An Introduction to Fossil Fuels; BioE 205: Bioengineering Thermodynamics; EnrE 250: An Introduction to Environmental Pollution Control; ChE 301: Chemical Engineering Thermodynamics; EnrE 495: Energy Conversion; EnrE 501: Advanced Thermodynamics, and other courses in the areas of fluid mechanics, heat transfer, mass transfer, engineering design, irreversible thermodynamics, mixtures, atomic and molecular nanotechnology, nanobiotechnology, phase transitions, statistical mechanics, transport phenomena, and applied mathematics). In addition Dr. Mansoori has delivered numerous short courses on matters related to energy and nanotechnology at the sites of many companies, professional societies, government laboratories and research centers, NGOs, etc.

Professor Mansoori has been active in research in the areas of renewable and non-renewable energies, thermodynamics, nanotechnology, statistical mechanics and molecular-based study of flow assurance in petroleum and natural gas industries. He has over 550 publications which includes authorship, co-authorship and editorship of ten books, including the present one. He has been a consultant to numerous energy-related companies and organizations.

Dr. Mansoori is the founder of the series of conferences on International Non-Renewable Energy Source Congress (INRESC), co-founder of Fluids and Thermal Energy Conversion (FTEC) Conferences, founder of Annual Midwest Thermodynamics and Statistical Mechanics Symposium Series, organizer of over 185 symposium series, workshops and short courses for various professional societies, industrial organizations and academic institutions on energy and related topics. He has given over 160 invited seminars at various universities and research centers on energy and related topics.

Dr. Mansoori has served as the editor-in-chief of Energy Sources Journal for six years, co-editor of Journal of Petroleum Science and Engineering for eight years, honorary editor-in-chief of International Journal of Nanoscience and Nanotechnology for ten years, as the series editor of books related to energy, and a member of editorial boards of numerous energy and nanotechnology related subjects, the list of which are too many to include here.

Nader Enayati

Dr. Enayati, is a consultant with chemical & energy industries. He holds degrees of BSc, MSc, and PhD in chemical engineering and has worked in energy conversion areas, specifically in the development of first and second generation biofuels through thermochemical and biochemical pathways. Prior to his work as a consultant, he was in charge of biofuels development programs at American Science and Technology, collaborating with University of Wisconsin, Superior; University of Wisconsin, Stevens Point and South Dakota State University. He was one of the organizers of alternative energy technology symposiums at the Chicago State University and University of Wisconsin, Superior. He has over nine years of experience in process research and development in chemical/biochemical engineering. Nader also served as an editorial assistant of Energy Sources Journal for four years.

Barnie L. Agyarko

L. Barnie Agyarko, is a chemical engineer and an instructor at Kennedy-King College, in Chicago, Illinois. He holds degrees of BSc and MSc, both in chemical engineering. In the past six years, he has been collaborating in engineering research at University of Illinois at Chicago, working on energy related projects. For four years, he worked for Gas Technology Institute in Des Plaines, Illinois, on natural gas processing technologies. Prior to his Master's degree, he spent one year at the Argonne National Laboratory, while developing novel materials for high energy density batteries for specialty and commercial applications. He has publications on both fuels processing and renewable energy utilization.

Contents

1

Energy: Sources, Conversion, Conservation and Sustainability

"*It is clear that there is some difference between ends: some ends are energeia* [*energy*], *while others are products which are additional to the energeia.*" [The first description of the concept of energy] **Aristotle**, 384BC–322BC.

"*Energie is the operation, efflux or activity of any being: as the light of the Sunne is the energie of the Sunne, and every phantasm of the soul is the energie of the soul.*" [The first recorded definition of the term energy in English] **Henry More** FRS (12 October 1614–1 September 1687) In Platonica: A Platonicall Song of the Soul (1642). Henry More was an English philosopher of the Cambridge Platonist School.

"*As the saying goes, the Stone Age did not end because we ran out of stones; we transitioned to better solutions. The same opportunity lies before us with energy efficiency and clean energy.*" **Steven Chu** (Noble Laureate and former U.S. Secretary of Energy), in letter (1 Feb 2013) to Energy Department employees announcing his decision not to serve a second term.

1.1. Introduction

In this book, we present various energy sources, conversions technologies and conservation possibilities. In every case, we have presented various options available for a country, for a state, or for a community to achieve its goal of energy sufficiency, clean environment, and as a result, sustainability. Variety of schemes related to each energy source and its related conversion technologies are presented and sustainability of renewable energy sources is discussed. All the possible energy sources including coal, natural gas, petroleum, nuclear, solar, wind, biofuels, and geothermal energy are presented in this book, as well as energy storage options. We have also presented various ways of dealing with carbon dioxide, which is produced from fossil fuels combustion, including its collection, transportation, storage and sequestration. The energy storage systems presented in this book will facilitate reliable and full integration of renewable power to the grid.

Energy is simply defined as "the ability to do work." Energy is a discrete measurable quantity which has units as defined by various means and at various scales.

Joule [J] is the basic energy unit of the metric system as well as the International System of Units (SI). One Joule is equal to the work done by a force of one Newton [N], when its point of application moves one meter in the direction of action of the force. One Newton is the force [F] that accelerates an object with a mass of one kilogram [kg] by one meter [m] per second [s] in each second ($1m/s^2$). Newton is the force exerted by Earth's gravity at sea level on one kilogram of mass.

$$1J = 1N.m = \frac{1kg.m^2}{s^2} \tag{1.1}$$

Watt is the basic unit of power in the metric system. It is the power produced from one Joule of work done in one second ($1\ J = 1\ W.s$).

The smallest unit of energy used in science is the electron volt [eV], which is a unit of energy equal to $1.60217657 \times 10^{-19}$ J.

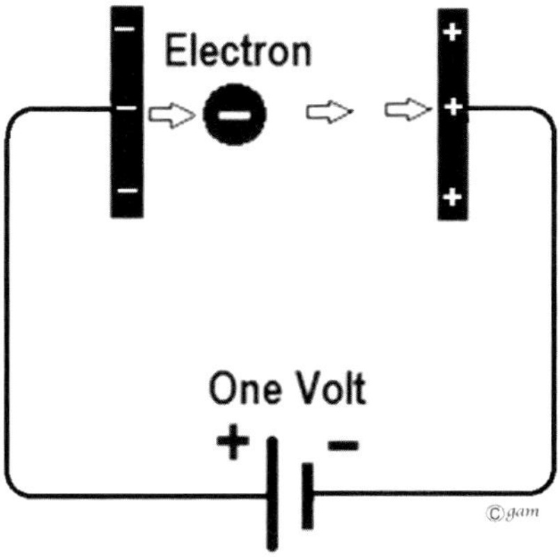

Fig. 1.1. Illustration of the electron volt (eV) definition.

One electron volt is the amount of energy gained (or lost) by the charge of a single electron moved across an electric potential difference of one volt.

An *erg* (short for ergon, a Greek word meaning "work/task") is the next smallest energy unit used in science after electron volt. An *erg* is the amount of work done by a force of one dyne exerted for a distance of one centimeter. In the CGS (centimeter-gram-second) base units, it is equal to one gram centimeter-squared per second-squared [$g \cdot cm^2/s^2$]. One *erg* is equal to 10^{-7} Joule.

In industry the more common units of energy used are the calorie (cal), British thermal unit (Btu), Kilowatt-hour (kWh) and Horsepower-hour (hph). Calorie is defined as the amount of heat required to raise the temperature of one gram of water by one degree Celsius at a pressure of one standard atmosphere. Similarly, Btu is defined as the amount of heat required at a pressure of one standard atmosphere, to raise the temperature of one pound of water by one degree Fahrenheit. In Table 1.1, we report all the energy units and their conversions in Joule.

There are vast and readily available conversion factors between various SI and British unit systems of length, mass, weight, force, energy, power, etc. in literature and in electronic media and we avoid presenting them here.

Table 1.1. Energy units and their conversions to Joule.

Energy Unit	Conversion to Joule
Electron Volt	$1.60217657 \times 10^{-19}$
Erg	10^{-7}
Joule	1
Watt. Second	1
Calorie	4.184
Btu	1, 055.05585
Kilowatt-hour (kWh)	3,600,000.
Horsepower-hour (hph)	2,684,519.536886

An entity or system that contains energy may have its energy in various forms which are categorized into three different types. They are potential energy, kinetic energy and intrinsic energy.

The potential energy ($P.E.$) of a system due to its elevated position from an external frame of reference is referred to as the gravitational potential energy. For example, water at the top of a water fall has gravitational potential energy which is released upon falling down. Mathematically, the gravitational potential energy of a system of mass m is shown by the following equation,

$$P.E. = m.g.\Delta Z \tag{1.2}$$

In this equation, m is the mass of the system, g is the gravitational field strength (acceleration of gravity), and ΔZ is the elevation/height of the system with respect to the frame of reference.

Kinetic energy ($K.E.$) of a system with mass m is the energy gained by a system with a velocity, $\dot{x} = dx/dt$, with respect to an external frame of reference at a distance x. The equation for kinetic energy is

$$K.E. = (1/2).m.\dot{x}^2 \tag{1.3}$$

Intrinsic energy ($I.E.$) of a system with mass m is the energy stored inside a system in various forms, mostly due to the nature and structure of the system. For example, a fossil fuel like coal has its intrinsic energy in the form of combustible materials like hydrocarbons, while uranium has its intrinsic energy in the form of its fissionability. The intrinsic energy that we have defined above is more general than the internal energy as defined in the science of thermodynamics, since we are including nuclear (fission and fusion) energies. Also, considering the fact that energies of a system are additive and that they can be converted into one another, we write the following equation for the total energy of a system:

$$E = P.E. + K.E. + I.E \tag{1.4}$$

Of course, in technological societies the most important application of various types of energies is their direct conversion to mechanical and thermal energies or to electric energy (electricity) for transmission. The remaining chapters of this book will address these issues in detail.

1.2. Energy Sources

Energy sources originate from our natural environment and they include the sun, wind, plants, geothermal, natural elements prone to nuclear processes like uranium, hydrogen, and fossil fuels such as coal, petroleum and natural gas. Some of these energy sources like the sun, wind, and geothermal are readily available to us, while others, such as plants, fossil fuels, and elements prone to nuclear processes, require chemical reactions, combustion, and nuclear reactions to produce them.

Energy sources are generally categorized as renewable and non-renewable energy sources. Renewable energy sources are those that are replenishable. On the timescale of human civilization, the major renewable energy sources that are readily available everywhere are solar energy, wind energy, plants/biofuel energy, and geothermal energy, which will be presented and discussed in detail in this book. There are other renewable energy sources which include hydropower

Fig. 1.2. Renewable energy resources. Left to right. (Top row): solar, wind, biomass/biofuel; (Bottom row): geothermal, hydro, ocean thermal gradient.

(dams, waves and tidal energy) and ocean thermal energy gradients. There are potentials for hydropower utilization in certain locations around the world and the related technologies are well developed and understood. However, utilization of ocean thermal energy gradients is limited to certain off-shore locations and requires advanced technology. In this book, we limit our presentations and discussions to the major renewable energy sources, i.e. solar energy, wind energy, plants/biofuel energy, and geothermal energy (Fig. 1.2).

Non-renewable (or non-replenishable) energy sources, which are known as fossil fuels, are known to deplete and release carbon dioxide and other pollutants to the atmosphere when used for energy production through combustion. Non-renewable energy sources are mostly made up of hydrocarbons and other compounds that are considered impurities from energy sources perspective. There exist seven well-known fossil fuel resources in nature (Mansoori 2009a, 2009b). In the order of their fluidity, they are natural gas, gas-condensate (also

Fig. 1.3. The normal conditions of the twelve naturally occurring fossil/hydrocarbon resources, in the order of fluidity from left to right. (Top row): natural gas, gas-condensate (NGL), light crude oil, intermediate crude oil, heavy oil, tar sand (Mansoori 2009a, 2009b); (Bottom row): oil shale, anthracite coal, bituminous coal, subbituminous coal, lignite and peat. Naturally occurring hydrocarbon energy sources vary in fluidity color, odor, and physicochemical properties, as shown in this figure.

known as NGL which stands for natural-gas liquid), light crude, intermediate crude, heavy oil, tar sand, and oil shale, as shown symbolically in Fig. 1.3. These are all naturally occurring complex mixtures, made up of hydrocarbons and other organic and inorganic compounds with varieties of molecular structures and sizes.

The technologies for utilization of coal, petroleum, natural gas, and other non-renewable energy sources were quite well developed in the course of the past two centuries. In this book, we devote Chapters 2–5 to brief histories and recent developments of such fossil fuels, from the point of view of environmental pollution control and governmental legislations and policy trends towards better utilization and/or reduction in their consumption.

Nuclear power technology can be quite diverse, depending on the type of nuclear reaction (fission or fusion) and nuclear fuel (uranium, thorium, plutonium, hydrogen, etc.) used. Conventional nuclear power stations (Fig. 1.4) use uranium as the element prone to nuclear fission reaction in order to produce thermal energy and its subsequent conversion to electricity. Uranium is a heavy metal which

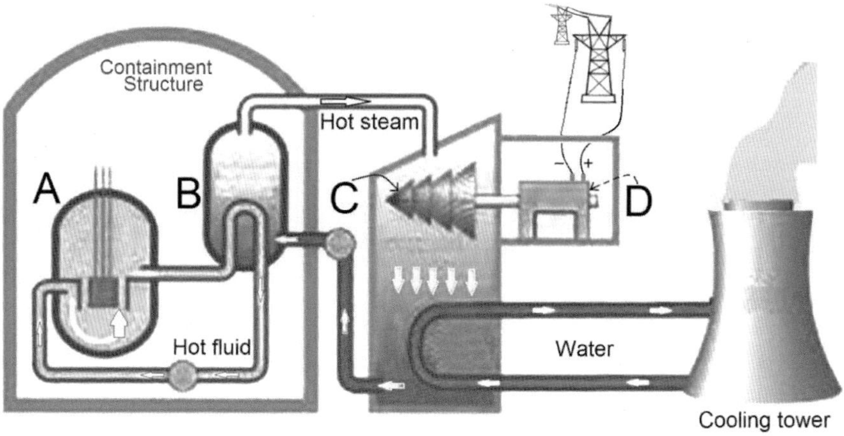

Fig. 1.4. General schematic of the conventional nuclear power system: A: main reactor, B: steam generator, C: steam turbine, and D: electricity generator.

is mined in certain parts of the world where it is available (Deffeyes and MacGregor, 1980). But uranium resource can be considered as non-renewable because the economically minable uranium ore with high enough concentration of uranium metal is finite. Therefore it is possible that this type of nuclear fuel will eventually be exhausted, if it is consumed at its current rate.

As we discuss in Chapter 6, nuclear fission "breeder" reactor technology can be considered as a renewable energy source, if it is well-designed and implemented, and provided it satisfies all the scientific requirements for safety and reliability. However, futuristic nuclear fusion technology which is in research stage, is a renewable energy source. This is because the fuel source for nuclear fusion reactor is hydrogen, an infinitely abundant natural resource that is produced through the dissociation of water (H_2O).

1.3. Energy Conversions and Efficiencies

The basic idea behind energy conversion is optimal transformation of one form of energy to another for our everyday use. In doing so, energy needs to be extracted from its source and transformed to various intermediate forms of energy, and subsequently, be converted into useful and controllable energies like electricity, mechanical energy and heat. Over the past centuries, a wide array of methods and devices has been developed for this purpose. The structures of early versions of these energy converters were quite simple, but the modern methods are complicated. The early hydro-powers, for example, transformed the potential energy of water into mechanical energy for grinding grain. Similarly, the early windmills utilized the kinetic energy of wind and converted it into mechanical energy for grinding of grains and pumping of water. However, modern energy-conversion systems for utilization of natural energy sources are rather complex. In order to achieve higher efficiencies, these systems require multiple steps or various processes, during which a natural energy source undergoes a whole series of transformations, prior to being used by humans. For example, in a conventional nuclear reactor the nuclear energy is first converted to thermal energy through fission reaction. The resulting high temperature heat generated from the fission reaction is extracted to boil water to produce steam with high thermal energy. The resulting steam is used to rotate the blades of a turbine which converts the thermal energy into mechanical energy. The rotor of the turbine is coupled to a shaft which drives an electric generator to produce electricity. The electric energy is then transferred through electric cables to residential, commercial, and industrial areas for conversion into all kinds of useful mechanical energies, or for lighting and space heating. Sometimes, the excess electric energy is converted into various forms of energy for storage, using pumped hydroelectric storage system, compressed air energy storage system or other storage systems using phase change materials.

In converting one kind of energy to another, there is always an efficiency conversion factor which needs to be considered. Majority of the energy conversion methods involve transformation of thermal energy into mechanical energy and then into electrical energy. The efficiency of such systems is, however, subject to fundamental

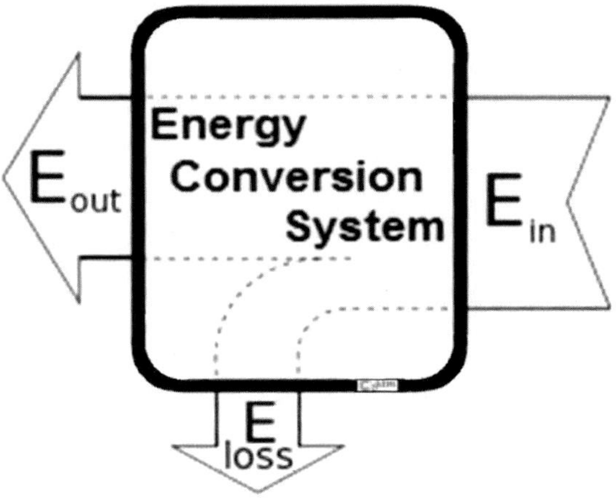

Fig. 1.5. General scheme of energy conversion and the fact that when we convert one kind of energy to another the efficiency of conversion is always less than 100%.

limitations set forth by thermodynamics principles, as dictated by the laws of thermodynamics and thermodynamic property relations. It can be demonstrated that none of the energy conversion methods is 100% efficient (Fig. 1.5).

The highest energy conversion efficiency which is close to 100% is the conversion of any kind of energy to thermal energy. But then most of the energy sources being utilized are in the form of thermal energy, and we need to convert those to more useful kinds of energy, which are either mechanical energy or electricity.

The origin of science of thermodynamics, which began about two centuries ago, was to understand and improve the efficiency of thermal energy conversion to mechanical energy. Originally, when the steam engine was developed, the highest conversion efficiency of thermal energy into mechanical energy was less than 5% (Goldemberg, 2012). Presently, with all the technological improvements made in steam power cycles, the highest conversion efficiency of thermal energy to mechanical energy is less than 45% (Ajith Krishnan and Jinshah, 2013). This means that more than 55% of the thermal energy sources which are available to us are wasted through the release of

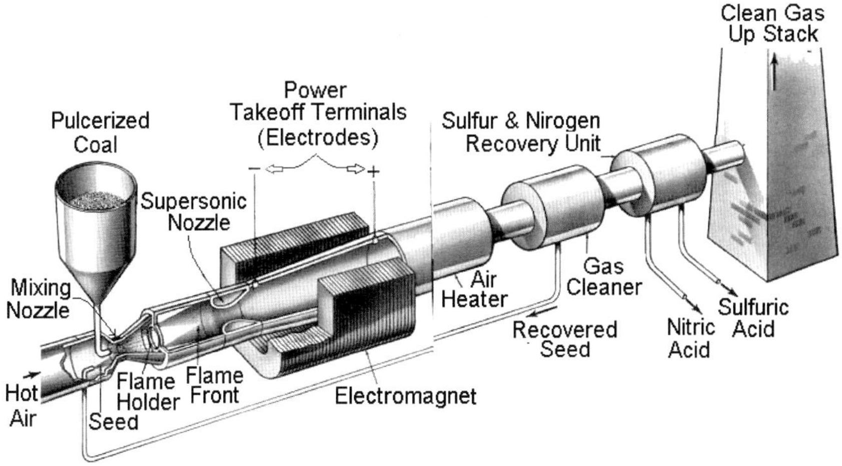

Fig. 1.6. Magnetohydrodynamics electric power plant principles [Modified from Freese, 1969].

heat to the environment. It should be mentioned that the efficiency of internal combustion engines used in automobiles for conversion of the intrinsic chemical energy of the fuel to mechanical energy is 20% or less (US-SCENR, 2007). The most advanced method of utilizing thermal energy to electric energy, which is expected to have much higher efficiency, is the magnetohydrodynamics (MHD). MHD transforms thermal energy and kinetic energy of a fluid flow directly into electricity (see Fig. 1.6). MHD generators operate at very high temperatures without any moving mechanical parts.

A U.S. DOE-funded feasibility study (1000 MWe Advanced Coal-Fired MHD/Steam Binary Cycle Power Plant Conceptual Design) of where the MHD technology could be used was published in June 1989 and showed that a large coal-fired MHD combined cycle plant could attain a HHV energy efficiency approaching 60%, well in excess of other coal-fueled technologies, so the potential for low operating costs exists (Freese, 1969).

Understanding of thermodynamics is necessary as it applies to conversion of heat to mechanical energy and it gives us information about the limiting efficiencies of thermal cycles including the

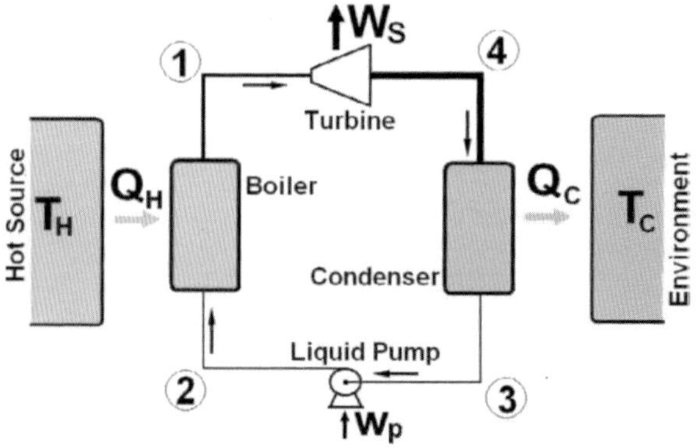

Fig. 1.7. A basic Rankine thermal power cycle (Modified from Mansoori, 2013).

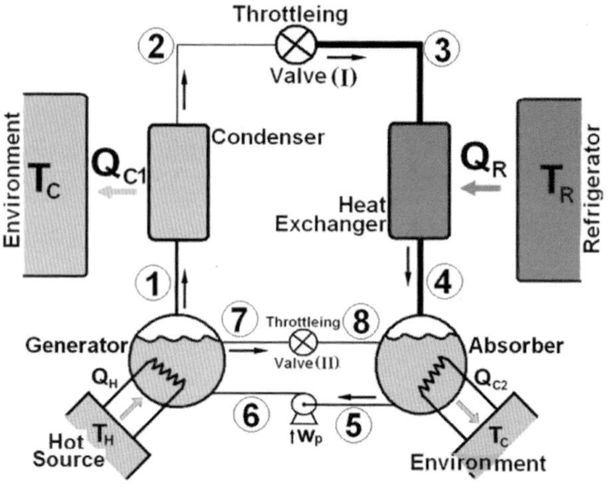

Fig. 1.8. Absorption cooling cycle (Modified from Mansoori, 2013).

Rankine cycle. Figures 1.7 and 1.8 show the conventional methods for producing power and refrigeration/air conditioning systems from thermal energy. In Fig. 1.7, we report the basic Rankine thermal power cycle for production of mechanical energy, a hot source such as burning a fossil fuel, a nuclear reaction or heat collected from solar radiation.

In addition, a basic absorption cooling and heat pump cycle which uses heat to produce cooling (refrigeration and air conditioning) is illustrated in Fig. 1.8.

The general practice in estimating the efficiency of thermal energy conversion to mechanical energy (Fig. 1.7) is by using the Carnot efficiency equation:

$$\eta_{Carnot} = 1 - \frac{T_C}{T_H} \tag{1.5}$$

Also, it has become a general practice in calculating the coefficient of performance (COP) of absorption cooling cycles (Fig. 1.8) using the following Carnot-based equation:

$$COP_{Carnot} = \frac{T_R}{T_H} \cdot \left[\frac{T_H - T_C}{T_C - T_R} \right] \tag{1.6}$$

In the above two equations, T_H is the temperature of the hot source, T_C is the temperature of the environment and T_R is the temperature of the refrigeration (or air conditioning). According to these equations, both the efficiency and COP as reported by Carnot are independent of the nature of the cycles and the working fluids used in the cycles, which makes these expressions valid for ideal cycles working with ideal gases. However, due to the ideal nature of the Carnot assumptions, both of the above equations are the highest upper bound to efficiency and COP of cycles and are quite far from the real efficiency and COP values.

These marvelously simple and highly cited Carnot cycle and its related efficiency and COP relations were proposed at the time when the principles of thermodynamics were at their infancy (Mansoori, 2013). The genius Nicolas Léonard Sadi Carnot, who proposed his cycle in 1823, recognized the need to develop his theory independent of the knowledge about properties of working fluids, especially at the time when accurate thermodynamic property data for such fluids were non-existent. Presently, extensive amount of accurate data and theoretical knowledge about thermodynamic properties of materials are available. Through the application of the 2nd law of thermodynamics for irreversible processes, which results in the general inequality relation for the entropy production, new upper- and lower-bounds to the efficiency of thermal power cycles and COP of absorption cooling and heat pump cycles have been developed. The resulting upper- and lower-bounds are closer to the actual efficiency and COP of cycles. Such relations allow a more precise design of cycles and the choice of working fluids for cycles. The available upper and lower bounds to efficiency and COP which are much closer to their actual values than those of Carnot cycle values are shown below (Mansoori, 2013):

$$\frac{1}{T_H}.\frac{(h_1 h_4)}{(s_1 - s_2)} \leq \eta_{Actual} \leq \frac{(h_1 - h_4)}{(h_1 - h_2)} \leq 1 - \frac{T_C}{T_H}.\frac{(s_4 - s_3)}{(s_1 - s_2)} \leq 1 - \frac{T_C}{T_H} \tag{1.7}$$

$$LL \leq COP_{Actual} \leq \frac{h_4 - h_2}{(h_1 - h_7) + \left(\frac{1 - x_G}{x_A - x_G}\right)(h_7 - h_5)} \leq UL \leq \frac{T_R}{T_H}.\left[\frac{T_H - T_C}{T_C - T_R}\right] \tag{1.8}$$

In the above expressions, h_i stands for specific enthalpy, s_i stands for specific entropy ($i = 1, 2, 3, \ldots$) of the points specified on the two cycles as shown in Fig. 1.8, and T_i's are temperatures as defined in the cycles. X_A is the mass fraction of refrigerant in the liquid phase coming out of absorber and x_G is the mass fraction of the liquid phase coming out of the generator. With the use of the above two expressions, it is possible to perform calculations to predict efficiencies and COPs more accurately in the course of design and choice of working fluids for energy conversion methods.

Conversion efficiencies of other kinds of energies to one another are generally better than the conversion of thermal energy to mechanical energy. For example, as we discuss in Chapter 8 of this book, a

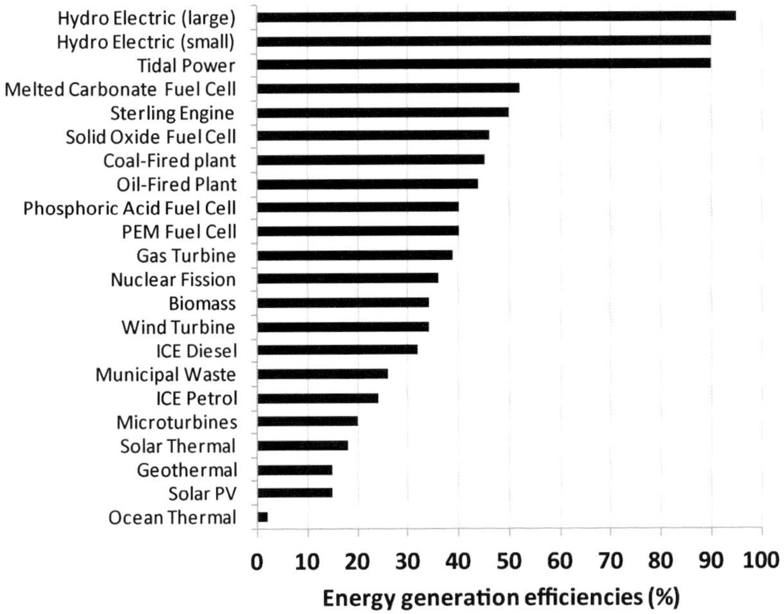

Fig. 1.9. The theoretical efficiency of converting various energy sources by a variety of methods into useful electrical energy (Lawson, 2005).

maximum of 59.3% of wind energy can be converted to mechanical energy. As another example, the conversion efficiency of mechanical energy to electric energy and *vice versa* can be as high as 90% (Chen *et al.*, 2009). This explains why the overall efficiency (from natural energy source to mechanical energy) of an electric car is better than the efficiency of a car with an internal combustion engine. Several literatures report and update the trends in efficiency improvements of various energy conversion devices. Figure 1.9 shows the theoretical efficiency of converting the various energy sources by a variety of methods into useful electrical energy (Lawson, 2005).

Presently, considerable attention has been focused on a number of direct energy-conversion technologies, notably solar cells and fuel cells that bypass the intermediate steps of conversion to heat for eventual mechanical and electrical power generation.

1.4. Energy Conservation

Energy conservation is the development of strategies and procedures to reduce energy consumption. Strategies such as driving fewer automobiles, turning off lights in areas when lighting is not essential, reducing the temperature of living and work spaces in winters and reducing the use of hot water for everyday living are examples of energy conservation. Local, state, and national laws, in the form of energy tax laws, carbon taxes, or economic incentives, have been developed in the U.S. and other countries to encourage the public and/or private sector to conserve and reduce energy use (Jaffe and Stavins, 1994).

In this book, we provide variety of examples on this subject of how the highly progressive State of Illinois, as a model state, has been developing ways of energy conservation/reduction in fossil fuels and electric energy. One interesting method of energy conservation which is implemented in Illinois is through a demand-response program. Demand Response is defined as "changes in electric usage by end-use customers from their normal consumption patterns in response to changes in the price of electricity over time, or to incentive payments designed to induce lower electricity use at times of high wholesale market prices or when system reliability is jeopardized" (US-DOE, 2006). An alternative approach to conserve energy and reduce emissions from coal-fired power plants is through increased energy efficiency and conservation programs. This means judicious use of energy for equipment, buildings, or products without altering the quality of their services. Among various options being considered for emission control in Illinois and several other states in the USA is energy efficiency resource standard (EERS), which sets overall energy savings targets for utility-sector efficiency programs (Prindle, 2009). Additionally, Illinois policy makers can implement the newly unveiled US Environmental Protection Agency (US-EPA) measures to force some of the older coal-fired power plants to retire and curb the emissions (SourceWatch, 2013). A new bill (HB 2607/SB 1485) introduced in the Illinois Senate in February 2015 calls to achieve a 20% reduction in energy demand through energy efficiency standards by 2025 and raising the energy generated by renewable sources to 35% by 2030, up from the current goal of 25% by 2025 (Sierra Club, 2015).

1.5. Energy Sustainability and Green Energy

The concept of sustainability is quite well discussed in recent years in the public media and in technology development circles.

The Brundtland Report (Brundtland, 1987) of the World Commission on Environment and Development of Organization for Economic Co-operation and Development of the United Nations defined sustainability as "*the ability of humanity to make development sustainable, by ensuring that it meets the needs of the present without compromising the ability of future generations to meet their own energy needs*". In this definition, sustainability includes availability of the energy resource relative to their rate of consumption, as well as economics, environmental and social criteria. The economics, ecological considerations and social acceptance are considered the three interconnected pillars of sustainability in any aspect of human activity (Gibson, 2006; Hansmann *et al.*, 2012) including energy use.

The methods and approaches that contribute to sustainability in energy are purely those that direct us to utilize renewable energy sources. There is no doubt that proper utilization of renewable energies, through improved efficiency of conversion technologies, can contribute greatly to energy and environmental sustainability.

Presently, our societies are using mostly non-renewable energy sources and nuclear energy. To reduce emissions of carbon dioxide and other environmental pollutants from the use of fossil fuels, certain trends have been developed towards the use of less polluting non-renewable energy sources such as natural gas and biofuels derived from agricultural food products and food wastes that are renewable.

However, sustainable energy is distinct from low-carbon and less polluting energies, which do not add appreciable amount of CO_2 and other pollutants to the environment.

In order to minimize the release of CO_2 and other pollutants to the environment and stop accumulation of hazardous radioactive waste, we need to use renewable energy sources which include agricultural waste-based biofuels, geothermal energy, hydro power, ocean thermal energy gradients, solar energy, wave and tidal power, wind energy, and hopefully fusion-based nuclear energy. Fusion-based nuclear energy is the way the sun produces its energy, and is a sustainable energy if the technology becomes practically available. We also need to advance our energy storage technologies for battery-based electrical energy storage, thermal energy storage, hydroelectric energy storage, compressed gas energy storage, etc., as well as energy transmission technologies for electrical power transmission (Panek *et al.*, 1998), in order to take better advantage of far-distance energy sources.

To achieve sustainability in energy, we need to use renewable energies and optimize the way a renewable energy source is utilized and consumed. This means that we need to increase energy conversion efficiencies to the maximum allowable by natural laws, while reducing energy waste (through frictions, heat losses, etc.). From the standpoint of energy sustainability, energy optimization from the consumption/demand side of energy utilization is equally important as energy optimization from the generating sources. Therefore, application of renewable energy, improvement of energy conversion efficiencies and energy conservation are certainly stepping stones in a path toward energy sustainability.

1.5.1. *The role of advanced technologies to achieve sustainability*

Advanced technologies include nuclear fusion, biotechnology, information technology and nanotechnology. These technologies which are being developed and advanced to the level of practical applications can help achieve sustainability in energy and environment.

In Chapter 6 of this book, we present nuclear technology and discuss future generation technologies which include nuclear fusion.

Chapter 7 of this book is on biofuels, and includes advantages and prospects of the applications of biotechnology in the production of 2nd and 3rd generation biofuels, which will help achieve sustainability in energy and environment.

The role of information technology for achieving sustainability in energy and environment includes the idea of using better and faster communications and exchange of data and information. Consequently, individuals and decision makers in our societies will be better informed and can communicate more effectively to make people become aware and educated about the different ways of achieving sustainability in energy and environment (McCollum *et al.*, 2011). Information technology allows renewable energy to be integrated into smart micro-grid systems to provide cheaper, environmentally friendly and a more reliable power supply to consumers.

1.5.1.1. *Role of nanotechnology in energy industry*

Advances in nanotechnology may help a society in energy savings, through improved power lines for energy distribution, improved energy storage devices and better catalysts for fossil and biofuels production.

There exist a great deal of recent publications and speculations about the role of nanotechnology in energy and environmental sustainability, see for example: (Mansoori *et al.*, 2008; Smith and Granqvist, 2010; Markovic *et al.*, 2012; Brinker and Ginger, 2011).

We are all very optimistic about the future of nanotechnology and the prospects it could bring to positively impact every discipline, including energy, environment, medicine etc., and in general, our quality of life (Mansoori, 2005a, 2005b; Mansoori *et al.*, 2007a,

2007b). Nanotechnology can have pronounced effect on energy and environmental sustainability, as it can contribute to smoother and faster shift towards higher energy conversion efficiency, better ways to conserve energy and appreciable advances in end-use energy storage systems, while minimizing the adverse effects of fossil energy use on the environment.

There is no doubt that the present energy technology has taken almost two centuries to be developed and that many organizations, companies, interest groups and individuals' livelihood depend on its *status quo*. As a result, there will be a major resistance to any changes that advanced technologies, including nanotechnology, may bring towards energy sustainability.

As regards the improvements in energy utilization, photovoltaic panels are expected to be built from nano-optimized cells (polymeric, quantum dot, thin film, multiple junctions) and antireflective coatings. Similarly, rotor blades of wind turbines may be constructed with nano-composite materials to make them lighter, stronger and resistant to corrosion and wear. Additionally, biomass energy yield may be optimized using nano-based precision farming, such as nanosensors, controlled release and storage of pesticides and nutrients.

Nanotechnology may also be used in the future for coal purification and improvements in petroleum refining and upgrading. In nuclear energy, nano-composites materials may be used in equipment for radiation shielding, and for futuristic fusion reactors as a long term option (Luther, 2008).

Besides construction of more efficient energy conversion systems, nanotechnology may lead to more efficient energy distribution. For example, by using nanofillers for electrical isolation systems and soft magnetic nano-materials for efficient current transformation, it may be possible to obtain high voltage transmissions lines. There is a great deal of power loss in existing power transmission wires which can be minimized using normal-temperature superconductive materials for wires, if they become available (Panek *et al.*, 1998). By using nanoscale interface design, it may be possible to produce normal-temperature superconductors. Also, research into superconducting materials based on optimized nano components may lead to efficient power transmission in the future (Elcock, 2007). Furthermore, communities may also benefit from smart-grid technology, which uses nanosensors for intelligent and flexible grid management that is capable of managing highly decentralized power feeds. Using nanotechnology, efficient heat transfer in industries and buildings based on "nano-optimized heat exchangers and conductors" may be possible in future.

In addition to energy distribution, nanotechnology may provide us with efficient energy storage systems (Luther, 2008). The stored energy can be of chemical, electrical, or thermal nature. In chemical energy storage, lithium ion battery is being optimized with nanostructured electrodes and flexible ceramic separator foils. Applications of such batteries can be found in mobile electronics and automobiles. Also, more efficient electrical energy storage (batteries) systems may be constructed using nanostructured capacitors with ultra-high power density and excellent charge and discharge rates (Luther, 2008). Nanomaterials can also be used for encapsulation of phase change materials for thermal energy storage and as adsorptive storage systems using nano-porous materials like zeolites. Materials like these can be used in air conditioning systems and for electricity conservation.

Besides better energy storage systems, nanotechnology may provide us with efficient energy usage systems, through applications of improved thermal insulation, more efficient air conditioning, light weight and robust construction materials, more efficient industrial processes and more efficient lighting. Currently, nanoporous foams and gels are used for thermal insulation, while electromagnetic windows, micro mirror arrays and IR-reflectors are used for air conditioning (Luther, 2008; Bell, 2009). Also construction of light weight buildings using nano composite materials, such as metal-matrix-composites, carbon nanotubes, nano-coated light metals, ultra performance concrete and polymer-composites are already in use and are being further improved.

In industry, energy intensive processes are being replaced with nanotechnology process innovations, such as molecular self-assembling processes that are less energy intensive (Elcock, 2007). Also, nanotechnology is playing an important role in the development of novel catalysts for commercial-scale cellulosic biofuels production. Additionally, nanotechnology can be applied to create the next generation light bulbs, such as organic light-emitting diodes for energy efficiency of buildings.

1.6. State of Illinois — Our Model State

In this book, we have tried to cover all aspects of energies of a state, a community, or a country from source to end use in a manner that will be educational to students of energy, while serving as a road map for energy planners and policy makers, energy-related engineers and scientists, citizens and community leaders, who are concerned about energy in general.

There are already numerous books on energy in various technical and scientific depths. However, the present book is designed to bring about a collective picture of technologies, policies and possible plans for energies that could help a society (a state, a country and any community) to learn more about how to deal with energy issues and achieve energy independence and sustainability.

State of Illinois (see Fig. 1.10) located in the central (Midwest) region of the U.S. is highly industrial, agricultural and truly a pioneer state in dealing with issues related to energy sources, conversion, conservation and sustainability.

Many of the energy technologies that are used around the world have their roots in Illinois. Presently, about 49% of the electric energy consumed in Illinois comes from nuclear energy, with most of the rest coming from coal combustion. Environmental concerns associated with mining and combustion of coal has prompted officials to enforce restrictive laws, encouraging use of less polluting energy sources. Illinois policy makers in all levels of government

Fig. 1.10. Map of continental USA in which the location of the State of Illinois and other Midwestern States (Iowa, Indiana, Kansas, Michigan, Missouri, Minnesota, Nebraska, North Dakota, Ohio, South Dakota and Wisconsin) are shown.

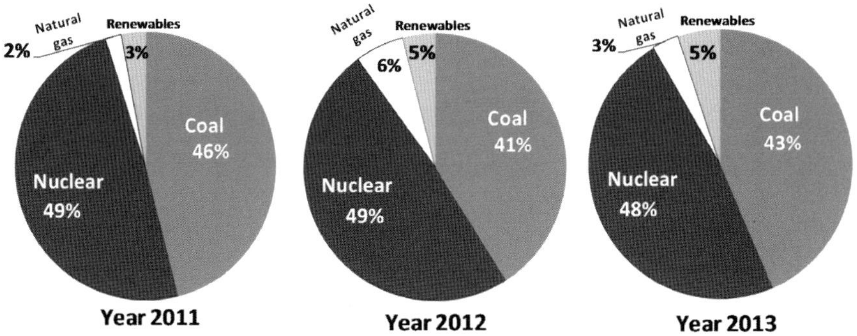

Fig. 1.11. Illinois electricity generation from various energy sources in 2011, 2012 and 2013 (US-EIA 2013a, US-EIA 2013b; US-EIA, 2014a).

and community organizations are quite concerned and proactive in dealing with energy, as well as the environmental matters related to energy. Figure 1.11 shows the electricity generation profile in Illinois from the various energy sources in 2011, 2012 and 2013.

Figure 1.11 indicates an increase in natural gas usage from 2% in 2011 to 3% in 2013, while the share of coal in generating electricity in Illinois reduced from 46% in 2011 to 43% in 2013. Also, according to a 2013 annual report issued by the Illinois Power Agency, over the past five years, the amount of electricity generation from coal-fired power plants represented a lower percentage of the total electricity generation in the state, while nuclear electricity generation has remained unchanged (IPA, 2013). This reduction in coal-fired power generation in Illinois is largely due to lower prices of natural gas, which the Energy Information Administration (EIA) has projected only a modest increase for the next 10 to 15 years (IPA, 2013). As this current boom of natural gas and petroleum production appears to provide some optimism in finding a solution to the U.S. energy problem, these economic gains can only be transient because neither energy sources is sustainable, especially with the world's population approaching the projected nine billion in the year 2050.

As a model state, we study and report various energy sources, conversion, conservation options, and technologies available for the State of Illinois to achieve its goal of energy sufficiency, clean environment and sustainability. While some of the data reported in the chapters are specific to Illinois, the options, technologies, implementation procedures, the guidelines, rules and regulations necessary to achieve sustainability in energy and environmental protection reported in this book are generally applicable to any state, province, or countries all around the world.

Many parts of the world, and specially the State of Illinois, is quite accustomed to using nuclear fission and fossil (non-renewable) energies to meet their energy needs. In recent years, Illinois, having the 5th largest population among the states in the U.S., ranked 5th in the U.S. in total energy consumption with about 10^{18} calories ($\sim 4 \times 10^{15}$ Btu) per year, behind Louisiana, Florida, California and Texas. In 2012, Illinois ranked 26th in the nation in total energy consumption per capita at 7.56×10^{10} calories (300 million Btu) and ranked 42nd in energy expenditure at $3,737 per capita (US-EIA, 2014b).

For over a century, coal has become the bedrock of Illinois energy source (see Chapter 2 of this book), feeding its electric power plants to generate electricity. Illinois has the largest overall, as well as the largest strippable bituminous coal reserves in the U.S.. However, the state cannot use much of its coal because of unfavorable geologic condition and surface development, coupled with the high sulfur content of Illinois coal which emissions wreak havoc on the environment (US-EIA 1994). This situation has caused a decline in the state's annual coal production, which was estimated at ~30 million metric tons per year in the 2000's, down from the ~60 million metric tons of coal per year during the 1970's (US-EIA, 2014c). But in 2012, Illinois coal production witnessed a dramatic increase to about 48 million metric tons, largely due to exports that reached a record of 13 million metric tons, a five-fold increase from 2.5 million metric tons in 2010 (Medine, 2013; US-EIA, 2014c). According to a 2013 report by the Illinois Department of Commerce and Economic Opportunity (IDCEO), in 2011 and 2012, Illinois coal was exported to at least 18 different countries around the world. These included Canada, Chile, Mexico and Dominican Republic in the Americas; Belgium, Denmark, Finland, France, Great Britain, Germany, Holland, Ireland, Portugal and Spain in Europe; and China, India, Pakistan and South Korea in Asia (Medine, 2013; Miller, 2014).

Despite the significant increase in Illinois coal exports in recent years, the state's current expenditure on low sulfur content coal imports, mostly from the Powder River Basin of Wyoming is more than \$1 billion per year, ranking the state 5th in the nation in net coal imports by weight, behind Iowa, Georgia, Missouri and Texas, respectively (Union of Concerned Scientists, 2014; US-EIA, 2014d). In 2012, Illinois imported 42.94 million tons of coal at the cost of \$1.45 billion, at the same time the state exported 20.63 million tons of coal worth \$1.33 billion, resulting in a net import of 22.31 million tons of coal worth \$120 million (Union of Concerned Scientists, 2014; Lydersen, 2014). To avoid net outflux of money from the state, Illinois is searching for clean coal technologies. This decision is purported to cut pollution and maximize the use of indigenous coal, while restoring the level of employment opportunities that used to be in the coal industry. Chapter 2 of this book will discuss the current clean coal technology projects in the state of Illinois that are designed to use Illinois coal.

Besides coal production, Illinois depends on nuclear energy to meet its energy needs as will be discussed in Chapter 6. As of 2014, the state has six nuclear power plants, representing about 12% of the nation's nuclear energy production, which makes it the largest nuclear energy producing state in the continental U.S. However expansion of nuclear plant fleet in the state has experienced stumbling blocks due to uncertainty in storage of nuclear spent fuels. Illinois has the largest spent nuclear fuel (a high risk radioactive waste) in the nation, amounting to approximately 8, 000+ tons. According to experts, the best current option for management and containment of spent nuclear fuel is storage in underground geologic formations. However, the use of this method has become very polemical, as every state in the U.S. tries to protect its territory from nuclear contamination. Alternative approach for reprocessing of nuclear spent fuel cannot be practiced in Illinois since this method has been banned in the U.S. since 1992 to avoid nuclear proliferation (Cerami, 2010).

Illinois is not a major producer, but consumes appreciable amount of petroleum and natural gas (see Chapters 3 and 4). As a result, the state imports crude oil from Canada, while natural gas is purchased from overseas through the U.S. Gulf Coast, the U.S. midcontinent regions, Western Canada, Colorado and Wyoming.

In 2012, Illinois consumed about 226 million barrels of crude oil at a cost of about $30 billion, representing 61.8% of the total amount of money on energy expenditure by the state. In the same year, Illinois consumption of natural gas totaled 26.56×10^9 standard cubic meters (~938 BSCF) at the cost of ~$6.2 billion, or 12.8% of the total money spent on energy by the state (US-EIA, 2014e).

Illinois is a major transportation hub for crude oil and natural gas distribution throughout North America, due to its central location and pipeline infrastructure. The 2010 energy industry economic output from the pipeline transportation sector was approximately

Fig. 1.12. Illinois Basin and its shares of its locations in Illinois, Indiana and Kentucky (Schaefer, 2013).

$662.6 million (Lewis and Bergeron, 2010). This revenue was based on services offered by the oil and gas transportation hubs, employment, as well as taxes paid by the pipeline companies to the state.

In recent years, there has been a shift in global energy supply, mainly due to new discoveries of oil and gas reserves in shale rock, particularly in the U.S. (Rosenthal, 2012). To reduce its oil and natural gas imports, Illinois legislators are debating on how to tap into the presumed vast resource of shale gas located in the Illinois Basin, shown in Fig. 1.12.

Such new technologies like hydraulic fracturing and horizontal drilling have opened up new natural gas and oil reserves that were otherwise technically challenging or economically prohibitive. Natural gas prices have kept decreasing in recent years and electricity generation from natural gas in U.S. would exceed that from coal in 2035 (IEA, 2013). With possible abundance of cheap natural gas in the state, Illinois could replace some or all of its coal-fired power plants with natural gas-fired power plants to reduce emission of harmful compounds like sulfur oxides, carbon oxides, etc. It is also predicted that there is 55% chance of more petroleum production and 45% chance of improved energy efficiency in both residential and transportation sectors (Rosenthal, 2012).

Additionally, Illinois is planning the infrastructure for carbon dioxide capture and sequestration from its new generation coal-fired power plants (see Chapter 5). Some of the carbon dioxide could probably be used for enhanced oil and coal-bed gas recovery in the Illinois Basin to boost the state's petroleum and natural gas production.

In recent years, Illinois has made investments in renewable energy production, such as biofuels (Chapter 7), wind energy (Chapter 8), and solar energy (Chapter 9) towards its goal of energy sufficiency, environmental cleanliness and energy and environmental sustainability.

In other states across the U.S., there are a few towns that use 100% renewable electricity. However, by the end of 2013 there were 91 towns in Illinois that purchased 100% renewable electricity, a number that exceeded anywhere in the country. The growth of the number of towns in Illinois using 100% electricity is depicted in Fig. 1.13.

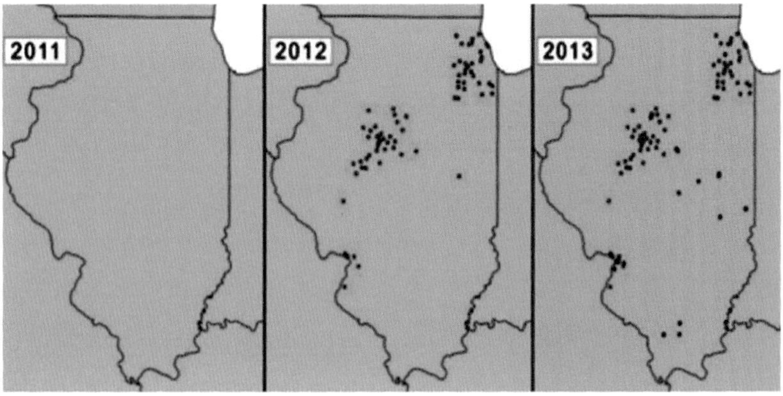

Fig. 1.13. Spread of towns with 100 percent renewable electricity in Illinois (WWF, 2014).

Some policies at the state level can indeed encourage use of renewable energy. For example, Illinois is one of the six states in the country that allows community choice aggregation, a system where residents can use their bulk purchasing to solicit bids from electricity providers.

In 2014, the state ranked 3rd place behind Nebraska and 2nd place behind Texas in ethanol and biodiesel production, respectively, in the nation (IER, 2014), and had over 150 wind energy utilization companies with a combined workforce of over 15,000 people. Chicago hosts at least 13 global and U.S. headquarters of major wind energy companies, more than any city in the U.S. (Craig *et al.*, 2011). The Illinois Power Agency Act mandated the Renewable (energy) Portfolio Standards (RPS). The RPS requires Investor-Owned Utility companies with more than 100,000 customers to supply 1.5% of their energy from wind energy as of 2008, reaching a goal of 18.75% (75% of all the renewables) by 2025. Highly variable weather in Illinois makes the geothermal and seasonal energy storage systems quite viable to reduce Illinois dependency on coal and other fossil fuels.

In essence, Illinois needs energy security which transcends the present needs, by maximizing production and efficient use of renewable energy resources which are indigenous and environmentally friendly. The reason is that long term energy security and management of the environment are important to the energy sector, since the impacts of energy production and consumption can have dire consequences on generations to come. Illinois offers a Property Tax Exemption (35 ILCS 200/Property Tax Code) to commercial, industrial, and residential sectors for on-site installations of renewable energy systems.

1.7. Our Motivations for Writing this Book

We finished writing this book in the year 2015, passing 2014 as the warmest year on record. Global annually-averaged temperature in 2014 was 0.69°C (1.24°F) higher than the 20th century average. The 2014 global average ocean and land surface temperatures were 0.57°C (1.03°F) and 1.0°C (1.8°F), respectively, above the 20th century average (NOAA, 2015). The year 2014 was not the warmest year in the U.S., but one of the warmest on record and the 18th consecutive year with an annual average temperature above the 20th century average for the country (NOAA, 2015). Greenhouse gas emissions are major contributors of global warming and in fact scientific studies have shown near-linear relationship between global warming and cumulative carbon dioxide (CO_2) emissions (Allen *et al.*, 2009; Matthews *et al.*, 2009; Meinshausen *et al.*, 2009). According to a very comprehensive report by Stocker *et al.*, (2013), global mean sea level rose by 0.19 meter (~7.5 inches) from 1901–2010. Similarly, increase in concentrations of carbon dioxide, methane and nitrous oxide was observed at levels unprecedented in the last 800,000 years. The rise in CO_2 concentration is primarily due to use of fossil fuels and secondarily from deforestation, reducing absorption of CO_2 from the air (Stocker *et al.*, 2013). The anthropogenic emission of carbon dioxide has caused ocean acidification and it is believed that more than half of the global average surface temperature increase is due to emission of greenhouse gases (Stocker *et al.*, 2013). It is very likely that extreme temperatures (high and low) and their frequency of occurrence are the result of anthropogenic forces.

To avoid dangerous climate change, more than 100 countries have adopted the average global surface temperature increase of 2°C (3.6°F) over the pre-industrial average levels to be a guiding principle (Meinshausen *et al.*, 2009). A recent study (Friedlingstein *et al.*, 2014) estimated that for a 66% probability of staying below the 2°C (3.6°F) threshold, global CO_2 and non-CO_2 emissions must be kept below the quota of 3,200 billion metric tons (or 3,200 gigatons, Gt). The same study projected that with 66% probability, and in order to stay below the 2°C (3.6°F) limit, the remaining emissions of CO_2 from 2015 onwards cannot be more than 1,200 billion metric tons.

With the current rate of emissions, it will take approximately 30 years to exhaust 2°C (3.6°F) limit.

Fossil fuels are one of the primary sources of greenhouse gas emissions, resulting in 57% of the global CO_2 emissions. Energy supply and transportation sector account for 39% of the global greenhouse gas emissions (US-EPA, 2007).

To avoid disastrous implications of global warming, substantial and sustained reduction of greenhouse gas emissions is necessary. In addition to environmental concerns, depletion of fossil fuels is yet another major factor forcing application of less-polluting renewable energy. Understanding of energy utilization processes is required to devise methods of increasing efficiency or to seek alternative sources of energy and routes of conversion processes. In presenting all the available energy sources and their utilization in this book, we have had the above facts in mind and we have tried to look at the possibilities of sustainability in energy and environmental quality.

We, the authors of this book, are quite fortunate to have lived most of our adult lives in the State of Illinois, a pioneer state in dealing with issues related to energy, and we have worked in a variety of energy-related fields. Indeed, it is the above-mentioned concerns about energy use, and as a service to our State, in particular, and sharing our studies and knowledge about energy and our State's pioneering experiences with the rest of the world, in general, that we undertook the task of writing this book. To make our writings based on facts of science and technology, we did neither solicit nor received any kind of financial support from any individual or organization in writing this book.

This book consists of eleven chapters and a glossary. After this introductory chapter, we have devoted the next three chapters to nonrenewable energies, which include coal, petroleum and natural gas respectively. Chapter 5 is on carbon dioxide, mainly generated from burning of fossil fuels, its capture and sequestration. In Chapter 6, we discuss nuclear energy production, consumption and its future prospects. Then the next four chapters (7–10) are devoted to biofuels, wind energy, solar energy, and geothermal energy respectively, which constitute the renewable energies. Chapter 11 is on energy storage technologies for energy management, backup and seasonal reserves services. In every chapter, we present the analysis of energy options from the point of view of individual and collective sustainability. Additionally, we have included an extensive set of glossary and notations at the end of this book which is necessary, due the interdisciplinary nature of energy technologies and the variety of terminologies presented in the book.

Bibliography

Ajith Krishnan, R. and Jinshah, B. S. (2013). Magnetohydrodynamics power generating technology. *Int'l J. of Scientific and Research Pub's*, **3**(6): 1–11.

Allen, M. R., Frame, D. J., Huntingford, C., Jones, C. D., Lowe, J. A., Meinshausen, M. and Meinshausen, N. (2009). Warming caused by cumulative carbon emissions towards the trillionth tone. *Nature*, **458**: 1163–1166.

Bell, T. E. (2009). Nanotechnology for energy: Is downsizing to atomic scale one way forward? *The Bent of TAU Beta Pi*, Knoxville, TN.

Brinker, C. J. and Ginger, D. (2011). Nanotechnology for sustainability: Energy conversion, storage, and conservation. *Sci. Policy Reports*, **1**: 261–303.

Brundtland, G. H. (1987). Report of the World Commission on environment and development: Our common future. United Nations, New York.

Cerami, J. S. (2010). Nuclear fuel reprocessing: Now we're ready (?) A presentation at the Kent College of Law, Chicago, IL.

Chen, H., Cong, T. N., Yang, W., Tan, C., Li, Y. and Ding, Y. (2009). Progress in electrical energy storage system: A critical review. *Progress in Natural Sci.*, **19**(3): 291–312.

Deffeyes, K. S. and MacGregor, I. D. (1980). World uranium resources. *Scientific America*, **242**(1): 50–60.

Elcock, D. (2007). Potential impacts of nanotechnology on energy transmission applications and needs. Report ANL/EVS/TM/08-3, Argonne National Lab., Env'l Sci. Div. Lemont, IL.

Freese, A. S. (1969). MHD: giant blow torches promise cheaper power. *Popular Sci.*, **194**(6): 84.

Friedlingstein, P., Andrew, R. M., Rogelj, J., Peters, G. P., Canadell, J. G., Knutti, R., Luderer, G., Raupach, M. R., Schaeffer, M., van Vuuren, D. P. and Le Quéré, C. (2014). Persistent growth of CO_2 emissions and implications for reaching climate targets. *Nature Geosci.*, **7**: 709–715.

Gibson, R. B. (2006). Beyond the pillars: Sustainability assessment as a framework for effective integration of social, economic, and ecological considerations in significant decision-making. *J. Env'l Assessment Policy and Management*, **8**(3): 259–280.

Goldemberg, J. (2012). Energy: What Everyone Needs to Know. *Oxford Univ. Press*.

Graig, A., Learner, H., Gay, P. and Dobies. J. (2011). The clean energy supply chain in Illinois: Wind, solar and geothermal. Env'l Law and Policy Center Organization, Chicago, IL.

Hansmann, R., Mieg, H. A. and Frischknecht, P. (2012). Principal sustainability components: Empirical analysis of synergies between the three pillars of sustainability. *Int'l J. of Sustainable Development & World Ecology*, **19**(5): 451–459.

[IEA] Int'l Energy Agency (2013). World energy outlook 2013 fact sheet. IEA, Paris, France.

[IER] Inst. for Energy Research (2014). Illinois: An energy and economic analysis. IER, Washington, DC.

[ISTC] Illinois Sci. and Tech. Coalition (2012). Illinois nanotechnology report: A road map for economic development. ISTC, Chicago, IL.

[ISTC] Illinois Sci. and Tech. Coalition (2014a). Nanotechnology. ISTC, Chicago, IL.

[ISTC] Illinois Sci. and Tech. Coalition.(2014b). Nano case studies. ISTC, Chicago, IL.

[IPA] Illinois Power Agency (2013). Annual report: The costs and benefits of renewable resource procurement in Illinois under the Illinois Power Agency and Illinois Public Utilities Acts. IPA, Chicago, IL.

Jaffe, A. B. and Stavins, R. N. (1994). The energy paradox and the diffusion of conservation technology. *Resource and Energy Economics*, **16**(2): 91–122.

Lawson, B. (2005). Battery and energy technologies: Energy efficiencies. *Electropedia, Woodbank Communications Ltd*, Chester, UK.

Lewis, J. and Bergeron, L. (2010). The economic impacts of the energy industry in Illinois. *Regional Development Inst.*, Northern Illinois University, IL.

Luther, W. (2008). Applications in nanotechnology in the energy sector. Vol. 9, Aktionslinie Hessen Nanotech of the Hessian Ministry of Economy, Transport, Urban and Regional Development, HA Hessen Agentur GmBh, Germany.

Lydersen, K. (17 January 2014). Despite decline, states still spending billions to import coal. *Midwest Energy News*, St. Paul, MN.

Mansoori, G. A. (2005). Principles of nanotechnology: Molecular-based study of condensed matter in small systems. *World Sci. Pub. Co.*, Hackensack, NJ.

Mansoori, G. A. (2009a). A unified perspective on the phase behavior of petroleum fluids., *Int. J. Oil, Gas and Coal Tech.*, **2**(2): 141–167.

Mansoori, G. A. (2009b). Phase behavior in petroleum fluids. Petrol. Eng'g Downstream section of Encyclopedia of Life Support Systems, UNESCO, Paris, France.

Mansoori, G. A. (2013). Entropy production gives rise to upper- and lower-bounds to efficiency and COP of cycles. *J. Thermodynamics & Catalysis*, **4**(2): 1–5.

Mansoori, G. A., George, T., Zhang, G. and Assoufid, L. (eds.). (2007a). Molecular Building Blocks for Nanotechnology: From Diamondoids to Nanoscale Materials and Applications, Springer, New York, *Topics in Applied Physics*, **109**.

Mansoori, G. A., Mohazzabi, P., McCormack, P. and Jabbari, S. (2007b). Nanotechnology in cancer prevention, detection, and treatment: Bright future lies ahead. *World Review of Sci., Tech. and Sustainable Development* (*WRSTSD*), **4**(2/3): 226–257.

Mansoori, G. A., Rohani-Bastami, T., Ahmadpour, A. and Eshaghi, Z. (2008). Environmental applications of nanotechnology. *Annual Review of Nano Research*, **2**.

Mansoori, G. A., Soelaiman, T. A. F. (2005). Nanotechnology — An introduction for the standards community. *J. ASTM Int'l*, **2**(6).

Markovic, D. S., Zivkovic, D., Cvetkovic, D. and Popovic, R. (2012). Impact of nanotechnology advances in ICT on sustainability and energy efficiency. *Renewable and Sustainable Energy Reviews*, **16**(5): 2966–2972.

Matthews, H. D., Gillett, N. P., Stott, P. A. and Zickfeld, K. (2009). The proportionality of global warming to cumulative carbon emissions. *Nature*, **459**: 829–833.

McCollum, D. L., Krey, V. and Riahi, K. (2011). An integrated approach to energy sustainability. *Nature Climate Change*, **1**: 428–429.

Medine, E. S. (2013). Realizing Illinois Coal Export Potential. Dep't of Commerce and Economic Opportunity, Office of Coal Development, Springfield, IL.

Meinshausen, M., Meinshausen, N., Hare, W., Raper, S. C. B., Frieler, K., Knutti, R., Frame, D. J. and Allen, M. R. (2009). Greenhouse-gas emission targets for limiting global warming to 2°C. *Nature*, **458**: 1158–1162.

Miller, J. W. (5 May 2014). Dirty U.S. coal finds a home in Europe. *The Wall Street J.*, New York, NY.

[NOAA] National Oceanic and Atmospheric Administration (2015). Global analysis-Annual 2014. National Climate Data Center, Asheville, NC.

Panek, D., Prando, G. and Mansoori, G. A. (1988). Superconductors: Past, present, and future. *Energy Sources*, **10**(3): 159–172.

Prindle, W. (2009). Energy efficiency as a low-cost resource for achieving carbon emissions reductions. National action plan for energy efficiency, ICF International Inc. Fairfax, VA.

Rosenthal, A. (12 November 2012). U.S. to be world's top oil producer in 5 years, report says. *New York Times*, NY.

Schaefer, K. (23 September 2013). Illinois Basin's New Albany Shale: The next big U.S. horizontal oil play? *Oil and Gas Investment Bulletin*, Vancouver, BC, Canada.

Sierra Club (2015). Lawmakers introduce bipartisan bill strengthening renewable energy, energy efficiency standards to create tens of thousands of jobs. content.sierraclub.org/press-releases/2015/02/. Sierra Club, San Francisco, CA.

Smith, G. B. and Granqvist, C. S. (29 September 2010). Green Nanotechnology: Solutions for Sustainability and Energy in the Built Environment. CRC Press.

SourceWatch (2013). Clean Air Act. *SourceWatch.org*, Madison, WI.

Stocker, T. F., Qin, D., Plattner, G.-K. *et al.* (eds.) (2013). Climate Change 2013: The Physical Science Basis. Cambridge University Press.

Union of Concerned Scientist (2014). Burning coal, burning cash: Ranking the states that import the most coal-2014 Update. Union of Concerned Scientist, Chicago, IL.

[US-DOE] U.S. Dep't of Energy (2006). Benefits of demand response in electricity markets and recommendations for achieving them: A report to the U.S. Congress Pursuant to Section 1252 of the Energy Policy Act of 2005. U.S. DOE, Washington, DC.

[US-EIA] U.S. Energy Info. Admin. (1994). State Coal Profiles. Office of Coal, Nuclear, Electric and Alternative Fuels, U.S. EIA, Washington, DC.

[US-EIA] U.S. Energy Info. Admin. (2011). State energy data system: Illinois. U.S. EIA, Washington, DC.

[US-EIA] U.S. Energy Info. Admin. (2013a). State energy data system: Illinois. U.S. EIA, Washington, DC.

[US-EIA] U.S. Energy Info. Admin. (2013b). Electric power annual 2012. U.S. EIA, Washington, DC.

[US-EIA] U.S. Energy Info. Admin. (2014a). Electricity: Detailed state data. U.S. EIA, Washington, DC.

[US-EIA] U.S. Energy Info. Admin. (2014b). State energy data system: Illinois. U.S. EIA, Washington, DC.

[US-EIA] U.S. Energy Info. Admin. (2014c). State energy production estimates, 1960 through 2012. U.S. EIA, Washington, DC.

[US-EIA] U.S. Energy Info. Admin. (2014d). State energy price and expenditure estimates 1970 through 2012. U.S. EIA, Washington, DC.

[US-EIA] U.S. Energy Info. Admin. (2014e). State profile and energy estimates: Illinois. U.S. EIA, Washington, DC.

[US-EPA] U.S. Env'l Protec. Agency (2007). Global greenhouse gas emissions data. U.S. EPA, Washington, DC.

[US-SCENR] U.S. Congress, Senate Committee on Energy and Natural Resources (2007). Transportation sector fuel efficiency: Hearing before the committee on energy and natural resources. U.S. Senate, 110th Congress.

[WWF] World Wildlife Fund (2014). Leading from the middle: How Illinois communities unleashed renewable energy. World Wildlife Fund Inc., Washington, DC.

2

Coal

2.1. Introduction

The abundance of coal in the world makes it one of the cheapest sources of fuel for base load electricity generation and possibly other uses. Demand for electric energy can be categorized as base load, intermediate load and peak load. Base load electric energy demand in U.S. and many other countries around the world is met mostly by coal and nuclear power plants. Base load power plants are typically larger (over 600 MW) than intermediate and peak load power plants, comparatively more efficient and reliable, and can continuously produce low-cost electric energy (Shaalan, 2003; PSCW, 2004; Lefton and Besuner, 2006). On the other hand, intermediate and peak load power plants are designed to be smaller and quickly brought online to respond to electric power demand during peak periods. These plants usually require natural gas as the fuel or a combination of fuels to meet the various load demands.

Coal-fired plants generate electricity by burning pulverized coal mixed with very hot air to produce steam (Osborne, 2013; Speight, 2013). The high pressure steam turns a turbine that drives an electric generator to produce electricity, somehow similar to the one shown in Fig. 2.1.

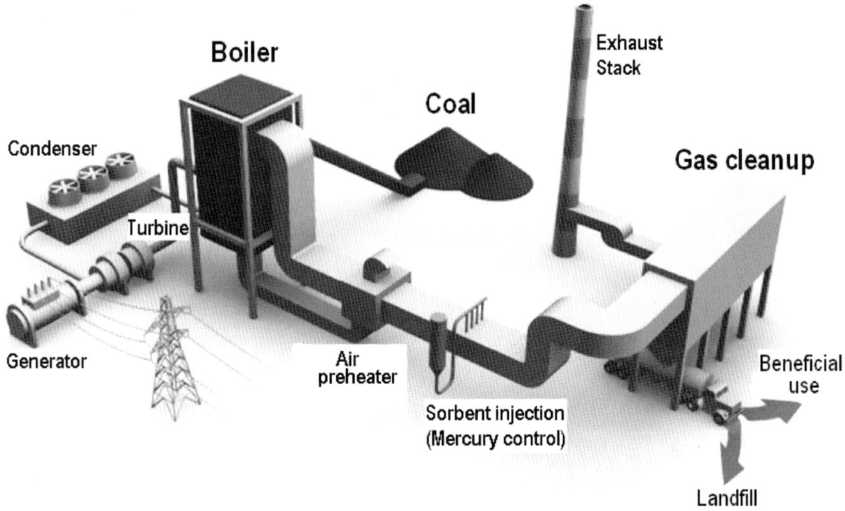

Fig. 2.1. Schematic of a conventional coal-fired power plant (www.grafport.com/illustration/ImageGalley.swf).

Despite the fact that coal is one of the cheapest sources of fuel, the U.S. and many other countries around the world face a dilemma about its use, due to environmental concerns for air and water quality. These environmental concerns associated with the use of coal include greenhouse effect due to release of significant levels of carbon dioxide (CO_2), direct health hazards posed by release of nitrogen oxides (NO_X), sulfur oxides (SO_X), particulate matter and mercury to the environment, as well as contamination of aquifers by coal mine slurry injections.

The U.S. policymakers in federal, states, and municipality levels have passed legislations that could curtail these pollutants with minimal financial burden on consumers. In these legislations, coal-fired power plant companies will have the option to decide between paying extra money for more efficient scrubbers and pollution control devices or have some of their plants closed as a means to reduce emissions (Wernau, 2011a). The implication of such policies is that consumers could face more expensive energy supply. For this reason, policymakers are introducing some measures for energy usage and exploring other alternative energy sources that are cost effective and environmentally benign. These include the use of energy efficiency programs, demand-response, operation of more natural gas plants, implementation of clean coal technologies, addition of generating capacity to ramp up underutilized generating plants, and utilization of biofuels, geothermal, solar and wind energy.

In the U.S., Illinois has the largest overall as well as the largest strippable bituminous coal reserves.

The total recoverable reserve (i.e., quantity of coal that can be recovered from existing coal reserves at reporting mines) is about 1.294 billion metric tons (IDCEO, 2008; Agyarko and Mansoori, 2013). The total demonstrated Illinois reserve base is 104.2 billion metric tons. This includes publicly available data on coal beds mapped and located underground with thicknesses considered technologically minable at the time of determinations. The total estimated recoverable reserve is 37.9 billion metric tons. These include the coal in the demonstrated reserve base considered recoverable, excluding coal estimated to be unavailable due to land use restrictions or currently economically unattractive for mining after applying the recovery rate assumptions. Out of these estimated recoverable reserves, 27.8 billion metric tons are exploitable by underground mining methods, and about 10.1 billion metric tons are exploitable through surface mining methods (IDCEO, 2012).

But Illinois cannot use much of its indigenous coal, partly due to unfavorable geologic conditions and surface developments and partly due to the high sulfur content of the coal. This situation has forced several coal mines in the state to shutdown, affecting coal production since early 1990s as shown in Fig. 2.2 (US-EIA, 2014a; Agyarko and Mansoori, 2013). As we also mentioned in Chapter 1 and according to Fig. 2.2, in 2012, Illinois coal production witnessed a dramatic increase to about 48 million metric tons, largely due to exports that reached a record of 13 million metric tons, a five-fold increase from 2.5 million metric tons in 2010.

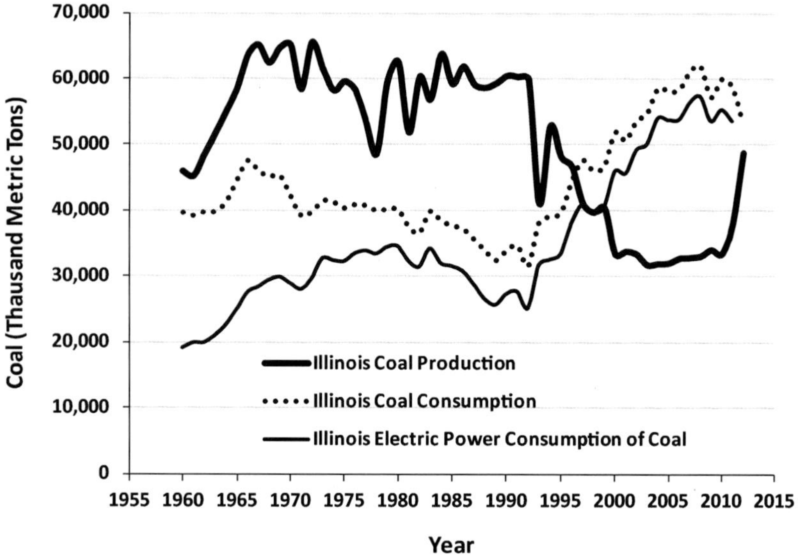

Fig. 2.2. Profiles of coal production and consumption in Illinois (Data from US-EIA, 2014a).

Power plant companies prefer using low sulfur content coal to reduce gas clean-up cost and meet the minimum environmental regulations (US-EIA, 1994). As mentioned earlier in Chapter 1, in 2012, Illinois imported 42.94 million tons of coal (85% of the coal burned in its power plants) at the cost of $1.45 billion, mainly from the state of Wyoming (Union of Concerned Scientist, 2014). Most of this expenditure was offset by selling electricity to other states and by shipping about $1.33 billion worth of Illinois' high sulfur content coal to other states. In 2013, electric energy demand in Illinois was mostly met by coal and nuclear power plants, with coal power representing 43% and nuclear energy representing 48% of electricity production in the state.

In 2008, there were 21 conventional coal-fired power plants similar to the one shown in Fig. 2.1 in Illinois operated by eight utility companies, generating a total of 16,240 MW of electricity (IDCEO, 2008).

These included Midwest Generation EME, LLC (six plants), Dynegy Midwest Generation (five plants), AmerenEnergy Generating (four plants), Dominion Energy Services Company (one plant), AmerenEnergy Resources (two plants), Electric Energy, Inc (one plant), Springfield City Water, Light and Power Company (one plant) and Southern Illinois Power Cooperative (one plant) (IDCEO, 2008).

In 2009, the Illinois electric power industry ranked 8th in both sulfur and nitrogen oxides (SO_x and NO_x) emissions in U.S., estimated at 237,000 and 78,000 metric tons, respectively (US-EIA, 2011). Based on the most recent data [US-EIA, 2014] and as of 2011, Illinois CO_2 footprint from the electric power industry ranked 6th in the nation, estimated at more than 225 million metric tons per year. The Illinois Environmental Protection Agency (IEPA) estimated that the state's coal-fired power plants emit 3.5 tons of mercury every year (IDCEO, 2008). The 2010 estimated emission of particulate matter (PM_{10}) in Illinois was 30,931 tons (IEPA, 2011).

In June 2014, the Illinois Governor announced the creation of Illinois Clean Energy Fund in the state. The revolving equity fund will be formed by the State of Illinois with $2.3 million from an allocation of federal funds approved for investing in clean energy businesses. This amount will be matched with an investment from Clean Energy Trust, a non-profit group created in Illinois about four years ago (Dudek, 2014). The purpose of the Fund is to help fill the gap in seed funding necessary for energy innovation in the state. The ultimate goal is to accelerate the commercialization and growth of small businesses and to create jobs in the state.

2.2. The Characteristics of Coal

There are many different kinds of coal available in different parts of the world and numerous research papers and books are already published about various coal characteristics (Berkowitz, 2012). However, since we used the State of Illinois for our case study, we report here on the characteristics of Illinois coal.

Fig. 2.3. Cross section of an Illinois coal seam (www.museum.state.il.us).

According to the Illinois State Museum (www.museum.state.il.us), Illinois coal occurs in seams (layers). These seams vary from less than 2 centimeters (0.75 inch) to more than 4.5 meters (about 15 ft) thick. Dozens of coal seams underlie Illinois (see Fig. 2.3).

The major seams have been given names to identify them. The largest is the Herrin Coal, named for a Southern Illinois town in Williamson County. In 1868, geologists numbered the coal seams by age. Number 1 is the oldest. The Springfield Coal is number 5 and Herrin Coal is number 6. The Springfield Coal seam averages 1.5 meters (about 5 ft) in thickness, while Herrin Coal seam averages 1.8 meters (about 6 ft). Because of the gradual down-warping of the basement rock, Illinois coal is near the surface in some areas and buried as deep as 365 meters (~1,200 ft) in other locations.

Illinois coal has a high heating value ranging from 6,111 kcal/kg (11,000 Btu per lb) in the northwestern part of the state to about 8,333 kcal/kg (15,000 Btu per lb) in the Southeastern part of the state (IDCEO, 2010). The ash content of Illinois coal may vary from place

Table 2.1. General characteristics of Illinois coal.

Organic sulfur[a]	0.4%–5%
Pyritic sulfur[a]	0%–5%
Mean moisture[a]	10.5%
Mean fixed carbon[a]	43.4%
Average ash[b]	10%
Heating value[b]	~6,111 kcal/kg

(a) IDCEO, 2012; (b) IDCEO, 2010

to place, with an average ash content of about 10%. In Table 2.1 the general characteristics of Illinois coal are reported.

According to IDCEO (2010), the high sulfur content of Illinois coal beds is attributed to the character of the strata laying underneath the coal. Usually, coals overlain by marine strata have sulfur content ranging from 3–5%. In certain non-marine areas where gray shale exceeds 20 ft in thickness above the coal, the sulfur content is less than 2.5%, commonly averaging 1.5%. In Illinois coal, as in other coals, sulfur exists in two forms: organic and pyritic.

The organic sulfur content of Illinois coals varies from a minimum of about 0.4% to a maximum of about 5%. Pyritic sulfur varies from nearly zero to as high as 5%. In Table 2.2, we report the range of typical analysis of Herrin and Springfield Coals in various Illinois counties.

Almost all the coal mined in Illinois are subject to cleaning at preparation plants in order to reduce the sulfur content of the final product by as much as one-third. Detailed analysis and data on characterization and quality of coals from the Illinois Basin can be obtained from Affolter and Hatch [2002].

2.3. "Clean Air Act" and its Impact on Coal-fired Power Plants

The "Clean Air Act" of 1970 authorized the U.S. Environmental Protection Agency (US-EPA) to set limits on how much of a pollutant can

Table 2.2. Range of typical analyses, Herrin coal and Springfield coal by counties (IDCEO 2010).

Counties	Moisture [%]	Volatile Matter [%]	Fixed Carbon [%]	Ash [%]	Sulfur [%]	Calorific Value [Btu/lb]	Rank Index*	Ash Fusion [°F]
			Herrin Coal					
LaSalle, Grundy	13–16	36–41	35–40	7–11	3–5	10,500–11,400	116–123	1,950–2,150
Bureau, Stark, Henry, Knox	6–20	31–35	37–40	8–13	3–5	9,700–10,300	111–118	1,900–2,120
Peoria, Fulton	15–19	32–35	37–43	8–13	2–4	10,000–10,700	111–120	NA
Sangamon, Macoupin	12–16	34–40	37–41	9–11	3–5	10,400–10,900	116–123	1,930–2,160
Christian, Montgomery, Bond, Madison	12–14	35–40	38–41	9–11	3–5	10,500–11,000	117–125	1,920–2,170
Douglas, Vermillion	4–16	32–36	38–41	8–12	1–3	10,400–11,100	118–128	2,080–2,220
Clinton, St. Clair	10–13	34–40	37–42	9–12	1–4	10,000–10,700	121–129	1,920–2,090
Marion, Washington, Randolph, Perry	8–12	35–39	38–44	9–13	1–4	10,800–11,300	124–133	1,920–2,610

Springfield Coal								
Peoria, Fulton, Tazewell, Schuyler	14–18	33–38	34–40	9–12	2–4	10,100–10,800	115–122	1,890–2,270
McLean, Logan, Menard, Sangamon	13–17	34–39	35–41	9–12	3–5	10,400–11,000	117–125	1,890–2,600
Macon, Shelby	12–16	36–39	35–40	8–12	3–4	10,500–11,100	119–127	NA
Edgar	10–12	36–40	37–43	8–10	3–4	11,200–11,500	NA	NA
Randolph, Perry	8–13	35–38	40–44	9–12	4–5	11,000–11,400	124–135	2,168–2,174
Jackson	8–9	35–36	44–55	1	3–4	11,600–11,800	130–135	1,940–2,010
Gallatin, Saline, Williamson	5–7	33–38	47–53	2–5	2–5	11,900–12,500	132–141	2,040–2,090
Gallatin (Eagle Valley)	4–5	34–37	48–52	3–4	3–4	12,400–12,700	130–147	NA

*Rank Index is defined as the calorific value of moist coal on a mineral-matter-free basis to the nearest 100 Btu/lb (Jacobson and Korose, 2003).

be released into the environment in the U.S. The law permits individual states to have stronger pollution control regulations than is stated in the Act. Under Phase I of the "Clean Air Act", power plants built after 17 August, 1971 were subjected to New Source Performance Standards (air pollution emission standards), which allowed a maximum emission rate of 2.16 kg sulfur dioxide (SO_2) per million kcal (1.2 lb SO_2 per million Btu) (IDCEO, 2008; Agyarko and Mansoori, 2013). Also, the Phase II of the "Clean Air Act" of 1970 required New Source Performance Standards to remove a certain percentage of SO_2 from flue-gases by January 2000.

The 1970 "Clean Air Act" was later extended by amendments in 1977 and later in 1990. It authorized US-EPA to establish standards for a number of atmospheric pollutants, including SO_2. The 1990 "Clean Air Act" Amendments encouraged the use of market-based principles and other innovative approaches, such as emission banking and trading and performance-based standards. Furthermore, the 1990 Act provided a framework from which alternative clean fuels would be used by setting standards in the fleet and a California pilot program that could be met by the most cost-effective combination of fuels and technology. Additionally, the 1990 Act encouraged the use of the following:

(i) Use of clean low-sulfur coal and natural gas, including innovative technologies to clean high-sulfur coal to reduce the acid rain.
(ii) Reduce energy waste and create sufficient market for clean fuels derived from agriculture and natural gas to cut reliance on oil imports by one million barrels per day.
(iii) Use innovative approaches and market-based principles, such as performance-based standards and emission banking and trading (IDCEO, 2008) to reduce pollution.

With regard to the Clean Air Mercury Rule, Illinois opted for a more stringent one than the US-EPA's limit because it wanted to cut emissions from coal-fired power plants by an average of 90% by 2009. For example, the Illinois mercury rule is expected to achieve over 90% of mercury reductions by 2015, while federal rules only require a 78% reduction by 2018. Again, the Illinois mercury rule is expected to achieve between 65–80% reduction in SO_2 by 2019, on the other hand, the federal rule only requires 34% reduction by 2019 (IDCEO, 2008). These additional rules for reduction of pollutants by Illinois power plants have made the state a leader of power plants pollution control in the U.S.

The "Clean Air Act" and its subsequent amendments in 1990 brought innovations into the energy industry in Illinois. In 2006, Illinois' second largest utility company, Ameren Corporation, reached an agreement with the state to curb emissions of the three most harmful power plant pollutants: mercury, SO_2 and NO_x (Environmental News Service, 2006). To reach this goal, Ameren Corporation agreed to invest $1.6 billion in technology upgrades at seven of its power plants in Central and Southern Illinois. This was intended to enable the company to use more Illinois coal, which has higher sulfur content than coal from other regions.

In April 2007, the U.S. Supreme Court authorized the US-EPA to regulate CO_2 as a pollutant. According to the ruling of the U.S. Supreme Court, US-EPA's reason for not regulating CO_2 was inadequate (IDCEO, 2008). The result of this ruling has had tremendous repercussions on the power plant industry, including those in Illinois, because utility companies need to spend more money to meet the new regulations.

In 2008, the U.S. government proposed the "Clean Air Interstate Rule" which was purported to cut pollution that spreads hundreds of miles and has enormous negative impacts on millions of Americans (Hawthorne, 2010). However, the U.S. Court of Appeals for the District of Columbia ordered the US-EPA to revise the rule, after deciding that the agency had overstepped its authority. This court ruling allowed many utility companies to withdraw their investments in pollution control. However in 2010, the U.S. government revisited the "Clean Air Interstate Rule" by urging the US-EPA to reduce pollution in areas around coal-fired power plants and in states downwind, where air quality is degraded by SO_2 and NO_x oxide emissions.

On 24 May, 2011, the US-EPA for the first time held a national debate in Chicago on its proposed standards for setting federal new limits for mercury, arsenic, lead, hydrochloric acid and other hazardous air pollutants (Learner & Rowe, 2011). Among the two key safety rules the US-EPA proposed under the Clean Air Act to protect public health and the environment were the Toxics Rule and the Clean Air Transport Rule.

On 21 December, 2011, the US-EPA unveiled its final rule on the new Toxic rules. There were no changes to the original proposals. However, there was a presidential memorandum attached to the new rules. This memorandum authorized the US-EPA to use the law to give energy companies more time beyond the three-year deadline for equipment installation or shut down old plants if they refuse. The US-EPA rule on power plant emissions for SO_2 and NO_x that crosses state borders was set to take effect in 2012; however, the rule was challenged in court by the energy companies (Drajem, 2011).

In January 2012, The U.S. Court of Appeals in the District of Columbia suspended the Cross-State Air Pollution Rule some days before it could be implemented, a decision that gave impetus to the claims of a group of power plant companies and states that the new rule is too austere and illegal. In April 2012, the court began a hearing on those challenges leveled against the U.S. Environmental Protection Agency regulations. According to legal fillings at the court, the challengers raised at least 10 questions about the legality of the new US-EPA rule. Among the key questions was the argument by the electric power companies that the US-EPA failed to justify the pollution limits of the new rules. Also, a group of 15 upwind states questioned as to whether US-EPA did not violate the law by not allowing them to develop their own plans for reducing emissions before the new rule was implemented. But not all power plant companies opposed the new rule, especially the owners of nuclear and natural gas power plants.

These electric power companies which include Exelon Corporation and Calpine Corporation joined nine downwind states in filing briefs in support of the new rule. Additionally, several environmental and public health groups supported the agency (Dow Jones Newswires, 2012).

In April 2014, the U.S. Supreme Court finally overturned the Appeals Court's decision and supported the 2011 EPA Cross-State Air Pollution Rule (CSAPR) that targeted harmful emissions from coal-fired power plants [Stohr, 2014]. Compared to 2005, EPA estimates this rule and other federal rules will lower power plant annual emissions of SO_2 in the CSAPR region by 6.4 million tons per year (a 73% reduction) and annual emissions of NO_X by 1.4 million tons per year (a 54% reduction), including 340,000 tons per year of NO_X during the ozone season [US-EPA, 2014].

This current turnaround of events is forcing several utility companies in Illinois to shut down and phase out some of their aging power plants. For example, some utility companies that own coal-fired electric power plants in the Chicago area are facing federal lawsuits, alleging that the companies failed to install pollution control devices, while keeping the plants operating far beyond the time they needed to be closed. Most of these power plants, like the Fisk electric generating plant (Fig. 2.4) in Chicago's Pilsen neighborhood, have been operating since 1903, before being forced to retire by the new EPA rules in August 2012 (Wernau, 2012).

The 1970 "Clean Air Act" exempted these older power plants under the assumption that they would soon be retired (Thorner, 2011). However, the utility companies continued to operate these power plants without retiring them or installing emission controls.

With the new US-EPA standards requiring that operators of aging coal-fired power plants either install costly new retrofits or shut down the plants, most Illinois utility companies are reviewing these new rules to decide the fate of their plants before 2018. The new standard for sulfur dioxide emission in Illinois is 75 ppb measured over one hour. Most SO_2 emissions in the state are from coal-fired electric

Fig. 2.4. The 326 net-MW coal-fired Fisk electric generating plant serving Chicago area, which started in 1903 at the Pilsen neighborhood [picture courtesy of Midwest Generation, An Edison International Company].

power plants and ethanol production facilities. The area around the City of Pekin, the county seat of Tazewell County in Illinois, has the highest level of SO_2 air pollution in the state (235 ppb over one hour), according to US-EPA, a level that now exceeds tightened Federal standards (Pekin Times, 2011).

It was estimated that compliance with the new clean air standards will bring over 122,000 capital investment jobs in Illinois. In addition, Illinois will benefit from approximately 1,500 permanent jobs in operation and maintenance of the new air pollution control equipment (Learner and Rowe, 2011). On the other hand, any delay to implement the new US-EPA's transport rule will mean about $450 million in lost revenue to Illinois every year. These losses are attributed to reduced productivity from air pollution caused to Illinois by other states.

Figure 2.5 shows the profile of electric power industry emissions estimates in Illinois for the period of 1990–2009. This diagram is produced using the data reported by the U.S. Energy Information

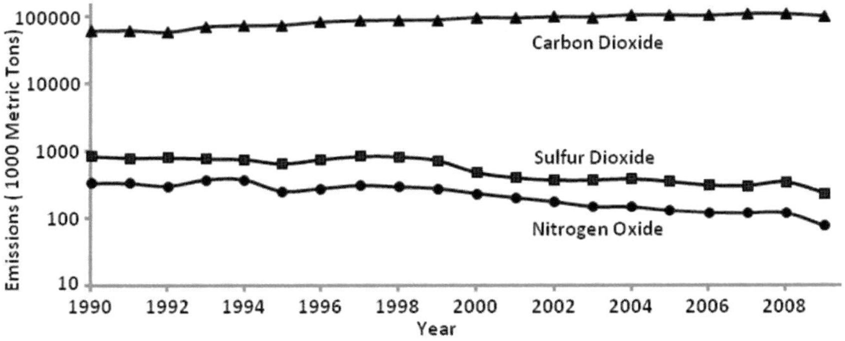

Fig. 2.5. Profile of electric energy industry emissions estimates, 1990–2009, Illinois (Data from US-EIA, 2011).

Adminstration (US-EIA, 2011). According to this figure, CO_2 emission from electric power plants increased in the past decade, while SO_2 and NO_x emissions decreased.

The new state and government rulings in regard to pollution reduction has brought about new activities towards using clean coal technology for electric power generation in Illinois, which are discussed in the next section.

2.4. Electric Power Plant Projects Using Clean Coal Technology

The clean coal technology (CCT) program which began in 1986 was the most ambitious government-industry initiative ever engaged to develop environmental solutions for U.S. abundant coal resources. The U.S. Department of Energy (DOE) generally defines clean coal technology (CCT) as a state-of-the-art technology which when implemented, improves the environmental performance and efficiency of coal-fired power plants, relative to the current coal-fired electric power plants. Clean coal technology is commonly referred to all the technologies being developed to reduce environmental impact of energy generation relative to older methods of burning coal (Buchan and Cao, 2004). However, when a technology is proven, it is no longer classified as CCT, but instead, it is considered as best available control technology or best available retrofit technology that can be used commercially. Several CCT options are currently available, but the economic viability of some of them remains questionable due to high energy taxes. There are many CCT projects currently being undertaken in the U.S., including the State of Illinois, to make the state a leader in clean and efficient use of coal in the foreseeable future. However, they are mostly based on four new technologies which include:

(i) Supercritical and ultra-supercritical steam cycle technology;
(ii) Oxy-combustion technology;
(iii) Integrated gasification combined cycle (IGCC) technology;
(iv) Hybrid integrated gasification combined cycle (HIGCC) technology.

In the following sections, we report on four important Illinois CCT projects and where these four new technologies have been proposed or are being used. They include: Prairie State Energy Campus (PSEC), FutureGen 2.0, coal-to-gas IGCC power plants, and Taylorville Energy Center, which are discussed briefly in the following sections.

2.4.1. *PSEC project (Supercritical steam cycle technology)*

A major CCT project in Illinois is the Prairie State Energy Campus (PSEC) and its adjacent Lively Grove New Coal Mine in Washington County (IDCEO, 2010). The PSEC belongs to Peabody Energy (the nation's biggest coal company) and eight not-for-profit public power companies. The eight companies are American Municipal Power, Illinois Municipal Electric Agency, Indiana Municipal Power Agency, Kentucky Municipal Power Agency, Missouri Joint Municipal Electric Utility Commission, Northern Illinois Municipal Power Agency, Prairie Power, Inc., and Southern Illinois Power Cooperative (PSGC,

2012). The projected total cost for the development and construction of the twin 810 MW electric unit power station was nearly $4 billion (IDCEO, 2009). The plant generates sufficient electricity to serve approximately 2.5 million families. This new power station uses the technology of pulverized coal to produce supercritical steam (see Sec. 2.4.5.1.) and up-to-date equipment for post-combustion flue-gas clean-up. The plant will use 6.3 million tons of coal per year from the adjacent newly constructed Lively Grove underground mine. The estimated carbon emissions will be 15% lower than the typical U.S. coal-fired power plant. The 1,600 MW power plant began producing commercial power on behalf of its owners in June 2012 (for Unit 1) and in November (for Unit 2) of the same year (PSGC, 2012). About 95% of the plant's power output is already dedicated to the eight not-for-profit Midwestern-based public power utilities, while Peabody Energy (the original developer) now retains a 5% interest in the campus (PSGC, 2012). The adjacent Lively Grove New Coal Mine current coal reserve is 200 million tons and is projected to last for nearly 30 years. The plant is estimated to add about $785 million annually to Illinois economy. Furthermore, nearly 450 full-time employments are estimated to be created by the project, which out of this number about three-fourths would be at the coal mine (Coal Age, 2011). The claimed emissions from the power plant compared to the EPA's projected Cross State Air Pollution Rule (CSAPR) regulations set for 2014 is shown in Table 2.3.

2.4.2. *FutureGen 2.0 project (oxy-combustion technology)*

The FutureGen 2.0 project is a private partnership between the U.S. DOE and the FutureGen Alliance, a consortium of companies including Babcock & Wilcox, Air Liquide Process & Construction and Ameren Energy Resources. The project is a revised version of the FutureGen Project which was cancelled by the DOE in January 2008, due to escalating cost. However, in September 2010, DOE reached

Table 2.3. Comparison of PSEC emissions (2011–2012) with the 2014 EPA's projected Cross State Air Pollution Rule (CSAPR) (PSGC, 2012).

	Pollutants			
	NO_X [lb/million Btu]	**SO_2 [lb/million Btu]**	**NO_X [kg/million kcal]**	**SO_2 [kg/million kcal]**
CSAPR	0.17	0.3	0.31	0.54
PSEC	0.07	0.182	0.126	0.33

CSAPR = Cross State Air Pollution Rule; PSEC = Prairie State Energy Campus

an agreement with the Alliance for the construction of FutureGen 2.0, a full-scale advanced oxy-combustion technology on an existing Ameren 200 MW coal-fired power plant in Meredosia, Illinois (McDonald *et al.*, 2011; Crooks, 2011; US-DOE, 2010).

The FutureGen 2.0 project consists of retrofitting the shuttered coal-fired power plant to demonstrate the advanced oxy-combustion technology, and transportation of carbon dioxide for underground storage (TCS, 2011). Principles behind oxy-combustion technology are presented in Sec. 2.4.5.2. The project will demonstrate a large-scale feasibility of the technology in meeting emission standards, confirm the cost basis for retrofitting and repowering existing coal-fired units, provide the information and experience that will reduce the cost for larger plants in future and establish operating and maintenance experience for future commercial plants. Furthermore, the project will confirm the technical and economic feasibility of efficiently producing low-cost electricity from coal, while nearly eliminating emissions. The cost of the new project was estimated at $1.3 billion and was supposed to generate 200 MW of gross electricity (140 MW of net electricity), capture 1.3 million tons per year of CO_2, and have near-zero emission of NO_X, SO_X, mercury, particulates and other hazardous air pollutants (McDonald *et al.*, 2011). A 32-mile pipeline will be constructed to pump the plant's CO_2 emissions into a 4,500 ft deep underground geologic formation with estimated capacity of over 1.3 million tons per year, located at Morgan County in Illinois.

It should be mentioned that following the passage of the American Recovery and Reinvestment Act (ARRA) of 2009, the Federal Government set aside about $1.1 billion federal funds for the FutureGen 2.0 project (Daniel, 2015). Since then the project made a slow start in obtaining permits and negotiating an arrangement with the State of Illinois that will require electricity consumers in Illinois to finance the rest of the costs of the plant which has a current estimate of $1.7 billion. But the project faced several lawsuits and challenges that made it impossible to raise the over $600 million additional non-federal funds. Besides, the Illinois landowners were challenging the Class VI permits obtained from U.S. EPA in September 2014 for underground injection of CO_2, while the Sierra Club filed a court appeal challenging the validity of the state's air permit for the plant. In the wake of these challenges, in February 2015, the DOE directed that the FutureGen 2.0 project be terminated, based on the following statement issued by the CEO of the FutureGen Alliance developing the project: "*The U.S. Department of Energy has directed the suspension of FutureGen 2.0 project development activities. The DOE has concluded that there is insufficient time to complete the project before federal funding expires in September 2015*" (Daniel, 2015).

2.4.3. *Southern Illinois & Chicago projects (IGCC technology)*

In July 2011, the Illinois governor signed a bill that could begin the construction of two coal-to-gas IGCC power plants to produce synthesis gas (syngas) in 2012 (Wernau & Karp, 2011; Wells, 2011). The principles behind IGCC technology is described in Sec. 2.4.5.3.

One of the plants will be constructed by Chicago-based Power Holdings, LLC near Waltonville in Jefferson County of Southern

Illinois and would cost approximately $2.3 billion. The plant will use nearly four million tons of Southern Illinois coal per year and create around 1,000 construction jobs, 300 permanent mining jobs and 250 permanent plant positions, while providing gas to 500,000 homes (Wells, 2011). The construction of the plant was supposed to begin in 2012, and be readied for testing by the end of 2013, followed by full operation by the year 2014. However, faced with the problem to attract investors due to low natural gas prices, the company applied for an extension of the project's construction permit, which expired in February 2012. But in June 2012, the Illinois EPA denied the permit request for the construction of the $2 billion plant. According to Illinois EPA, the company's request neither indicated the new or revised standards and operations required for utility plants and boilers, nor did it show the Best Available Control Technologies (BACT) for emissions, particulate matter and greenhouse gases. Nevertheless, Power Holdings LLC can reapply for the permit, should the company decide in the future to continue with construction of the plant (Gregerson, 2012).

The other plant will be constructed by New York-based Leucadia National Corp at the Southeast Side of Chicago at the cost of nearly $3 billion (Wernau & Karp, 2011). This modern coal gasification plant will produce syngas and will be developed on a brownfield site within the Chicago-Calumet industrial corridor. The plant will use about 1 million tons of Illinois coal per year, together with petroleum coke to produce syngas. The proposed Illinois coal-fueled facility will use General Electric quench gasifier to produce substitute natural gas and electricity from the waste stream. About 85% of the CO_2 from the plant will be captured and sequestered.

These projects are expected to create thousands of jobs, spur economic growth in the communities, replace the 1950s coal-fired power plants and demonstrate clean use of Illinois coal. In August 2012, the governor of Illinois vetoed the legislation that would have allowed Leucadia National Corp to move forward with construction of the coal gasification plant (Bzdak, 2012). It should be mentioned that in 2011, the governor supported a bill for the construction of the same project based on the fact that the state's four major gas utilities would buy the plant's synthetic natural gas for 30 years at a price set by a formula. Also, consumers would have paid for the gas through utility bills, and at a guaranteed price to save them about $100 million over 30 years (Wernau, 2011b). But, two utilities companies (Peoples Gas and North Shore Gas) whose names were identified in the original bill as buyers of the gas failed to sign the necessary 30-year purchase contracts. As a result, Illinois policymakers had to revise the bill to enable Ameren and Nicor to buy the gas produced by the plant (Tribune Staff, 2012). In essence, the revised bill would have required the customers of Ameren and Nicor Gas to pay 95% of the costs of building and operating the plant, a situation that compelled the governor to veto the bill to protect consumers (Bzdak, 2012). Below we report the statement from the spokesperson of Nicor (a provider of natural gas to most residents in the suburbs) on this issue.

"*We estimate the costs for synthetic gas that would be produced at the Leucadia plant to be more than three times current market prices. If this legislation is passed, Nicor Gas customers would bear an additional $126 million in gas costs each year. With the abundance of low cost natural gas, the Leucadia plant is neither necessary nor a cost-effective way to supply gas to Illinois residential, commercial, and industrial customers.*" (Schutz, 2012).

2.4.4. *Taylorville Energy Center project (HIGCC technology)*

A proposed project waiting for approval from the Illinois General Assembly is the construction of the Taylorville Energy Center at Christian County in Southern Illinois. The project is a joint venture between Tenaska Inc., an energy company based in Omaha, Nebraska, and MDL Holding Co. of Louisville, Kentucky. The project would use HIGCC technology to convert Illinois coal into pipeline-quality synthetic gas to generate wholesale electricity

(Flores *et al.*, 2010). However, the full scale 602 MW HIGCC electric power plant at a projected cost of \$3.5 billion was opposed by the Illinois Senate, despite being earlier approved by the Illinois House of Representatives. The Illinois Commerce Commission (ICC) has also expressed concerns about the exorbitant cost relative to the cost of other energy generation technologies and "uncertain future benefits" of the project (Flores *et al.*, 2010). For example, the cost per MWh of electricity for traditional coal-fired power plants is \$141.08–\$153.03, for nuclear power plants is \$101.45–\$128.03, while for Taylorville HIGCC is \$212.73.

The proposed Taylorville HIGCC power plant is expected to capture more than 50% of CO_2 (about 1.9 million tons per year) that will be transported via pipelines for storage. The project is expected to generate enough electricity to power about 600,000 homes in the Taylorville area and create more jobs. The Illinois Senate Bill (SB 1987), created the Clean Coal Portfolio Standard, requiring utility companies to purchase 5% of the electricity from coal-fired plants that capture at least half of the CO_2. But the law limited the rate impact from the Taylorville project to 2% for residential customers. By 2017, all new coal-fired power plants built in Illinois will be required by law to sequester at least 90% of total CO_2 emissions.

Opponents of the Taylorville project claim that the technology is too expensive and not proven. Besides, there are claims that capturing 90% of the emissions from this HIGCC power plant could increase the total cost of electricity because the processes involved in capturing the emissions are energy intensive.

In late November 2011, the Illinois Senate finally passed a bill (SB 678), granting permission for construction of the \$3.6 billion Taylorville coal-to-gas electric generating plant. The bill would make the Taylorville plant eligible for a tax credit estimated at \$8.7 billion over 30 years period and includes provisions that require the major state utility companies to purchase electricity from the plant for 20–30 years (Yeagle, 2011). However, in July 2012, the Illinois EPA withdrew the pollution permit for the construction of the Taylorville

Energy Center. The Illinois EPA earlier issued the pollution permit allowing the Taylorville Energy Center project to operate as a natural gas plant. But, a request from the US-EPA, demanding that the state add coal gasification and carbon sequestration privileges in the project, necessitated the state agency to cancel the permit (MITEI, 2013).

In Table 2.4, we compare the compiled characteristics of the three new energy generation power plant projects reported by various sources as referenced.

In Table 2.5, we compare the cost per megawatt-hour (MWh) of electricity generation of the new energy generation projects (power plants) (as also reported in Table 2.4) with traditional coal-fired power plants, nuclear power plants, IGCC power plants and wind energy power plants.

Obviously, the reported data for the Prairie State Energy Campus does not seem to be reasonably accurate in general. This low cost per megawatt-hour electricity is the Average Costs to Participants of the project. This cost is far less than the wholesale electricity price because the 68 member participants of the project are locked in to a contract for several years.

2.4.5. *Science and engineering basis of electric power plant projects using clean coal technology*

In this section, we briefly report on the basic technologies of the new Illinois electric power plant projects.

2.4.5.1. *Supercritical and ultra-supercritical steam cycle*

a. *Supercritical steam cycle*

The technology of pulverized coal producing supercritical steam takes advantage of higher boiler design pressures to raise the water

Table 2.4. Comparison of the three new power plant (PP) projects.

—	Prairie State Energy Campus	FutureGen 2.0 Project	Taylorville PP
Power plant (PP) type	Supercritical[1]	Oxy-combustion[8]	HIGCC[4]
Gross PP capacity (MW)	1600[1]	200[2]	716[4]
Product	Electricity[1]	Electricity[2]	Electricity + SNG[4]
Projected cost ($ billion)	4–5[1,7]	~1.3[3]	~3.6[4]
Electricity cost ($/MWh)	58–63.45 for PPs 1st operation [7]	NA	212.73[4]
Annual coal consumption (million tons)	6.3[1]	NA	1.5–2.4
CO_2 emission	15% less than typical US coal-fired PP[1]	>90% captured[2]	>50% captured facility-wide & >90% captured from gasification island[6]
Estimated annual economic Impact ($ million)	$785[1]	NA	>$350[9]
Estimated no. & kind of jobs creation	450[1]	900 construction or direct + 1000 indirect[5]	16,000 direct/indirect construction-related[6]
Projected No. of customers	2.5×10^6 homes[1]	NA	6×10^5 homes[4]
Proposed completion date	2013[1]	April 2016[8]	2017[9]

(1) Coal Age, (2011); (2) McDonald *et al.* (2011); (3) Crooks, (2011); (4) Flores *et al.* (2010); (5) US-DOE, (2010); (6) Lundy and Salmi, (2012); (7) Lydersen, (2012); (8) McDonald and Sturm, (2012); (9) Jana Martin of Tenaska, Inc. (Private communication); NA — not available.

Table 2.5. Comparison of cost per MWh electricity generation by various kinds of power plants.

Kind of Power Plant	(Cost/MWh)
Prairie State Energy Campus	\$58.18–68.01
The FutureGen 2.0 project	NA
Taylorville HIGCC [(1)]	\$212.73
Traditional coal-fired [(1)]	\$141.08–153.03
Nuclear[(1)]	\$101.45–128.03
Combined cycle combustion turbine (IGCC)[(1)]	\$154.05–160.78
Wind[(1)]	\$88.80–121.97

(1) Flores *et al.*, 2010.

boiling temperature and the average temperature of the heat input. For bituminous coal (such as Illinois coal), about 70–75% of the fine coal particles size is below 75 μm, while less than 2% of the particles are over 300 μm (IEA, 2011). This finely powdered pulverized coal is blown with a portion of the combustion air into the furnace through a series of burner nozzles. The combustion process is enhanced by the use of secondary and tertiary air injection into the combustion chamber. Depending on the rank of coal, combustion would occur at temperatures ranging from 1300–1700°C, with typical residence time of particles ranging from 2–5 seconds (CIAB, 2010).

There are two types of boiler designs used in supercritical steam generation: the traditional two-pass layout boiler and the tower type boiler. The two-pass layout boiler type consists of a furnace chamber, topped by heat transfer tubings, which recovers the flue-gas heat content and, as a result, it further reduces the flue-gas exit temperature. In the tower boiler, nearly all the heat transfer sections is mounted vertically above each other over the combustion chamber (CIAB, 2010). This system allows increase of water pressure and temperature above its critical point values (221 bar and 374°C, respectively) which is known as the supercritical condition (CIAB, 2010; Park *et al.*, 1987; Touba and Mansoori, 1998).

Table 2.6. Summary of performance of pulverized coal power plants using anthracite/hard coal (IEA, 2008).

			Ultra-Supercritical	
Steam Cycle	**Subcritical**	**Supercritical**	**(Best available)**	**(AD 700)**[(a)]
Steam conditions	180 bar (540°C)	250 bar (560°C)	300 bar (620°C)	350 bar (700°C)
Net output [MW]	458	458	456	457
Net efficiency [%][(b)]	40.2	42.0	43.4	45.6
CO_2 emission [ton/MWh-net][(c)]	0.83	0.80	0.77	0.73

(a) **AD700** = Advanced (700°) PF Power Plant: Clean Coal Technology. Industry-focused technology program initiated in 1990 to develop advanced materials which could allow increasing the main and reheat steam temperatures in pulverized fuel (PF) boilers up to 700°C or higher.
(b) **Net efficiency** [%] = the net electric energy output as a percentage of the fuel energy input of a thermal power plant.
(c) **CO_2 emission [ton/MWh-net]** = the amount of carbon produced by a facility per net amount of energy produced by the facility.

This results in corresponding increase in average temperature of steam entering the turbines. Higher temperature steam contributes to increase in thermal (Rankine) cycle efficiency of the power plant by almost 2%, which results in less fuel usage and less greenhouse gas production (IEA, 2011). Boilers operating at supercritical conditions do not require hot reservoirs or steam drums because supercritical steam is highly dry and there is no need to separate water and steam (CIAB, 2010). The advantages of supercritical steam coal-fired electric power plants include higher net efficiency and lower greenhouse gas emissions compared to those of conventional subcritical steam coal-fired electric power plants (see Table 2.6).

On the negative side, supercritical-steam coal-fired electric power plants are often perceived as unproven and costly technology unsuitable for local coal, especially coal with high ash content (Burnard

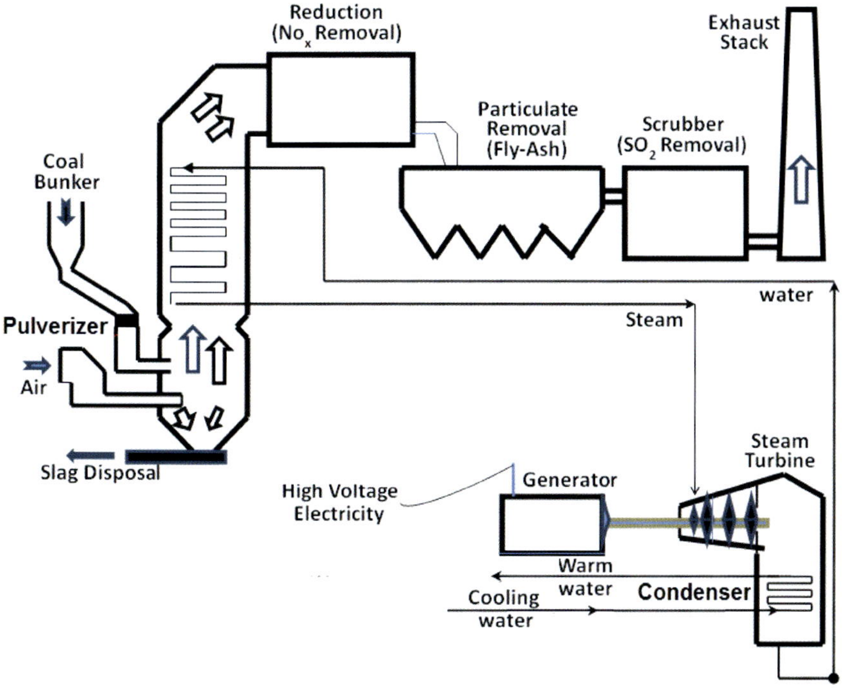

Fig. 2.6. Typical supercritical and ultra-supercritical coal-fired electric power plant (Environmental Engineering Solutions, 2008).

and Bhattacharya, 2011). Figure 2.6 shows the schematic diagram of the supercritical-steam coal-fired electric power plant technology.

b. *Ultra-supercritical steam cycle*

Boilers operating at supercritical pressure above 240 bars and supercritical temperature above 593°C are classified as ultra-supercritical boilers (CIAB, 2010). Increasing the temperature and pressure of steam to ultra-supercritical conditions will further increase the Rankine cycle efficiency of the power plant that use steam as the working fluid with improved environmental performance. The ultra-supercritical steam cycle is a clean-coal technology that can be considered by Illinois as an alternative solution to reduce power plant emissions. Figure 2.6 shows the schematic diagram of the ultra-supercritical steam coal-fired electric power plant technology.

The ultra-supercritical steam coal-fired electric power plant technology is currently the most efficient technology for electricity production using pulverized coal, with efficiencies hovering around 42% (IEA, 2008; Viswanathan *et al.*, 2010). Supercritical- and specially ultra-supercritical steam is highly reactive with inorganic materials, including many metals (Touba and Mansoori, 1998; Touba *et al.*, 1998). The world's first ultra-supercritical electric power plant (Philo Unit 6) in Ohio, built by American Electric Power (AEP) could not operate at ultra-supercritical conditions constantly, due to failure of the material of construction when subjected to prolonged high pressures and temperatures and especially when it was in contact with ultra-supercritical steam (Sigmon, 2008). Consequently, the power plant was finally operated at supercritical conditions until it was decommissioned. With recent metallurgical advancements in the design of high pressures and temperature materials with higher creep-strength-enhanced ferritic steels, advanced austenitic alloys and nickel-based alloys, ultra-supercritical power plant has become practical (Viswanathan *et al.*, 2010).

Lately, almost two dozen ultra-supercritical coal-fired power plants have been commissioned globally. Among those is the Waigaoqiao III ultra-supercritical power plant in Shanghai, China, which reportedly has been saving some 900,000 tons of raw coal and reducing CO_2 output by 1.9 million tons since start-up in 2008 (Patel, 2010). Also, among the ultra-supercritical power plants currently under construction is the state-of-the-art €1.4 billion, 800 MW (equivalent to an installed cost of $2,547/kW (2009 $)) Lünen plant located at the outskirt of Lünen in Germany (Probasco & Ruhlman, 2011). The net efficiency of this plant is around 45.6% and the boiler outlet steam conditions are 280 bars pressure and temperature range of 600–610°C. Late in the year 2012 was defined as the expected date of service of this plant using low-sulfur bituminous coal. In the U.S., AEP recently completed the construction of a 600 MW ultra-supercritical plant in Arkansas. The boiler outlet steam conditions for this plant are 250 bars and 602–610°C. This $1.5 billion project completed in late 2012 is the only ultra-supercritical power plant in the USA (Viswanathan *et al.*, 2010).

The latest ultra-supercritical steam generators can operate at supercritical pressures above 241 bars and at steam temperatures around 593°C. These operating pressure and temperature of ultra-supercritical steam cycle allow less consumption of coal, fewer emissions, and more efficient operation of the turbine cycle. Furthermore, the new construction materials can sustain high temperatures for extended periods of time. Besides improved environmental performance, ultra-supercritical plants are equipped with state-of-the-art emission control technologies, such as dry flue-gas desulfurization system for SO_2 reduction and baghouse technology for particulate removal and selective catalytic reduction system for NO_X reduction (AEP, 2011). Currently, the U.S. DOE in collaboration with Ohio Coal Development Office (OCDO) and other institutions is researching advanced materials for ultra-supercritical coal-fired boilers that can operate at a temperature of 760°C and an unspecified pressure (Viswanathan *et al.*, 2010). Also, ultra-supercritical steam cycle capable of operating at steam temperatures of 700°C (AD700) and a pressure of 300 bars are in the research and development stage. Table 2.6 shows the performance of pulverized fuel power plants using hard coal (anthracite). The net specific CO_2 emission in tons per megawatt hour-net (ton/MWh-net) is defined as the amount of carbon produced by a facility per net amount of energy produced by the facility. The net (%) efficiency is the net electric energy output as a percentage of the fuel energy input of a thermal power plant. The advantages of ultra-supercritical steam coal-fired electric power plants include lower fuel consumption, higher net efficiency and lower greenhouse gas emissions relative to those of supercritical and conventional subcritical steam coal-fired electric power plants (see Table 2.6). On the negative side, like supercritical power plants, ultra-supercritical steam coal-fired electric power plants are often wrongly perceived as unproven and costly technologies unsuitable for local coal, especially coal with high ash content (Burnard and Bhattacharya, 2011). The misguided perception about ultra-supercritical coal-fired electric power technology makes it less attractive in the U.S., despite being demonstrated to be commercially viable and competitive. Instead, the utility companies prefer the conventional subcritical technology because of their familiarity with the technology (Burnard and Bhattacharya, 2011).

2.4.5.2. *Oxy-combustion technology*

Figure 2.7 shows the oxy-combustion technology used for the FutureGen 2.0 project in Illinois (WNA, 2011). The oxy-combustion plant consists of a conventional coal-fired power plant, air separator and gas clean-up equipment.

The oxy-combustion separates oxygen from air and combines it with coal for the combustion to proceed in a mixture of CO_2 and oxygen medium (Phillips and Maxson, 2011). The resulting flue-gas

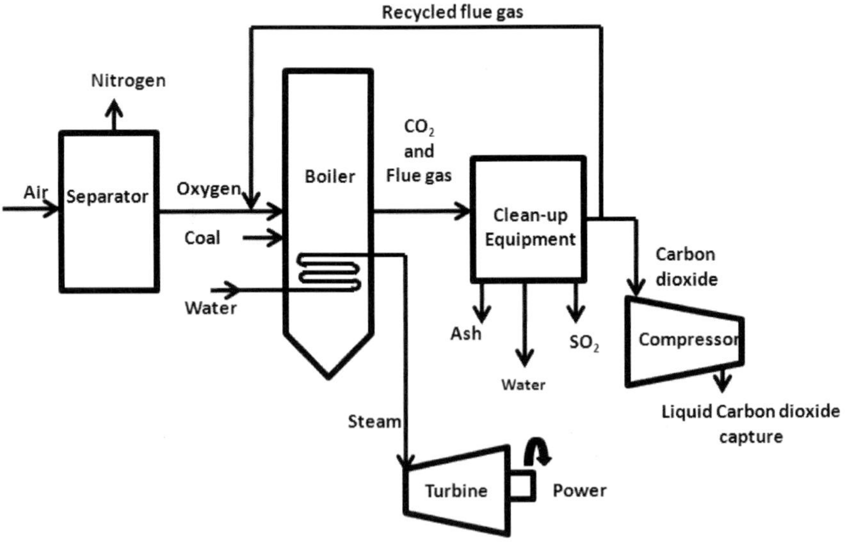

Fig. 2.7. Schematic diagram of the oxy-combustion process (McDonald *et al.*, 2011).

has about 70–90% (dry basis) CO_2. The high purity of CO_2 facilitates its easy collection, transportation and storage. Oxy-combustion plants are constructed to be highly reliable because of their proven commercial components. They are also attractive to the utility industry because their operation is very similar to pulverized coal-fired power plants. In addition, their operating scheme is developed to maximize revenue during times of dramatic price fluctuations. They also have design flexibility for partial capturing of carbon dioxide whenever CO_2 regulations start below 90%, or full capturing when it is above 90%. Besides, use of air separating unit reduces NO_x and other emissions in the flue-gas to near-zero levels (Perrin *et al.*, 2013; SRI Consulting, 2009).

The variable operation and maintenance (O&M) cost for an oxy-combustion supercritical technology with carbon capture is $6.15/MWh. On the other hand, the variable O&M cost for an oxy-combustion ultra-supercritical technology with carbon capture is $5.69/MWh. The major advantage of the latter technology is that it facilitates CO_2 capturing by producing high concentration of CO_2 in the flue gas, and by providing additional flue gas purification, dehumidification and pressurization (Global CCS Institute, 2009).

2.4.5.3. *Integrated gasification combined cycle (IGCC)*

The basic principle of the IGCC technology is to convert coal into syngas at high temperature and pressure, clean-up the gas, then combust it in a two-stage process to produce steam for electricity generation. A simplified block flow diagram of the process is shown in Fig. 2.8. The pollutants from the process are converted into reusable products to reduce emissions. It is estimated that the levelized-cost of energy using the IGCC and carbon capture and sequestration technology will be $147.4 per MWh (US-EIA, 2014b). Additionally, the plant can reduce emission levels to 1.0% of that of traditional coal-fired power plants. By creating the Illinois Clean Coal Portfolio Standard Law (SB

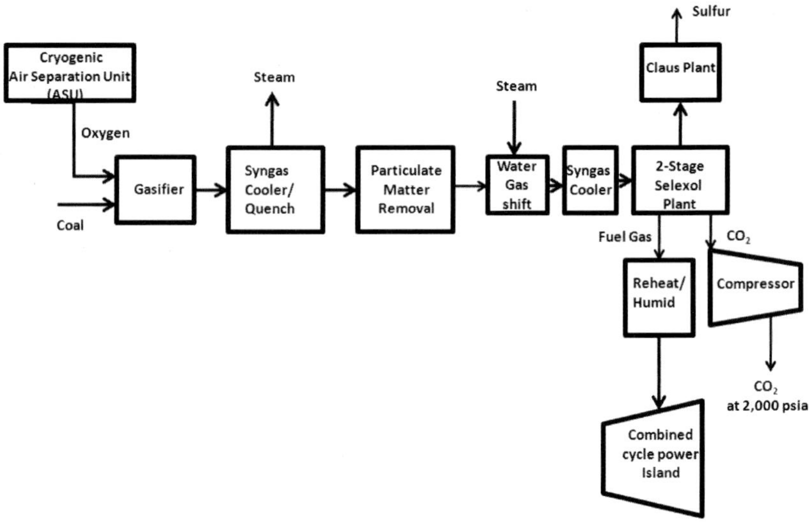

Fig. 2.8. IGCC with carbon capture and sequestration plant design configuration (US-EIA, 2010).

1987), Illinois lawmakers hope that the use of IGCC power plants in the state would curtail greenhouse gas emissions, while making the high sulfur content coal in Illinois attractive for the state's electric power plants. Consequently, in July 2011, the governor of Illinois signed into law two coal-to-gas bills to pave the way for the construction of two IGCC plants in Illinois (Wernau & Karp, 2011; Wells, 2011).

The first step in a gasification process is pyrolysis, which occurs at a temperature of over 400°C and in the absence of oxygen/air. In this step, the hydrogen-rich volatile organic gases are driven out of coal in the absence of oxygen to reduce it to carbon-rich char. In the second step, the char is gasified, using appropriate amounts of pure oxygen from a cryogenic air distillation process (see Chiu and Newton, 1980) and steam to produce carbon monoxide (CO) and H_2 gases at a temperature of over 700°C, leaving behind ash as by-product. The overall reactions for this step are (WNA, 2011).

$$2C(s) + O_2(g) \rightarrow 2CO\,(g) \tag{2.1}$$

$$C(s) + H_2O(g) \rightarrow CO(g) + H_2(g) \tag{2.2}$$

The resulting syngas, which is essentially hydrogen and CO, is cleaned up before combustion in a gas turbine. The hot gases exiting the gas turbine are used to produce steam to generate more electricity. A typical IGCC power plant based on this system of reactions has a thermal efficiency of about 45%. Consequently, current research work on the IGCC technology is focused on adding a water-gas shift reactor to oxidize CO with water, such that the exiting gas stream will be essentially hydrogen and CO_2, with a small quantity of nitrogen and other impurities such as hydrogen sulfide and mercury (WNA, 2011).

Water-gas shift reaction:

$$CO(g) + H_2O(g) \rightarrow CO_2(g) + H_2(g) \tag{2.3}$$

After CO_2 and other impurities are removed from the reaction product, hydrogen becomes the only fuel for energy generation. The lower heating value (known as LHV) of the hydrogen produced is about 28,900 kcal/kg (i.e., five times higher than coal in terms of energy density). By using hydrogen in a gas turbine for mechanical energy generation, the overall system efficiency in the long-term is expected to be about 60% (WNA, 2011).

Approximately, 85% of the CO_2 is captured for sequestration, while the hydrogen sulfide is reduced to elemental sulfur using the Claus process. However, the energy requirement for the air separation plant used to produce oxygen for partial combustion of coal leading to the water-gas shift reaction is roughly 20% of the gross power generated by the plant. For this reason, there is currently no commercial-scale IGCC power plant using hydrogen fuel. One way to reduce this power loss, with corresponding reduction in capital cost of the IGCC plant, is to replace the cryogenic air distillation unit with ion-transport membrane oxygen technology (WNA, 2011). This technology uses a ceramic membrane, a complex metal oxide which under pressure and temperature, ionizes and separates oxygen molecules from air (Dyer, *et al.*, 2000). The technology does not require external source of electrical energy. Therefore, compared to traditional cryogenic air separation units, ion transport membrane technology could reduce the IGCC power plant internal power requirement by as much as 30%, while reducing the capital cost by almost 30% when installed with either oxy-combustion or IGCC power plants. However, these membranes need frequent replacements which will increase the operating cost of the IGCC power plant.

2.4.5.4. *Hybrid integrated gasification combined cycle (HIGCC)*

As mentioned in Sec. 2.4.4, HIGCC technology is an IGCC technology with a methanation unit added to the process for converting the syngas into pipeline-quality substitute-natural-gas (SNG) or methane. Part of the SNG produced in this process can be used for electric power generation, while the balance can be used for households. Hence, this technology is called 'hybrid', since it can supply both chemical and electrical energy simultaneously. Figure 2.9 shows the block diagram of the HIGCC technology process.

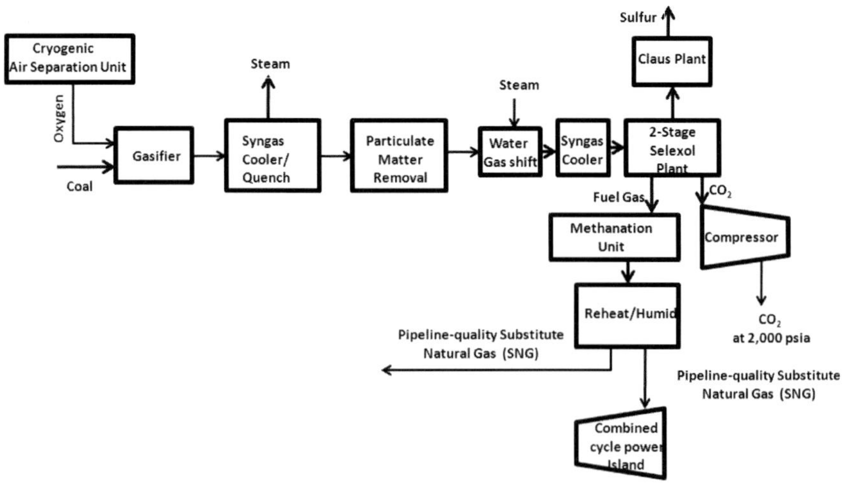

Fig. 2.9. HIGCC plant configuration (Flores *et al.*, 2010).

The syngas produced from the gasification process is cleaned in the particulate scrubber, prior to conversion into SNG in the methanation unit as shown by the equation below.

$$CO + 3H_2 \rightarrow CH_4 + H_2O \tag{2.4}$$

The higher energy density of the SNG, relative to that of syngas improves the efficiency of the power generation (Flores *et al.*, 2010).

2.5. Potential of Underground Coal Gasification (UCG)

The unmineable coal seams have long been recognized as good targets for UCG as shown in Fig. 2.10.

The early efforts of pre-commercial scale application of UCG technology foundered on several requirements. These requirements included the proximity of the UCG mine to populated areas in need of the products of gasification and the availability of dedicated surface facilities that could process and use the large volume of gas and energy produced from the UCG.

In 1981, DOE funded Hammesfahr, Winter & Associates, Inc. to determine the existence of adequate surface infrastructure in any U.S. states with UCG amenable coal reserves. These infrastructures would provide an industrial application, as well as revenues for the synthesis gases produced from the pre-commercialization project, using a multi-well field test. Illinois was selected among five states for the project because of its important coal reserves and its proximity to major markets. New Mexico, Washington State, and Wyoming were included on the list based on their broad database on western coals, while Texas was selected because of its local interest in UCG. Since the objective of the project was to reduce the construction of new surface facilities, the selection criteria was focused on three major fuel consumers: the utility industry, the refinery industry and natural gas processors. However, an exception was granted to states with facilities which could produce transportation grade methanol for

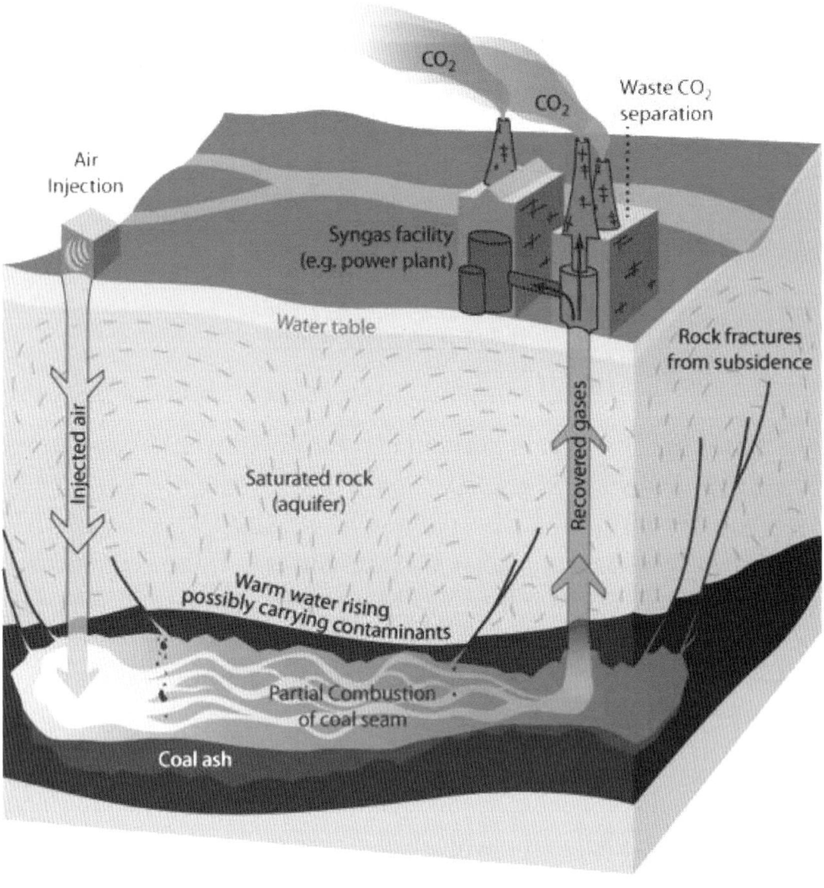

Fig. 2.10. Schematic of an underground coal gasification process adapted from "Alaska's Wild Resource Web" (www.groundtruthtrekking.org/WildResource/Issues/UndergroundCoalGasification.html).

local use, using UCG products. The estimated rate of production for the pre-commercialization test was 1.21×10^{12} cal (4.8 billion Btu) per day, an equivalent of the energy required for 20 MW of power production (Hammesfahr and Winter, 1983).

Table 2.7 shows the number of potential users of the UGC products by state. According to Table 2.7, within the 50 miles radius, Illinois has the highest number of potential users (12) compared with the other four states reported.

Table 2.7. Number of potential users of UGC products by state in 50-mile radius around the coal resource (Hammesfahr and Winter, 1983).

States	**No. of users**
Illinois	12
New Mexico	7
Texas	3
Washington	11
Wyoming	9

Coal reserves of the other 4 states shown in Table 2.7 are quite smaller than that of Illinois, according to Fig. 2.11.

In 1982, Williams Brothers Engineering Company conducted a study on UCG amenable bituminous coal resources in U.S. states. In their study, they included Illinois, Kentucky, Ohio (which obviously have more coal reserves than New Mexico), Texas, Washington, and Wyoming and their potential users of UCG products within 50 miles radius of their coal reserves. Table 2.8 shows the projected target resources in the states of Illinois, Kentucky, and Ohio which were evaluated by Williams Brothers Engineering Company as having the required potential test sites for continued research and development in bituminous coal (Martin *et al.*, 1983).

In 1983, Morgantown Energy Technology Center of the U.S. Department of Energy proposed a research and development project for an UCG technology that could lead to the extraction of energy from the eastern U.S. bituminous coal, which included Illinois coal. The overall goal of this project was to establish a database for the recovery of energy from deep, thin seam, swelling, high sulfur bituminous coals that were less economical to surface mine. In this proposal, again Illinois, Kentucky, and Ohio were recommended as in Williams Brothers Engineering Company study, as having the best potential sites for the project. The selection criteria for the UCG

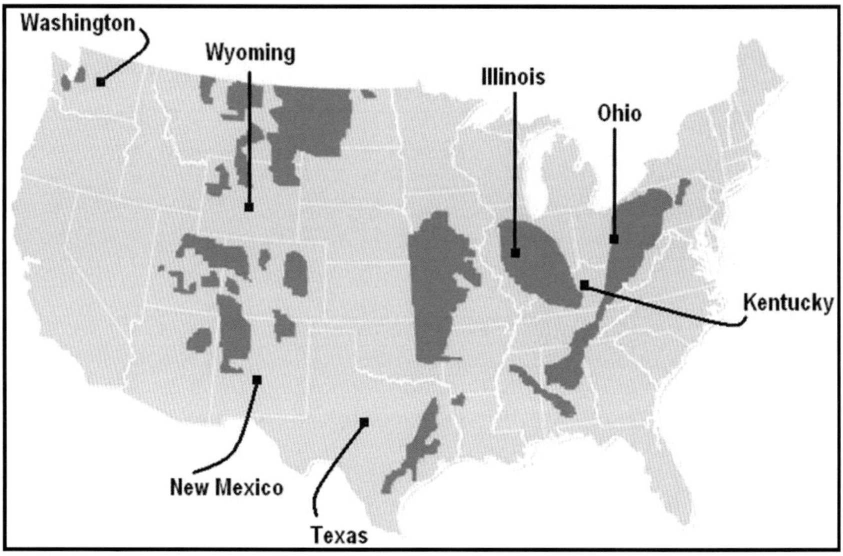

Fig. 2.11. U.S. coal reserves (US-EIA, 2007). States of Illinois, Kentucky, New Mexico, Ohio, Texas, Washington, and Wyoming which are identified in this map were considered in various studies as potential states for underground coal gasification sites.

Table 2.8. UCG amenable bituminous (in metric tons) coal resources in selected U.S. states (Martin *et al.*, 1983).

States	**[10^6 tons]**
Illinois	9,134
Kentucky	1,645
Ohio	82

deep bituminous coal seam site(s) was based on factors, such as coal seam greater than 1.22 meters (4 ft) in thickness, location within a commercially viable quantity of coal, target coal seam of at least 91.44 meters (300 ft) below the water-table, a competent roof and floor structure for the targeted coal seam, and initial coal properties amenable to UCG testing (Martin *et al.*, 1983).

In 1986, a series of laboratory experiments were conducted by Western Research Institute in Laramie, Wyoming to simulate the underground gasification of Herrin bituminous coal from Illinois (Illinois coal number 6). The test program was conducted under a cooperative agreement with the U.S. DOE. The coal samples were provided by Peabody Coal Company from its River King 3 mine near New Athens in Illinois. The purpose of this test was to address some specific concerns, regarding the use of UCG technology for eastern bituminous coals. These concerns included low permeability, swelling tendency, tar production, coal reactivity and geologic setting. The conclusions drawn from these performance tests indicated that the weakly swearing eastern bituminous coal could readily produce 1,780 kcal per standard cubic meter (200 Btu/scf) of gas in a bore hole link system with good lateral growth. Besides, the swelling tendency of the eastern bituminous coal (with free swelling index of 4 — a value to indicate the coking of coal) does not adversely affect the UCG process and that a sustained gasification is achievable when the coa l is gasified under ash slagging conditions. Also noted during the performance test was that sidewall reactions and gas flow patterns appeared to control the rate and shape of cavity growth, and therefore lateral cavity growth in thin eastern coal seams may be greater than originally expected (Glaser and Johnson, 1986).

Despite the potential for UCG, there has never been any UCG project taking place in the U.S. (Bradshaw and Dance, 2005). The reason is that most of the UCG projects in the U.S. which began in the early 1960s were terminated by mid-1980s, due to the decline of oil prices around the world. Nevertheless, from 1974 to 1989, there was a surge in research and development of UCG in the U.S., with 33 UCG pilot projects located in West Virginia, Wyoming, Alabama, Texas and Washington State. Out of these projects, the most successful one was the Rocky Mountain 1 project, conducted in the 1980s in Carbon County, Wyoming and organized by the U.S. DOE, the Electric Power Research Institute (ERPI), Amoco Production Company, Union Pacific Resources and the Gas Research Institute (GRI).

In 2009, a research report on the potential for UCG cited two coal fields in the Illinois Basin that would meet the important criteria (coal thickness, depth, and other characteristics, such as moisture content and heating value for these two coals) for UCG. The coal fields are the Springfield and Seelyville coal beds in Southwest of Indiana, located in the Illinois Basin. The report recommended initiating activities on the possibility of locating an UCG pilot project in those coal fields in the future (Ag Mohamed *et al.*, 2009). The total capital cost of the future UGC project in Indiana, based on capital expenditures (in $2009) of an air-fired UCG was estimated as $130.6 million. This estimate included the cost of drilling, air compression, piping, accessories, gas-cleaning, site facilities, land acquisition, salaries during plant construction, permitting and legal fees and working capital (Ag Mohamed *et al.*, 2009).

Currently, there are no UCG facilities operating in the United States and there are no major U.S. private companies committed to UCG process. However, there is a host of institutions in the country involved in various research and development activities in promoting UCG's potential (Friedmann, 2009). It is argued that the potential impact of UCG application in the U.S. is to drive down the price of oil, a situation similar to the effect of shale gas utilization on natural gas price. This is because a large portion of the UCG products may be converted into gasoline and other petroleum products using the Fischer–Tropsch/gas-to-liquid processes.

2.6. Environmental Concerns About Coal and Clean Coal Technology

Even with the claimed advances in clean coal technology, environmental activists around the world object to the use of coal in any form, due to environmental and health impacts associated with mining and burning of coal, especially high-sulfur coal, as well as the problems and adverse effects of storing large amounts of CO_2 underground

(Agyarko and Mansoori, 2013). In what follows we enumerate the problems associated with the use of coal.

2.6.1. *Coal mining fatalities*

Since coal, petroleum and natural gas reservoirs are found in the same locations in the Illinois Basin (shared by three States of Illinois, Indiana and Kentucky), it appears that the simultaneous exploitation of these resources can be very problematic. As we will discuss in Chapter 3 on petroleum production and consumption, there are many oil and gas wells already drilled in the Illinois Basin during the past several years. The large number of oil and gas wells drilled through or near coal mines may penetrate the mines, leading to explosions during coal mining and possible loss of life (Rice *et al.*, 1913). As a result, there are huge amounts of coal left around the oil wells that could not be mined due to explosion hazard. In the past, the methods for protecting coal mines from gas and oil wells leakages have included open workings and pillars or coal that may be left as pillars. But the use of coal pillars for separation of coal mines from oil and gas leaks results in the loss of coal. These problems would exasperate with a dramatic increase in mining of coal, which is envisioned for application in clean coal technology to produce electricity.

There have been several reported cases of deaths as a result of explosions and fires, possibly due to leakages of oil and gas in coal mines in the Illinois Basin. In Table 2.9, we report the deadliest of these coal mining accidents, as reported by the U.S. Department of Labor.

2.6.2. *Adverse health effects of coal mining*

There is a major concern about the health of coal miners and that of the residents of coal mining areas. Underground coal miners frequently suffer from diseases, such as pneumoconiosis (black lung disease) caused by breathing of coal dust. The effects of this disease include shortness of breath, as well as the risk of having emphysema,

Table 2.9. Deadliest Illinois Basin coal mining accidents (U.S. Department of Labor: *www.msha.gov/MSHAINFO/FactSheets/MSHAFCT8.htm*).

Location	Year	Fatalities	Type
Cherry	1909	259	Fire
Centralia	1947	111	Explosion
West Frankfort	1951	119	Explosion
Rosiclair	1971	7	H_2S gas

bronchitis and fibrosis diseases. According to the National Institute for Occupational Safety and Health (NIOSH), in 2005–2006, 9% of miners with 25 or more years of mining experience tested positive for the black lung disease, as opposed to only 4% in the late 1990s. The reason for such an increase can be attributed to higher production rates and use of mining machinery that churn up more dust.

2.6.3. *Mountaintop removal mining*

Apart from coal miners' health problems, the pollutants from coal mining are associated with chronic illnesses among the residents of coal mining communities. In 2008, a study conducted by West Virginia University Health Sciences Center indicated that the residents of coal mining communities have 70% increased risk of having kidney disease, 64% increased risk of developing chronic obstructive pulmonary diseases (COPD) and 30% more likely to have high blood pressure (hypertension). Also, the same study indicated that human mortality rates are higher in coal mining communities compared to other areas in U.S. (Energy Justice Network, 2012). It is envisioned that such problems will be more severe with mining of high sulfur coal.

Furthermore, there are concerns about public safety and the environment since the public are against coal mining at mountaintops (mountaintop removal mining like Fig. 2.12), a process which entails

Fig. 2.12. A mountaintop mining operation in West Virginia. (www.epa.gov/region3/mtntop/).

the removal of up to 305 meters (1,000 feet) of soil at the top of mountains, in order to gain access to the underlying coal seams (Jasmine, 2009).

Mountaintop removal mining has serious repercussions on the ecological balance of the environment because it degrades the natural landscape. An example of such a scene is in the Appalachian Mountains, where an estimated one million acres of forests have been destroyed in coal mining. According to the U.S. Environmental Protection Agency, an estimated 1,931 kilometers (1,200 miles) of rivers and streams in U.S. have already been buried permanently, as a result of mountaintop removal of soil for coal (Chua, 2012). Also, coal mining at mountaintops produces slurries that contain heavy metals and carcinogens which are discharged into open lagoons. The pollutants from these lagoons can contaminate underground drinking waters. In 1972, a flood from a waste pond at Buffalo Creek in Logan County, West Virginia killed 125 people, injured 1,000 others, and left 4,000 people homeless. In addition, in 2000, a waste pond in Inez,

Kentucky spilled over 300 million gallons of pollutants into local waterways (Reece, 2006). Furthermore, mountaintop removal of soil in coal mining threatens the livelihood of many people. For example, there are claims that some families living close to the Appalachian Mountains were threatened by the mining companies who wanted to dislocate them in order to facilitate their mining activities (Jasmine, 2009).

2.6.4. *Carbon dioxide causing climate change*

Climate change due to excessive amount of carbon dioxide in the air is happening now and happening fast, and there is a need for urgent action to cut greenhouse gas pollution. There is a concern that CO_2 capture and sequestration (CCS) technology may contribute to the onslaught of climate change which is fast approaching. It is estimated that the earliest possible time for commercial-scale deployment of the CCS technology would be after 2030 (Jasmine, 2009). Furthermore, other studies have indicated that there are potential problems associated with the integrity of CO_2 storage reservoirs, due to chemical reactions from the CO_2 injection. For example, the injected CO_2 has the tendency to acidify the water in the reservoir, which can dissolve minerals such as calcium carbonate and cause formation damage by increasing the permeability of the underground rock, the repair of which would be very expensive (Paktinat and Mansoori, 1993). Enhanced rock permeability can cause the escape of CO_2-rich fluids from the reservoir through different pathways, which may pollute underground drinking waters. Besides, since CO_2 is heavier than air, it could cause affixation in case of any leakage into enclosed living areas of humans and livestock. In 1986, a large quantity of CO_2 emissions from a volcanic crater in Lake Nyos in Cameroon, West Africa killed 1,700 people (Kling *et al.*, 2005). Also, there are concerns about transportation of CO_2 in pipelines, especially in populated areas, because it has similar environmental risks in case of accidental leakage (Folger, 2008).

2.6.5. *Direct environmental problems of clean coal technology*

It is generally accepted that with the use of advance environmental pollution control methods for removal of pollutants and their safe disposal, together with CO_2 sequestration, the net efficiency of coal-fired power plants would be drastically reduced. For CCT to have a dominant role in the future of any state's energy supply, their net positive efficiency and environmental performance must be improved.

The major problems confronting the development of CCTs are mounting costs of projects, marginal improvements in technology efficiencies due to high cost of environmental cleanup of pollutants, and CO_2 sequestration difficulties. Also, the high capital cost of CCTs make it difficult for it to compete with other alternative fuels for electricity generation (Philibert and Podkanski, 2005). But this efficiency improvement is often hampered by factors such as high combustion temperatures, which have the tendency to breakdown the combustion turbines, resulting in frequent plant downtime and loss of revenue (Viswanathan *et al.*, 2010).

2.7. Environmental Problems Associated with Coal Power Generation

Coal has long been one of the major fossil fuels used to produce electricity and one of the largest sources of environmental pollution in the world. Coal mining and power generation using coal has been a major contributor to many economies since it has provided employment, generated revenue, helped develop infrastructure and improved the livelihoods of people (Mansoori, 1994; Agyarko and Mansoori, 2013). Despite its importance as an energy source, coal mining and power generation has become the focus of attention in regards to emissions of greenhouse gases (GHG) and toxic chemicals (Shibli, 2014). In Chapter 5, we present the GHG emissions from

Fig. 2.13. Plumes of a small coal-fired power plant.

coal and other fossil fuels (petroleum, natural gas, etc.). In this section, we discuss toxic chemical emissions from using coal in power generation.

Environmentally, coal emits more pollutants than other energy sources. A picture taken by one of the authors of this book (GAM) from a small coal-fired power plant is depicted in Fig. 2.13, showing the size of plumes being released to the atmosphere.

EPA has indicated that in order to meet the goal of GHG emissions reduction, CCS technology must be an integral part of future coal-fired power plants. For example, coal combustion in IGCC power plant without carbon capture and sequestration emits 0.06 lb of NO_X per MMBtu, 0.025 lb of SO_2 per MMBtu and 206 lb of CO_2 per MMBtu (US-EIA, 2013). These emissions contribute to global warming and cause health problems. Illinois Basin coal has high sulfur

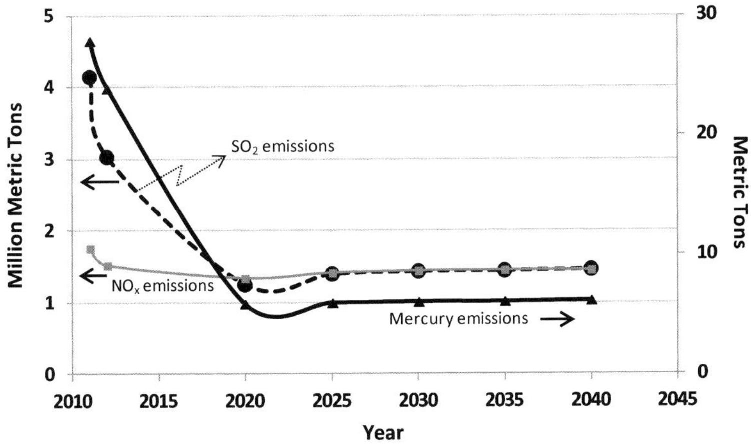

Fig. 2.14. Estimation of electric power sector emissions in the U.S. (US-EIA, 2014c).

content (about 5% weight), and its large scale combustion will be a major source of pollution.

Coal-fired electric generating units are subject to the Mercury and Air Toxics Standards (MATS), which requires fossil-fuel steam electric generators with capacities greater than 25 MW to greatly reduce emissions of mercury, acid gases and toxic metals. The standards were scheduled to take effect in April 2015 and EIA estimates that emissions of SO_2 and Mercury from electric power plants will be reduced significantly by the year 2040, as shown in Fig. 2.14.

2.7.1. *Adverse health effects of coal combustion*

Presently, coal-fired power plants produce electricity for many countries but, according to the American Lung Association, they also produce more hazardous air emissions than any other industrial pollution sources. In the U.S. 386,000 tons of 84 separate hazardous air pollutants spew from over 400 plants in 46 states. Such emissions threaten the health of people who live near these plants, as well as those who live hundreds of miles away. Also, according to the American Lung Association, despite the concentration of these plants largely in the Midwest and Southeast parts of the U.S., their toxic emissions threaten the air in communities nationwide (Am Lung Assoc, 2011). According to the Clean Air Task Force (CATF, 2010), scientific findings concerning the link between fine particle air pollution emissions from the U.S. coal-burning power sector and public health indicate tens of thousands of premature deaths each year and hundreds of thousands of heart attacks, asthma attacks, emergency room visits, hospital admissions and lost workdays.

2.7.2. *Corrective actions to reduce environmental and health hazards of coal*

To comply with the MATS, it is assumed that all coal-fired plants have control equipment such as flue gas desulfurization scrubbers or dry sorbent injection systems installed by 2016. The control equipment needed to reduce mercury is specific to each plant configuration and coal type. According to EPA, complying with MATS mandate would cost the electric generating industry $9.6 billion annually, however, the health benefits of the rule far exceed its cost and prevent up to 11,000 premature deaths and provide $90 billion in health benefits (Jacobs, 2014).

In the U.S., coal-fired power plants have been under significant economic pressure in recent years, mainly due to low price of natural gas and low electricity demand and also because they are forced to comply with the new environmental mandates (see Fig. 2.15).

From 2009 to 2013, 18 GW of coal-powered electricity generators have retired and 28 GW announced their retirement by 2025 (Fleischman, *et al.*, 2013). Overall, a total of 51 GW coal-fired capacity will be retired between 2012 and 2040.

Water consumption is an integral part of power generation technologies and an aspect of coal energy extraction that does not get required attention. In coal processing, water is used for cooling and, depending on coal quality, may be used to reduce ash and sulfur content of coal. Water consumption for coal extraction is on the

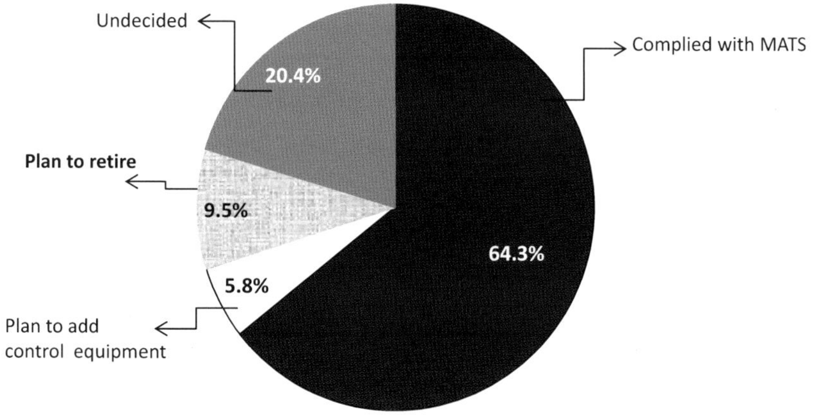

Fig. 2.15. MATS compliance strategies for U.S. coal-fired electric power sector as of end of 2012 (Data from (US-EIA, 2014d)).

average 2.6 gal/MMBtu (219 liter per metric ton) for mining and 0.8 gal/MMBtu (67 liter per metric ton) for washing (Mielke *et al.*, 2010). Water scarcity is one of the main challenges facing the international community. It is estimated that global freshwater demand will increase by 30%, while supply will decrease by 40% (Baowen, 2014). A study by the International Energy Agency (IEA, 2012) indicated that world energy demand will grow 30% by 2035, but that water consumption for energy generation and production will grow at higher rate and increase by 85%.

There are reports that indicate existence of technologies that can incrementally increase the efficiency of a coal-fired power plant, but such incremental increase in efficiency may not be enough to markedly reduce GHG and toxic chemicals to an accepted level set by the EPA. It seems that substantial improvements in reduction of such emissions would not come to fruition, unless major improvements in technology are realized.

Predictions indicate that despite its negative connotation, coal will be one of the main sources of power generation in the U.S. It is estimated that between 2012 and 2040, U.S. coal consumption will increase by an average of 0.3% per year and reach 22.6 quadrillion

Btu (more than one billion metric tons, assuming coal's heat content of 19.5 million Btu per short ton) in 2040 (US-EIA, 2014c). Hopefully, the tightened emission rules which are set to prevent release of greenhouse gasses by fossil fuel power generating plants will be effective as predicted.

2.8. Concluding Remarks

In this chapter, we reported on problems of conventional methods of coal production and consumption, the characteristics of coal, the impact of "Clean Air Act" on the future of coal-fired power plants and electric power plant projects using CCT. We also reported on the underground coal gasification possibilities and the environmental concerns about coal and CCT. There exist major difficulties and tremendous costs involved in true implementation of the CCT, which includes major reduction in pollutants released to the environment, i.e. collection and sequestration of CO_2 and other emissions in keeping the air clean.

Globally, coal resource is forecasted to last for about 113 years, and vast reserves that are expected to last for over 100 years exist in U.S., Russia, India and Australia (BP, 2014; Agyarko and Mansoori, 2013; Mansoori, 1994). Consumption of coal is currently growing in the world as an energy source, due to increasing population in the world and growing economies like China and India. Consequently, there are predictions that the world could be running out of coal sooner than was previously predicted (Heinberg and Fridley, 2010; Nassos, 2011). State of Illinois is endowed with the largest overall bituminous coal and the largest strippable bituminous coal reserves in U.S., enough to feed its coal-fired power plants for about 250 years, based on current rate of consumption (IDCEO, 2010).

In 2011, Illinois announced a total of \$5.3 billion investments for the construction of two coal-to-gas plants in the state. This is in addition to four clean coal technology construction projects in the state based on supercritical steam cycle technology, oxy-combustion technology, IGCC technology and HIGCC technology (McDonald *et al.*, 2011, Wernau and Karp, 2011; Wells, 2011).

Economically, coal is one of the cheapest sources of energy for power generation, with projected levelized cost of electricity for the year 2015 estimated at \$50–90 per MWh, making it one of the most competitive energy sources (NEI, 2011). However, with projected average U.S. levelized cost of electricity for coal-fired power plants entering service in 2019 estimated at \$95.6 per MWh (for conventional coal) and \$115.9 per MWh for IGCC CCT cycle, coal power will soon be less competitive as compared with wind or natural gas power (US-EIA, 2014b).

According to the National Energy Technology Laboratory (NETL), the average efficiency of coal-fired plants in the U.S. was 32% in 2007 (Campbell, 2013). In order to achieve the goal of reduction in GHG and other emissions, increasing heat rate efficiency of coal-fired plants is necessary. However, as Campbell (2013) indicated in his report, efficiency improvements in an individual area of a coal-fired power plant are marginal, as shown in Table 2.10.

The relatively low cost of Illinois Basin coal, which is mainly due to its higher sulfur and ash content, may make it an economically attractive option for power generation. However, this economic advantage will come at the cost of higher rate of environmental

Table 2.10. Actual efficiency improvements of coal-fired electric generating units (Campbell, 2013).

Efficiency Improvement Technology	Percentage (%) Efficiency Increase
Sootblower Optimization	0.10–0.65
Low-Rank Coal Drying	0.10–1.70
Combustion Control Optimization	0.15–0.84
Cooling System Heat Loss Recovery	0.20–1.00
Flue Gas Heat Recovery	0.30–1.50
Steam Turbine Design	0.84–2.60

pollution or higher cost of emission-prevention technology. Characteristics of coal and environmental problems of using coal as a source of energy makes it very challenging to use coal as a contributor to energy sustainability in any state or country. Considering the difficulties of achieving the goals of clean coal technology and the strong public opinion against burning coal to produce electricity, it is unlikely that coal will remain a major economically attractive source of energy in the U.S.

As the renewable energies replace fossil fuels and nuclear energy use, the non-fuel uses of fossil fuels, including coal, may become more important, as the source of chemical feedstock in the future. Coal, in addition to carbon, contains many different chemical compounds which have many non-combustion applications in chemicals, pharmaceutical, materials and construction industries. The ways in which coal constituents may be separated and utilized include leaching, supercritical extraction, carbonization, etc. For example, conventional carbonization is heating of coal in the absence of air in a coking oven at about 900–1200°C (1650–2200°F). Liquid coal tar is the major liquid product resulting from the carbonization of coal with solid coke as the by-product. Coal tar contains a combination of simple and complex aromatic hydrocarbons including anthracite, naphthalene, phenanthrene derivatives and generally poly aromatic hydrocarbons (PAHs), as well as diamondoids, phenols, heterocyclic oxygenated compounds, sulfur compounds and nitrogen compounds. There are extensive numbers of literature about chemistry, separations of constituents and non-fuel applications of coal-derived compounds (see for example the two book by van Krevelen (1948 and 1993); the study by Schobert and Song (2002) and the book on the chemistry and technology of coal by Speight (2012). If non-fuel uses of coal can be proven as economically viable and environmentally friendly technologies, then there are tremendous opportunities that can be explored in the U.S. and specifically in Illinois.

Bibliography

[AEP] American Electric Power (2011). Ultra-supercritical generation — increased efficiency with improved environmental performance, fact-sheet. American Electric Power, Columbus, OH.

Affolter, R. H. and Hatch, J. R. (2002). Characterization of the quality of coals from the Illinois basin, Chapter E, in Hatch, J. R. and Affolter, R. H. (eds.), *Resource Assessment of the Springfield, Herrin, Danville, and Baker Coals in the Illinois Basin*. U.S. Geological Survey Professional Paper No. 1625-D.

Ag Mohamed *et al.* (2009). Viability of underground coal gasification with carbon capture and storage in Indiana, V600 Capstone Course, Bloomington School of Public and Environmental Affairs, Indiana University, IN.

Agyarko, L. B. and Mansoori, G. A. (2013). A Review of Non-Renewable Energy Options in Illinois. *Int. J. Oil, Gas and Coal Technol.*, **6**(3): 288–347.

[Am Lung Assoc] American Lung Association (2011). Toxic Air — The case for cleaning up coal fired power plants. Lung.Org., Chicago, IL.

Baowen, L. (2014). Water crisis calls for common action. Editor's note. *Cornerstone Magazine*, **2**(1), p. 1.

Berkowitz, N. (2012). *An Introduction to Coal Technology*. Elsevier.

[BP] British Petroleum. (2014). Statistical Review of World Energy, 63rd Edn.

Bradshaw, J. B. and Dance, T. (2005). Mapping geological storage prospectivity of CO_2 for the world sedimentary basins and regional source to sink matching. In *Proccedings of the 7th Int'l Conf. on Greenhouse Gas Technologies* (*GHGT-7*), 1, pp. 583–591. Vancouver, 5–9 September, BC.

Buchan, B. and Cao, C. (2004). Coal-fired generation: proven and developing technologies, Office of Market Monitoring and Strategic Analysis, Florida Public Service Commission, Tallahassee, FL.

Burnard, K. and Bhattacharya, S. (2011). Power Generation from coal: ongoing developments. Int'l. Energy Agency, OECD/IEA, pp. 5–40.

Bzdak, Z. (2012). Quinn and consumers: this contrarian gamble on natural gas prices is a risk. Opinion section, *Chicago Tribune*, Chicago, IL.

Campbell, R. J. (2013). Increasing the efficiency of existing coal-fired power plants, Congressional Research Service Report No. R4334.3.

[CATF] The Clean Air Task Force (2010). The toll from coal — an updated assessment of death and disease from America's dirtiest energy source. The Clean Air Task Force, Boston, MA. *catf.us*.

Chiu, C. -H. and Newton, C. L. (1980). Second Law Analysis in Cryogenic Processes, *Energy*, 5(8–9): 899–904.

Chua, J. M. (2012). Find out your connection to mountaintop removal. *science.howstuffworks.com*.

[CIAB] Coal Industry Advisory Board (2010). Power generation from coal: measuring and reporting efficiency performance and CO_2 emissions, Report No. OECD/IEA2010, CIAB, Int'l Energy Agency, Paris, France.

Coal Age (2011). Prairie state energy campus startup just a few months away. *Coal Age News,* **117**(8): p. 6.

Crooks, E. (2011). FutureGen Alliance selects Morgan County, Ill. as the site for the FutureGen 2.0 carbon storage facility, Press release, FutureGen Alliance Organization, Jacksonville, IL.

Daniel, S. (2015). FutureGen 'clean-coal plant' is dead. News Section, *Crains Chicago Business*.

Dow Jones Newswires (2012). U.S. court to scrutinize EPA's delayed power — sector rule. Fox Business News.

Drajem, M. (2011). EPA issues air toxics cap for U.S. coal-fired power plants. *Bloomberg Business Week*, New York, NY.

Dyer, P. N., Richards, R. E., Russek, S. L. and Taylor, D. M. (2000). Ion Transport Membrane Technology for Oxygen Separation and Syngas Production, *Solid State Ionics*, **134**(1–2): 21–33.

Dudek, M. (2014). State to invest $2.3 million in clean-energy companies. Chicago Sun-Times, Chicago, IL.

Energy Justice and Network (2012). Fuels and Technologies: Coal: Dangerous Power, *Energy Justice and Network*, **3**(1): 1–5.

Environmental News Service (2006). Ameren to slash emissions at Illinois coal-burning power plants. *Environment News Service*, Lincoln City, OR.

Environmental Engineering Solutions (2008). High performance coal-fired supercritical power plants-promotes cleaner environment, *Environmental Engineering Solutions,* **2**(15): 1–23.

Fleischman, L., Cleetus, R., Deyette, J., Clemmer, S., Frenkel, S. (2013). Ripe for Retirement: An Economic Analysis of the U.S. Coal Fleet. *The Electricity J.*, **26**(10): 51–63.

Flores, M., Colgan, J., Ford, L., Elliot, S. and O'Connell-Diaz, E. (2010). Analysis of the Taylorville energy center facility cost report, Illinois Commerce Commission (ICC), TEC Report No. 1, pp.1–45.

Folger, P. (2008). Carbon capture and sequestration (CCS), Congressional Research Service, CRS Report No. RL33801, pp. 1–31.

Friedmann, S. J. (2009). Coal without carbon: accelerating development of underground coal gasification. *Clean Air Task Force*, Lawrence Livermore National Lab., **2**(37): 1–17.

Glaser, R. R. and Johnson, L. A. (1986). Physical simulations of underground coal gasification in eastern bituminous coal, University of Wyoming Research Corp., Laramie, Western Research Inst., Report No. DOE/FE/60177-2249.

Global CCS Institute (2009). Strategic analysis of the global status of carbon capture and storage — Report 2: Economic assessment of carbon capture and storage technologies, Final report, Global CCS Institute, Canberra, Australia.

Gregerson, J. (2012). Ill. EPA says no to coal-to-gas project's construction permit, ENR Midwest, Chicago, IL.

Hammesfahr, F. W. and Winter, P. L. (1983). Underground coal gasification: The leading source for lower-priced gas. In *Proceedings of 9th Ann. Underg. Coal Gasify. Symp*, pp. 68–74. Bloomingdale, 7–10 August, IL.

Hawthorne, M. (2010). EPA targets coal-fired power plants: New rule would cut smog, soot in Chicago, nationwide. *Chicago Tribune*, Chicago, IL.

Heinberg, R. and Fridley, D. (2010). The End of Cheap Coal. *Nature*, **468**: 367–369.

[IDCEO] Illinois Dep't of Commerce and Economic Opportunity (2008). The Illinois coal industry. Report, Office of Coal Development, Springfield, IL.

[IDCEO] Illinois Dep't of Commerce and Economic Opportunity (2009). Prairie State Energy Campus & Lively Grove Mine. Annual Report, Office of Coal Development, Springfield, IL.

[IDCEO] Illinois Dep't of Commerce and Economic Opportunity (2010). The Illinois coal Industry. Office of Coal Development, Springfield, IL.

[IDCEO] Illinois Dep't of Commerce and Economic Opportunity (2012). The Illinois coal Industry. Office of Coal Development, Springfield, IL.

[IEA] Int'l Energy Agency (2008). Energy technology perspectives 2008-scenarios and strategies to 2050, in support of the G8 plan of action. IEA, Paris, France.

[IEA] Int'l Energy Agency (2011). Clean coal technologies. IEA Clean Coal Center, Database Section, Clean Coal Technologies, London, UK.

[IEA] Int'l Energy Agency (2012). World energy outlook 2012. IEA, Paris, France.

[IEPA] Illinois Environmental Protection Agency, (2011). 4th Illinois Annual Air Quality Report. IEPA, Springfield, IL.

Jacobs, J. (2014). Federal judges uphold EPA mercury and toxics standards. *Midwest Energy News*, St. Paul, MN.

Jacobson, R. J. and Korose, C. (2003). Coal Geology in Illinois, Keystone Coal Industry Manual, Primedia Intertec Pub. Co.

Jasmine, G. (2009). The many problems with clean coal, the environment and wildlife cause, Redwood City, CA. *care2.com.*

Kling, G. W., Evans, W. C., Tanyileke, G., Kusakabe, M., Ohba, T., Yoshida, Y. and Hell, J. V. (2005). Degassing Lakes Nyos and Monoun: Defusing Certain Disaster. *PNAS*, **102**(40): 14185–14190.

Learner, H. and Rowe, J. (2011). Clearing the air: when opposites actually agree — in search of cleaner air. *Chicago Tribune,* Chicago, IL.

Lefton, S. A., and Besuner, P. (2006). The cost of cycling coal fired power plants. *Coal Power Magazine*, pp. 16–20.

Lundy, D. and Salmi, H. (2012). Illinois EPA issues Taylorville energy center final air permit; slashes mercury emissions following request by Tenaska. *Reuters*, New York.

Lydersen, K. (2012). Ohio towns have buyer's remorse over Prairie State Energy Campus. *Midwest Energy News*, St. Paul, MN.

Mansoori, G. A. (1994). Where Do We Stand with Fossil Fuels? (A Global Perspective), FLASHPOINT, AIChE-F&PD Publication, American Inst. of Chemical Eng., New York, **5**(1): 4–5.

Martin, J. W., Byrer, C. W., Brummert, A. C. and Wieber, P. R. (eds.)(1983). Potential for development of underground coal gasification in the bituminous coal resource. In *Proceedings of 9th Annual Underground Coal Gasification Symp.* pp. 150–158. Bloomingdale, 7–9 August, IL.

McDonald, D. K., Moorman, S., Darde, A., and de Limon, S. (2011). Oxy-combustion: A promising technology for coal-fired plants. *Power Magazine*, Houston, TX.

McDonald, D. K. and Sturm, E. (2012). FutureGen 2.0: oxy-coal's opportunity. Technical Paper No. BR-1853. Presented at the 36th Int'l Technical Conf. on Clean Coal & Fuel Systems, Clearwater, FL.

Mielke, E., Diaz Anadon, L., and Venkatesth, N. (2010). Water consumption of energy resource extraction, processing, and conversion, A review of the literature for estimates of water intensity of energy-resource extraction, processing to fuels, and conversion to electricity. Energy Technology Innovation Policy Discussion Paper No. 2010-15, Belfer Center for Science and Int'l Affairs, Harvard Univ., Boston, MA.

[MITEI] MIT Energy Initiative (2013). Taylorville energy center fact sheet: carbon dioxide capture and storage project, Carbon Capture and Sequestration Technologies Program at MIT, CCS Project Database, MIT, Boston, MA.

Nassos, G. P. (2011). A sustainable environment: our obligation to protect God's gift, can renewable energy truly replace non-renewables? *georgepnassos.com*.

[NEI] Nuclear Energy Institute. (2011). The cost of new generating capacity in perspective, White paper, Nuclear Energy Institute, Washington, D.C.

Osborne, D. (2013). *The Coal Handbook: Towards Cleaner Production: Coal Production*. Woodhead Publishing.

Paktinat, J. and Mansoori, G. A. (1993). Reduction of formation damage, emulsion and adsorption through the use of ethoxylates in sandstone. In the *Proceedings of the 1st Int'l Non-Renewable Energy Sources Congress* (*INRESC '93*), M. Edalat and G. A. Mansoori (eds.) pp. 167–180.

Park, S. J., Kwak, T. Y. and Mansoori, G. A. (1987). Statistical Mechanical Description of Supercritical Fluid Extraction and Retrograde Condensation. *Int'l J. of Thermophysics*, **8**(4): 449–471.

Patel, S. (2010). China: A world powerhouse. *Power Magazine*, Houston, TX.

(2011). Pekin has worst sulfur dioxide air pollution in state. *Pekin Daily Times*, Pekin, IL.

Perrin, N., Dubettier, R., Lockwood, F., Court, P., Tranier, J. P., Bourhy-Weber, C. and Devaux, M. (2013). Oxycombustion for carbon capture on coal power plants and industrial processes: advantages, innovative solutions and key projects. *Energy Procedia*, **37**: 1389–1404.

Philibert, C. and Podkanski, J. (2005). International energy technology collaboration and climate change mitigation: case study 4: Clean coal technologies, Int'l Energy Agency, Paris, France.

Phillips, J. and Maxson, A. (2011). Research and development for future coal generation. *Power Magazine,* Houston, TX.

Probasco, D. and Ruhlman, B. (2011). Coal-fired generation cost and performance trends. *Power Magazine*, Houston, TX.

[PSCW] Public Service Commission of Wisconsin (2004). Electric Power Plants. PSCW, Madison, WI.

[PSGC] Prairie State Generating Company (2012). Yes. We are the future of energy. PSGC Annual Report, Marissa, IL.

Reece, E. (2006). Moving mountains: the battle for justice comes to the coal fields of Appalachia. *Orion magazine*, pp. 1–8.

Rice, G. S., Hood, O. P. *et al.* (1913). Oil and gas wells through workable coal beds, U.S. Dep't of Interior, Bureau of Mines, Petroleum Technology, Vol. 7, Bulletin 65.

Schobert, H. H. and Song, C. (2002). Chemicals and Materials from Coal in the 21st Century. *Fuel*, **81**(1): 15–32.

Shaalan, H. E. (2003). Generation of electric power. *Handbook of Electric Power Calculation*, 3rd Edn. McGraw-Hill Book Co.

Shibli, A. (2014). Coal Power Plant Materials and Life Assessment: Developments and Applications. Woodhead Publishing

Sigmon, W. L. (2008). The lure of ultra-supercritical: exploring the future of coal-burning. *Energybiz Magazine*, **5**(5): 90–91.

SRI Consulting. (2009). Disruptive Technologies-Global Trend 2025 — Clean coal technologies (Background). Appendix D, SRI Consulting Business Intelligence.

Stohr, G. (2014). Obama power-plant pollution rule upheld by top U.S. Court. *Bloomberg Business News*, New York, NY.

Schutz, P. (2012). Chicago clean energy project. Chicago Tonight. *Public Media Organization, WTTW*, Chicago, IL.

Speight, J. G. (2012). *The Chemistry and Technology of Coal*, 3rd Edn. CRC Press.

Speight, J. G. (2013). *Coal-Fired Power Generation Handbook.* Wiley-Scrivener.

[TCS] Taxpayer for Common Sense (2011). The Department of Energy FutureGen Initiative, fact sheet. Taxpayer for common Sense, Washington, D. C.

Thorner, N. (2011). New EPA regulations for coal-fired plants are economy busters in Illinois. *illinoisreview.typepad.com.*

Touba, H. and Mansoori, G. A. (1998). Structure and Property Prediction of Sub- and Supercritical-Water. *Fluid Phase Equilibria*, **150–151**: 459–468.

Touba, H., Mansoori, G. A., and Matteoli, E. (1998). Subcritical and Supercritical Water Radial Distribution Function. *Int J. Thermophys*, **19**(5): 1447–1471.

[Tribune Staff] (2012). Quinn vetoes coal-to-gas plant on Southeast Side. *Chicago Tribune Business*, Chicago, IL.

Union of Concerned Scientist (2014). Burning coal, burning cash: Illinois dependence on imported coal. Fact Sheet, Union of Concerned Scientist, Cambridge, MA. *ucsusa.org.*

[US-DOE] U.S. Dep't of Energy (2010). FutureGen industrial alliance announces carbon storage site selection process for FutureGen 2.0, National Energy Technology Laboratory, Technologies Database, News Release. *energy.gov.*

[US-EIA] U.S. Energy Inf. Adm. (1994). State coal profiles. US-EIA, Washington, DC.

[US-EIA] U.S. Energy Inf. Adm. (2007). US Coal Reserves Regions. US-EIA, Washington, DC.

[US-EIA] U.S. Energy Inf. Adm. (2010). Updated capital cost estimates for electricity generation plants. Independent Statistical Analysis, US-EIA, Washington, DC.

[US-EIA] U.S. Energy Inf. Adm. (2011). State electricity profiles 2009. Office of Electricity, Renewables, and Uranium Statistics, U.S. EIA, Washington, DC.

[US-EIA] U.S. Energy Inf. Adm. (2013). Updated capital cost estimates for electricity generation plants. US-EIA Independent Statistical Analysis, U.S. EIA, Washington, DC.

[US-EIA] U.S. Energy Inf. Adm. (2014). State-level energy-related carbon dioxide emissions, 2000–2011. Analysis and Projections, US-EIA, Washington, DC.

[US-EIA] U.S. Energy Inf. Adm. (2014a). State energy data system: Illinois. Independent Statistics and Analysis, Technical Notes and Documentation, US-EIA, Washington, DC.

[US-EIA] U.S. Energy Inf. Adm. (2014b). Annual Energy Outlook 2014: Levelized cost and levelized avoided cost of new generation resources in the Annual Energy Outlook 2014. U.S. EIA, Washington, DC.

[US-EIA] U.S. Energy Inf. Adm. (2014c). Annual Energy Outlook 2014 — with projections to 2040. Office of Integrated and Int'l Energy Analysis, US-EIA, Washington, DC.

[US-EIA] U.S. Energy Inf. Adm. (2014d). Coal-fired power plant operators consider emissions compliance strategies. U.S. EIA, Washington, DC.

[US-EPA] U.S. Environmental Protection Agency. (2014). Cross-State Air Pollution Rule (CSAPR). www.epa.gov/airtransport/CSAPR/basic.html. Accessed on 15 October, 2014.

van Krevelen, D. W. (1948). *Elements of Coal Chemistry*. Nijgh & Van Ditmar, Rotterdam, the Netherlands.

van Krevelen, D. W. (1993). Coal, Third Edition: Typology — Physics — Chemistry — Constitution (Coal Science & Technology). Elsevier Science.

Viswanathan, R., Shingledecker, J. and Purget, R. (2010). Evaluating materials technology for advanced ultra supercritical coal-fired plants. *Power Magazine*, Houston, TX.

Wells, L. (2011). Southern Illinois coal-gasification plant will bring hundreds of jobs. Business section. *Evansville Courier & Press*, Evansville, IN.

Wernau, J. (2011a). Illinois options for replacing coal-fired power plants. Business section. *Chicago Tribune*, Chicago, IL.

Wernau, J. (2011b). Gov. Pat Quinn OKs coal-to-gas plant for Chicago's Southeast side: After earlier veto, addition of consumer protections leads to change of heart. Business section. *Chicago Tribune*, Chicago, IL.

Wernau, J. and Karp, G. (2011). Quinn signs bill opening door to Southeast Side coal-to-natural gas plant. Business section. *Chicago Tribune*, Chicago, IL.

Wernau, J. (2012). Closure of Chicago's Crawford, Fisk electric plants ends coal era: Plants were the last coal-fired electric generating plants in a major U.S. city. Business section. *Chicago Tribune*, Chicago, IL.

[WNA] World Nuclear Association (2011). Clean coal technologies, carbon capture and sequestration. WNA, Public Inf. Service Database: Climate Change and Sustainability Development, **7**(1): 1–9.

Yeagle, P. (2011). Coal gasification plant approved by senate. *Illinois Times*, Springfield, IL.

3

Petroleum

3.1. Introduction

Petroleum and natural gas are natural hydrocarbon resources in fluid (gas and liquid) states which are considered as the non-renewable energy sources. They are mostly made up of hydrocarbons and other compounds which are considered impurities from energy source perspective. There exist seven well-known hydrocarbon fluids in nature (Mansoori, 2009a, 2009b). In the order of their fluidity, they are natural gas, gas-condensate (also known as NGL standing for natural-gas liquid), light crude, intermediate crude, heavy oil, tar sand and oil shale as shown symbolically in Figure 3.1. These are all naturally occurring complex mixtures made up of hydrocarbons and other organic and inorganic compounds with a variety of molecular structures and sizes (see Fig. 3.1, which is part of Fig. 1.3 in Chapter 1).

The hydrocarbons and most other organic compounds present in all these seven naturally occurring fluids are generally polydispersed, having a range of size, shape and molecular weight distributions (see for example Mansoori *et al.*, 2007; Speight, 2014). As a result, continuous functions may be used for their mole fraction distributions (see for example (Du and Mansoori, 1986, 1987; Mansoori *et al.*, 1989; Vakili-Nezhaad *et al.*, 2001)). Figure 3.2 shows the approximate relative composition of various petroleum fluids with respect to the number of carbon atoms of their organic contents.

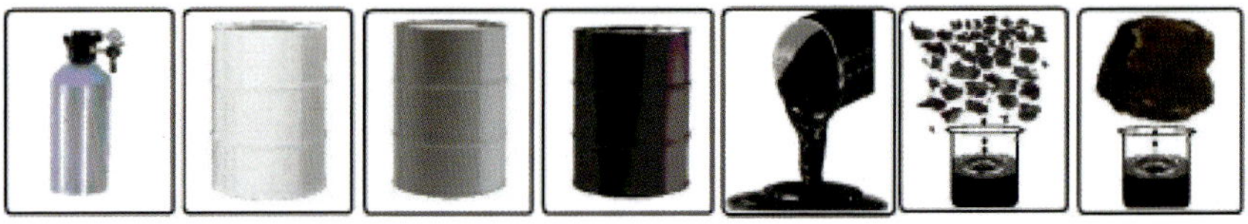

Fig. 3.1. The normal conditions of the seven naturally occurring hydrocarbon fluids, in the order of fluidity from left to right: natural gas, gas-condensate (NGL), light crude oil, intermediate crude oil, heavy oil, tar sand and oil shale (Mansoori, 2009a, 2009b).

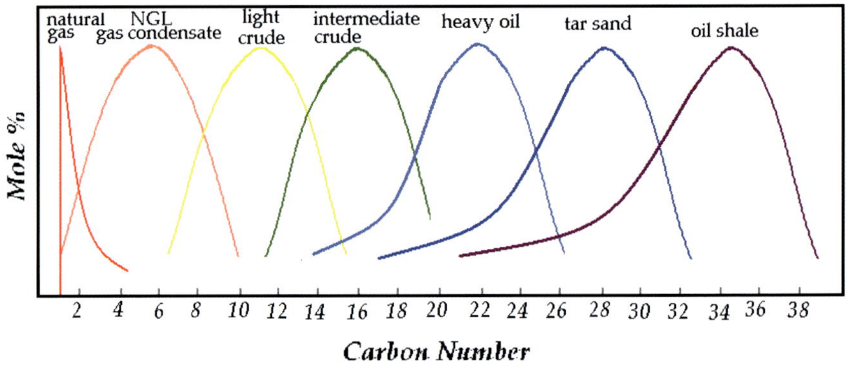

Fig. 3.2. Various categories of natural gas and liquid hydrocarbon reserves and their approximate hydrocarbon molecular weight distributions according to their carbon numbers (Mansoori, 2009a, 2009b).

This novel classification (Mansoori, 2009a, 2009b) will allow investigators to look at all the naturally occurring hydrocarbon fluids from a unified basis, which is their hydrocarbon contents and their relative size distributions. It should also be mentioned that heavy oil, tar sand and oil shale contain appreciable amounts of resins and asphaltenes, which are not hydrocarbons (Branco *et al.*, 2001; Speight 2012, 2013), while the percentages of these compounds are generally low or non-existent in the lighter naturally occurring hydrocarbon fluids.

Generally, many compounds are known to be present in these fluids. The number of carbon atoms in the hydrocarbons and other organic compounds present in naturally occurring hydrocarbon fluids can vary from one (as in methane) to over 100 (as in heavy oil, tar sand and oil shale) (Mansoori *et al.*, 2003; 2007; Speight, 2012, 2013).

During their production, transportation and processing, naturally occurring hydrocarbon fluids may go through a number of phase changes, which include evaporation (separation of gases from liquid streams), retrograde condensation (separation of liquids from gas streams), and solids formation and deposition (separation of crystalline solids, colloids and aggregates from liquid or vapor streams) (Jones and Pujadó, 2006; Park *et al.*, 1987; Mansoori *et al.*, 1988; Mansoori and Savidge, 1989; Schulz *et al.*, 1991). The latter is mostly due to precipitation and/or deposition of diamondoids (Mansoori *et al.*, 2012), wax (Mansoori *et al.*, 2003), asphaltenes (Mansoori, 1988) and inorganic compounds. In view of the complexity of such fluids, study and understanding of their properties and phase behavior is still a challenging and an industrially important task (Mansoori, 2009a, 2009b). Such an understanding will help us design a more economical route for the related production, transportation and refining projects (Macdonald, 2011; Meyers, 2003; Speight, 2010; Speight and Ozum, 2001). The complexity of the phase behavior in naturally occurring hydrocarbon fluids is due to the existence of the variety and polydispersity of hydrocarbons and other organic molecules.

The predominant hydrocarbons present in lighter naturally occurring hydrocarbon fluids (natural gas, gas-condensate, light-crude and intermediate-crude oil systems) are alkanes (also known as paraffin hydrocarbons) as shown in Fig. 3.3.

As we look at the physical properties of these fluids, going from natural gas to oil shale, their alkanes content decrease while their viscosity and density increase. Other hydrocarbons present in naturally occurring hydrocarbon fluids are aromatic hydrocarbons (see Fig. 3.4) at somehow lower concentrations than paraffins, depending on the underground natural reservoir source.

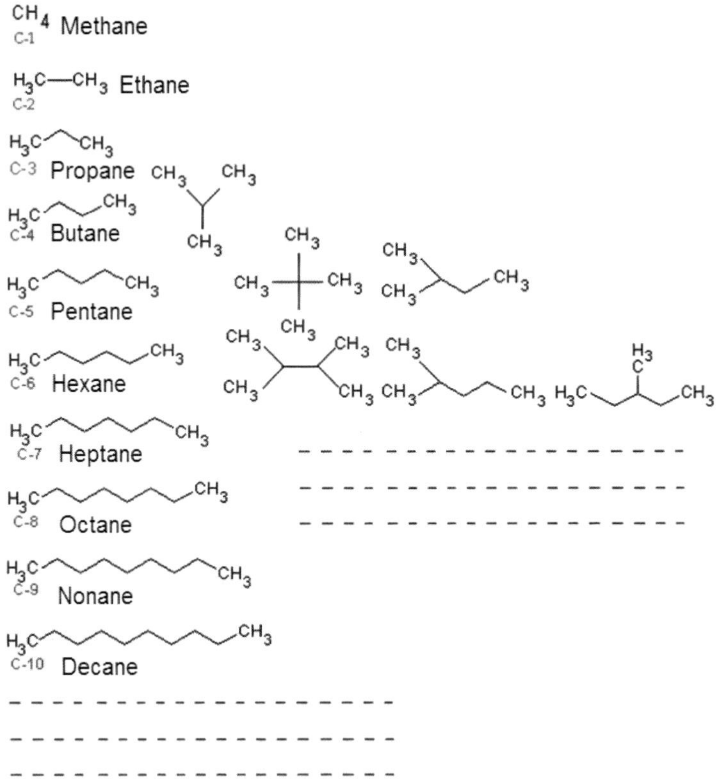

Fig. 3.3. Lighter paraffin hydrocarbons present in naturally occurring hydrocarbon fluids (Mansoori, 2009a, 2009b).

Other families of hydrocarbons like alkenes, alkynes, cycloalkanes, alkadiene, diamondoids, etc. may also be present in naturally occurring hydrocarbon fluids, but their concentrations would be generally low. Quite frequently, these reservoir fluids contain various amounts of other organic and inorganic compounds, which are usually termed as impurities from the petroleum industry point of view (Mansoori, 2009a, 2009b).

Molecular structure of organic impurities is mostly composed of elements such as carbon, hydrogen, metals, nitrogen, oxygen and sulfur. Molecules of inorganic impurities are generally composed of carbon, metals, oxygen and sulfur.

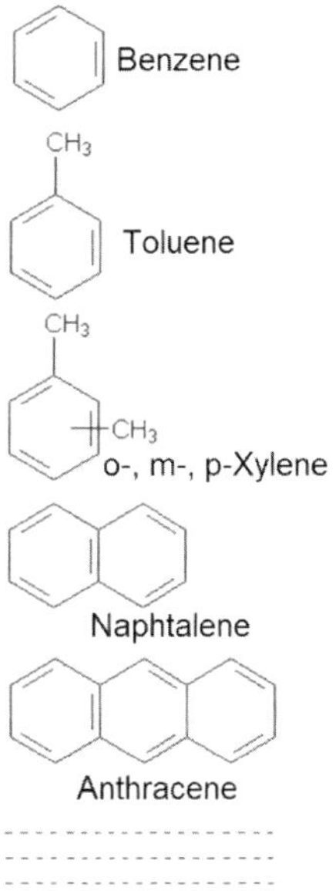

Fig. 3.4. Lighter aromatic hydrocarbons present in naturally occurring hydrocarbon fluids (Mansoori, 2009a, 2009b).

Petroleum fluid impurities can be categorized into five groups of compounds (Mansoori, 2009a, 2009b)

(i) The major low molecular weight impurities, which include carbon dioxide (CO_2), hydrogen sulfide (H_2S), metal oxides (Al_2O_3, Fe_2O_3, SiO_2, etc.), nitrogen (N_2), oxygen (O_2), salts (NaCl, $CaCO_3$, etc.), sulfur (S) and water (H_2O).

(ii) High molecular weight impurities, which could be present in the naturally occurring hydrocarbon fluids heavy fractions. They include asphaltenes, asphaltogenic acids, diamondoids and derivatives, mercaptans, metal carbenes, organometallics, petroleum resins and wax. In Table 3.1, the general closed chemical formulae of high molecular weight impurities of petroleum fluids are presented (Mansoori, 2009a, 2009b).

(iii) There may also be other organic and inorganic compounds present in naturally occurring hydrocarbon fluids, which depend on their underground natural reservoir sources. The

Table 3.1. General closed chemical formulae of high molecular weight impurities of petroleum fluids. In this table, C is carbon, H is hydrogen, N is nitrogen, O is oxygen, S is sulfur, X is metals and R represents a hydrocarbon segment (Mansoori, 2009a, 2009b).

High MW Impurity	General Closed Formula
Asphaltenes	$C_m H_n N_i O_j S_k$
Asphaltogenic Acids	$C_m H_n N_i O_j S_k - COOH$
Diamondoids	$C_{4n+6} - H_{4n+12}$
Mercaptans	$C_m H_n N_i O_j S_k X$
Metal Carbenes	$R_1 HC = CHR_2 X$
Organometallics	$Ex.: R_1 - X - R_2$
Petroleum Resins	$C_m H_n N_i O_j S_k$ $i, j, k = 0$ or 1.0
Wax	$C_m H_n$ $18 \leq m \leq 60$ $n \leq 2m + 2$

concentration of other organic matter is generally quite low and they have little or no effect on petroleum fluids phase behavior.

(iv) Another category of impurities which may be present are compounds which have been added to these reservoir fluids during their production, transportations and storage stages for various reasons. These include but not limited to acids, alcohols, aromatic hydrocarbons, detergents and polymers.

(v) Naturally occurring hydrocarbon fluids often contain compounds, which are the result of physical association of hydrocarbons and the above-mentioned impurities and contaminants (Petrov, 2012). They include clathrates (Mahajan *et al.*, 2007), colloids, crystalline solids, flocs and slugs.

The literature on petroleum and natural gas is rich with numerous data banks, correlations and nomograms for the analysis and treatment of such reservoir fluids and their liquid-vapor phase behavior (*www.api.org*; *www.nist.gov*; Chorn and Mansoori, 1988; Mansoori *et al.*, 2012). The technologies for petroleum exploration, production, transportation, refining, upgrading and use are well developed and advanced (Macdonald, 2011; Meyers, 2003; Speight, 2010; Speight and Ozum, 2001) and they will not be discussed in detail in this chapter. What are presented in this chapter are the brief historical and developmental stages and trends in petroleum exploration, production, transportation and consumption in a representative state, i.e. state of Illinois.

3.2. Petroleum Production

In Southeastern Illinois, there are numerous natural underground petroleum reservoirs. Illinois oil production reached its peak around 1941, when several major oil and gas companies were operating in the Illinois Basin (Finley *et al.*, 2005; Agyarko and Mansoori, 2013). Until 1970, Illinois was one of the major oil-producing states in the U.S. Currently, Illinois is a minor oil producer, ranking a distant 16th among states in the U.S., with 2014 production of 9.54 million barrels of crude oil. Illinois' high quality, light sweet southern crude oil (with the API gravity of about 36.8) is attractive to refiners, despite having a lower gasoline yield of 27.50 wt% or 2% less gasoline per unit volume than West-Texas intermediate crude oil (with the API gravity of about 39.6) (RCF&REAL, 2009). Approximately 16,000 oil wells have been drilled in the Illinois Basin, representing about 3% of all the U.S. wells in oil production. Out of the 16,000 oil wells, over 90% are marginal wells, each producing a mere 1–2 barrels a day (Lewis and Bergeron, 2010).

In 2014, the U.S. Energy Information Administration (US-EIA) reported Illinois recoverable proven crude oil reserves at 42 million barrels (US-EIA, 2014). Oil exploration activities in the Illinois Basin in the 1970's led to the discovery of a more mature resource base (i.e. subsurface regions with organic matter which has undergone complete thermal transformation) rich in hydrocarbon deposits, with presumably 60% of the original-oil-in-place (assessed in 2006) still unrecovered (Finley *et al.*, 2005). However, the relatively shallow depths of the reservoirs of most of these oil reserves in the Illinois

Basin makes the long term prospects of oil production unlikely (IOGA, 2014). The current Illinois oil resource will require more advanced technology than what was required during the first one hundred years of oil production in the state. For this reason, the major oil companies abandoned the mature oil fields in the Illinois Basin, while prospecting for oil fields in the U.S. Gulf Coast or Texas and in Alaska which are less expensive to develop. Currently, oil prospecting in the Illinois Basin is mostly undertaken by small local companies (IOGA 2014).

In 2006, there were about 602 oil fields in Illinois with estimated 2.4 billion barrels of cumulative production potential (Finley *et al.*, 2005). These 602 oil fields included the 43 largest oil fields in the state located in the coal-rich areas of the Illinois Basin (see Fig. 3.5).

Because many of these fields may be reaching the end of their economic life and have depths of about 600–1500 meters, they may require new methods to boost their production rates. Among the options being considered to boost oil production in Illinois

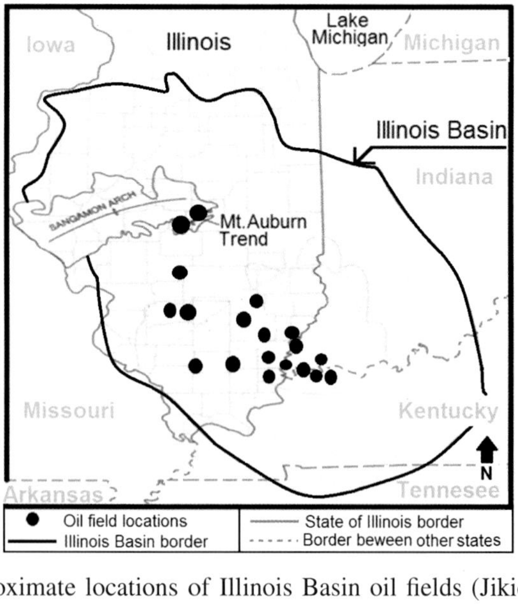

Fig. 3.5. Approximate locations of Illinois Basin oil fields (Jikich *et al.*, 2012; Lasemi 2009, 2014).

is horizontal drilling, as well as enhanced oil recovery (EOR) technology using CO_2, nitrogen, or surfactant plus polymer flooding of the reservoirs.

Application of artificial intelligence and expert systems (Gharbi and Mansoori, 2005) would also be necessary for a more efficient petroleum exploration and production. Some limited test results from the oil fields using carbon dioxide injection to bring the reservoir to minimum miscibility condition (Benmekki and Mansoori, 1986) for 31–41 days, indicated an improvement in oil recovery from 3–60 barrels of oil per day in a single well. Another oil well at the Manttoon field in Illinois showed a similar progress in oil recovery from 9–45 barrels a day. Out of the 602 individual oil fields identified in Illinois, 43 contained at least one reservoir that satisfies minimum criteria for carbon dioxide miscible flooding (Finley *et al.*, 2005). Currently, the safest approach for commercial scale exploitation of oil in the Illinois Basin, without compromising the integrity of the coal mines, is through the use of horizontal and directional drilling. The horizontal and directional drilling technologies results in drilling of fewer wells in the oil fields, while reducing the risk of penetrating a coal mine. In Table 3.2, we report the Illinois petroleum resources profile.

3.2.1. *Enhanced oil recovery (EOR)*

The EOR projects in Illinois include Loudon Oil Field in Fayette County, Markham City North Oil Field in Jefferson and Wayne Counties and Lawrence oil Field in Lawrence County. The overall estimated EOR potential for Illinois is estimated at 632–979 million barrels (Finley *et al.*, 2010; Agyarko and Mansoori, 2013) as reported in Table 3.2.

Since its discovery in 1937, the Loudon oil field in Illinois has an estimated cumulative production of 400 million barrels of oil (ARI, 2006). In March 2007, Midwest Geological Sequestration Consortium led by the Illinois State Geological Survey (ISGS), together with

Table 3.2. Illinois petroleum resources profile.

API Gravity	36.8
Gasoline yield[a]	27.50 wt%
Estimated number of oil wells[b]	16,000
Estimated number of marginal wells[b]	90% of total oil wells
Estimated number of oil fields[c]	~602
Unrecovered movable crude oil[d]	4.1×10^9 barrels
Recoverable proven crude oil reserves[e]	42×10^6 barrels
Estimated EOR potential[f]	$(632 - 979) \times 10^6$ barrels
Estimated cumulative production potential[c]	2.4×10^9 barrels
2014 oil production[e]	9.54×10^6 barrels

(a) Data from M. F. Scott, Illinois Geological Survey (private communication);
(b) Lewis and Bergeron, (2010); (c) Finley *et al.*, (2005); (d) IOGA, (2014); (e) US-EIA, (2014); (f) Finley *et al.*, (2010).

the Indiana and Kentucky Geological Surveys and industry partners, including Petco Petroleum Corp., conducted the first EOR pilot study in Loudon Field, in Fayette County in Illinois (Rodosta *et al.*, 2012).

The pilot formation was the Cypress Sandstone with average permeability of 31 mD and average porosity of 16%. The Loudon field test involved injection of 40 metric tons of CO_2 in gaseous state into a producing well, at a depth of approximately 475 meters. The CO_2 injection continued for 3–5 days at the rate of 5–10 metric tons per day (MGSC, 2011). The gas was allowed to mix and dissolve in the oil for approximately one week, before placing the well back on production to measure the amount of petroleum fluids produced. After the 1-week soaking period, daily oil production initially increased from 2 to 8 bbl/day (or fourfold) for a few days, but subsequently decreased to 5–6 bbl/day with sustained oil rate of 1–2 bbl/day (Finley, 2007). Although the initial injection volume target was just 40 metric tons of CO_2, it was a critical first step in testing geological sequestration and EOR in the Illinois Basin (Rodosta *et al.*, 2012). In Chapter 5, application of CO_2 in enhanced oil (petroleum) recovery is further discussed.

In February 2012, Strategic American Oil Corporation (SAOC) made an announcement that it had begun water injection at its V. People No. 1 well located in Markham City North Oil Field in Jefferson and Wayne Counties in Illinois. The location of this pilot project is near the center of the potential water flood area, which is surrounded by additional leasehold held by SAOC and its partner. From 1943–2009, the Markham City North Oil Field produced an estimated 1,381,300 barrels of oil during the primary phase of production. However, it is believed that a significant amount of oil still remains in the reservoir with a portion that is recoverable using water flooding. For several years, similar water-flood projects close to the location of this pilot project have performed well, with some reporting secondary to primary recovery ratio of 1 to 1. Hence, it is anticipated that this project may yield potential gross recoverable reserves in excess of one million barrels of oil. The results from this pilot phase may be used to determine if further water flood development plans for the entire field would be productive (Schor and Foy, 2012).

Oil production at the Lawrence oil field in Illinois started in 1906. Since then, over 410 million barrels of oil have been produced by several companies in this field. In April 2012, DOE reported that EOR was successfully used to boost oil production in excess of its peak (by more than 300%) from a well located in Lawrence oil field in Illinois. The Lawrence oil field well production was increased from 16 barrels per day to between 65 and 75 barrels per day, using an innovative alkaline surfactant polymer (ASP) flooding technique. According to the DOE, the ASP flooding raised the overall oil cut in the 15-acre project area from 1% to 12%. The DOE sponsored project was operated by Rex Energy Corporation headquartered in State College, Pennsylvania. The project was an integral part of a larger effort in progress at the ISGS supported by DOE's Fossil Energy Office. Because of the oil potential in this field, Rex Energy may expand testing to cover another 58-acre portion of the field. The company apparently planned to invest in ASP flooding technology on 351 acres of the field just south of the current operations in 2013. Currently, ISGS is characterizing the Mississippian Cypress sandstone reservoirs, as well as the Pennsylvanian Bridgeport sandstone underlying Lawrence field. This project aims to define the remaining oil reserves, with the view of projecting how additional oil could be economically recovered. While the two formations are the most prolific ones in Illinois, they are structurally complicated. As a result, ISGS is expecting that the ASP flooding technique could be used to recover the extra 130 million barrels of oil estimated to be technically recoverable at the Lawrence oil field. The original-oil-in-place in the reservoirs of Lawrence oil Field was estimated at one billion barrels. The success of the ASP technology could be extended to similar oil fields in U.S. (Snow, 2012; Jikich *et al.*, 2012).

3.2.2. *Recent oil exploration and production activities*

Activities in the Illinois Basin have started to change because of advanced technology. Since 2011, there has been continuous influx of larger oil companies into Southern Illinois, in the Illinois Basin, buying mineral rights for oil prospecting, a situation which is reminiscent of what happened in the early 1900s regarding coal rights (AP, 2013). In April 2012, Imperial Petroleum Inc. announced that it had formed a lease pool with a private group to begin acquisition of acreage on certain oil and natural gas drilling prospects in the Illinois Basin (Wilson, 2012; Agyarko and Mansoori, 2013).

The reason for the renewed interest in oil exploration in the Illinois Basin is the discovery of a very significant oil and gas resource area in the southeast margin of an old oil production area in central Illinois, known as Mt. Auburn Trend (see Fig. 3.5).

The Mt. Auburn Trend is located along the southeastern flank of the Sangamon Arch and includes several oil fields (Lasemi, 2009). The area extends to about 64.7 kilometers (40 miles) long, spanning from central Macon County and across Northern Christian County,

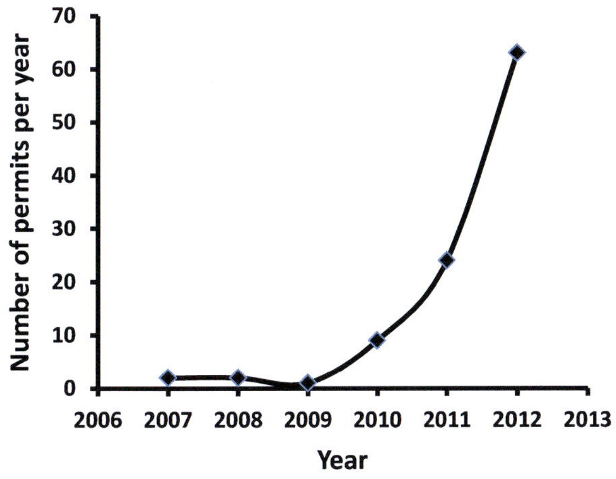

Fig. 3.6. Trend of Mt. Auburn oil and gas permits 2007–2012 (ISGS, 2014).

and extending over several oilfields that have produced over 18 million barrels of oil. In essence, the area is entering its third developmental phase of oil exploration, after the oil booms from late 1940s to mid-1950s and subsequently from late 1960s to early 1980s (ISGS, 2014). From early 1980s to 2010, oil exploration in the area was sporadic, due to lack of a predictive model in identifying the approximate locations of the oil and gas reservoirs (ISGS, 2014). For example, prior to 2010, the annual number of permits issued for oil and gas exploration along the Mt. Auburn Trend was just one or two compared to over 60 in 2012 (see Fig. 3.6).

Such dramatic increase in the number of permits is attributed to a publication of depositional and dolomitization model by the ISGS. The model serves as an important predictive tool for oil and gas exploration in the Illinois Basin because it explains the formation and the deposition of the oil-producing rocks (porous dolomites). Additionally, it explains the percolation of sea water into the sediments which occurred during the Silurian period, and how the process was linked to fluctuations of sea-level (ISGS, 2014). It should be mentioned that on the geological time scale, the Silurian and Devonian periods refer to the divisions of the Paleozoic Era (542–251 million years ago). During the Silurian period (438–408 million years ago), North America (including the present day Illinois) was attached to Europe in a large landmass situated at the equator. The first half of the Silurian period is characterized by rising sea levels, which subsequently fell or retreated during the rest of the period. Also the Silurian period is characterized by the appearance of large coral reefs, including diversification of fish species. Additionally, primitive life forms of plants and insects began on land. The Silurian period was preceded by the Devonian period (from 408–360 million years ago). During the Devonian period, the first forest was formed as land plants grew and matured, followed by the first emergence of vertebrate animals to colonize land (USGS, 2002). Geologists point out that it was the slow decomposition of the remains of these plants and animals that were buried under layers of rocks and ancient sea several millions of years ago that led to the formation of our fossil fuels, including the Mt. Auburn Trend oil reservoir in Illinois. According to ISGS, the 2010 model has led to the completion of a number of prolific wells during the years 2011 and 2012. For example, from April 2011 to December 2012, four wells produced over 295,000 barrels of oil in Macon County. Also during the year 2012, two wells produced over 160,000 barrels of oil in Christian County. Furthermore, it is claimed that the potential for new oil discoveries in the region appears to be promising. This is because the approximate locations of potential reservoirs in unexplored areas of Mt. Auburn Trend was assessed using this model (ISGS, 2014). Below, we report some of the recent activities in this significant oil and gas resource area in the Illinois Basin.

In April 2011, Production Energy LLC, completed the No. 1 Bowman well, a Silurian test well located in Macon County. Although there was no report of the initial oil production from the well, the Illinois Pipeline Production Report in December 2012 indicated a cumulative oil recovery of over 114,000 bbl, an amount of oil typical of a prolific well. Also during April 2011 to December 2012, the oil recovery of four wells on Bowman and Butterfield leases owned by Production Energy LLC was over 290,000 bbl of oil. The wells are located at 1.93 kilometers (1.2 miles) Southeast of Mt. Auburn Trend at Harristown South field. It is claimed that the Harristown South field that was discovered in 1975 has cumulative oil production of approximately 150,000 bbl from nine wells (Lasemi, 2009).

In January 2012, Belken Oil LLC completed its No. 1 Sheldon well, a Silurian test well located at Christian County. The claimed initial oil production rate from the 602 meters (1,975 ft) well of a dolomitic sandstone and dolomite reservoir was 1,440 barrels per day. By December 2012, the claimed oil production from the well was over 82,000 bbl. In addition to the No. 1 Sheldon well, the company completed its Sheldon No. 2 well in March 2012 to accumulate over 160,000 bbl of oil from the two wells in that year.

In June 2014, Kansas-based Oil and Gas Exploration Company announced its plans to establish an Illinois Oil Basin headquarters on a land located at South of Fairfield in Illinois. The company expects to drill as many as 15 test wells in the Knox formation in the Illinois Basin at the depth of 2440 meters (8,000 feet), but prefers a formation at a much shallower depth of 1372 to 1524 meters (4,500 to 5,000 feet) located in the oil and gas-rich New Albany Shale. Besides the ongoing drilling program, the company has begun a large-scale three-dimensional seismic study of land across the Illinois Basin. In Franklin County, the company has tested 95.8 km^2 of land (37 square miles) using three-dimensional seismic technology. Also in Wayne County the company has tested 147.6 km^2 (57 square miles) of land, and has projected to test a total of 595.7 km^2 (230 square miles) of Southern Illinois farmland (Wells, 2014).

In June 2014, Galilee Energy Limited, an Australia-based energy company entered into a joint venture to carry out exploration drilling targeting high return shallow oil prospects in the Illinois Basin, in Illinois. Galilee and its joint venture partners include a group of experienced U.S. based oil and gas professionals. The company has acquired almost 3000 acres of leases in an active and ongoing leasing program in the Illinois Basin. These include areas in Sangamon County, Shelby County, Bond County, Logan County, Macon County, Christian County and Clark County. As of November 2014, Galilee Energy had completed the first four wells in the Illinois Basin shallow oil exploration program.

3.3. Petroleum Transportation and Refining

The world's first and largest multinational corporation (Standard Oil of Ohio) was established in Ohio in 1862. In 1890, the company set up an oil-marketing organization (Standard Oil of Illinois) in Chicago, and later owned the majority shares of P.C. Hanford Oil Co., an Illinois company established in 1861 dealing in standard illuminating and lubricating oils, axel grease, white lead and various varnishes (Andreas, 1886; US-DOJ, 1909; Agyarko and Mansoori, 2013). Standard Oil of Illinois was subsequently purchased by Standard Oil of Indiana. But following an antitrust lawsuit in 1911, Standard Oil was forced to splinter into several subsidiaries. The refined oil from Standard Oil of Indiana refinery located outside Whiting in Indiana along the border with Chicago was transported to Chicago and other Midwestern cities through two pipelines originating in Lima, Ohio. Also, oil from Standard Oil of Indiana was transported by land using the railroad terminal belonging to the Chicago & Calumet Terminal Railroad. In 1985, the Standard Oil Company in Indiana changed its name to Amoco Corporation with headquarters in Chicago. But in 1998, the company merged with British Petroleum and was renamed as BP-Amoco plc, a name which was maintained till 2000, when the company changed the name to BP Plc (Grant, 2007).

Illinois leads the Midwest in crude oil refining capacity with four refineries with average combined production of 950,000 barrels per day (US-EIA, 2014). Two of these refineries are ExxonMobil in Joliet and Citgo in Lemont, both near Chicago; the third one is

ConocoPhillips, located in Illinois suburb of St. Louis and the forth is Marathon Petroleum Company in Robinson, near the Indiana border with Chicago. However, none of these refineries process any meaningful amount of Illinois petroleum. Instead, approximately 67% of the crude oil produced in Illinois is processed at CountryMark Refinery in Mt. Vernon, Indiana, while the balance is refined in Michigan and Ohio (RCF&REAL, 2009).

Illinois is a key transportation hub for crude oil and natural gas for North America because it is centrally located and has well-developed pipeline infrastructure. Illinois refineries rely on crude oil from Canada and the U.S. Gulf Coast, with much of the oil transported from Canada. Much of the Canadian heavy crude oil imported to Illinois is refined and consumed in the state, especially in transportation and industrial sectors, or transported to other states after being refined. Illinois imports more Canadian crude oil than any other state in the U.S. In 2007, Illinois import from Canada totaled $12 billion of crude oil and $91 million of refined products (Grossman, 2011).

In May 2014, the Illinois Commerce Commission approved the recommendation of an administrative law judge, authorizing Enbridge Inc., a Canadian Pipeline Company, to use "eminent domain" to acquire easements across 127 tracts of land for construction of a proposed $800 million oil pipeline in Illinois (Tomich, 2014). In legal terms "eminent domain" means the right of a government or its agent to expropriate private property for public use, with payment of compensation. However, the company still needs to obtain the remaining 120 parcels of land from court, as some landowners intend to challenge construction of the project. The Southern Access Extension pipeline that was proposed in 2007 was scheduled for construction in summer of 2014, and it is planned to be operational by mid-2015 (Enbridge, 2013). The 167-mile pipeline would start from the company's Flanagan oil terminal at Pontiac, Illinois and would cross eight counties (Livingston, McLean, DeWitt, Macon, Christian Shelby, Fayette, Marion) to an oil terminal and pipeline hub at Patoka in Marion County, Illinois (AP, 2014).

Also faced with a growing market demand for petroleum, Enbridge Energy Partners LP (a subsidiary of Enbridge Inc.) is proposing to expand its pipeline infrastructure alongside existing pipelines in Indiana and Illinois. The proposed project (Line 78 pipeline) would involve construction of about 79-mile long and 36-inch diameter pipeline. The pipeline would start from Enbridge's Flanagan terminal near Pontiac in Illinois, and travel Northeast to Enbridge's terminal near Griffith in Indiana. The completion of this project would expand Enbridge's capacity to transport the increasing supplies of light and heavy crude oil produced in Western Canada, including the crude oil produced in the Williston Basin region around North Dakota (Starling, 2014).

3.4. Petroleum and Petroleum Products Consumption

Globally, it is estimated that approximately 32 billion barrels of oil and liquid fuels were consumed in 2012 (US-EIA, 2013), with the U.S. consuming nearly 25% (8 billion barrels) of this global oil consumption. In 2012, Illinois consumed about 226 million barrels of petroleum at the cost of $29.74 billion, representing 76.6% of the total energy expenditure by the state. These included 43.7 million barrels of distillate fuel oils, 24.7 million barrels of jet fuel, 21.1 million barrels of liquefied petroleum gas (LPG), 109.8 million barrels of motor gasoline (includes fuel ethanol blended into motor gasoline), 34,000 barrels of residual fuel oil and others (includes asphalt and road oil, aviation gasoline, kerosene, lubricants) (US-EIA, 2014; Agyarko and Mansoori, 2013). Figure 3.7 shows the profiles of crude oil production and consumption in Illinois.

According to Fig. 3.7 (a semi-log graph), consumption of petroleum in Illinois has not changed appreciably, while petroleum production in Illinois has decreased even with bigger rate in the period of 1960–2009. The use of petroleum for electric power generation in Illinois has been a function of the price of petroleum as compared with

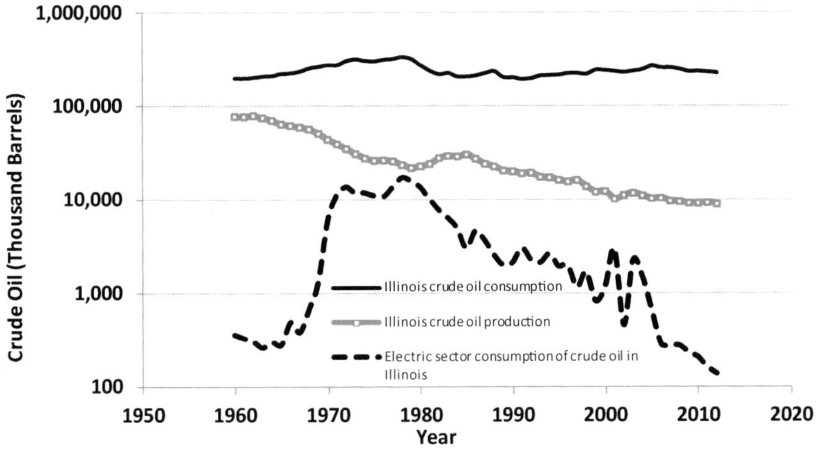

Fig. 3.7. Profiles of petroleum fuel production and consumption in Illinois (Data from US-EIA, 2014).

natural gas and coal. As seen in Fig. 3.7, in recent years the amount of petroleum fuel used for electricity generation in Illinois has not been significant. For example, in 2007 the Illinois electric energy sector consumed only 136,000 barrels of petroleum in the form of residual fuels, distillate fuel oils and petroleum coke (US-EIA, 2014).

In Table 3.3, we report the consumption of various petroleum products in Illinois in commercial, industrial, transportation and residential sectors in 2007.

According to Table 3.3, in 2012 the Illinois transportation sector consumed most of the petroleum fuels with 36.342 million barrels of distillate fuels, 24.668 million barrels of jet fuels and 107.539 million barrels of motor gasoline, representing about 75% of the total petroleum consumed in the state. The transportation sector also consumed 109,000 barrels of aviation gasoline, in addition to jet fuel and motor gasoline in 2012. The industrial sector in Illinois is the major consumer of LPG and residual fuel oils and the commercial sector is the least consumer of petroleum products.

Table 3.3. Consumption of various petroleum products (in thousand barrels) in Illinois in commercial, industrial, transportation and residential sectors in 2012 (Data from US-EIA, 2014).

	Sector				
Petroleum Product	**Commercial**	**Industrial**	**Residential**	**Transportation**	**Electric power**
Distillate fuel oil	1,009	6,158	65	36,342	136
Kerosene	2	0	7	0	0
LPG[(1)]	554	13,045	4,877	717	0
Motor gasoline	250	2,005	0	107,539	0
Residual fuel oil	0	12	0	21	0
Lubricants	0	0	0	1,086	0
Aviation fuel[(2)]	0	0	0	24,777	0
Others[(3)]	0	27,257	0	0	0
TOTAL	1,815	48,477	4,949	170,482	136

(1) LPG: liquefied petroleum gases, includes ethane and olefins.
(2) Aviation fuel includes jet fuel (24.668 million barrels) plus aviation gasoline (109,000 barrels), totaling 24.777 million barrels as reported in this table.
(3) Others: include asphalt and road oil, kerosene, lubricants and 16 other petroleum products.

According to Table 3.4, Texas is a far bigger consumer of petroleum products than the other three states, while Illinois is the smallest consumer.

3.5. Environmental Problems Linked with the Use of Petroleum

The environmental problems associated with the use of petroleum include hydrocarbon emissions from refineries, motor vehicles, pipelines, piles of petroleum coke (pet coke) found around refineries that pose health problems to residents, CO_2 emissions in the refineries and from internal combustion engines and other toxic gases from the refineries, incomplete combustion of hydrocarbons and gasoline

Table 3.4. Comparison of petroleum products consumption (in thousand barrels) and barrels per capita (in parenthesis) in Illinois with three other major states in the U.S. (California, New York and Texas) in 2012 (Data from US-EIA 2014 and the U.S. Census Bureau 2014).

	State			
Petroleum Product	**Illinois**	**California**	**New York**	**Texas**
Distillate fuel oil	43,575 (3.4)	89,754 (2.4)	60,638 (3.1)	160,384 (6.2)
Jet fuel	24,668 (1.9)	94,474 (2.5)	25,823 (1.3)	62,428 (2.4)
LPG[(1)]	19,193 (1.5)	15,524 (0.4)	6,983 (0.4)	521,771 (20)
Motor gasoline	109,794 (8.5)	343,071 (9)	128,275 (6.6)	293,796 (11.3)
Residual fuel oil	34 (0.003)	26,576 (0.7)	9,802 (0.5)	21,321 (0.8)
Others[(2)]	28, 461 (2.2)	50,392 (1.3)	9,445 (0.5)	220,289 (8.5)
TOTAL	225,724 (17.5)	619,791 (16.3)	240,966 (12.3)	1,279,988 (49.1)

(1) LPG: liquefied petroleum gases, includes ethane and olefins.
(2) Others: include asphalt and road oil, kerosene, lubricants and the 16 other petroleum products.

additives. Below we report three of the major environmental pollution cases related to the use of petroleum in the state (Agyarko and Mansoori, 2013).

(i) Following decades of operation, pipeline and other refinery emissions from refineries around Hartford, a village in Madison County, Illinois has resulted in the formation of a layer of gasoline and other petroleum products floating on top of the groundwater underneath much of the northeast part of the village. The clean-up of the gasoline has become difficult, due to natural events such as rain, drought, or the level of the Mississippi River that can subject the groundwater level to fluctuations and trap the gasoline on top of groundwater. High levels of gasoline vapors emitted from the polluted ground water in an enclosed space have been causing explosions and fires. In 1969, several fires at the Northeast part of the village were attributed to such gasoline vapors. Also during the 1970s and 1980s, there were reports about odor, health concerns and vapor intrusions into homes in the area. To resolve the gas intrusion problem in Hartford, in late 1970s and subsequently in the early 1990s, the Illinois EPA and the Illinois Attorney General's Office (AGO) required Clark Oil and Refining Company (currently Premcor Refining Group) to install petroleum recovery systems in the affected areas (IEPA, 2003). In response to EPA's and AGO's requirements, in the 1970s, Clark Oil and Refining Company installed three recovery wells in North Hartford to pump petroleum from the groundwater. However, these three old recovery wells are currently not operating. In the early 1990s, Clark Company installed a vapor recovery system (consisting of 12 boreholes drilled below the ground in the area) to capture petroleum vapors in soils before they could escape to home basements. According to Premcor Refinery's report, these systems have recovered about 7000 m^3 ($\sim$ 247,000 ft^3) of vapor and about 4,400 m^3 ($\sim$1.16 million gallons) of liquid gasoline since the installations. It was initially assumed that the emission levels in Hartford had been restored to normalcy and the installations were functioning adequately to protect the community. But in spring 2002, the Illinois Department of Public Health (IDPH) and the Illinois Environmental Protection Agency (Illinois EPA) again received several reports from the residents of Northeast Hartford, complaining about odors, chronic headaches and sporadic nausea linked with vapors in their homes (IEPA, 2003). As a result, the IDPH took steps to measure and perform an evaluation test of the levels of volatile organic compounds (VOCs) in the air around the basements of selected homes in Hartford. The VOCs are chemicals with high vapor pressures at room temperature found in gasoline vapors. The results of the samples found in the Hartford basements indicated that long-term and intermittent exposure to the levels of VOCs posed no apparent increased risk of cancer. Nevertheless, according to the evidence from past events, those conditions that led to the May 2002 vapor intrusions could resurface and cause short-term exposures and health effects. For this reason, the IDPH concluded that the gasoline vapor intrusions in Hartford pose a public health hazard (IEPA, 2003). Since the spring 2002 complaints, the Illinois EPA and the Illinois AGO have asked the oil companies that were operating in the area or used these pipelines to help in searching for a temporary solution (for the short term), and a long-term solution to this gasoline vapor intrusion problem. The current owners of these refineries are Premcor Refining Group (formerly Clark Oil and Refining; Apex Oil Company was another former owner), BP Products of North America (formerly Amoco) and ConocoPhillips (former owner of Shell). Of these companies, the Illinois EPA and AGO have filed lawsuit against Premcor, and Apex Oil Company, the company which became Clark Oil, because these companies have been identified as responsible parties for the pipeline emissions (IEPA, 2003). In essence, the state is bringing pressure to bear on these oil companies that have operated in the area to fund and perform the investigative work needed to resolve many of these outstanding issues.

The EPA is planning to conduct a test to learn more about how to remove the trapped gasoline underneath the affected area at Hartford. The test, referred to as “focused pumping”, is designed to pump a significant amount of groundwater out of the ground at the time when water levels are naturally low. In this process, the water level around the wells will be forced to drop further, allowing the gasoline to rise up through the soil and be pumped, rather than remaining trapped underground to cause environmental and public health problems (EPA-R5, 2014). Also, separate wells will be used to recover the gasoline and transfer it directly to a set of storage tanks designed specifically for safe keeping. Once the tanks are full, the gasoline will be transferred to recycling facilities for reuse (EPA-R5, 2013).

(ii) Another problem linked with petroleum utilization in Illinois is the growing piles of petroleum coke (pet coke) around the refineries in the Chicago area. Pet coke is similar in appearance to coal. It is a carbonaceous solid material formed after thermal decomposition of petroleum. It is also one of the valuable products formed in oil refining and its applications include electricity generation and a range of other industrial uses. As mentioned in Chapter 2, the proposed coal-to-gas IGCC power plant that was to be constructed by New York-based Leucadia National Corp at the Southeast Side of Chicago was planned to use about one million tons of Illinois coal per year, together with petroleum coke to produce synthetic gas that could be burnt to produce electricity. In late summer of 2013, the residents of Southeast Chicago, Illinois began complaining about windblown dust and water pollution claimed to have been caused by pet coke storage and handling facilities near the Calumet River (see Fig. 3.8).

The pet coke is generated by an Indiana refinery that refines tar sand imported from Alberta in Canada, and is transported to Chicago for storage. The major concern to the residents is the fear that the piles of pet coke could expand in the coming years, especially with the $3.8 billion expansion of BP's oil refinery in nearby Whiting, Indiana, which could wreak further havoc on their neighborhood (Arbogast, 2014). Of the three storage facilities in Chicago, two are owned by KCBX Terminals, Inc., and one is owned by Beemsterboer Slag Corporation (US-EPA, 2014b). In response to residents' complaints, on 15 November 2013, the US-EPA requested that air monitors be installed and operated by KCBX around its facilities. In December 2013, the City of Chicago drafted new regulations governing the storage of pet coke in Chicago. Also on 13 January 2014, the Governor of Illinois announced his own emergency administrative rules for pet coke storage in the state. Then on 16 January 2014, the Illinois EPA filed its proposed rules for pet coke storage in the state. However, on 23 January, 2014, those rules were overturned by the Illinois Pollution Control Board and returned to the Illinois EPA for further deliberations, due to lack of emergency (Arbogast,

Fig. 3.8. In this 25 October, 2013 photo, the Willis Tower in downtown Chicago provides a backdrop to a huge mound of pet coke in the a residential area Southeast part of the city. An increasing volume and size of pet coke piles is causing environmental concerns for residents in this working-class neighborhood. (AP Photo/Charles Rex Arbogast) (Webber, 2013).

2014). According to the board, the governor should have gone through the normal process of rulemaking for the passage of ordinances or laws in dealing with the pet coke. As a result, the EPA failed to show there was an imminent threat to public health posed by the piles (Moskowitz, 2014). Nevertheless, on 3 June, 2014, the US-EPA finally issued a notice of violation (NOV) to KCBX, based on air monitoring data submitted by KCBX for the monitoring period beginning February 18, 2014 through May 10, 2014, including dust wipe sampling taken from homes in the neighborhood adjacent to KCBX. According to US-EPA, "*KCBX caused the emission of* PM_{10} *into the air, so as, either alone or in combination with contaminants from other sources, to cause or tend to cause, air pollution in Illinois and/or to prevent the maintenance of the revised NAAQS for* PM_{10} *in violation of the Illinois SIP at 35 Ill. Admin. Code §201.141*"

(US-EPA, 2014a). Particulate matter concentrations (PM_{10}) refer to fine suspended particulates of size less than 10 microns in diameter. Excess emissions of particulate matter contain microscopic solids or liquid droplets, which may get deep into the lungs and cause serious health problems. Particulate matter exposure contributes to irritation of the airways, coughing, difficulty in breathing, decreased lung function, aggravated asthma, chronic bronchitis, irregular heartbeat, non-fatal heart attacks, and premature death in people with heart or lung disease (US-EPA, 2014a). In addition to EPA's issuance of NOV, the State of Illinois filed lawsuits against all the three facilities for violations of various air and water pollution regulations. In response to the state's lawsuit, Beemsterboer Slag Corporation, owner of one of the three pet coke storage facilities has removed all bulk materials, including pet coke from its southeast Chicago facility. The Chicago Department of Public Health has new rules requiring that all large bulk material storage facilities enclose their bulk material within two years. This measure was taken to protect residents around the storage area from inhaling pollutants from the bulk material (US-EPA, 2014b).

(iii) In September 2010, a leak from a pipeline owned by Enbridge Energy Partners LP (an affiliate of Enbridge Inc., based in Calgary, Canada) spilled oil into vegetation, a retention pond and a sewage treatment plant in Romeoville, a suburb of Chicago, Illinois. Following the oil spill, the pipeline was shut off for about a week, which disrupted the delivery of about 670,000 barrels of oil products per day to the refineries located in the U.S. Midwest. Federal officials estimated the leak from the pipeline to be at least 256,000 gallons of crude oil, which caused temporary displacement of some businesses in Romeoville, as well as increased crude oil and gasoline prices in the Midwest (AP, 2011). As a result, the Illinois Attorney General's Office filed a lawsuit against Enbridge Energy partners LP, on the allegation that the company violated the state's air and pollution laws and caused danger to public health and welfare. The lawsuit required Enbridge Energy partners LP to test the groundwater, soil, and sediment for possible contamination and also pay the cleanup cost estimated at $60 million (Manchir, 2010).

(iv) In March 2014, the Whiting refinery owned by British Petroleum (BP) located in Northwest Indiana (near the border of Chicago South side) spilled between 2.38 and 6.20 m^3 (630 and 1,638 gallons) of crude oil into Lake Michigan, after a claimed malfunction of one of the refinery's crude oil distillation units. The initial assessment of the oil spill was claimed to be appearing minimal, however, the company informed the officials of the Coast Guard, EPA, and Indiana Department of Environmental Management to help in containing the oil spill to prevent it from spreading (Jackson, 2014; Horn, 2014). The claimed area of the lake covered by the oil spill was about 465 m^2 (5,000 square feet). But according to officials from the US-EPA, the spill is unlikely to pose any long-term risk to Lake Michigan, which is a source of drinking water for the seven million people living in Chicago and its suburbs. This is because the 68th Street water intake crib is located at about eight miles Northwest of the oil spill site and will not be contaminated (Hawthorne, 2014).

3.6. Illinois' Refineries, Their Capacities, Products and Hazardous Air Pollutant Emissions

Table 3.5 shows the current operating capacities, types of products, and the hazardous air pollutants emitted by the petroleum refineries in Illinois.

(i) As mentioned in Sec. 3.3, the Joliet Refinery located in Joliet, in Will County, Illinois is owned by ExxonMobil Company. The refinery was commissioned in 1972 and has current operating capacity estimated at 250,000 barrels per day. The refinery processes Canadian crude oil supplied via the Enbridge Line 14 Oil Pipeline. The products from the refinery are diesel and gasoline,

Table 3.5. Illinois refineries capacity, products and hazardous air pollutant emissions.

Name of Refinery	**Joliet**	**Lemont**	**Robinson**	**Wood River**
Ownership	ExxonMobil	Citgo (PDVSA)	Marathon	Conoco/ Cenovus
Location in Illinois	Joliet	Lemont	Robinson	Roxana
		Capacity(1)		
bbl/day	250,000	167,000	206,000	356,000
Million tons/Yr	12.5	8.35	10.3	17.8
		Refining Units capacity(1) **(bbl/day)**		
Atm. Distillation	250,000	167,000	206,000	306,000
Vac. Distillation	126,700	75,000	71,500	130,200
Delayed Coker	59,400	46,420	29,000	18,000
Fluidized Cat Cracker	98,000	69,000	55,000	101,000
Hydrocracking	NA	NA	28,000	41,000
Catalytic Reformer	52,600	NA	78,000(3)	78,800(3) 17,700(4)
Naphtha Hydrotreat.	NA	95,700	NA	NA
Kerosene Hydrotreat.	NA	15,000	NA	NA
Diesel Hydrotreating	NA	35,500	NA	NA
Naphtha Desulphurization	110,600	NA	66,500	69,300
Diesel Desulfurization	85,100	NA	78,500	46,000
Gasoline Desulfurization	NA	NA	39,000	80,000
Kerosene Desulfurization	NA	NA	NA	53,250
Sulfur Recovery Unit	NA	NA	NA	32,000

(Continued)

Table 3.5. (*Continued*)

Name of Refinery	Joliet	Lemont	Robinson	Wood River
	Hazardous air pollutant emissions in 2010[(2)]			
[lb]	663,297	244,565	590,860	692,529
	Products Produced			
	Diesel, gasoline, propane, butane, industrial fuels, asphalt	Gasoline, diesel, aviation fuel, heating oil, ultra low sulfur diesel fuel (ULSDF)	Gasoline, diesel, kerosene, jet fuel, petroleum coke, propane, propylene, sulfur, slurry, butane	Clean transportation fuels

(1) Wikidot (2012); (2) Earthjustice (2011); (3) Low pressure; (4) High pressure; Atm-Atmospheric; Vac- Vacuum; NA-not applicable; hazardous air pollutant emissions include VOC benzene, toluene and xylene.

propane, butane, industrial fuels and asphalt (Wikidot, 2012). In 2005, ExxonMobil Company agreed to invest in emissions reductions technologies after a Consent Decree with the EPA, and in 2008, completed the installation of a wet flue gas scrubber at the facility. In 2010, 663,297.48 lb of hazardous air pollutant emissions were reported at the Joliet facility (Earthjustice, 2011). These included VOCs benzene, toluene and xylene.

(ii) The Lemont Refinery located in Lemont, Will County, Illinois is owned by Citgo Petroleum Corporation, a subsidiary of the National Oil Company of the Bolivarian Republic of Venezuela. The original refinery was built in 1920 with crude oil processing capacity of 25,000 barrels per day (CITGO, 2011). But in 1928, the refinery was purchased by Globe Oil & Refining Company and its capacity was expanded to 38,000 barrels of crude oil per day (Wikidot, 2012). In 1954, the refinery was acquired by Pure Oil Company and the crude oil processing capacity was increased to 53,000 barrels per day, reaching over 100,000 barrels per day in 1970. In 1997, the company became 100% subsidiary of Petróleos de Venezuela, S.A., (PDVSA) and processes heavy Venezuelan crude oil produced by CITGO's parent company. The crude oil is supplied from the Shell Lockport Tank Farm via the Buckeye pipeline. The current crude oil processing capacity of the refinery is 167,000 barrels per day (8.35 million tons per year). The products from the refinery include motor gasoline, diesel fuel, aviation turbine fuel and home heating oil. Additionally, the company processes 100% Ultra Low Sulfur Diesel Fuel, due to diesel desulfurization unit commissioned in 2010 at the facility. In 2010, 244,565 lb of hazardous air pollutant emissions was reported by the facility (Earthjustice, 2011).

(iii) The Robinson refinery mentioned in Sec. 3.3 is located in Robinson in Crawford County, Illinois and is currently owned by Marathon Petroleum Company LLC. The original refinery was built in 1906 by the Lincoln Oil Company, with a capacity of 750 barrels of crude oil per day. In 1904, the refinery was acquired by the Ohio Oil Company. In 1940, the capacity was increased to 40,000 barrels of crude oil per day, and was nearly doubled to 125,000 barrels of crude oil per day in 1970 (Wikidot, 2012). The current crude oil processing capacity of the refinery is 206,000 barrels per day, and uses both sweet and sour crudes supplied by the Patoka Oil Terminal Hub. Also, various Canadian crude oils are used in the refinery, including ultra-heavy and sour. The refinery's products include gasoline, diesel fuel, kerosene, jet fuel, petroleum coke, propane, propylene, sulfur, slurry and butane. In 2010, the Robinson refinery reported 590,860 lb of hazardous air pollutant emissions at the facility (Earthjustice, 2011).

(iv) The Wood River Refinery is located in Roxana, in Madison County, Illinois. The refinery was originally built by Shell Oil

Company in 1917 and began operating as the Roxana Petroleum Company until it was owned by ConocoPhillips. The refinery uses heavy crude oil from Canadian oil sands and has crude oil processing capacity of 356,000 barrels per day. The products from the refinery are clean transportation fuels (Wikidot, 2012). In 2005, the refinery installed its S Zorb™ Sulfur Removal Technology and completed its coker and refinery expansion (CORE) project in 2011. In 2010, ConocoPhillips reported 692,528 lb of hazardous air pollutant emissions by the facility (Earthjustice, 2011).

3.7. Concluding Remarks

According to the latest short-term estimate by the US-EIA (2015) and depicted in Fig. 3.9, the world's total production of petroleum and other liquids will increase from 87.38 million barrels per day (MMbpd) in 2010 to 93.50 MMbpd in 2016. For the same period, petroleum consumption would increase from 87.34 MMbpd to 93.42 MMbpd, representing average annual growth of more than 1%. For the long-term, production of petroleum and lease condensate is estimated to grow 0.5% annually from 2012–2040. Consumption of

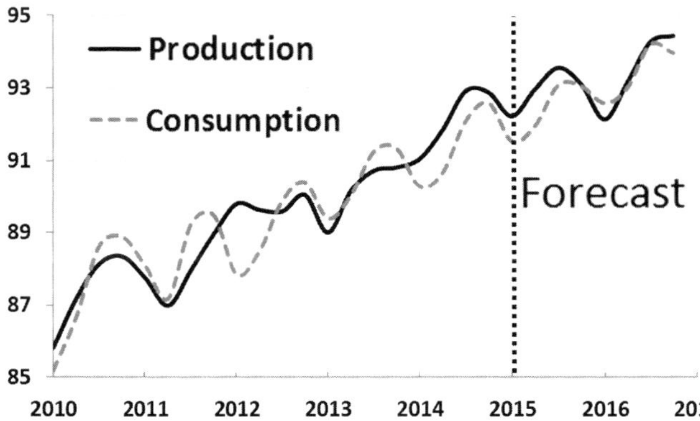

Fig. 3.9. World petroleum and other liquids production and consumption (US-EIA, 2015).

Table 3.6. Production and consumption estimates of petroleum and other liquids in the U.S. (50 states) (Million barrels per day) (US-EIA 2014a, 2015).

	Short-Term			Long-Term				
	2014	**2015**	**2016**	**2020**	**2025**	**2030**	**2035**	**2040**
Production	13.98	14.38	15.37	14.2	13.9	13.2	12.9	12.4
Consumption	19.06	19.32	19.43	17.28*	17*	16.72*	16.58*	16.58*

(*)Data were calculated based on 1 barrel of oil = 5.85 million Btu.

petroleum and other liquids on the other hand, is estimated to decrease by 0.1% annually from 2012–2040 (US-EIA, 2014a).

For the U.S., the short-term production of petroleum annual growth is estimated at 4.9%, while consumption is estimated to increase by 1% annually, as shown in Table 3.6.

Currently, Illinois is a minor oil producer, ranking a distant 16th among states in the U.S., with production of 9.54 million barrels of crude oil in 2014.

However, petroleum is mostly imported to Illinois from Canada because of high cost of its production in the state. According to the U.S. Energy Information Administration, as of 2012 Illinois proven petroleum oil reserves was 42 million barrels. Additionally, the state's unrecovered movable oil in reservoirs was estimated at 4.1 billion barrels, which is about one half of its original volume in 1905, when oil was first discovered in the state (IOGA, 2014). Based on this data, it can be inferred that within the span of about 100 years, Illinois depleted almost 50% of its original oil reserves.

Currently, there is renewed interest in oil exploration in the Illinois Basin, due to the discovery of a very significant oil and gas resource area in the southeast margin of an old oil production area in central Illinois, known as Mt. Auburn Trend (Lasemi, 2009). This situation has led to a dramatic increase in the annual number of oil and gas exploration permits issued by the state. According to ISGS, a 2010 model has led to the completion of a number of prolific wells during the years 2011 and 2012. For example, from April 2011 to December 2012, four wells produced over 295,000 barrels of oil in Macon County. Also during the year 2012, two wells produced over 160,000 barrels of oil in Christian County (Lasemi, 2009).

The state's high quality, light sweet southern crude oil (with the API gravity of about 36.8) is attractive to refiners, despite having a lower gasoline yield of 27.50 wt% or 2% less gasoline per unit volume than West-Texas-intermediate crude oil (with the API gravity of about 39.6) (RCF&REAL, 2009; Agyarko and Mansoori, 2013).

There have been new discoveries of oil reserves, such as the Canadian oil sands with volumetric capacity of about 73 billion barrels, which is second only to Saudi Arabia (Government of Alberta, 2008). But economically, crude oil prices can be very volatile, due to geopolitical issues around the world (Mansoori, 1994; Speight 2011), which can adversely impact Illinois' economy.

Illinois consumes more petroleum every year than its current annual production. As a result, the state is considering the use of horizontal drilling and EOR techniques to harness the estimated 4.1 billion barrels of unrecovered movable oil in the state's oil reservoirs in the Illinois Basin. Based on the average annual petroleum oil consumption of Illinois, which is estimated at 247.5 million barrels, the 4.1 billion barrels unrecovered movable oil in the state's portion of the Illinois Basin may last for up to 16 years (US-EIA, 2011; IOGA, 2014). This means that in the long term, it will be difficult for the state to become independent in petroleum consumption without an alternative solution besides oil importation. One suggested solution to this problem is conversion of some of the state's abundant coal resource into petroleum products using coal-to-liquid technology, provided the sulfur contents of the Illinois coal could be dealt with.

Currently, Illinois leads the Midwest in crude oil refining capacity with four refineries combining to produce 958,000 barrels per day (US-EIA, 2014). Two of these refineries are ExxonMobil in Joliet and Citgo in Lemont all near Chicago; the third is the ConocoPhillips refinery in the Illinois suburb of St. Louis, Missouri, and the forth refinery is the Marathon Petroleum Company refinery in Robinson, Illinois near the Indiana border. However, none of these refineries process any meaningful amount of Illinois petroleum. Instead, approximately 67% of the crude oil produced in Illinois is processed at Country Mark Refinery in Mt. Vernon, Indiana, while the balance is refined in Michigan and Ohio (RCF&REAL, 2009).

Environmentally, emissions from crude oil pipelines and refineries are also a major concern in the state, since they pollute the air and contaminate land and water resources and cause health problems. The environmental problems associated with the use of petroleum in Illinois and elsewhere include carbon dioxide, hydrocarbons, and other gaseous emissions from vehicles, pipelines, refineries, and oil producing wells, as well as piles of petroleum coke around refineries, posing health problems to residents living close to these emission sources.

The average emissions rates in the U.S. from oil-fired electric power plants are 490 lb of CO_2 per MMBtu (1672 lb of CO_2 per MWh), 3.51 lb of SO_2 per MMBtu (12 lb of SO_2 per MWh) and 1.17lb of NO_x per MMBtu (4 lb of NO_x per MWh) (US-EPA, 2013). Consequently, the finite nature of petroleum oil coupled with its environmental and geo-political problems make this energy resource unsustainable. In Chapter 5, we present CO_2 emission, capture, sequestration and utilization.

Bibliography

Agyarko, L. B. and Mansoori, G. A. (2013). A Review of Non-Renewable Energy Options in Illinois. *Int. J. Oil, Gas and Coal Technol.*, **6**(3): 288–347.

Andreas, A. T. (1886). *History of Chicago: From the Earliest Period to the Present Time, 1871–1885*, Vol. 3. A. T. Andreas Pub. Co., Chicago, IL. *archive.org*

[AP] Associated Press (2011). Romeoville oil spill cost: Enbridge estimates cleanup at $60 million. *Huffpost Alive. Huffington Post*, Washington, DC.

[AP] Associated Press (2013). Southern Illinois braces for oil rush as 'fracking' regulations considered by lawmakers. Politics section. *Fox News*.

[AP] Associated Press (2014). Illinois regulators OK eminent domain for pipeline. News section. *The Washington Times*, Washington, DC.

Arbogast, C. R. (2014). For pet coke solutions, look to how state handles fracking. *Chicago Sun-Times*, Chicago, IL.

[ARI] Advanced Resources Int'l (2006). Basin oriented strategies for CO_2 enhanced oil recovery: Illinois & Michigan Basins. Prepared for the U.S. DOE-Office of Fossil Energy-Office of Oil and Natural Gas, Advanced Resources Int'l, Arlington, VA.

Benmekki, E. H. and Mansoori, G. A. (1986). Accurate vaporizing gas-drive minimum miscibility pressure prediction, SPE Paper No. 15667. In the *Proceedings of 61st Ann. Tech. Conf. and Exhib., Soc. of Petrol. Eng.*, pp. 1–12. New Orleans, LA.

Branco, V. A. M., Mansoori, G. A., De Almeida Xavier, L. C., Park, S. J., Manafi, H. (2001). Asphaltene Flocculation and Collapse from Petroleum Fluids. *J Petrol Sci. Eng.* **32**: 217–230.

Chorn, L. G. and Mansoori, G. A. (eds.) (1988). *C7+Fraction Characterization*. Taylor & Francis Pub. Co., New York.

[CITGO] CITGO Petrol. Corp. (2011). Refinery: Lemont. *www.citgorefining.com.*

Du, P. C. and Mansoori, G. A. (1986). Phase Equilibrium Computational Algorithms of Continuous Mixtures. *Fluid Phase Equilibr.*, **30**: 57–64.

Du, P. C. and Mansoori, G. A. (1987). Phase Equilibrium of Multicomponent Mixtures: Continuous Mixture Gibbs Free Energy Minimization and Phase Rule. *Chem. Eng. Commun.* **54**(1–6): 139–148.

Earthjustice (2011). Emissions data from EPA's Petroleum Refining Sector ICR Component 2. *earthjustice.org*.

Enbridge (2013). Southern Access Extension Pipeline Project. *Project Update*, (NRG/RIP/6-13/0.0M), Enbridge Corp. Houston, TX.

[EPA-R5] Envron. Prot. Agency, Region 5 (2013). Test will help EPA remove gas from beneath village. EPA, Chicago, IL. *www.epa.gov/region5.*

[EPA-R5] Envron. Prot. Agency, Region 5 (2014). Hartford area hydrocarbon plume site. Chicago, IL. *www.epa.gov/region5.*

Finley, R. J., Gustison, S. R., Leetaru, H. E. (2005). Geological considerations for CO_2 sequestration in a cratonic basin: An assessment of options in the Illinois Basin, USA. In *The Proceedings of the 7th Int'l Conf. on Greenhouse Gas Control Technologies*, pp. 1339–1344. Vancouver, 5–9 September 2004, BC, Canada.

Finley, R. J. (2007). Field test overview, and assessment of the geological carbon sequestration options of the Illinois Basin. In *The Proceedings of 6th Ann. DOE Conf. Midwest Geological Sequestration Consort.* UIUC, Champaign, IL.

Finley, R. J., Greenberg, S. E. and Frailey, S. M. (2010). *Carbon Sequestration Atlas of the United States and Canada*, DOE/NETL, 3rd Edn (Atlas III), pp. 48–59. Midwest geological sequestration consortium.

Gharbi, R. B. C. and Mansoori, G. A. (eds.) (2005). Artificial Intelligence Applications in Petroleum Exploration and Production. *J Petrol Sci Eng*, **49**(3) 93–96.

Grant, T. (ed.) (2007). *Int'l Directory of Co. Histories*, Gale Virtual Reference Library (eBook), Vol. 14. St. James Press, Chicago, IL.

Grossman, I. (2011). Question of the month — July/August 2011: How much Canadian oil and gas are imported into the Midwest. The Council of State Gov.'s, CSG Midwest. *csgmidwest.org.*

Hawthorne, M. (2014). BP raises estimate of Lake Michigan oil spill. News section. *Chicago Tribune*, Chicago, IL.

Horn, S. (2014). BP doubles initial size estimate of Lake Michigan Oil Spill, pp. 1–3. *Desmogblog.com.*

[IEPA] Illinois Envron. Prot. Agency (2003). Petroleum vapor problems in Northeast Hartford Basements — Fact Sheet No 1. IEPA, Springfield, IL.

[IOGA] Illinois Oil and Gas Assoc. (2014). IOGA-Brochure. *www.ioga.com.*

[ISGS] Illinois State Geol. Survey (2014). Publication spurs oil production in central Illinois. ISGS, Champaign, IL. *isgs.illinois.edu.*

Jackson, A. (2014). Up to 755 gallons of crude oil spilled into Lake Michigan from BP refinery. News section. *The Grand Rapids Press*, Ann Arbor, MI.

Jikich, S., Grube, J., Seyler, B., Webb, N. and Damico, J. (2012). Illinois Basin applications demonstrating potential of ASP EOR technology. Cover Story. *The American Oil and Gas Reporter*, Derby, KS.

Jones, D. S. J. and Pujadó, P. P. (2006). *Handbook of Petroleum Processing*. Springer.

Lasemi, Y. (2009). Oil Potential Seen in Silurian Reef-related Reservoirs in Illinois Sangamon Arch. *Oil & Gas J.*, **107**(29): 36–40.

Lasemi, Y. (2014). Central Illinois Shows Siluro-Devonian Oil Potential. *Oil & Gas J.*, **112**(3): 49–60.

Lewis, J and Bergeron, L. (2010). The economic impacts of the energy industry in Illinois, ISCC Energy Industry Report. Regional Development Institute, Northern Illinois Univ., DeKalb, IL.

Macdonald, S. (2011). Petroleum Conservation in the United States: An Economic Analysis. Earthscan.

Mahajan, D., Taylor, C. E. and Mansoori, G. A. (2007). An Introduction to Natural Gas Hydrate/Clathrate: The Major Organic Carbon Reserve of the Earth. *J Petrol Sci Eng*, **56**(1–3): 1–8.

Manchir, M. (2010). Energy company to pay for cost of Romeoville oil spill Illinois Attorney General's office sues Enbridge to ensure follow-through. News Section. *Chicago Tribune,* Chicago, IL.

Mansoori, G. A. (1994). Where Do We Stand with Fossil Fuels? (A Global Perspective). *FLASHPOINT*., AIChE-F&PD Publication, Fuels & Petrochem. Div. of AIChE, New York, **5**(1):4–5.

Mansoori, G. A., Schulz, K. and Martinelli, E. E. (1988). Bioseparation Using Supercritical Fluid Extraction/Retrograde Condensation (Feature Article). *Nature Biotechnol*, **6**(4): 393–396.

Mansoori, G. A. and Savidge, J. L. (1989). Predicting retrograde phenomena and miscibility using equation of Sate. In *Proceedings of SPE Ann. Tech. Conf. and Exhib.*, Soc. of Petrol. Eng., pp. 383–398.

Mansoori, G. A., Du, P. C. and Antoniades, E. (1989). Equilibrium in Multiphase Polydisperse Fluids. *Int J Thermophys*, **10**(6): 1181–1204.

Mansoori, G. A, Barnes, H. L. and Webster, G. M. (2003). Petroleum Waxes. In *Fuels and Lubricants Handbook*, Chapter 19, pp. 525–558. ASTM Int'l, West Conshohocken, PA.

Mansoori, G. A., Vazquez, D. and Shariaty-Niassar, M. (2007). Polydispersity of Heavy Organics in Crude Oils and Their Role in Oil Well Fouling. *J Petrol Sci Eng*, **58**(3–4): 375–390.

Mansoori, G. A. (2009a). A Unified Perspective on the Phase Behavior of Petroleum Fluids. *Int. J. Oil, Gas and Coal Technol.*, **2**(2): 141–167.

Mansoori, G. A. (2009b). Phase behavior in petroleum fluids. In *Petrol. Eng.*, Downstream section of Encyclopedia of Life Support Systems, UNESCO, UN, Paris, France.

Mansoori, G. A., de Araujo, P. L. B, and de Araujo, E. S. (2012). *Diamondoid Molecules: With Applications in Biomedicine, Materials Science, Nanotechnology & Petroleum Science*. World Sci. Pub. Co., Hackensack, NJ.

Meyers, R. (2003). *Handbook of Petroleum Refining Processes*. McGrah-Hill Pub. Co.

[MGSC] Midwest Geol. Sequestration Consortium (2011). Enhanced oil recovery — EOR I pilot-Loudon Oil Field Huff 'n' Puff. MGSC Projects, MGSC, Univ. of Illinois, Champaign, IL.

Moskowitz, P. (2014). Illinois pollution board rules against new limits on pet coke. Environment section. *Aljazeera America*. america.aljazeera.com.

Park, S. J., Kwak, T. Y. and Mansoori, G. A. (1987). Statistical Mechanical Description of Supercritical Fluid Extraction and Retrograde Condensation. *Int J Thermophys*, **8**(4): 449–471.

Petrov, A. A. (2012). Petroleum Hydrocarbons. Springer.

[RCF&REAL, 2009] RCF E&FC, Inc. and REAL, UIUC (2009.). The Oil and Gas Industry in the Illinois Economy — Q report prepared for IOGA. Regional Economics Applications Lab., UIUC, Champaign, IL.

Rodosta, T., Damiani, D. and Finley, R. J. (2012). Midwest Geological Sequestration Consortium-Validation Phase. National Energy Technology Lab. Morgantown, WV.

Schor, T. and Foy, J. (2012). Strategic American oil begins water injection in its Illinois waterflood project. Duma Energy Corp., Houston, TX.

Schulz, K., Martinelli, E. E. and Mansoori, G. A. (1991). Supercritical Fluid Extraction/Retrograde Condensation (SFE/RC) with Applications in Biotechnology. In T. J. Bruno and J. F. Ely (eds.) *Supercritical Fluid Technology*, pp. 451–478 CRC Press, Boca Raton, FL.

Snow, N. (2012). DOE Reports EOR Success at Lawrence Field in Illinois, News Section. *Oil & Gas J.*, Tulsa, OK.

Speight, J. G. (2010). The Refinery of the Future. William Andrew.

Speight, J. G. (2011). An Introduction to Petroleum Technology, Economics, and Politics. Wiley-Scrivener.

Speight, J. G. (2012). Shale Oil Production Processes. Gulf Pub. Co.

Speight, J. G. (2013). Heavy Oil Production Processes. Gulf Pub. Co.

Speight, J. G. (2014). *The Chemistry and Technology of Petroleum*, 5th Edn. CRC Press.

Speight, J. G. and Ozum, B. (eds.) (2001). Petroleum Refining Processes. CRC Press.

Starling, R. (2014). Enbridge to Expand Pipeline System in Illinois and Indiana. *Energy Global*, Farnham, Surrey, UK.

Tomich, J. (2014). Enbridge pipeline moves a step closer to reality in Illinois. Midwest Energy News, St. Paul, MN.

U.S. Census Bureau (2011). State & County Quick Facts. USCB, Washington, DC.

[US-DOJ] U.S. Dept. of Justice (1909). United States of America, Petitioner, V. Standard Oil Company of New Jersey *et al.*, Defendants, Vol. 17, p. 3481. Gov. Printing Office, Washington, DC.

[US-EIA] U.S. Energy Inf. Admin. (2011). State energy data system: Illinois. Independent Statistics and Analysis, Tech. Notes and Documentation, U.S. EIA., Washington, DC.

[US-EIA] U.S. Energy Inf. Admin. (2013). World petroleum use sets record high in 2012 despite declines in North America and Europe. US-EIA, Washington, DC.

[US-EIA] U.S. Energy Inf. Admin. (2014). State energy data system: Illinois. *Independent Statistics and Analysis*, Tech. Notes and Documentation, U.S. EIA, Washington, DC.

[US-EIA] U.S. Energy Inf. Admin. (2014a). Annual energy outlook 2014. Analysis and Projections, U.S. EIA, Washington, DC.

[US-EIA] U.S. Energy Inf. Admin. (2015). Short-term energy outlook. Analysis and Projections, U.S. EIA, Washington, DC.

[US-EPA] U.S. Environ. Prot. Agency (2013). Clean energy. US-EPA, Washington, DC.

[US-EPA] U.S. Environ. Prot. Agency (2014a). Petroleum coke in Chicago: KCBX notice of violation — June 3, 2014. US-EPA, Washington, DC.

[US-EPA] U.S. Environ. Prot. Agency (2014b). What is EPA doing about pet coke in Chicago? US-EPA, Washington, DC.

[USGS] U.S. Geol. Survey (2002). Tapestry of time and terrain: The union of two maps — geology and topography. U.S. Dept. of the Interior, Washington, DC.

Vakili-Nezhaad, G. R., Modarress, H. and Mansoori G. A. (2001). Continuous Thermodynamics of Petroleum Fluids Fractions. *Chem Eng Process: Process Intensification*, **40**(5): 431–435.

Webber, T. (2013). Piles of Midwest 'pet coke' raising residents' ire. *Associated Press, Yahoo News*.

Wells, L. (2014). Kansas based oil company plans Illinois headquarters. Business section. *Courier & Press*, Evansville, IN.

Wikidot (2012). US refineries: A barrel full. *abarrelfull.wikidot.com*.

Wilson, J. T. (2012). Imperial begins leasing program in the Illinois Basin. News section. *Business Wire*, San Francisco, CA.

4

Natural Gas

4.1. Introduction

Natural gas is mostly made up of methane (80% or more) and other light hydrocarbons (ethane, propane, butane and pentane) and traces of heavier hydrocarbons. Natural gas is mostly used in households for heating and cooking. Natural gas was used in Illinois and some other states in the U.S. for electric energy generation, mostly during peak electricity consumption periods for peak-shaving purposes until late 1990s. The regular Illinois natural gas-fired power plants construction boom started in 1997 and continued until 2007, when the electricity market in the U.S. was going through a period of deregulation (Finley, 2011).

The natural gas which is produced in Illinois is mostly from deep underground formations, coal-bed sources, and in minute quantities from anaerobic digestion of organic solid waste buried in landfills. In older times, there was also natural gas produced from petroleum reservoirs in Southern Illinois oil fields, but those oil fields are now depleted of volatile light hydrocarbons (separated gas) which contributed to the natural gas resources of the state. There are also natural gas reserves around the world in solid-state gas-hydrate (clathrate) forms (Mahajan *et al.*, 2007), but such resources are not known to exist in Illinois in major quantities.

The natural gas produced from the above mentioned sources contain impurities which include carbon dioxide, nitrogen, oxygen, sulfur, sulfur oxides and water (in various quantities depending on the source), which need to be removed from the gas before its use. Too much nitrogen and carbon dioxide in natural gas will reduce its caloric value beyond acceptable use. Besides carbon dioxide, sulfur and sulfur oxides are of health and environmental concerns (Colborn *et al.*, 2010). Minute amounts of water in natural gas will prevent its transportation through pipeline, due to fouling of the pipeline resulting from gas-hydrate crystals formation (Mahajan *et al.*, 2007).

Natural gas industry in the U.S., in general, and in Illinois, in particular, is quite mature and several natural gas related research and development organizations, including the Gas Technology Institute, have been active in Illinois. Also, natural gas production, storage, transmission and distribution system in Illinois is well established and linked to the national system. Meanwhile, a technology for natural gas production from the New Albany Shale in the Illinois Basin is being developed and is discussed in this section. In Fig. 4.1, an illustration of the conventional natural gas production, treatment, storage, transmission and distribution system is shown.

It is worth mentioning that the field of natural gas engineering is a well-established field and many outstanding books, such as the ones by Lyons and Plisga (2004), Mokhatab and Poe (2012) and Speight (1990, 2007, 2010) are available. Also the database (aga.org; api.org; nist.gov), correlations and prediction methods (Agahi *et al.*, 1995; Araujo *et al.*, 2012; Edalat *et al.*, 1993; Mansoori 1994; Mansoori and Edalat, 1996; Mansoori *et al.*, 2012; Mohsen-Nia *et al.*, 1994, 2002; Pires *et al.*, 2001; Riazi and Mansoori, 1993) for the thermo physical properties of natural gas and impurities are known and are available for design of separation, transportation and storage facilities of natural gas.

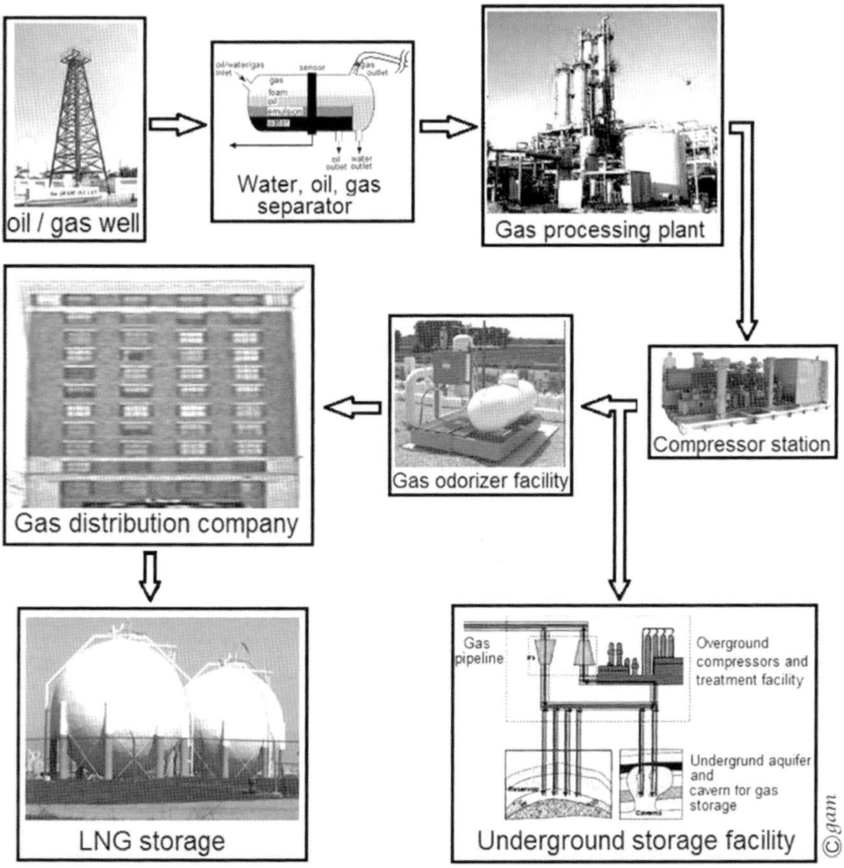

Fig. 4.1. The general scheme for natural gas production, treatment, transmission, storage and distribution system.

4.2. Natural Gas Production

Illinois is not a major producer of natural gas in the U.S. and ranked 27th in 2012. The 2013 production of natural gas from the Illinois Basin was 81.75×10^6 standard cubic meters (2,887 MMscf) as shown in Fig. 4.2 (US-EIA, 2014a).

Because of low production, natural gas for household use and electric power generation in Illinois is purchased from the U.S. Gulf Coast, U.S. mid-continent regions, Western Canada, Colorado and

Wyoming. From these locations the gas is transported through major pipelines to the Illinois natural gas market centers (Chicago Hub and ANR Joliet Hub). These centers also serve as major hubs for natural gas distribution in North America.

In 2013, the electric power sector in Illinois consumed 1.49×10^9 standard cubic meters of natural gas (52.6×10^9 scf) (see Fig. 4.2). In the same year, the residential sector consumed most of the natural gas (12.8×10^9 standard cubic meters or 452×10^9 standard cubic feet) imported to Illinois, with more than 80% of Illinois households relying on the fuel as their primary energy source for home heating and cooking. To meet the winter peak demand, Illinois stores natural gas in underground natural gas storage facilities.

In Fig. 4.3, we report the Illinois natural gas electric power price estimates for the period 1997–2013.

The declining natural gas price from 2008 to 2009 shown in Fig. 4.3 was due to increase in production of natural gas in Illinois and also shale gas in other parts of the U.S., as Illinois had not started producing shale gas. It is projected that shale gas production in the U.S. will represent about 20% of the U.S. natural gas consumption by the year 2020 (US-EPA, 2010).

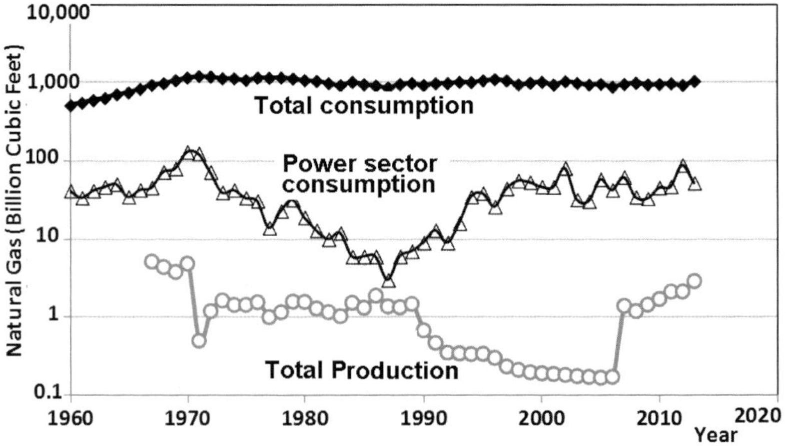

Fig. 4.2. Illinois natural gas production, total consumption and the electric power sector consumption (Data from US-EIA, 2014a).

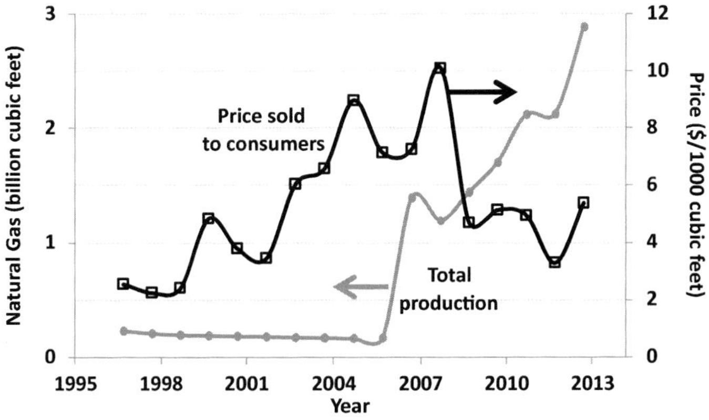

Fig. 4.3. Illinois natural gas production and price sold to electric power generators, 1997–2013 (Data from US-EIA, 2014a).

Table 4.1. ECBM resource potential in the Illinois Basin by state (Finley *et al.*, 2010).

	ECBM	
State	**[10^9standard cubic meters]**	**[10^9 standard cubic feet]**
Illinois	76.46–277.5	2,700–9,800
Indiana	4.25–17.0	150–600
Kentucky	3.68–13.30	130–470
Total	**84.39–307.80**	**2,980–10,870**

Illinois natural gas reserve is currently unknown, but the state's enhanced coal-bed methane (ECBM) potential in the Illinois Basin is estimated at $76.46 - 277.51 \times 10^9$ standard cubic meters (Finley *et al.*, 2010), as reported in Table 4.1, along with the data for Indiana and Kentucky from the Illinois Basin.

ECBM is a method for producing additional coal-bed methane from a source rock, and is similar to enhanced oil recovery applied to petroleum fields. Below, we present a comparison of the ECBM potential of the three states located in the Illinois Basin. While Illinois ECBM potential reserves may seem significant when compared with the annual natural gas consumption reported in Fig. 4.2, under the best scenario it would only be sufficient for approximately 10 years of the state's natural gas consumption.

4.3. Shale Gas Exploitation

Recent estimates indicating large quantities of natural gas reserves and resources in shale formations have resulted in a renewed interest in natural gas as a relatively low carbon energy option. Shale gas is extracted through hydraulic fracturing (known for short as fracking). Fracking involves high-pressure injections of water, chemicals and sand into shale formations to open seams that enable hydrocarbons to flow. Through this process, millions of gallons of water mixed with additives are left in the ground, which could lead to pollution of groundwater aquifers. Furthermore, it has been claimed that shale gas emissions of greenhouse gases are even higher than the conventional natural gas emissions, and may be equal or higher than the emissions caused by using coal or petroleum (Howarth *et al.*, 2011).

Based on advances in horizontal well drilling, hydraulic fracturing and recent high gas prices, Illinois policymakers passed a bill (SB 0664) in November 2011, allowing fracking in Illinois. This bill would require owners of shale drilling companies to submit specific information about their fracking plans to the Illinois Department of Natural Resources (IDNR). This information would include the geological names and description of the wells, nature of additives used in the hydraulic fluid and the depth of the formation where the fluid is applied. The bill could also make it illegal to apply any volatile organic compounds like ethyl benzene, benzene, toluene and xylene, including any use of petroleum distillates for the specified types of shale gas extraction based on hydraulic fracturing. Additionally, well owners would be required to declare information about the disposal of the used hydraulic fluid.

In June 2013, the Illinois Governor signed the legislation allowing regulation of fracking in the state. The legislation which overwhelmingly passed both the State Senate and the House of Representatives was hailed as the strictest for oil and gas drilling in the U.S. This is because while similar regulations exist in states such as Texas, Colorado, Arkansas, Wyoming and Pennsylvania, those regulations neither require fracking companies to disclose the specific chemicals used prior and after the fracking process, nor do they mandate companies to conduct water testing during the entire process of fracking (Kasper and Maloney, 2013). Therefore, since Illinois became the first state in the U.S. to introduce these additional rules, it could compel other states to follow. In essence, the legislation will force oil and gas companies to register with the Department of Natural Resources (Kasper and Maloney, 2013). Also, the oil and gas companies must indicate in the permitting process details of how the wells will be drilled, the amount of fluid used and the applied pressure, how it will withdraw water, contain waste, and disclose the chemicals used (Kasper and Maloney, 2013). In addition, a 30-day public comment period is needed, seven days after the Department of Natural Resources receives a permit application. Residents who are suspicious about their water being polluted due to fracking activities, have the right to request the IDNR to investigate the pollution within 30 days and reach a determination within 180 days (Kasper and Maloney, 2013).

Meanwhile, environmentalists are concerned about the problems in regards to drafting of the rules, which undermined what they have accomplished for several years. One of the problems is the loophole that allows 80,000-gallon threshold of fluid to be injected into rock before the law becomes effective. This practice is likely to pollute ground water sources in the state (ISWS, 2014). It should be mentioned that in 2010, researchers working with oil and gas in the New Albany shale informed the state of Illinois about the potential use of gas, foam, mist or gel for fracking. This is because the New Albany shale formation is shallow and too absorbent to use water in fracking (Wernau, 2014a). As a result, there are claims that many potential fracking firms will not use water, but instead will be using large quantities of fluids that are not measurable in gallons, such as gas, foam or gel in the fracking process (Wernau, 2014a). For this reason, environmentalists demanded that the state should include provisions for non-water-based fracking technology in the fracking rules. Consequently, Illinois lawmakers gave the IDNR up to November 15, 2014 as deadline to complete the writing of these rules to reflect the law. But the IDNR was required to consider public grievances of between 30,000–35000 responses (Hagerty, 2014). In November 2014, the IDNR published on its website (www.dnr.illinois.gov/adrules/documents/62-245.pdf) the final rules approved by the Illinois lawmakers in 7 November 2014. The rules clarify the quantification of non-water based fracking fluids, such as foams and gases. The fracking fluid or "based fluid" as defined by the rules covers any kind of fluid used in the hydraulic fracturing (fracking) process above the 300,000 gallon threshold. As defined by the rules, "base fluid" is a continuous phase fluid type, including, but not limited to, water used in a high volume horizontal hydraulic fracturing operation. Furthermore, the rules require chemical disclosure by permit holders, trade secret disclosure to a health professional, managing of natural gas and hydrocarbon fluids during production, produced water disposal or recycling, transportation and reporting requirements and restoration of the well site and production facility (IDNR, 2014).

But not everyone is satisfied with the final rules approved by Illinois lawmakers. For example, a lawsuit filed on November 10, 2014 in Madison County, Illinois by a group of environmentalists sought to block the rules from publication in the Illinois Register. According to the group, the IDNR failed to give the public enough notice during the rule-making process. Furthermore, the group claims that the IDNR did not use scientific studies in crafting its first draft of the rules and "left members of the public scrambling", to review and comment on the rules before publication (Wernau 2014b).

4.3.1. *Shale hydraulic fracturing, its benefits and problems*

In the oil and gas industry, hydraulic fracturing is used to stimulate wellbores drilled into unconventional reservoirs, such as shale rock or coal-beds, with the view to maximize extraction of underground resources (petroleum and natural gas) (US-EPA, 2010). Hydraulic fracturing allows the production of natural gas and oil from subterranean natural reservoirs, generally at a depth of 1,500–6,000 meters ($\sim$ 5000–20,000 ft) below the surface of the earth. Hydraulic fracturing wells are drilled either vertically all the way or first vertically and then horizontally or directionally (Cardott, 2008). Vertical wells may reach depths of 300–2,500 meters ($\sim$ 1,000–8,000 ft), but horizontal sections of the well may reach thousands of feet away from the production pad on the surface. Hydraulic fracturing is required for wells drilled in low permeability reservoir rock to ensure permeability (see Fig. 4.4).

According to some estimates, hydraulic fracturing has increased the U.S. recoverable reserves of petroleum by at least 30% and of natural gas by approximately 90% (Montgomery and Smith, 2010). The types of equipment used for hydraulic fracturing in the oil and gas industry are generally slurry blenders, combined with high pressure and high volume fracturing pumps (powerful triplex, or quintiplex pumps) and a monitoring unit. Fracture propagation during stimulation is monitored using seismic acoustic fracture mapping, tiltmeter measurements and treatment pressure analysis. The auxiliary equipment consists of high pressure treating iron, fracturing tanks, a chemical additive unit, low pressure pipes and gauges for flow rate, fluid density and pressure. The operating pressure of the fracturing equipment is approximately 15,000 psi (103,421 kPa) at injection rates of up to 100 barrels per minute.

The hydraulic fracture is created by pressurizing fracturing fluid through the wellbore into the geologic formation. The nature of

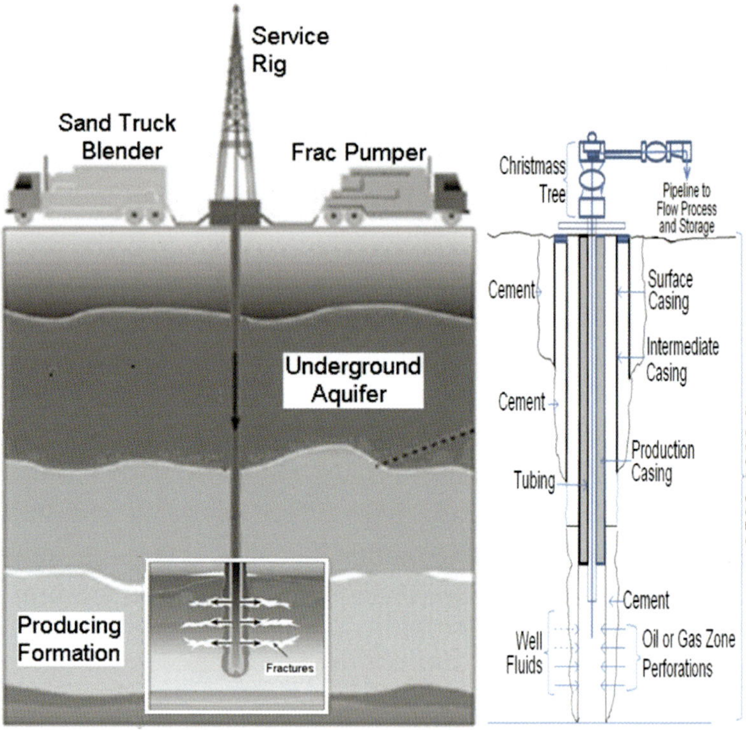

Fig. 4.4. Single-stage hydraulic fracturing (Modified from Arthur and Seekins, 2010).

additives in fracturing fluid will enhance the fractures through chemical and physical interactions with the wellbore rocks. The fluid is pressurized at a rate high enough to raise the downhole pressure to a value greater than the fracture gradient of the formation rock. This will cause the geologic formation to crack. The crack is then propagated as more fluid enters the rock. To keep the fracture open, a proppant (sieved round sand or ceramic beads) is injected into the fracture before releasing the pressure (see Fig. 4.5). This allows the escape of gas from the fracture into the well.

Approximately 15–30% of the fracturing fluids are recovered from the well to the surface (Arthur and Seekins, 2010). Hydraulic fracturing is normally recommended to be performed in short

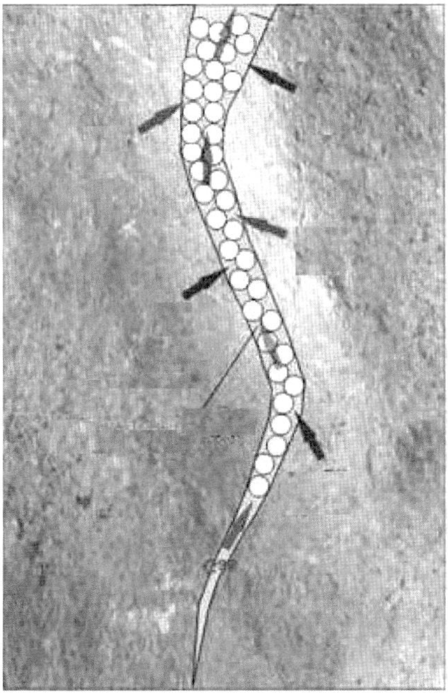

Fig. 4.5. The role of proppants in enhanced gas recovery, Adapted from: http://thorsoil.com/main.html.

durations of time away from the shallow underground sources of drinking water (Arthur *et al.*, 2008; Arthur and Seekins, 2010). Extra protection is provided using multiple casing strings and stringent construction requirements. For the process to be cost-effective, it is important that all fractures created remain in the targeted zone. Fracturing of horizontal wells require 3–5 million gallons of water. The required water is delivered by trucks, via temporary pipeline and from stored tanks or local/centralized impoundments (Arthur and Seekins 2010). For high volume hydraulic fracturing operation, a comprehensive up-front modeling is used for designing stimulation jobs to ensure safety.

The components of a fracturing fluid are proppant, water and additives. The selection of additives depends on the type of wellbore and the nature of the geologic formation. They generally may include acids, biocides or disinfectants, compressed gases, corrosion inhibitors, cross linkers, friction reducers, foams, gels, gelling agents, iron control/stabilizing agents, oxygen scavengers, pH adjusting agents, scale inhibitors and surfactants.

The environmental and human health risk factors linked with hydraulic fracturing include contamination of ground water, risk to air quality, migrating chemicals including hydraulic fracturing chemicals to the surface and potential improper waste disposal (US-HR, 2011). The U.S. Congress asked the US-EPA office of Research and Development to carry out a study on the relationship between hydraulic fracturing and drinking water resources (Arthur and Seekins, 2010). The US-EPA study areas encompasses topics such as quality and quantity of water source, the type of chemicals injected into the well bore, geological or man-made features in the vicinity of the well, methods of waste water handling, factors that influence choice of chemicals and operating conditions and methods of fluids handling on and around the well pad.

In March 2010, the US-EPA announced its intention to conduct a study in response to the request from the Congress. In November 2011, the US-EPA announced its final research plan on hydraulic fracturing (US-EPA, 2011). The progress report was released in 2012 and the final report was estimated to be delivered in 2014 (US-EPA, 2012a). However, the final study is not expected to be released sooner than 2016.

From 2005 to 2009, it was reported that 14 companies who render services to the oil and gas industries used more than 2,500 hydraulic fracturing products containing 750 chemicals (US–HR, 2011). During this period, 780 million gallons of hydraulic fracturing products were used, excluding the water added at the well site. Among the chemicals used were some harmless compounds like salt and citric acid and interestingly, some unanticipated compounds such as instant coffee and walnut hulls. The chemicals used included benzene, lead, methanol (applied in 342 hydraulic fracturing fluids cases), isopropyl alcohol (applied in 274 cases), 2-butoxyethanol (applied in 126 cases) and ethylene glycol (applied in 119 cases). While benzene and lead are considered toxic, methanol is listed as a potential candidate for regulation under the Safe Drinking Water Act (US-HR, 2011). At least one of the service companies involved in the high volume hydraulic fracturing business was subpoenaed by the US-EPA before the company declared the composition in its products. Because of this, a number of lawmakers are arguing that some companies may be hiding certain information about their products, which could be harmful to the environment (US-EPA, 2010).

According to the American Petroleum Institute, hydraulic fracturing fluids generally comprise of 90% water, 9.5% sand and 0.5% chemicals (API, 2015). Low-volume vertical fracking uses less than 80,000 gallons of water or foam equivalent per well, while unconventional hydraulic fracturing (high-volume fracturing) uses an average 3–7 million gallons of water, equating to 150,000–350,000 gallons of chemicals per well. Low-volume vertical fracking has been legal in Illinois and Illinois policymakers passed rules for high-volume hydraulic fracturing on 6 November 2014. As of 15 March 2015 only one company; Denver-based Strata-X Energy (strata-x.com), was registered under fracking rules at the IDNR (dnr.Illinois.gov).

4.3.2. *The New Albany shale gas in the Illinois basin*

The New Albany shale rock is the only natural gas bearing shale rock identified in Illinois. For over a century, the New Albany Shale located in the Illinois Basin of Southeastern Illinois has been identified to have hydrocarbon deposits. However, lack of technical know-how has hampered efforts to harness this energy resource on the commercial scale (Cluff and Dickerson, 1982). The shale rock was named after its outcrop New Albany in Indiana. The shale covers much of Illinois, Western Kentucky and Southwestern Indiana.

Gas has been produced from the New Albany Shale since 1858, from at least 40 fields in Kentucky, 19 in Indiana and 1 in Illinois. To date, more than 600 wells have been drilled into the New Albany Shale in the Illinois Basin, principally in Southern Indiana and Western Kentucky.

The Illinois Basin remains largely undeveloped, due to poorly developed pipeline infrastructure and, until very recently, public policy that permitted meaningful gas production data to remain obscure from public inspection. However, changes in public policy and higher gas prices have spurred a renewed interest in Shale development in the basin.

Kentucky Natural Gas Corporation (KNG) has core assets situated at the southeastern tip of the Illinois Basin in an area referred to as the Rough Creek Fault Zone. Shale in this area is extensively fractured by proximity to the Rough Creek Fault System, the largest fault system in the Eastern Interior U.S.

The elevation of the top of the New Albany Shale stretches from nearly 230 meters (750 ft) above sea level near the outcrop to over 1,370 meters (4,500 ft) below sea level in Southeastern Illinois. The formation estimated area in the Illinois Basin is 137,269 kilometers square (53,000 square miles) and has its thickest depth of over 140 meters (460 ft) in Southeastern Illinois and Western Kentucky (KNG, 2005). The estimated gas-in-place of the New Albany Shale is $(2.44–4.53) \times 10^{12}$ standard cubic meters (86–160 Tscf) with technically recoverable volumes ranging from 1.9 to 19.2 Tcf ($53.8 \times 10^9 - 544 \times 10^9$ m^3) (Durham, 2010; GTI, 2009; Dahaghi and Mohaghegh, 2009; Mastalerz *et al.*, 2010).

In mid-1976, the Illinois State Geological Survey undertook a comprehensive study of the geology and geochemistry of the New Albany Shale to evaluate its potential for shale gas and other hydrocarbon compounds. Through organic geochemistry, existence and nature of petroleum and natural gas can be determined (Carpentier *et al.*, 2007). Subsequent research work conducted in 1982 indicated that the most favorable place in Illinois for shale gas resources, based on four major geologic criteria (depositional facies, shale thickness, thermal maturation and known faulting and fracturing) is the 19-county area of Southeastern Illinois (Cluff and Dickerson, 1982). In 1979, the first quantitative assessment data of gas content of core samples from New Albany Shale gas wells in Wayne County of Illinois became available. In addition to the 2.44 – 4.53 × 10^{12} standard cubic meters (86 – 160 × 10^{12} scf) of shale gas-in-place in 2007, the U.S. Geological Survey estimated that the Illinois Basin contains the following quantities of undiscovered, but technically recoverable oil and gas resources: a mean of 214 × 10^6 barrels of oil, a mean of 132 × 10^9 standard cubic meters (4.66 × 10^{12} scf) of natural gas and a mean of 24 × 10^6 barrels of natural gas liquids (US-GS, 2007).

The New Albany Shale is the most carbon-rich geologic formation among all the gas shale basins in the U.S. The formation is of Devonian and Mississippian age and is a major source of hydrocarbons. The formation was deposited under anoxic marine conditions and is made up of brown, black and green shale with minor beds of dolomite and sandstone. The organic-rich content of the New Albany Shale is 4% by weight and pyrite is a common accessory mineral. Also, present in the black shale layers are phosphorus, heavy metals and uranium, which results in an unusually high radioactivity level (Ettensohn and Barron, 1981). The formation is similar to the Ohio Shale of Ohio and Eastern Kentucky, the Antrim Shale of the Michigan Basin and the Chattanooga Shale of Tennessee and Central Kentucky. The New Albany Shale consists of the following stratigraphic (rock layering) members with distinctive physical properties: the Blocher which is slightly calcareous and dolomites black shale (brownish-black to grayish-black), the Selmier (greenish-gray to olive-gray shale), the Morgan Trail (brownish-black to olive-black fissile siliceous pyritic shale), the Camp Run (a greenish-gray to olive-gray shale interbedded with brownish-black shale), the Clegg Creek (a brownish black pyritic shale is the most organic-rich portion of the formation) and the Ellsworth (composed of a lower part of interbedded brownish-black shale and an upper part of greenish-gray shale) (Stevenson *et al.*, 1981).

To access gas production potential of New Albany Shale, it is important to know the correct properties of the reservoir for appropriate completion strategy of the well to increase gas recovery. These properties are organic richness, thickness, thermal maturation, permeability, porosity, pore pressure, gas-in-place, mineralogy and brittleness. In what follows, these characteristics of New Albany Shale are reported.

4.3.3. *Properties of New Albany Shale*

4.3.3.1. *Organic richness*

In general, the organic carbon content of shale rock is related to the hydrocarbon-bearing potential. Black shale enriched with elements such as arsenic, molybdenum and nickel is known to contain hydrocarbons (Bergstrom *et al.*, 1980). The total organic carbon (TOC) in the New Albany Shale is quite variable, ranging from near 0% by weight in greenish-gray beds to 25% by weight in brownish-black organic-rich beds and averaging about 4% by weight, which is comparable to other shale of the same family (Bergstrom *et al.*, 1980; GRI, 1994).

4.3.3.2. *Shale thickness*

There is a close relationship between areas of thick black shale accumulation and gas production in gas shale basins. Thick black shales normally indicate the presence of rich hydrocarbon reservoirs. The Illinois Basin has a couple of broad regions with moderately thick shale accumulation (Bergstrom *et al.*, 1980). The thickness of the New Albany Shale is more than 90 meters (295 ft) in parts of South-eastern Illinois, western Kentucky and southwestern Indiana, including West-Central Illinois. But the shale measures less than 30 meters (98 ft) in thickness towards the margins, where it is bounded by the

Cincinnati Arch in the East, the Kankakee Arch in the Northeast, the Wisconsin-LaSalle Arch in the North, the Mississippi River Arch in the Northwest, the NE Missouri Arch and the Ozark Dome in the West and the Pascola Arch in the South (Lazar and Schieber, 2004).

4.3.3.3. *Thermal maturation*

Thermal maturation is the time-temperature dependency of the transformation of sedimentary organic matter to mobile hydrocarbons. Most hydrocarbons, including oil and natural gas, evolved from thermal maturation of sedimentary organic matter. Therefore, measurement of thermal maturation of the organic matter in the shale samples (vitrinite reflectance) allows identification of oil and gas locations in the geologic formations (Bergstrom *et al.*, 1980). In Southeastern Illinois in the Illinois Basin, the vitrinite reflectance values for the hydrocarbon deposits are locally well over 0.80% (GRI, 1994). The deepest part of the Illinois Basin in Hamilton and Wayne Counties, in Illinois, has average vitrinite reflectance values between 0.63 and 0.76% (GRI, 1994). The recent vitrinite reflectance reported from the New Albany Shale in the Illinois Basin of Kentucky ranges from 0.55–1.01%. These recent data agree with all the previous data reviewed, thus placing the New Albany Shale in the Illinois Basin largely in the oil window (i.e., Type II to Type II–III kerogen range) (GTI, 2009).

4.3.3.4. *Permeability*

Permeability is defined as the capacity of rock to allow fluids to pass through it. The most commonly used unit of permeability is the Darcy, which may be defined as the rate of flow (in milliliters per second) of a fluid with a viscosity of one centipoise through a porous medium with cross-sectional area of one square centimeter, under a pressure gradient of one atmosphere (760 mm Hg) per centimeter. The measured permeability of the New Albany Shale in the Illinois Basin ranges from 1–100 microdarcies. While it was acknowledged that these permeability data are confusing, the reason for the discrepancy was not known by the time these data were being documented (GTI, 2009).

4.3.3.5. *Porosity*

Porosity is defined as the void (empty) spaces in a material that can be filled by fluids (gas or liquid) and is often expressed as a percentage ratio of the total pore volume to the rock or bulk volume. Porosity is a direct measurement of compactness of the shale. The three factors that can influence the porosity of shale rocks are the depth of burial, carbonate content and organic content (Bergstrom *et al.*, 1980). In the case of the New Albany Shale, most part of the total pore volume of the shale is linked with ultra-micropores with diameters less than 5Å (angstrom). As a result, the diffusion of gas within the ultra-micropores becomes activated and much dependent on temperature rather than pressure, and since the diffusion coefficient is a function of temperature, this explains the low-permeability of the shale. In general, the porosity of shale rock ranges from 4–6 porosity unit (p.u.) The porosity unit (p.u.) is the percentage of pore space in a unit volume of rock. The porosity of the New Albany Shale in the Illinois Basin in a core from Christian County in Kentucky ranges from 0.5–3.1%, averaging 1.8% (Kalantari-Dahaghi and Mohaghegh, 2011). The porosity of the New Albany Shale in the Illinois Basin in a core sample from Sullivan County, Indiana, varies from 0.6–9.3% and averaging 4.0% (GRI, 1994).

4.3.3.6. *Pore pressure*

Pore pressure is defined as the estimated amount of force exerted in the borehole by fluids or gases within the formation that has been penetrated per unit area of borehole. The estimated pore pressure is used to determine the amount of rock and fluid weight required to prevent the fluid or gas from escaping and causing a blowout or wellbore failure. As a result, pore pressure needs to be known to have a better understanding of subsurface environment to ensure safety in designing, drilling and construction of gas wells. A virtual multi-dimensional internal bond (VMIB) model was used to estimate the pressure gradient around a fracture in a well and was found that the pore pressure distribution elliptically decreases from the bottom-hole pressure at the fracture surface to the original reservoir pore pressure at far field (GTI, 2009).

4.3.3.7. *Gas-in-place*

The prediction of recoverable gas in shale rock basins requires an accurate estimate of gas-in-place. The gas-in-place is a combination of the free gas within the pores and natural fractures of the shale, including the gas adsorbed on the surface of the organic matter (kerogen) in the shale. The gas-in-place of the New Albany Shale ranges from 87.5–218.6 million cubic meters per kilometer square (8–20 Bscf per square mile), depending on locations and depths (Kalantari-Dahaghi and Mohaghegh, 2011). The amount of free gas depends on the pressure, porosity and gas saturation and is estimated to be from 15–80% for most gas shale basins (Lewis *et al.*, 2004).

4.3.3.8. *Mineralogy*

Mineralogy is the study of the mineral constituents of rocks, as well as the systematic geological changes that led to the deposition of the rock. The mineralogy of shale rock explains whether the shale is gas bearing reservoir. The types of minerals in the New Albany Shale are quartz, clay minerals, muscovite mica, minor amounts of potassium feldspar and plagioclase (GRI, 1994). In general, shale rich in swelling clay minerals and quartz is a potential source of gas reservoir of significant economic value (Nicolas and Bamburak, 2009). Also, shale rock rich in quartz grains enhances the brittleness of the rock, the effectiveness of reservoir fracturing techniques and consequently the gas production rate. However, swelling clay can be detrimental to potential gas reservoirs during drilling, reservoir stimulation and gas production. The reason is that swelling clay may reduce the porosity and permeability of the shale, resulting in low gas flow. Mineralogy can also provide important clues to facilitate the planning and drilling operations of the shale because of the impact of the minerals on the drilling fluids and the drilling techniques (Nicolas and Bamburak, 2009).

4.3.3.9. *Natural gas content*

The original gas-in-place estimates for the New Albany Shale is about 4.53×10^{12} standard cubic meters (160 Tscf), while technically recoverable volumes are estimated to be about 0.544×10^{12} standard cubic meters (19.2 Tscf).

In Table 4.2, we report the value and/or nature of New Albany Shale properties, which are discussed in this section.

4.3.4. *Technical challenges in hydraulic fracturing*

Recent increasing attention on the numerous shale gas fields in the U.S. is the result of advanced technologies, such as horizontal drilling and efficient hydraulic fracturing, and stimulation that have led to expected cost-effective gas production. Since shale rocks characteristics could vary from basin to basin, it is rather difficult to transfer technologies from one shale rock basin to another. A typical example is the occurrence of natural fractures which vary from basin to basin, as regards the intensity distribution, size, and porosity and occlusion patterns. Occlusion is the disruption of porosity by shale infiltration as carbonate disintegrates into rubble. These natural fractures may not even exist at certain shales, as is the case at Eagle Ford Shale in Texas (Durham, 2010). Consequently, the economic exploitation of these shale rocks remains a major challenge.

At the New Albany Shale in the Illinois Basin, it is necessary to develop specific techniques and methods to boost gas production at commercial levels. This will require careful planning of well drilling

Table 4.2. Characteristics of the New Albany Basin shale gas reservoir.

Property	Value/Nature
Mineralogy[(a)]	Quartz, clay, muscovite mica, little potassium feldspar and plagioclase
Estimated Basin area[(b)]	112,664 km^2 (43,500 $mile^2$)
Depth[(b)]	152.4–609.6 m (500–2000 ft)
Shale thickness[(c)]	30–90 m (98–295 ft)
Net thickness[(b)]	15.2–30.5 m (50–100 ft)
Depth to base of treatable water[(b)]	121.9 m ($\sim$ 400 ft)
Rock column thickness between top of pay and bottom of treatable water[(b)]	30.5–487.7 m (100–1600 ft)
TOC[(b,d)]	1–25 %
Thermal maturation (vitrinite reflectance)[(d,e)]	0.6–1.2%[(i)]
Total porosity[(b)]	10–14%
Permeability[(e)]	1.0–100 microdarcies
Total gas-in-place[(h)]	$(87.5 - 218.8) \times 10^6$ m^3/km^2 ($(8 - 20) \times 10^9$ $ft^3/mile^2$)
Free gas-in-place[(f)]	15–80% of total gas-in-place
Gas content[(b)]	1.13–2.26 std m^3/metric ton (40–80 standard ft^3/ton)
Water production[(b)]	795–79,500 liter/day (5–500 barrels/day)
Well Spacing[(b)]	32.4 hectare (80 acre)
Original gas in place[(b)]	4.53×10^{12} std. m^3 (160×10^{12} std. ft^3)
Recoverable Resources[(b)]	0.544×10^{12} std. m^3 (19.2×10^{12}std. ft^3)
Estimated initial prod. per horizontal well[(g)]	7.78×10^6 m^3/day (275×10^6 ft^3/day)
Targeted gas prod. per horizontal well[(g)]	14×10^6 m^3/day (500×10^6 ft^3/day)

(a) Bergstrom *et al.* (1980); (b) Arthur *et al.* (2008); (c) Lazar and Schieber (2004); (d) GRI (1994); (e) GTI (2009); (f) Lewis *et al.* (2004); (g) Durham (2010); (h) Kalantari-Dahaghi and Mohaghegh (2011); (i) Salehi (2010).

geometries, precise formation characterization and completion practices that would ensure optimum gas recovery. Exploitation of the New Albany Shale in the Illinois Basin began in the late 1800's, when oil exploration was at its peak. In the course of those activities, gas was detected as the drill bits pierced through the shale. However, any further development to exploit the gas foundered on several limitations, including very low permeability of the shale (Durham, 2010).

Over the years, none of the data from the New Albany Shale has translated into commercial scale quantities of gas. This raised concerns about the inherent technical problems (i.e., problems arising from the use of inaccurate data about the shale, which could compromise gas well simulation designs) linked with this gas shale basin. The New Albany gas wells are mainly from natural fractures with an estimated cost ranging from \$1–1.8 million.

The New Albany Shale rock has extremely low matrix permeability, coupled with few open natural fractures. Therefore, proper positioning of horizontal wells relative to the dominant fracture orientation will be needed in commercial scale production of gas. In this case, the fractures may be linked through the hydraulic fracture stimulation, while exercising caution with the water bearing rocks. The water-bearing rock in the New Albany Shale is expected to be one of the major problems hampering efforts to boost gas production rate (Durham, 2010).

In general, horizontal wells at the New Albany Shale basin have indicated much higher production rates than vertical wells with fractures. For example, the initial gas production rate of a typical New Albany Shale horizontal well with a lateral measurement of 762 meters (2,500 ft) is 7,787.13 cubic meters per day (275×10^3 scf per day), reducing to 2831.68 cubic meters per day (100×10^3 scf per day) in the first 24 months of operation (Durham, 2010). The peak production volume for these wells happens in the first 30 days, accompanied by a considerable gas production decline pattern following a shallow hyperbolic curve profile. The cumulative gas production over a 40-year period is estimated to be 28.32×10^6 standard cubic meters per well (one Bscf per well). In general, it is very common to have initial high production rates, accompanied by steep declines, and the New Albany Shale wells are no exception. Also, for some gas wells it is not uncommon for the steep decline in gas production to continue for several years, before realizing any significant volumes of gas production.

4.3.5. *Advanced technologies in shale gas and oil exploitation*

The advance technologies used in shale gas drilling include horizontal drilling, multi-stage fracturing, multilaterals drilling, simultaneous hydraulic fracturing drilling, three dimensional (3D) seismic monitors and real-time microseismic monitoring (Cardott, 2008).

The (3D) high-resolution seismic imaging instrumentation allows the exploration of hydrocarbon deposits in areas with complex structures lying below complex overburden. With this instrumentation, the driller of underground geologic formations can point with precision to the locations of high gas concentrations. This technique produces higher drilling success rates with improved economics of gas production. The environmental impact is also mitigated because fewer emissions and less waste are produced from drilling activities (Cardott, 2008).

Microseismic monitoring, which is a borehole seismic technique, can be used for detecting the micro-acoustic energy caused by changes in the stress of the shale rock matrix during hydraulic fracturing. The device allows accurate measurement of the hypocenter of acoustic emissions during the injection of fluids, gas, proppant and other materials in the borehole. The device can record an image of the fractures by sensing micro-earthquakes that are caused by shear slippage on bedding planes or natural fractures adjacent to the hydraulic fracture. The microseismic data acquisition device is used to improve oil and gas production economics by enhancing reservoir productivity and reducing oil and gas well completion costs (Taleghani and Lorenzo, 2011).

4.4. Underground Natural Gas Storage

Illinois underground natural gas storage began in 1941, when the first known experiments were performed by Superior Oil Company at New Harmony in Indiana. In the course of these experiments, 424,760 standard cubic meters (15 MMscf) of gas was injected into a reservoir, which resulted in gas exit when the well was opened (Buschbach and Bond, 1967). However, there was plenty of salt water in the gas, leading to low gas flow and eventual cut off of the flow and the experiment was abandoned. In 1950, Mississippi River Fuel Corporation at Waterloo, Illinois began the first practical use of underground gas storage in Illinois. But, it was not until 1952 that Natural Gas Pipeline and Panhandle Eastern Pipeline Companies began large scale projects at Herscher and Waverly, respectively (Buschbach and Bond, 1967). As of 2013, Illinois ranked second only to that of Michigan with total underground natural gas storage capacity of 28×10^9 standard cubic meters (1000 Bscf), as seen from the data reported in Fig. 4.6 (US-EIA, 2014b).

Not every state in the U.S. has natural gas storage systems as indicated in Fig. 4.6, and therefore the states without gas storage systems have to depend on others to meet their energy storage needs.

The three basic types of reservoirs for underground storage of natural gas are depleted gas and oil fields, aquifers (a typical example is shown in Fig. 4.7) and salt caverns. Only depleted oil and gas fields and aquifers are used in Illinois because there are more aquifers in Illinois than any other state in the U.S.

An example of depleted oil and gas fields is the Herscher natural gas storage site in Illinois, owned by Natural Gas Pipeline Company (Kinder Morgan) that was part of the Trenton oil fields operating in the 1880s, but was abandoned in early 1900s (Schill, 2009). Since

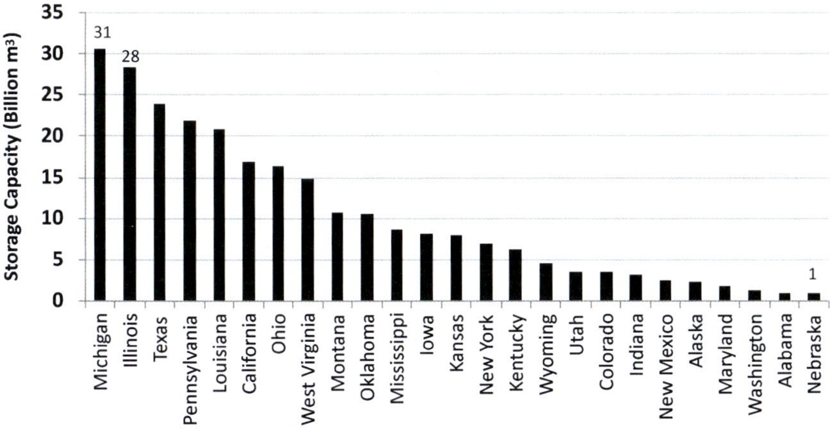

Fig. 4.6. Underground natural gas storage capacity of the top 25 U.S. states (US-EIA, 2014b).

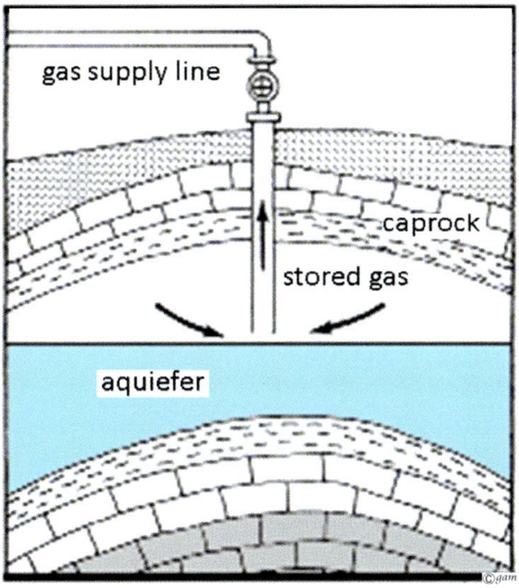

Fig. 4.7. A typical aquifer gas storage field showing injected gas displacing water.

1950s the Herscher site has been used to store natural gas, because the Cambrian-age sandstone makes it ideal for this purpose.

An underground storage facility for natural gas in the Illinois Basin requires an aquifer (see Fig. 4.7), or a storage porous rock layer with sufficient permeability and porosity, to accept and hold the gas, an impermeable caprock overlaying the storage area to prevent upward migration of gas, and a geologic trap to keep the gas from moving in a horizontal direction (Buschbach and Bond, 1967).

A reservoir originally filled with oil or gas and is referred to as depleted reservoir, usually contains porous rock layers that may be used for gas storage. At high pressures, gas also dissolves in liquids (Mansoori and Edalat, 1996). Natural aquifers, which were originally filled with fresh or salt water, are another possibility for gas storage in Illinois Basin. About 60% of Illinois natural gas is stored in salt water aquifers, while the remaining is stored in depleted gas and oil reservoirs. However, gas storage in depleted oil and gas fields represent the most widespread method of storing natural gas in large quantities (around 76% of gas storage volume) worldwide (Evans and West, 2008). In Illinois, natural gas storage facilities are used to meet traditional seasonal load variations, emergency situations and hourly swings.

The underground natural gas storage facilities in Illinois are currently used for the following purposes, most of which are well specified by the U.S. Federal Energy Regulatory Commission (FERC 2004):

i. To ensure the lowest cost of natural gas purchase from gas producers by the gas storage company.
ii. To provide constant gas supply to customers and meet related regulatory obligations.
iii. To provide lower gas prices to consumers through maintenance of specific levels of storage inventory.
iv. To ensure liquidity at market centers to help control price volatility and maintain orderly gas markets.
v. To offset the reduction in traditional supplies to meet winter demand, increase the comfort inventory level of working gas or top gas; serve to offset the growing summer peak impacts from electric generation through the injection of more gas during the shoulder months and support electric generation loads.

(a)

(b)

Fig. 4.8. Manlove gas storage complex in northwest Champaign County, Illinois. In the foreground is one of the numerous wells in the area with the LNG plant and tanks in the background. (a): Actual picture. (b): Vertically expanded picture to view facilities in the background (en.wikipedia.org/wiki/Newcomb_ Township,_Champaign_County,_Illinois, accessed on 29 January 2015).

Gas storage in underground storage field operations consists of several components and interdependent facilities. These include observation wells, injection and withdrawal wells, water disposal wells, dehydration facilities, gathering lines, compressors, gas measuring facilities and, at least, one major transportation pipeline for gas delivery or reception with proper design (Edalat and Mansoori, 1988). An example of underground gas storage in Illinois is the Manlove gas storage complex in Northwest Champaign County (see Fig. 4.8), which can hold approximately 4.25×10^9 standard cubic meters (150 Bscf) of gas at depths of about 1,200 m ($\sim$ 4,000 ft) (Finley *et al.*, 2005).

In Manlove gas storage complex, the gas is stored under about 27,500 leased acres of farmland, but there is a central location with offices, gas treatment facilities, pumps and pipeline connections (Finley *et al.*, 2005). Natural gas is injected into the reservoir (depleted oil and gas or aquifer formation) to build up a volume of compressed gas, which can later be withdrawn through operating wells. The gas storage can be cycled between a maximum and minimum operating pressures of the reservoir. The maximum pressure is the highest operable pressure that the caprock can support, and tends to be the original pressure of the reservoir at the time of discovery (Buschbach and Bond 1973). Below the minimum operating pressure is the cushion gas (base gas), which represents around 50% of the total gas in place in the reservoir and is physically unrecoverable gas. The working gas which can be made available to the market is the volume of gas in the reservoir above the level of base gas. The working gas capacity refers to the total gas storage capacity less than the base gas. The total gas in storage is the volume of storage in the underground facility at a particular time. The total gas storage capacity is the maximum volume of gas that can be stored in an underground storage facility in accordance with its design, which comprises the physical characteristics of the reservoir, installed equipment and operating procedures particular to the site.

In the 70s, over 16.4×10^9 standard cubic meters (580 Bscf) of natural gas was stored in underground reservoirs at 37 locations in Illinois. Out of this volume of gas, about one third or 5.47×10^9 standard cubic meters (193.1 Bscf) was working gas and two-third or 10.93×10^9 standard cubic meters (386 Bscf) was cushion gas (Buschbach and Bond, 1973).

Out of the 37 underground reservoirs in Illinois, 29 are with working gas capacity of 8.69×10^9 standard cubic meters (307 Bscf) and daily withdrawal capability of 173.6×10^6 standard cubic meters (6,129 MMscf) (Tobin, 2006). Out of these 29 locations, 11 are depleted reservoirs with working gas capacity of 1.44×10^9 standard cubic meters (51 Bscf) and daily withdrawal capability of 23.64×10^6 standard cubic meters (835 MMscf), and 18 are aquifer storage sites with working gas capacity of 7.25×10^9 standard cubic meters (256 Bscf) and daily withdrawal capability of 149.9×10^6 standard cubic meters (5,293 MMscf) (Tobin 2006). The remaining 8 (of the total 37) storage locations in Illinois remain inactive because gas companies can now store more gas using few storage sites than it was in the 70s, due to improved methods of storage.

In the past years, some of the aquifer storage facilities experienced leak problems and were shut down. The leaks were the result of inadequate sealing of caprock and problems associated with faulting. In Northern Illinois, the Cambrian Mt. Simon underground water-filled sand/sandstone aquifers are extensively used to store natural gas for peak seasonal usage in the Midwest. About 60% of the total gas storage inventory in Illinois is held at the Mt. Simon formations, which is an important supply for peak winter deliveries for Northern Illinois cities and specially Chicago (US-DOE, 1999a). Since 1968, there have been several modifications to the storage fields to meet changing market conditions. In Illinois, operators of underground gas storage facilities have added new designs and built surface and underground facilities. These changes include relocating wells within reservoirs, drilling of large diameter wells, adding coil tubing drilling, horizontal wells drilling and completing larger diameter wells. Furthermore, storage operators are adding dehydration units, compressors and new gathering lines (FERC, 2004). Besides, they are employing new technology to better understand reservoir geology confinement and reservoir flow behaviors, while adding new storage operational procedures and abandoning unprofitable facilities.

4.4.1. *Advanced technologies to improve gas storage facilities*

Currently, there are several available new technologies that may be used to improve natural gas storage field efficiency in Illinois. One technology allows debris in the well pipe to be removed using a low/high frequency sound wave device. The device can be used to provide lasting improvements in storage wells by vibrating scales off the well pipes to unclog them. The conventional way of removing debris is by washing, injection of acids and creating new perforations in the well pipe. Additionally, there are innovative fracturing technologies, such as injecting high pressure liquid carbon dioxide instead of water or other liquids, to prevent clays from sticking and sealing off parts of the reservoir (FERC 2004).

For storage of large quantities of gas using the same volume of reservoir, the gas may be converted into solid clathrate, also known as gas-hydrate (Mahajan *et al.*, 2007), at reservoir temperature and high pressures in the presence of water, to make it more compact. The advantage is that as much as 181 standard cubic meters ($\sim$ 6400 scf) of natural gas can theoretically be stored in a single cubic meter of hydrate (US-DOE, 1999b; Mahajan, *et al.*, 2007; Mansoori and Edalat, 1996).

Also, a gas storage facility in Sweden called "lined rock caverns" is being studied by DOE to expand the geographic diversity of gas storage in U.S. The system consists of a steel tank installed in a cavern carved into the rock of a hill. Around the steel tank is a cast concrete which transfers the pressure build-up from the tank to the rock. According to DOE, although the technique is more expensive than conventional method of gas storage (in aquifers, depleted oil or gas fields or salt formations), it allows multiple injections and withdrawals of gas several times during the year, a situation which is not always practical with the other systems of gas storage. This could make the service cost of the "lined rock cavern" comparable to conventional storage methods. In the U.S., a couple of sites have already been studied as possible locations for applying the "lined rock cavern" technology. These sites include a 113.3×10^6 cubic meters (4 billion cubic foot) working gas facility near Atlanta, Georgia, and a 56.6×10^6 cubic meters (2 billion cubic foot) working gas facility near Boston, Massachusetts (US-DOE, 1999b).

4.5. Natural Gas Consumption

Natural gas is mostly used for households in Illinois and other states in U.S. In 2013, from a total of 1.057×10^{12} cubic feet of natural gas consumption, only 5.1% was used for Illinois electricity production. However, electricity generation using natural gas could reach 31% with full utilization of the existing gas-fired electric power plants through more production or more import of natural gas (Wernau, 2011).

Other uses of natural gas in Illinois include industrial, transportation and commercial sectors of the economy as shown in Fig. 4.9 (US-EIA, 2014a). Natural gas pricing can be very volatile, which could put financial burden on consumers.

It is believed that an option for reducing carbon dioxide emissions would be to replace coal-fired power plants with natural gas-fired power plants. During early 1990s until 1996, some natural gas-fired power plants were built in Illinois to meet peak-load shaving demand because natural gas was cheaper than fuel oil (Kaplan, 2010). As mentioned in Sec. 4.1, the natural gas-fired power plants construction boom in Illinois started in 1997 and continued until 2007, when the electricity market in the U.S. was going through a period of

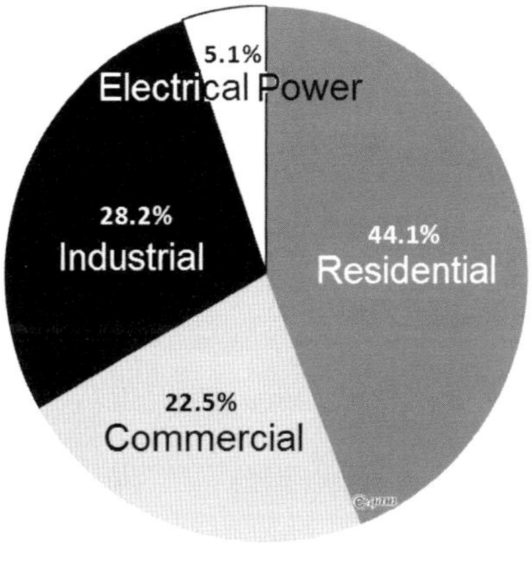

Fig. 4.9. Illinois natural gas consumption by end use (Data from US-EIA, 2014a).

deregulation. Around that period, more natural gas-fired power plants were built in Illinois by non-utility companies known as Alternative Retail Electric Suppliers, while Investor-Owned Utility companies nearly halted the construction of new power plants. Construction of the new power plants was partly motivated by the desire to reap quick financial gains from the newly deregulated power market, as well as by the cheap natural gas prices at that time. During 1998, about 16 new gas-fired power plants totaling 5,305 MW of capacity became operational or were being built in Illinois (Finley, 2011). By the end of 1998, Illinois was a net exporter of electricity.

By August 2000, over 50 new gas-fired power plant permits had been granted in the State of Illinois or were in the pending status. Furthermore, 10 new natural gas-fired power plants received a permit by the Illinois-EPA, while 20 plants were in the permit process (Finley, 2011). One example is the Exelon's Southeast Chicago Energy Project which was built in 2002 (see Fig. 4.10). The 8-unit

Fig. 4.10. The Exelon's Southeast Chicago Energy Project which was built in 2002. It is considered a Level-1 Blackstart Facility — meaning it can be restarted without outside power. In the event of a major power system failure, Southeast Chicago Energy Project will be used to reenergize the electric system and bring other regional power plants back online.

facility provides much needed peak generation to the city of Chicago during periods of high electric demand in the summer and winter months. Each unit of this power plant is rated at 44 MW power and runs on natural gas.

According to Illinois-EPA, by the year 2005, out of the 55 proposed small size gas-fired power plants, only 22 were built, while out of the 21 proposals for larger gas-fired power plants, just five were constructed (Orrick, 2005). According to Exelon Corporation, while the construction of new natural gas combined cycle (NGCC) power plants is the most economically viable option at the moment because of cheap natural gas prices, they are simply not required. This is because the current demand for electricity is low and many power markets are still saturated with generating capacity (Exelon, 2015).

4.5.1. *New generation of natural gas power plants*

As the state plans to phase out aging coal-fired power plants which we discussed in Sec. 2.3, there may be a need to operate underutilized natural gas plants to fill the lost capacity. Carbon dioxide emissions is 1,925 kg/1,000 m^3 of natural gas ($\sim$ 120 lb/1,000 ft^3) (US-EIA, 2013c) and 29% of total methane in the U.S. is emitted by natural gas and petroleum systems. Methane emissions are 9% of the total GHG emissions in the U.S. however, methane has over 20 times greater potential to trap radiation than CO_2 over 100-year period (US-EPA, 2012b). On the other hand, burning natural gas would emit much less sulfur, mercury and particulate matter than coal. A comparison of fossil fuels emissions are shown in Table 4.3.

Recent developments in high-temperature and more efficient natural gas combustion turbine (NGCT) plants by Mitsubishi Heavy Industries and NGCC plant by General Electric and Siemens, could further help Illinois in switching to natural gas-fired power plants in electric power production (Patel, 2011) and reduce the environmental pollution caused by coal fired power plants.

Table 4.3. Fossil fuels emission levels (Smith and Sikorski, 2014).

	Natural Gas		**Petroleum**		**Coal**	
Pollutant	**(lb/10^9 Btu)**	**(kg/10^9cal)**	**(lb/10^9Btu)**	**(kg/10^9cal)**	**(lb/10^9Btu)**	**(kg/10^9cal)**
Carbon Dioxide	117,000	211	164,000	296	208,000	375
Carbon Monoxide	40	0.072	33	0.06	208	0.375
Nitrogen Oxides	92	0.166	448	0.808	457	0.824
Sulfur dioxides	1	0.0018	1,122	2.02	2,591	4.67
Particulates	7	0.0126	84	0.151	2,744	4.95
Mercury	0	0	0.007	1.26×10^{-5}	0.016	2.89×10^{-5}

In 2011, the Mitsubishi Heavy Industries announced their J-Series NGCT with a claimed thermal efficiency exceeding 60%, and with the world's highest turbine inlet temperature of 1, 600°C (2, 912°F).

Also in 2011, the General Electric Company launched its new NGCC power plant known as FlexEfficiency-50. This new 510 MW power plant is claimed to be capable of starting under 30 minutes with a push of a button and could offer an efficiency greater than 61%.

In Germany, Siemens also announced setting a new world record in power plant efficiency by launching the SGT5-8000H NGCC power plant. It is claimed this new generation H-class gas turbine is the result of 10 years of research effort and is capable of reaching over 60% of efficiency. This Siemens turbine was designed for 400 MW in simple-cycle duty and for 600 MW in combined-cycle duty (Patel, 2011).

NGCT plant and NGCC plant are the two categories of plants that burn natural gas and compressed air to produce steam and generate electricity (see Fig. 4.11).

Example of an NGCT power plant operating in Illinois is the 640 MW natural gas-fired Lee Energy facility located at approximately 145 kilometers (90 miles) west of Chicago, in Lee County.

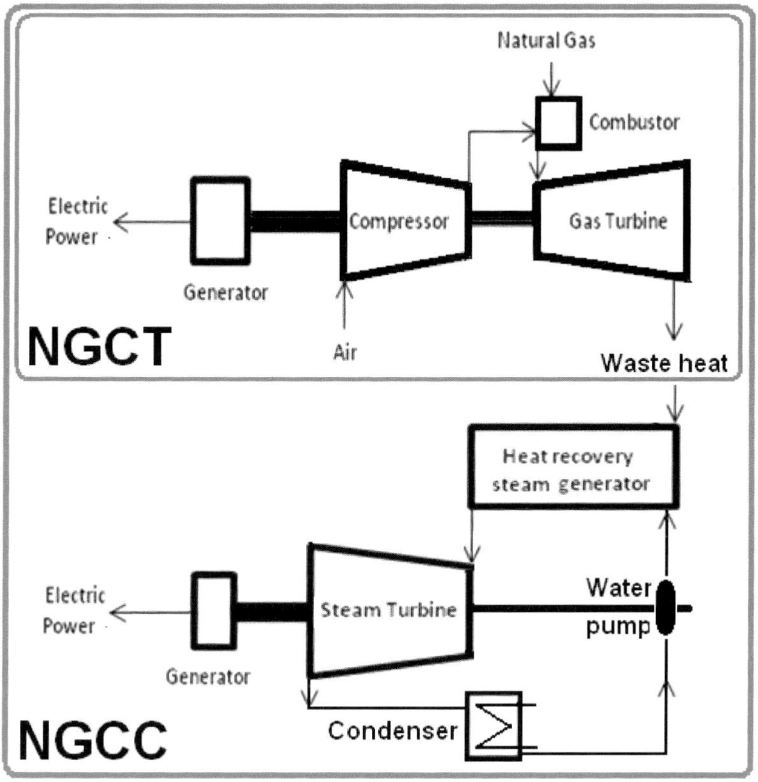

Fig. 4.11. Principles of NGCT plant, which is a subset of NGCC plant (Modified from US-EIA, 2010).

The Lee facility is owned by Duke Energy and consists of eight 80 MW natural gas-fired combustion turbines, operating in simple-cycle mode (IEPA, 2004).

An example of NGCC power plant operating in Illinois is the Minooka combined cycle power plant in Kendall County. The 1,160 MW facility owned by Dynegy Inc., consists of four-train GE 7FA combustion gas turbines with four three-pressure, reheat heat recovery steam generators and four GE steam turbine generators. The facility is duct-fired in the summer to boost power during local peak periods (SNC-Lavalin, 2005).

The NGCC plant is generally more efficient than the NGCT plant because it makes maximum use of the energy in the fuel through a two-step generating process that captures waste heat that would otherwise be lost. The NGCC plants are typically operated to meet intermediate load demands, while the NGCT plants operate in peak-load shaving demand periods or much fewer operating hours.

The NGCC plants may be constructed within two to three years at the cost of roughly $1,200 per kW (Kaplan, 2010), as opposed to coal-fired plants which take five to six years at the cost of approximately $3,500 per kW. The annual operation and maintenance costs of NGCC and NGCT plants are estimated at 4% of the investment costs per year (IEA, 2010; Seebregts, 2010). It is estimated that the levelized cost of energy for the NGCC plant is between $64.4–91.3 per megawatt hour (MWh) (US-EIA, 2014d). The efficiency of NGCT plants is about 35–42%, compared with 52–60% for NGCC plants (IEA, 2010; Seebregts, 2010).

There are certain factors that need to be dealt with in order to increase the likelihood of replacing coal with natural gas in Illinois. These include the amount of excess natural gas-fired generating capacity available in Illinois; the current operating patterns of coal and gas plants, the amount of flexibility power system operators may have for changing those patterns, and whether or not the transmission grid could deliver power from existing gas power plants to loads currently served by coal plants. Additionally, it is important to have sufficient natural gas supply pipeline and gas storage capacity to deliver large amounts of additional fuel to gas-fired power plants (Kaplan, 2010).

However, due to the prevailing uncertainties about natural gas prices and CO_2 emission tax that could rise in the future, it may be difficult to adopt any robust strategy for natural gas- fired power plants in electric power production deployment in Illinois. This situation may result in a changing economic balance between gas and coal-fired power plants; unless Illinois can find a solution to exploit its shale-gas deposits in the Illinois Basin.

4.6. Concluding Remarks

Natural gas is another non-renewable fossil energy resource. According to the International Energy Agency (IEA) and based on the current rate of consumption, the current global conventional natural gas reserve could last up to 130 years. Also, the technically recoverable natural gas reserve in the U.S. is estimated at 2,431 trillion standard cubic feet (Tscf) (US-EIA, 2012; US-EIA, 2014c) and there are claims that it will last for over 100 years, provided new gas supplies continue to be found in shale deposits (Nelder, 2011). But, in spite of these vast reserves of gas, this finite energy resource cannot be extended to meet the needs of indefinite generations ahead.

Currently, Illinois imports natural gas from Canada, Gulf of Mexico and other states in the U.S, including Wyoming and the Rocky Mountains Region. To boost natural gas production, Illinois policy makers are deliberating on how to exploit its huge shale gas deposit in the Illinois Basin, which could provide a steady gas supply with guaranteed price to the state's gas-fired power plants for up to 40 years or more (Oil-Gas-News, 2011). The gas content of the New Albany Shale in the Illinois Basin is estimated at 86 trillion cubic feet (Agyarko and Mansoori 2013).

Economically, the cost of electricity from gas-fired power plant has been very sensitive to natural gas price, which has seen more market volatility in recent years. For example, at the average price of natural gas in 2008, the levelized cost of electricity was 8.6 cents per kWh, compared to 5.5 cents per kWh based on the average 2009 gas prices (Flavin and Kitasei, 2010). However, recent discoveries of shale gas in the U.S. are expected to bring price stabilization to the gas market. This price reduction is reflected in the projected average U.S. levelized cost of electricity for advanced combined cycle natural gas-fired power plants entering service in 2019, which is estimated to be $64.4 per MWh (US-EIA, 2014d).

Environmentally, natural gas burns cleaner than other fossil fuels. Electric power plants burning natural gas based on advanced generation natural gas-fired combined cycle technology emit 0.0075 lb of NO_x per MMBtu, 0.001 lb of SO_2 per MMBtu and 117 lb of CO_2 per MMBtu (US-EIA, 2013b). These emissions of NO_x, SO_x, and carbon dioxide from natural gas-fired power plants per unit of electrical energy produced are less than those of oil-fired or coal-fired power plants, while leaving almost no ash particles. However, extraction and transportation of natural gas produces local air and water pollutants, including emissions of greenhouse gases. To reduce these emissions, Illinois may ramp up the capacity of its underutilized gas-fired plants or acquire high performance NGCC power plants or ultra-supercritical pulverized coal power plants rated at about 60% efficiency (Buchan and Cao, 2004). Despite being cleaner than coal, natural gas also does not meet the sustainability criteria because it is not renewable.

Bibliography

Agahi, R. R., Ershaghi, B., Lin, M. C. and Mansoori, G. A. (1995). Thermodynamic behavior of hydrogen/natural gas mixtures. In *Proceedings Of the* 74th *Ann. Conv. Gas Processors Assoc.*, pp. 23–27. GPA, Tulsa, OK.

Agyarko, L. B. and Mansoori, G. A. (2013). Non-renewable Energy Options in Illinois. *Int. J. Oil, Gas and Coal Tech.*, **6**(3): 288–347.

API (2015). Unlocking America's natural gas resources. Available at *api.org/hydraulicfracturing*. Am. Pet. Inst., Washington, DC.

Araujo, P. L. B., Mansoori, G. A. and Araujo, E. S. (2012). Diamondoids: Occurrence in Fossil Fuels, Applications in Petroleum Exploration and Fouling in Petroleum Production. *Int. J. Oil, Gas and Coal Tech.*, **5**(4): 316–367.

Arthur, J. D., Langhus, B. and Alleman, D. (2008). An Overview of Modern Shale Gas Development in the U.S. A Publication of ALL Consulting, Tulsa, OK.

Arthur, J. D. and Seekins, J. (2010). Water and shale gas development. In *Proceedings of National Assoc. of Royalty Owners Ann. Conf*, pp. 1–39. Pittsburgh, PA.

Bergstrom, R. E., Shimf, N. F. and Cluff, R. M. (1980). Geologic and geochemical studies of the New Albany shale group (Devonian-Mississippian) in Illinois. Report No. ORO-E-(40-1)-5203-12, Ill. State Geol. Survey, Urbana, IL.

Buschbach, T. C., Bond, D. C. (1967). Underground Storage of Natural Gas in Illinois. Illinois Petroleum, Ill. State Geol. Survey, 86: p. 54.

Buschbach, T. C., Bond, D. C. (1973). Underground Storage of Natural Gas in Illinois. Illinois Petroleum, Ill. State Geol. Survey, 101: p. 80.

Buchan, B. and Cao, C. (2004). Coal-fired Generation: Proven and Developing Technologies. Office of Market Monitoring and Strategic Analysis, Florida Public Service Commission Reports and other Sources: Electric and Gas, No. 6, pp. 1–25.

Cardott, B. J. (2008). Understanding Gains on 'New' Reservoir Shale Closing 'Conventional' Gap. *AAPG Explorer*, **29**(11): 78, Am. Assoc. Petrol. Geol., Tulsa, OK.

Carpentier, B., Wilhelms, A. and Mansoori, G. A. (2007). Reservoir Organic Geochemistry: Processes and Applications. *J. Petrol. Sci. & Eng.*, **58**(3–4): 341–343.

Cluff, R. M. and Dickerson, D. R. (1982). Natural Gas Potential of the New Albany Shale Group (Devonian-Mississippian) in Southeastern Illinois. *SPE J.*, **22**(2): 291–300.

Colborn, T., Kwiatkowski, C., Schultz, K. and Bachran, M. (2010). Natural Gas Operations From a Public Health Perspective. *Human and Ecological Risk Assessment*, **17**(5): 1039–1056.

Dahaghi, A. K. and Mohaghegh, S. D. (2009). Economic Impact of Reservoir Properties, Horizontal Well Length and Orientation on Production from Shale Formations: Application to New Albany Shale, SPE paper No 125893, pp. 1–11. In *Proceedings of 2009 SPE Eastern Reg. Meet.*, Charleston, WV.

Durham, L. S. (2010). Illinois Basin shale gets tech focus. News section, AAPG *Explorer*, Am. Assoc. Petrol. Geol., Tulsa, OK.

Edalat, M., Bozar-Jomehri, R. B., and Mansoori, G. A. (1993). Generalized Equation Predicts Vapor Pressure of Hydrocarbons, *Oil & Gas J.*, **91**(5): 39–40.

Edalat, M. and Mansoori, G. A. (1988). Buried Gas Transmission Pipelines, *Energy Sources J.*, **10**(4): 247–252.

Ettensohn, F. R., Barron, L. S. (1981). Depositional model for the Devonian-Mississippian Black-Shale sequence of North America: A tectono-climatic approach. Report No. DOE/METC/12040-2, U.S. Dept. of Energy.

Evans, D. J. and West, J. M. (2008). An appraisal of underground gas storage technologies and incidents, for the development of risk assessment methodology. Report No. OR/07/023, British Geol. Survey, Open Report.

[Exelon] Exelon Corp. (2015). A growing portfolio of natural gas generation. Exelon Energy, Chicago, IL. *exeloncorp.com.*

[FERC] Fed. Energy Regulatory Commis. (2004). Current state of and issues concerningunderground natural gas storage. FERC, Washington, D.C.

Finley, R. J., Gustison, S. R., Leetaru, H. E. (2005). Geological consideration for CO_2 sequestration in a cratonic basin: An assessment of options in the Illinois basin, Greenhouse Gas Control Technologies 7. In *Proceedings of the* 7th *Int'l Conf. on Greenhouse Gas Control Technologies*, Vol. II, Part 1, pp. 1339–1344.Vancouver, CA.

Finley, R. J., Greenberg, S. E. and Frailey, S. M. (2010). *Carbon Sequestration Atlas of the U.S. and Canada*, DOE/NETL, 3rd Edn (Atlas III), pp. 48–59. Midwest Geol. sequestration consortium.

Finley, R. J. (2011). Illinois in the global energy marketplace. Ill. State Geol. Survey, Prairie Research Inst., Short Tech. Topics.

Flavin, C. and Kitasei, S. (2010). The role of natural gas in a low-carbon energy economy. Briefing Paper, World Watch Inst.

[GRI] Gas Research Inst. (1994). Gas Potential of the New Albany Shale (Devonian and Mississippian) in the Illinois Basin. Final Report No. GRI-92/0391/, Illinois Basin Studies 2, GRI (now GTI), DesPlains, IL.

[GTI] Gas Tech. Inst. (2009). New Albany Shale Gas Project. A Joint Industry Project Sponsored by Research Partnership to Secure Energy for America (RPSEA), Ann. Progress Report, GTI, Des Plaines, IL.

Hagerty, J. (2014). Fast-track fracking bill dead, companies wait on IDNR. News section, *The Rock River Times*, Rockford, IL.

Howarth, R. W., Santoro, R. and Ingraffea, A. (2011). Methane and The Greenhouse-Gas Footprint of Natural Gas from Shale Formations, *Climatic Change*, **106**(4): 679–690.

[IDNR] Illinois Dep't of Natural Resources (2014). Hydraulic Fracturing Regulatory Act, *IDNR*, Springfield, IL.

[IEA] Int'l Energy Agency (2010). Gas-Fired Power. IEA Report. No. ETSAP-Tech. Brief E02, I.E.A., Paris, France.

[IEPA] Illinois Environ. Protection Agency (2004). Fossil fuel-fired power plants. Report to the House and Senate Env. and Energy Committees, pp. 1–115, I.E.P.A., Springfield, IL.

[ISWS] Ill. State Water Survey (2014). Ground water science. Ill. State Water Survey, UIUC, Champaign, IL.

Kalantari-Dahaghi, A. and Mohaghegh, S. D. (2011). A New Practical Approach in Modeling and Simulation of Shale Gas Reservoirs: Application to New Albany Shale, *Int. J. Oil, Gas and Coal Technology*, **4**(2): 104–133.

Kaplan, S. M. (2010). Displacing coal with generation from existing natural gas-fired power plants. Congressional Research Service Report No. R41027, U.S. Congress, Washington, DC.

Kasper, M. and Maloney, P. (2013). Illinois adopts nation's strictest fracking regulations. Climate section, *Think Progress*, Washington, DC.

[KNG] Kentucky National Gas Corp. (2005). New Albany Shale background. *kentuckynaturalgas.com.*

Lazar, O. R. and Schieber, J. (2004). New Albany and Ohio Shales: An introduction. *Devonian black shales of the Eastern U.S.: New insights into sedimentology and stratigraphy from the subsurface and outcrops in the Illinois and Appalachian basins*, Field Guide for the 2004 Great Lakes Section SEPM Ann. Field Conf., J. Schieber and O. R. Lazar (eds.), Indiana Geol. Survey Open File Study 04–05, pp. 31–34. Bloomington, IN.

Lewis, R., Ingraham, D., Pearcy, M., Williamson, J. and Sawyer, W. (2004). New evaluation techniques for gas shale reservoirs. In *Proceedings of 2004 Schlumberger Reservoir Symp*. Schlumberger Ltd., Houston, TX.

Lyons, W. C. and Plisga, G. J. (2004). *Standard Handbook of Petroleum and Natural Gas Engineering*. Gulf Pub. Co.

Mahajan, D., Taylor, C. E., and Mansoori, G. A. (2007). An Introduction to Natural Gas Hydrate/Clathrate: The Major Organic Carbon Reserve of the Earth, *J. Petrol. Sci. & Eng*., **56**(1–3): 1–8.

Mansoori, G. A. (1994). Principle of Throttle/Phase Change (T/PC) Process for Separation of H2S from Natural Gas, *IPI J.*, **37–38**: 12–16.

Mansoori, G. A., Araujo, P. L. B., and Araujo, E. S. (2012). *Diamondoid Molecules: With Applications in Biomedicine, Materials Science, Nanotechnology & Petroleum Science*. World Sci. Pub. Co., Hackensack, NJ.

Mansoori, G. A. and Edalat, M. (1996). Storage of natural gas in liquids. In *Proceedings of* 5th *World Congress of Chemical Engineering*, 14–18 July. San Diego, CA.

Mastalerz, M., Drobniak, A., Rupp, J. A, Strąpoć, D., Schimmelmann, A. and Hasenmueller, N. R. (2010). Constraints on the origin and volume of gas in the New Albany shale (Devonian-Mississippian, Eastern Illinois Basin. In *Proceedings of AAPG Eastern Section Conf.* Vol. 94, pp. 1713–1740. 26–29 September, Kalamazoo, MI.

Mohsen-Nia, M., Moddaress, H., Mansoori, G. A. (1994). Sour Natural Gas and Liquid Equation of State, *J. Petrol Sci. Eng.*, **12**: 127–137.

Mohsen-Nia, M., Modaress, H. and Mansoori, G. A. (2002). New Equation of State is Accurate for VLE, PVT Calculations of Oil Mixtures, *Oil & Gas J.*, **100**(50): 50–52.

Mokhatab, S. and Poe, W. A. (2012). *Handbook of Natural Gas Transmission and Processing*. Gulf Pub. Co.

Montgomery, C. T., Smith, M. B. (2010). Hydraulic Fracturing: History of Enduring Tech., *J. of Petrol. Tech.*, **62**(12): 26–32.

Nelder, C. (2012). What The Frack? Is There Really 100 Years' Worth of Natural Gas Beneath the U.S.? Health and Science section, *Slate*, New York, NY.

Nicolas, M. P. B. and Bamburak, J. D. (2009). Geochemistry and Mineralogy of Cretaceous shale, Manitoba (parts of NTS 62C, F, G, H, J, K, N): Preliminary Results. Report of Activities, Manitoba Innovation, Energy and Mines, Manitoba Geol. Survey, pp. 165–174, Manitoba, Canada.

Oil-Gas-News (2011). New Albany Shale gas play, Illinois Basin. News section, *Oil-Gas-News.com*.

Patel, S. (2011). Pushing the 60% efficiency gas turbine barrier. *Power Magazine*, Houston, TX.

Pires, A. P., Mohamed, R. S., Mansoori, G. A. (2001). An Equation of State for Property Prediction of Alcohol–Hydrocarbon and Water–Hydrocarbon Systems, *J. Petrol. Sci. & Eng.*, **32**: 103–114.

Orrick, D. (2005). Peaker plants return, for the moment controversial power plants struggle after being all the rage. News section, *Daily Herald*, Arlington Heights, IL.

Riazi, M. R., Mansoori, G. A. (1993). Simple Equation of State Accurately Predicts Hydrocarbon Densities. *Oil & Gas J.*, 108–111.

Salehi, I. (2010). New Albany shale gas project. Final report. RPSEA Project No. 07122-16, Gas Technol. Inst., Des Plaines, IL.

Schill, J. (2009). Natural gas storage field expansion uses static pipe bursting, *Pipeline & Gas J.*, **236**(5): 54–57.

Seebregts, A. J. (2010). Gas-Fired Power. Energy Tech. System Analysis Program (IEA-ETSAP), Agency Energy Tech. Network, ETSAP-Tech. Brief No. E02IEA, Int'l Energy Agency, Paris, France.

Smith, K. C. and Sikorski, M. (2014). Energy security in Central and Eastern Europe science-based decisions and U.S. best practices in hydraulic fracturing. Center for European Policy Analysis (CEPA), Washington, DC.

[SNC-Lavalin] Surveyer, Nenniger & Chenevert Consulting Engineers (2005). Minooka combined cycle power plant. Report No. WASH-0068-EN-2005, SNC-Lavalin Group Inc., Montreal, Quebec, Canada.

Speight, J. G. (2007). *Natural Gas: A Basic Handbook*. Gulf Pub. Co.

Speight, J. G. (1990). *Fuel Science and Technology Handbook*. CRC Press.

Speight, J. G. (2010). *Handbook of Industrial Hydrocarbon Processes*. Gulf Pub. Co.

Stevenson, D. L., Whiting, L. L. and Cluff, R. M. (1981). Geologic structure of the base of the New Albany Shale Group in Illinois, Illinois Inst. of Nat. Vol. 121, Res., Ill. State Geol. Survey, Champaign, IL.

Taleghani, A. D. and Lorenzo, J. M. (2011). An alternative interpretation of microseismic events during hydraulic fracturing. In *Proceedings of SPE Hydraulic Fracturing Tech. Conf. and Exhibition held*, SPE Paper No. SPE140468-PP, pp. 1–11. Woodlands, TX.

Tobin, J. (2006). Underground Natural Gas Storage, Developments: 1998–2005. Report No. DOE/EIA-0131, U.S. EIA, Office of Oil and Gas.

[US-DOE] U.S. Dept. of Energy (1999a). 3-D Modeling of Mt. Simon gas storage fields could enhance winter gas availability from the Illinois Basin. *DOE Fossil Energy Techline*, pp. 1–2, U.S. DOE, Washington, DC.

[US-DOE] U.S. Dept. of Energy (1999b). U.S. Dept. of Energy Fiscal Year 1998 Accountability. Report No. DOE/CR-0067, U.S. DOE, Washington, DC.

[US-EIA] U.S. Energy Information Administration (2009). "Underground Natural Gas Storage Capacity," Developments: 1998–2011, *U.S. Energy Information Administration* (http://www.eia.gov/dnav/ng/ng_stor_cap_dcu_SIL_a.htm).

[US-EIA] U.S. Energy Inf. Admin. (2010). Updated capital cost estimates for electricity generation plants. Independent Statistical Analysis, Report No. AEO2011 cycle, U.S. EIA, Washington, DC.

[US-EIA] U.S. Energy Inf. Admin. (2012). Natural gas annual. State Energy Data, U.S. EIA, Washington, DC.

[US-EIA] U.S. Energy Inf. Admin. (2013b). Updated capital cost estimates for electricity generation plants. Independent Statistical Analysis, EIA, Washington, DC.

[US-EIA] U.S. Energy Inf. Admin. (2013c). Carbon dioxide emissions coefficients. U.S. EIA, Washington, DC.

[US-EIA] U.S. Energy Inf. Admin. (2014a). State profiles and energy estimates. U.S. EIA, Washington, DC.

[US-EIA] U.S. Energy Inf. Admin. (2014b). Underground natural gas storage capacity. U.S. EIA, Washington, DC.

[US-EIA] U.S. Energy Inf. Admin. (2014c). Shale oil and shale gas are globally abundant. *Today in Energy*, U.S. EIA, Washington, DC.

[US-EIA] U.S. Energy Inf. Admin. (2014d). Ann. energy outlook 2014: Levelized cost and levelized avoided cost of new generation resources. U.S. EIA, Washington, DC.

[US-EPA] U.S. Environ. Prot. Agency (2010). Science in action: hydraulic fracturing research study. EPA Office of Research and Development, Doc. No. EPA/600/F-10/002, Washington, DC.

[US-EPA] U.S. Environ. Prot. Agency (2011). Plan to study the potential impacts of hydraulic fracturing on drinking water resources. EPA

Office of Research and Development, Doc. No. EPA/600/R-11/122, Washington, DC.

[US-EPA] U.S. Environ. Prot. Agency (2012a). Study of the potential impacts of hydraulic fracturing on drinking water resources. EPA Progress Report No. 601/R-12/011, Washington, DC.

[US-EPA] U.S. Environ. Prot. Agency (2012b). Overview of greenhouse gases. Methane Emissions, Washington, DC.

[US-GS] U.S. Geol. Survey (2007). Assessment of undiscovered oil and gas resources of the Illinois Basin, 2007. National Assessment of Oil and Gas Fact Sheet No. 2007-3058, USGS, Dept. of Interior, Washington, DC.

[US-HR] U.S. House of Representatives (2011). Chemicals used in hydraulic fracturing, pp. 1–32. U.S. HR Committee on Energy and Commerce Minority Staff, Washington, DC.

Wernau, J. (2011). Illinois options for replacing coal-fired power plants. Business section, *Chicago Tribune*, Chicago, IL.

Wernau, J. (2014a). Illinois fracking law has big loophole, environmentalists say. Business section, *Chicago Tribune*, Chicago, IL.

Wernau, J. (2014b). Fracking rules for Illinois published after long battle. Business section, *Chicago Tribune*, Chicago, IL.

5

Carbon dioxide: Emission, Capture, Sequestration and Utilization

"The slight percentage of carbonic acid in the atmosphere may, by the advances of industry, be changed to a noticeable degree in the course of a few centuries." **Svante August Arrhenius**, 1903 Noble Laureate in Chemistry.

"CO_2 is the exhaling breath of our civilization, literally... Changing that pattern requires a scope, a scale, a speed of change that is beyond what we have done in the past." **Al Gore**, the 2007 Nobel Peace Prize Laureate for his work in climate change activism, 45th Vice President of the U.S. (1993–2001) under President Bill Clinton.

"I think the Caribbean countries face rising oceans and they face increase in the severity of hurricanes. This is something that is very, very scary to all of us. The island states in the world represent — I remember this number — one-half of one percent of the carbon emissions in the world. And they will — some of them will disappear." **Steven Chu**, the 1997 Nobel Laureate in Physics and the 12th U.S. Secretary of Energy.

5.1. Introduction

One of the major problems associated with the use of fossil fuels is production and release of carbon dioxide (CO_2) to the atmosphere. CO_2 is a naturally occurring chemical compound, but concentrations of CO_2 in the atmosphere are increasing at an accelerating rate from decade to decade as shown in Fig. 5.1.

According to Fig. 5.1, during 2005–2014, the average annual rate of increase of carbon dioxide was 2.07 parts per million (ppm) which is more than double the increase in the 1960s.

In 2011, the State of Illinois ranked 6th in CO_2 emission among all the states in the U.S., with annual output estimated at about 225 million metric tons (MMT). The major emitters of CO_2 in the state are the electric power plants, followed by the transportation industry (see Table 5.1).

CO_2 emissions from electric power sector are mainly from coal- and natural gas-fired power plants, with coal being the most significant source and emitting twice as much CO_2 per megawatt hour (MWh), compared with natural gas used as fuel in combined-cycle power plants (US-EIA, 2014b).

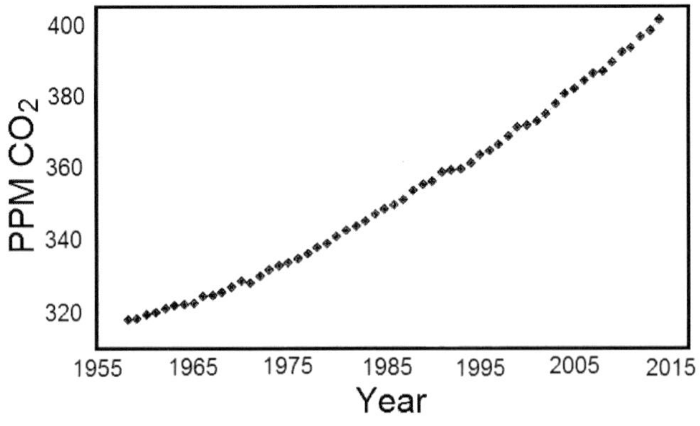

Fig. 5.1. Scripps Data atmospheric CO_2 concentration (1958–2014) from National Oceanic and Atmospheric Administration (NOAA); measuring location for this data is Mauna Loa Observatory, Hawaii: Latitude 19.5Â°N; Longitude 155.6 Â°W; Elevation 3397m (Tans, 2014).

Table 5.1. Illinois CO_2 emissions (in MMT) in 2011 (US-EIA, 2014a).

Sector	10^6 Metric Tons	% of Total
Commercial	12.4	5.50
Residential	23.6	10.47
Industrial	34.2	15.18
Transportation	63.9	28.36
Electric Power	91.2	40.48
Total	**225.3**	**100**

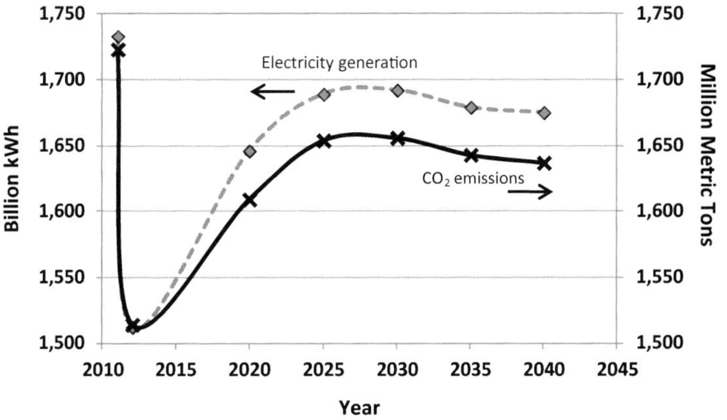

Fig. 5.2. Estimation of total electricity generation and CO_2 emission by coal in the U.S. (US-EIA, 2014b).

In 2011, the U.S. coal-powered power plants generated 1,733 billion kilowatt-hours (kWh) of electricity. However, due to retirement of some of the coal-powered plants, the share dropped to 1,512 billion kWh in 2012. EIA estimates that the share of electricity generation from coal in the U.S. will increase by 10.8% and reach 1,675 billion kWh by 2040, as shown in Fig. 5.2 (U.S. EIA, 2014b). The same figure shows that CO_2 emissions followed the same pattern as electricity generation and it is estimated that these emissions will increase by 8.1% from 2012 levels and reach 1,637 MMT by 2040.

Table 5.2. Carbon dioxide emission coefficients from different coal ranks (Schneider and Boggs, 2014).

Coal Rank	**Emission Co-efficient [kg CO_2/MMBtu]**
Bituminous	103.69
Anthracite	103.69
Lignite	97.72
Sub-Bituminous	97.17
Waste Coal	95.52
Coal-Derived Synthesis Gas	95.52
Coal-Based Synfuel	92.91

As mentioned in Chapter 2, Illinois has the largest overall and the largest strippable bituminous coal reserves in the U.S. Using Illinois coal for power generation will result in high amount of CO_2 emissions, since bituminous coal has the highest emission coefficient (see Table 5.2).

Despite the fact that Illinois power plants partially use cleaner coal from other states, Illinois still ranks 6th in CO_2 emission among all the states in the U.S. (US-EIA, 2014a). A report by Schneider and Boggs (2014) shows that the top five polluting power plants in Illinois emitted 39 MMT of CO_2 in 2012 (Table **??**).

Figure 5.3 shows the locations of CO_2 producing stationary point sources within the State of Illinois, as compiled by the Midwest Geological Sequestration Consortium (MGSC).

5.2. Global Warming Due to Excessive CO_2 in the Air

Carbon dioxide is a greenhouse gas and its increase in the atmosphere has been contributing to global warming (Raich and Schlesinger, 1992). The evidence that support global warming includes the

Table 5.3. Share of Illinois electricity sector carbon dioxide pollution contributed by the top 5 most-polluting power plants in 2012 [Schneider and Boggs, 2014]

Operator Name	Plant Name	Total CO_2 Emissions of Top 5 Plants [MMT of CO_2]	Top 5 as a Share of Power-Sector Total	Power-Sector as a Share of Statewide Total	Share of Statewide CO_2 by Top 5
Dynegy Midwest Generation Inc	Baldwin Energy Complex				
Midwest Generations EME, LLC	Powerton				
Electric Energy Inc.	Joppa Steam	39	43%	38%	18%
Ameren Energy Generating Co.	Newton				
Ameren Energy Generating Co.	Coffeen				

alarming rate at which continental glaciers in Greenland and West Antarctica are melting (Perlman, 2008; Pidcock, 2013). Also, killer heat waves have been attributed to global warming and were blamed for hundreds of human deaths in Illinois in 1995 and over 70,000 deaths in Europe in 2003 (Changnon *et al.*, 1996; Robine *et al.*, 2008). According to climatologists, the 5-day heat wave which engulfed Illinois and other Midwestern states during mid-July 1995 was one of the worst weather-related disasters in the history of Illinois. It caused 581 deaths in Illinois while the City of Chicago suffered the most with 525 deaths (Changnon *et al.*, 1996). According to NOAA (*www.ncdc.noaa.gov/sotc/global/*), January 2014 was the fourth warmest January around the world since global temperature records began in 1880, while the coolest was January1893.

Measurements indicated an increase of 0.3–0.6°C (0.54 – 1.08°F) in global temperatures during the 20th century. If the trends in emissions of greenhouse gases continue, there is a possibility that the global mean temperature by the end of the 21st century will be 3°C higher than its current figure (Rochelle *et al.*, 2009).

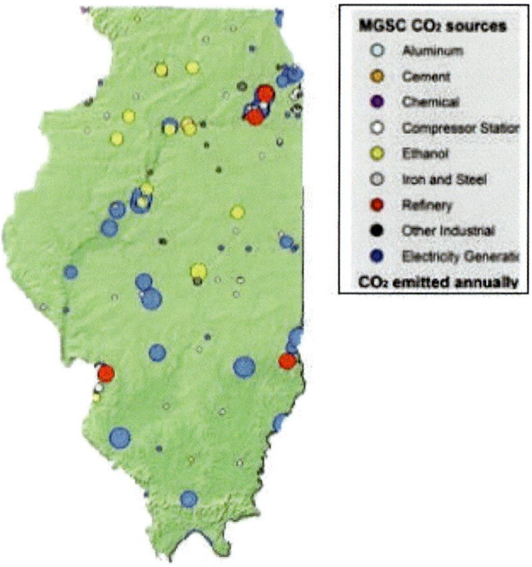

Fig. 5.3. Map showing the location of CO_2 producing point sources within the State of Illinois, according to MGSC (USDOE/NETL, 2012).

Vegetation (plants and trees) play a critical role in the carbon cycle, maintaining safe levels of CO_2 and oxygen (O_2) in the air, water and environment for millions of years (see Fig. 5.4).

The balance of O_2 and CO_2 in the environment has been maintained through respiration and photosynthesis in trees and plants, which played an important role in addressing climate change. Also, the natural formation of carbonate minerals in the earth crust, as well as CO_2 uptake by the oceans are the additional processes sustaining the long-term global carbon cycle (Folger, 2013; Oelkers *et al.*, 2008; Agyarko and Mansoori, 2013). According to a 2013 Congressional Service Report (Folger, 2013), during the 1990s about 45% of the CO_2 emissions from fossil fuels combustion and land use activities has remained in the atmosphere, with the balance taken up by the oceans, vegetation or soils on land surface. Because of excessive amount of CO_2 being released to the atmosphere from fossil fuels combustion, the role of plants and trees in maintaining the global climate is minimized.

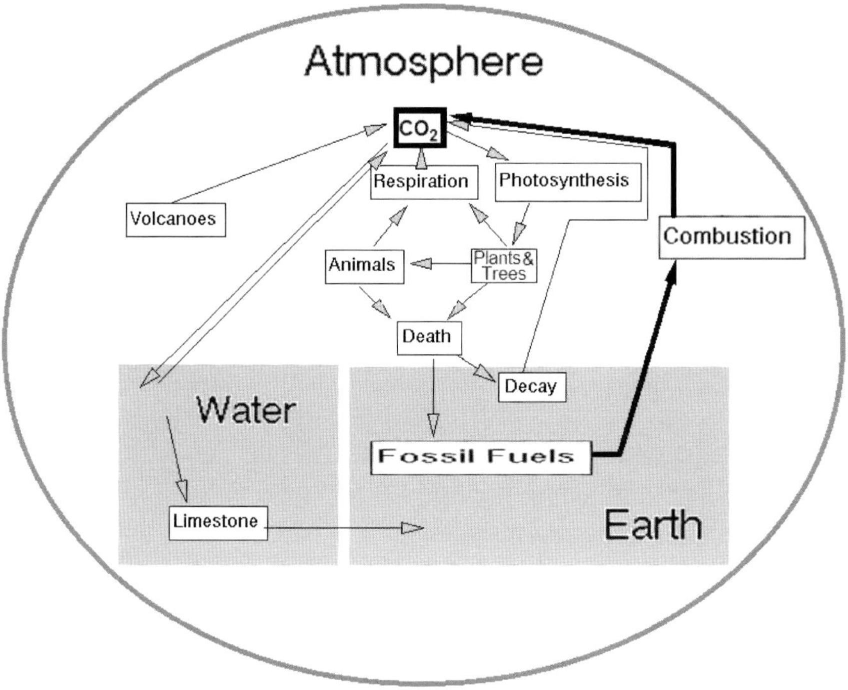

Fig. 5.4. The carbon cycle and the role of fossil fuels in excessive accumulation of CO_2 in the air.

To avoid environmental disaster, human intervention is required to improve the way energy is produced and consumed (Chen and Seiner, 2011). These include scientific and engineering approaches that are intended to significantly reduce emissions of greenhouse gases, more utilization of renewable energy sources and CO_2 capture and sequestration, with the aim of stabilizing atmospheric CO_2 concentrations at 550 ppm by 2025 (Fauth *et al.*, 2002; Yamin, 2005). Reducing CO_2 released into atmosphere would reduce, and perhaps reverse, the effects of global warming that we have been experiencing in the last several years.

Although electricity generation using conventional coal-fired power plants is a matured industry, there remain several obstacles when connecting this conventional process with CO_2 capture technology. Among the obstacles are the high capital cost of CO_2 capture technologies (with approximately 75% increase in capital cost for adding CO_2 capture equipment), a large footprint required for CO_2 capture equipment, operational problems and a high energy penalty for CO_2 removal from flue gases in absorption columns using liquid solvents and regeneration of absorbing solvents (Phillips and Maxson, 2011).

In the following sections, we present the various CO_2 capture technologies that can help a state meet its future environmental obligations. These include pre-combustion, oxy-fuel combustion and post combustion technologies. We also present the different types of CO_2 sequestration options and CO_2 potential utilization technologies, which can substantially reduce CO_2 emissions.

5.3. CO_2 Capture

According to the data reported in Table 5.1, electricity generation power plants must be the primary target for CO_2 capture. This is because they produce the largest fraction of CO_2 released to the environment through fossil fuels combustion and contrary to the transportation sector, they are stationary. From an economic point of view, it is best to concentrate on collecting CO_2 from large stationary combustion sources. Also, the larger the combustion process, the more economical will be to capture/collect CO_2 (Agyarko and Mansoori, 2013).

The three categories of Carbon Capture and Storage (CCS), as well as their related technologies being pursued are referred to as the 1st generation, 2nd generation and transformational (US-DOE/NETL, 2013a). The 1st generation technologies are those that are commercially available or are currently under demonstration. The 2nd generation technologies include those that will be ready for demonstration in 2020–2025 timeframe and are currently in research and development (R&D) stages. The transformational technologies, however, include those that offer the potential for improvements in cost and performance beyond what is expected from the 2nd generation technologies, and are in the early stages of development or are in conceptual design. The development and scale-up of the transformational technologies are expected to occur in the 2016–2030 timeframe, while their demonstration projects are expected to be initiated in the 2030–2035 time period (US-DOE/NETL, 2013a). The status of the 1st generation CO_2 capture technologies with reference to power generation are classified as high, medium and low, and are presented in Table 5.4.

Below, we provide a brief description of the three 1st generation routes for CO_2 capture as listed in Table 5.4.

5.3.1. *Pre-combustion capture*

In the pre-combustion CO_2 capture process, the primary fuel (coal or biomass) is first converted into synthetic gas (syngas) through a gasification process. The syngas is converted into hydrogen and carbon dioxide-rich fuel mixture using steam reforming (water gas shift) process (see Fig. 5.5).

Table 5.4. Readiness and development potential of the CO_2-capture techniques (CSLF, 2013).

	Readiness for Demonstration		**Development Potential**	
Technology	**Coal**	**Natural gas**	**Coal**	**Natural gas**
IGCC w/CCS*	Medium-High	N/A	High	N/A
Oxy-combustion	Medium-High	Low	High	Medium-High
Post-combustion	High	High	Medium-High	Medium-High

* Integrated gasification combined cycle (IGCC) power plant with carbon capture and sequestration (CCS) (a pre-combustion decarburization power plant).

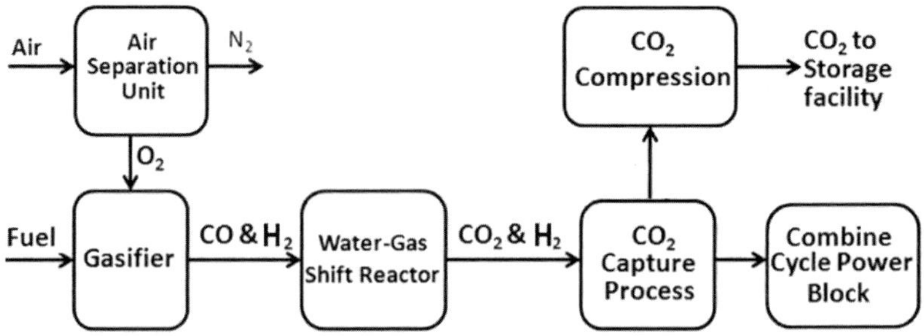

Fig. 5.5. A block diagram illustrating power plant with pre-combustion CO_2 capture (US-DOE/NETL, 2013a).

The CO_2 concentration in the fuel mixture can range from 15–50% and can be separated, while the hydrogen can be used as fuel or chemical feedstock. The pre-combustion technologies currently being developed are expected to capture more than 90% of CO_2 in flue gas. The current cost of commercially available pre-combustion carbon capture technologies (based on physical or chemical adsorption processes) using IGCC power plant is about \$60 per metric ton of CO_2. Therefore current research effort is focused on three key separation technologies (advanced solvents, sorbents and membranes) which aims to reduce this cost to \$40 per ton of CO_2 (US-DOE, 2014a).

5.3.2. *Oxy-fuel combustion capture*

We have already discussed the principles behind the oxy-combustion carbon capture technology in Chapter 2, with specific reference to the FutureGen 2.0 project in Meredosia, Illinois. The major problem associated with the deployment of the 1st generation oxy-combustion technology is its high cost (US-DOE/NTEL, 2013a). This high cost requires development of the 2nd generation and transformational technologies to lower the costs and energy requirements linked with the 1st generation systems. This cost reduction will require efforts to develop integrated components of oxy-combustion systems, which can result in more efficient operations and lower capital costs (US-DOE/NTEL, 2013a).

5.3.3. *Post-combustion capture*

The CO_2 generated as a result of fossil fuels combustion in electricity power plants is primarily a mixture of CO_2, water vapor and nitrogen. The majority of the post-combustion technologies for CO_2 capture include passing the flue-gas through a packed bed absorber column, where the sorbent (such as amines) selectively absorbs the carbon dioxide (Fig. 5.6) before the flue gas is released to the atmosphere (US-DOE/NETL, 2010).

The CO_2-rich solvent then enters a regenerating column or stripper, where the CO_2 is released through application of heat and the solvent is returned to the absorber. It is estimated that energy requirements for CO_2 capture and compression may reduce the net plant power output by 20–30% (Phillips and Maxson, 2011). Several methods have been designed, with the view to optimize the process and reduce the capital cost involved through testing of new solvents.

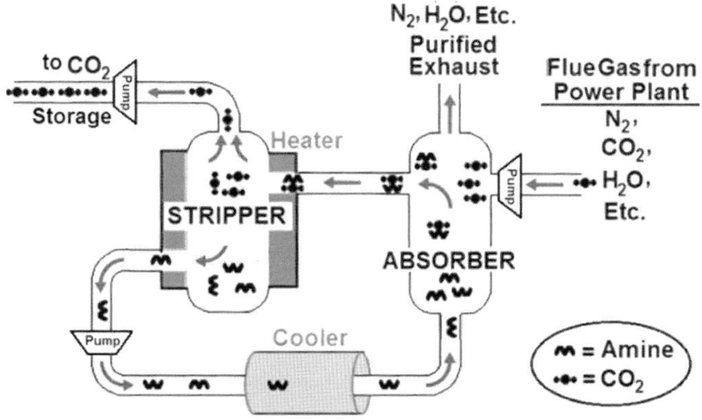

Fig. 5.6. Principle of post-combustion CO_2 capture using amines as the absorbent of CO_2.

The conventional approach for post-combustion CO_2 capture is the use of monoethanolamine (MEA) solvent. However, there are claims about development of advanced amine-based solvents (Mitsubishi Heavy Industries, 2009; Johnson *et al.*, 2009) that could capture more CO_2 than MEA.

Another post-combustion technology for capturing carbon dioxide is the use of chilled ammonia (Cerimele and McMillian, 2011). In this process, it is claimed that the chilled concentrated ammonia solution would chemically bind to CO_2, followed by application of heat to release CO_2 for compression and storage. The flue-gas is chilled to about 2°C before entering the absorber. In the absorber, ammonium carbonate, $(NH_4)_2CO_3$, and ammonia-based solvent absorb CO_2 to form ammonium bicarbonate NH_4HCO_3. Then the ammonium bicarbonate slurry is pumped to a regenerator to liberate CO_2 and recover the ammonium carbonate for reuse. It is claimed that the efficient operation of the process is ensured by heat recovery.

The initial demonstration of the later process took place at Wisconsin-Energy's power plant in Pleasant Prairie, Wisconsin, across the border from Illinois, using the exhaust flue-gas stream from a 1.7 MW electric power plant. The process was reported to be capable of 90% CO_2 removal from a slipstream of the plant's flue-gas (Cerimele and McMillian, 2011). The claimed environmental benefits using this technology included lower energy for CO_2 capture relative to other technologies, high removal of other pollutants such as sulfur trioxide and particulate matter of size 2.5 μm and higher. Similarly, a research group in California recently claimed to have developed a new metal-organic framework (MOF) material modified with nitrogen compounds called diamines, capable of capturing CO_2 at different temperatures. These materials are designed to release CO_2 at lower temperatures, thus potentially reducing the energy requirement by half when compared with the current carbon-capture materials (McDonald *et al.*, 2015).

5.4. CO_2 Sequestration

Currently, the three proposed options suitable for CO_2 sequestration are mineralization, geological storage and deep ocean storage. In what follows, we briefly describe these technologies.

5.4.1. *Mineral carbonation*

This method is used to convert CO_2 into stable carbonate minerals, such as calcite, $CaCO_3$, dolomite, $CaMg(CO_3)_2$, magnesite, $MgCO_3$ and siderite $FeCO_3$ for storage (Oelkers *et al.*, 2008). The process involves reaction of CO_2 with naturally occurring silicate minerals to produce geologically stable carbonate minerals and silica, according to the following generalized chemical equation (Goldberg *et al.*, 2001)

$$\begin{aligned}&(\mathrm{Mg, Ca})_x\mathrm{Si}_y\mathrm{O}_{x+2y+z}\mathrm{H}_{2z} + x\mathrm{CO}_2\\ &\quad \rightarrow x(\mathrm{Mg, Ca})\mathrm{CO}_3 + y\mathrm{SiO}_2 + z\mathrm{H}_2\mathrm{O}\end{aligned} \tag{5.1}$$

The products from the above exothermic reactions are highly stable, which can be disposed of without release of CO_2 to the atmosphere. However, the reactions are slow and must be carried out at high temperatures and pressures, and *in situ*, to avoid transportation of a large quantity of silicate minerals rocks (US-DOE, 2014b). *In situ* refers to the process whereby CO_2 is injected into geological formations rich in potential carbonate formers, and is similar to underground coal gasification mentioned in Chapter 4. In contrast, *ex situ* minerals carbonation occurs in processing plants away from the location of the silicate minerals. The first step in *ex situ* minerals carbonation is preparation of reactant minerals, which involves mining, crushing and milling, including transportation of the minerals to the processing plant. The second step is reaction of concentrated CO_2 stream with the prepared minerals, while the third and final step involves transportation of the carbonate products for storage

(Folger, 2013). These processes are energy and capital intensive, with an estimated total cost (in 2008) of ~ \$54 per ton of CO_2 sequestered (Oelkers *et al.*, 2008).

There is currently no project in Illinois using this technology. However, in 2002, the U.S. Department of Energy (DOE) tested samples of industrial solid residues (fluidized bed combustion ash and flue-gas desulfurization spray dryer ash) from Freeman United Coal Mining Company, in Farmersville, Illinois to determine their potential application for permanent sequestration of CO_2 as mineral carbonates. It was determined that 63,200 metric tons of CO_2 per year could be sequestered at the Farmerville, Illinois complex (Fauth *et al.*, 2002). But while quantitatively these industrial solid residues may not be enough to have a significant impact on Illinois CO_2 emissions, they can be easily obtained at low cost. Additionally, they are easily found around industrial sites already in fine particles (without the need for grinding or crushing) which facilitate their fast reactions with CO_2 (Fauth *et al.*, 2002).

In Australia, Calera Corporation is building a plant (Calera Yallourn) which will be the first commercial-scale facility in the world using mineral carbonation. The facility will produce 2.1 metric tons of building materials for every metric ton of CO_2 captured, at the cost of \$45–60 per metric ton of CO_2 (Kolstad and Young, 2010). It is estimated that the costs associated with the facility (including CO_2 capture and building materials) is \$300–380 million (PB and GCCSI, 2011; Kolstad and Young, 2010).

Current research on mineral carbonation is focused on how to reduce the energy requirements for grinding the feedstock materials needed for the mineralization process. There is more interest in developing chemicals or catalysts to speed up the reaction rates, as well as reducing the thermal and pressure requirements, including how to meet industrial standards for building materials, using the products from mineral carbonization (US-DOE, 2014b).

5.4.2. *Geological sequestration/deep welling*

This option for CO_2 storage uses geological formations such as depleted oil fields or saline aquifers through different trapping mechanisms. It is the best option for Illinois because of the abundance of these sites.

In 2003, the MGSC was launched as a Phase I DOE Regional Carbon Sequestration Partnership, which continued in 2005 as a Phase II effort. The objective of the MGSC was to study the geological carbon sequestration potential in the Illinois Basin (Frailey and Finley, 2006). MGSC was headed by the Illinois State Geological Survey, in collaboration with Indiana and Kentucky Geological Surveys and a group of subcontractors. The results of this project are reported in the subsequent paragraphs.

Since many states in the U.S., including Illinois, are currently transitioning to clean-coal-technologies, adequate CO_2 sequestration infrastructure development is assumed to be the key to success of this transition. In all CO_2 capture technologies, the cleaned flue-gas, containing mainly nitrogen, water vapor and low amount of CO_2 would flow to the stack, while the captured CO_2 would be transported to geologic storages. There are three types of subsurface formations in the Illinois Basin which have been identified for CO_2 storage. These are deep coal seams considered uneconomical for mining in the near future, depleted or mature oil and natural gas reservoirs and saline aquifers containing non-potable water (see Fig. 5.7).

As we discussed in Chapter 4, Sec. 4.4, depleted or mature oil and natural gas reservoirs and saline aquifers containing non-potable water are already used for natural gas storage in Illinois. It will be necessary to identify additional similar geological sites for CO_2 sequestration (see Fig. 5.8).

Coal-seams that are too thin and laying deeper than about 300 meters ($\sim$1000 ft) underground and less than about 100 cm ($\sim$ 40 inches) thick are considered to be technically and economically unmineable, as compared to shallow and thick coal seams (Frailey

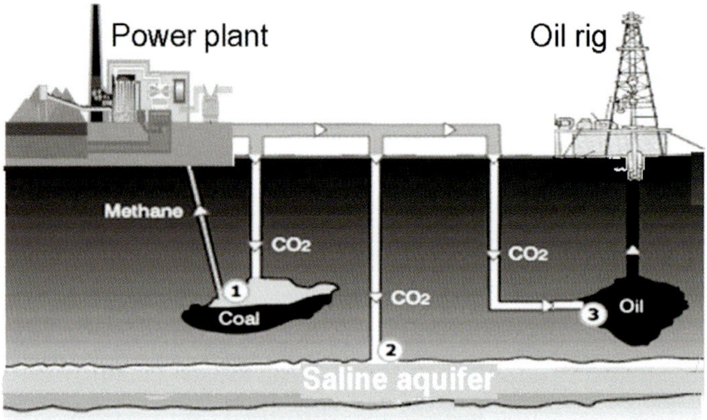

Fig. 5.7. The three types of geological CO_2 sequestration options that can be used to store CO_2 in Illinois (Modified from a figure by World Coal Association, formerly world Coal Institute (World Coal Institute, 2009).

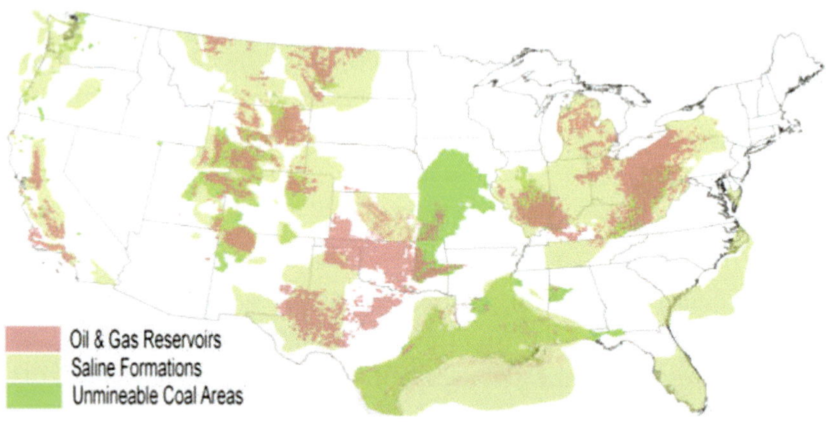

Fig. 5.8. Locations of possible CO_2 geological storage sites in U.S. (US-EPA, 2015).

and Finley, 2006). These unmineable coal seams in the Illinois Basin contain adsorbed methane gas that could be displaced using CO_2, due to the high affinity of CO_2 to coal relative to methane. An estimated 566.34×10^9 standard cubic meters (20 Tscf) of methane is trapped in the coal seams that cover Illinois and Indiana regions of the Illinois Basin (Finley *et al.*, 2005). At Saline County in Illinois,

BPI Industries, owner of coal-bed methane (CBM) project has been selling gas produced from CBMs into pipelines (Rodvelt and Oestreich, 2005).

Injecting CO_2 into coal seams may boost commercial level natural gas production in Illinois. The CO_2 could be separated from the methane using membrane separation technology or formulated amine solvents depending on the outlet gas specification of the contactor (Agyarko, 2005). Currently, CO_2 retention characteristics of coal seams are being studied as part of the DOE Regional Carbon Sequestration Partnership using new data from test wells.

As part of the DOE Regional Carbon Sequestration Partnership, a comprehensive evaluation of the oil fields in the Illinois Basin for CO_2 storage was carried out by the Midwest Geological Sequestration Consortium (MGSC). This evaluation was based on specific parameters of the oil fields, such as original-oil-in-place, porosity, permeability and water saturation, in addition to temperature and pressure (Finley *et al.*, 2005).

Deep saline formations in the Illinois Basin are the other options for large scale CO_2 sequestration in Illinois. Because the water in these formations has depth ranging from 1,200–1,500 meters (3936–4920 ft) and with two to three times the salinity of sea water, it is not drinkable (Finley *et al.*, 2005). In Table 5.5, we report the CO_2 storage capacity of unmineable coal areas, oil fields, saline formations and their total (per category) in the Illinois Basin by state.

An example of saline formations in Illinois is the Mt. Simon Sandstone, which is a major saline reservoir lying below the oil reservoirs in Southeastern Illinois and Southwestern Indiana. The nature of the geology of the Mt. Simon sandstone makes it an ideal location for CO_2 storage. The Mt. Simon sandstone has a depth ranging from about 600 to 4,300 meters (approximately 2,000 to 14,000 ft) below the surface of the earth (Finley *et al.*, 2010). Furthermore, the caprock seal of the Eau Claire Shale has demonstrated as an excellent seal for natural gas containment.

Table 5.5. CO_2 Storage capacity of unmineable coal areas, oil fields, saline formations and total (per category) in the Illinois Basin by state (Finley *et al.*, 2010).

	CO_2 Storage Capacity ($\times 10^6$ **metric tons**)			
State	**Unmineable Coal Areas**	**Oil Fields**	**Saline Formations**	**Total**
Illinois	1,470–2,900	106–358	8,400–116,000	9,976–119,258
Indiana	86–170	20–47	2,900–39,000	3,006–39,217
Kentucky	68–134	14–35	400–5,600	482–5,769
3-state Total	1,624–3,204	140–440	11,700–160,600	13,464–164,244

The St. Peter sandstone may also be suitable for CO_2 sequestration. The St. Peter Sandstone is extended, porous and permeable quartz sandstone which in general, includes fine-grained with good lateral continuity. Above the St. Peter sandstone are seals made up of several hundred feet of dense limestone and overlain by about 46–76 meters (150–250 ft) of Maquoketa Shale (Finley *et al.*, 2010). But, obtaining data for assessing the Mt. Simon Sandstone in the Southern Illinois area is a major challenge. As a result, a comprehensive data compiled from studies of natural gas storage fields was used for the first time to assess the feasibility of the deep Mt. Simon sandstone in Southern Illinois for CO_2 sequestration (Finley *et al.*, 2005).

The most widespread and prolific petroleum bearing sandstone in the Illinois Basin is the Cypress sandstone. These areas with thick Cypress sandstone tend to have a large water bearing zone that may be considered a saline storage target for CO_2. The porous and permeable sandstone is generally less than 30.48 meters (100 ft) thick and shows a significant variation in thickness and lateral extent, but it can also reach a thickness of 60.96 meters (200 ft). The Cypress sandstone is the shallowest among the saline reservoirs examined in this project, with estimated depths approaching approximately 914.40 meters (3,000 ft) in some parts of the Illinois Basin (Finley *et al.*,

Table 5.6. CO_2 storage capacity of saline formations in the Illinois Basin by reservoir (Finley *et al.*, 2010).

Reservoir	**CO_2 Storage Capacity [10^6 metric tons]**
Cypress sandstone	200–2,300
St. Peter sandstone	600–7,800
Mt. Simons sandstone	11,000–151,000
Total (rounded)	**11,800–161,100**

Fig. 5.9. The State of Illinois now boasts of the first large scale carbon sequestration project in the U.S., the ADM Project in Decatur (US-DOE/NETL, 2011; Marshall and Wire, 2011).

2010). The quantification of CO_2 storage resource of these saline formations is presented in Table 5.6.

In 2008, as part of the DOE Regional Carbon Sequestration Phase I project, the MGSC in conjunction with Archer Daniels Midland Company (ADM) in Decatur, Illinois and Schlumberger Carbon Services began a couple of projects to evaluate the safety and efficiency of CO_2 storage in the Illinois Basin (McKinney and Finley, 2008).

The first project (Illinois Basin-Decatur project) confirmed that the saline-water-bearing Mount Simon Sandstone rock formation running below an ADM complex at Decatur in Illinois is a suitable location for geologic carbon sequestration (see Fig. 5.9).

Consequently, in November 2011, the project began injecting CO_2 into the Mount Simon sandstone formation at a depth of 2,135 meters (7000 ft) below the ground (Kuipers, 2011). The injected CO_2 is a byproduct from ADM's biofuels plant near the storage site in Decatur, Illinois. The Decatur experiment will continue injecting CO_2 for three years, and has drawn significant interest from other industrial and scientific countries around the world.

The second project is the Illinois-Industrial Carbon Capture and Sequestration project (IL-ICCS). This project aims to demonstrate the commercial-scale sequestration of carbon dioxide within the Mount Simon formation. The project was scheduled to begin CO_2 injection into the Mount Simon formation in the third quarter of 2013. The project consists of a 2,000 metric tons per day compression/dehydration facility and a pipeline capable of transporting the CO_2 from ethanol plant to the sequestration and monitoring site (McDonald, 2011; US-DOE/NETL, 2011a).

The above-mentioned two projects will combine to inject a total of 3,000 tons of CO_2 per day when they become operational (McDonald, 2011). Data from these projects are intended to provide baseline information for future development of CCS opportunities not only for the Illinois Basin, but for the entire U.S.

Apart from the DOE-funded project, in 2009, Denbury Resources Inc. concluded a feasibility study on the construction of the Midwest CO_2 pipeline. The purpose of this project was to assess information and data regarding the possibility of creating an infrastructure to assist in transportation of captured CO_2. The captured CO_2 is expected to be transported from coal gasification plants in Illinois, Indiana and Kentucky to the Gulf Coast, for a safe and permanent underground sequestration through enhanced oil recovery(EOR) operations (IDCEO, 2010). More than 200 miles of the Midwest CO_2 pipeline would be located in Illinois and will be linked with one or more of the proposed gasification projects to Denbury's existing pipeline network in the Gulf States. A project like this is expected to spur economic growth in the region.

5.4.3. *Deep ocean storage*

The proposed methods for deep ocean storage of CO_2 include direct injection, biological sequestration and chemical sequestration. Below, we provide a brief description of each of these methods.

5.4.3.1. *Direct injection*

This option for CO_2 storage involves capturing and separation of CO_2 from flue gas produced by fossil-fueled power plants. The separated CO_2 is liquefied by compression, followed by transportation of the liquid CO_2 through a submerged pipeline for injection in deep oceans. It is estimated that over 90% of all the CO_2 released to the atmosphere by human activities could ultimately end up in the ocean through natural process. However, it would take thousands of years for complete natural absorption of all the CO_2 in the atmosphere by the oceans (Folger, 2013). For this reason, a proposed approach of making use of the vast CO_2 storage potential of the oceans is by direct injection of liquefied CO_2 at ocean depths ranging from 1,000–3,000 meters ($\sim$ 3,000–10,000 ft). At such depth, the CO_2 is claimed to remain in the liquid state due to high pressure. Also since liquefied CO_2 is heavier than sea water, the liquid will settle at the bottom of the ocean and could remain there for thousands of years as a pool of CO_2. However, not much information is known about the impact of ocean acidification on marine ecosystems (US-DOE/NETL, 2006; Feely *et al.*, 2006). Moreover, the assumption that liquid CO_2 will remain intact at the ocean bottom without dispersion into water and again getting released into the atmosphere is quite hypothetical. Therefore, because the full impact of this method on the environment is still unknown, the technology is not yet considered as a practical option. Furthermore, since the oceans already contain about 50 times as much of CO_2 stored in the atmosphere, injecting more CO_2 into the ocean could bring it to its saturation point, which can change its storage capacity in the future (Folger, 2013).

5.4.3.2. *Biological sequestration*

This alternative approach enhances the rate of CO_2 sequestration in the ocean and includes ocean fertilization. Ocean fertilization involves application of micronutrients (such as iron) to certain areas of the ocean known as "high nutrients, low chlorophyll" areas. The aim is to improve the primary production of biomass (phytoplankton) that depends on micronutrients for maintenance of their chlorophyll. The ultimate goal is to increase CO_2 uptake by the ocean through photosynthesis. However, it has been demonstrated that to some extent, iron enrichment can bring some modifications to the biogeochemistry of the ocean, which can lead to destabilization of the oceans ecology (Rackley, 2009). Also, instead of using external sources of micronutrients for ocean fertilization, use of wave-powered pumps have been proposed as an alternative to stimulate "upwelling" in the "high nutrients, low chlorophyll" areas of the ocean to achieve the same end. The "upwelling" process involves drawing of cool, micronutrient-rich waters from below the "thermocline" of the ocean to the surface (Rackley, 2009). The "thermocline" is a region of the ocean depth where sudden temperature changes occur more frequently than it does in the layers of water below or above it. The problem with the "upwelling" process is that since the deep ocean is rich in dissolved inorganic carbon (HCO_3^-), it may end up releasing more CO_2 to the surface of the ocean than it absorbs.

5.4.3.3. *Chemical sequestration*

This approach for CO_2 sequestration involves bubbling of CO_2 contained in flue gas (from electric power or cement manufacturing plants) through water and subsequently through limestone, in accordance with the chemical reaction steps below (Rackley, 2009).

$$H_2O + CO_2 \rightarrow H_2CO_3 \tag{5.2}$$

$$CaCO_3 + H_2CO_3 \rightarrow Ca^{2+} + 2HCO_3^- \tag{5.3}$$

Here, the bicarbonate-rich effluent stream produced in Eq. (5.2) can be discharged into the ocean to neutralize the effect of ocean acidification, which occurs through the direct dissolution of atmospheric CO_2 in the ocean. Although such a practice may yield some positive environmental impact, the major problem with this approach will be the sheer amount of limestone and water needed for large-scale application of the process (Rackley, 2009).

5.4.3.4. *CO_2 clathrate (gas hydrate) formation*

Clathrates (gas hydrates) are formed when a guest molecule, like CO_2, is trapped within a cage of water molecules at certain composition, temperature and pressure (Mahajan *et al.*, 2007). The cage of water molecules takes the form of a 12-or 14-faced hedron, (dodecahedron or tetrakaidecahedron) having pentagonal or hexagonal faces with an oxygen atom at each vertex, similar to normal ice, as illustrated in Fig. 5.10. The stoichiometry of this complex is approximately (CO_2: $6H_2O$), depending on the temperature and pressure of

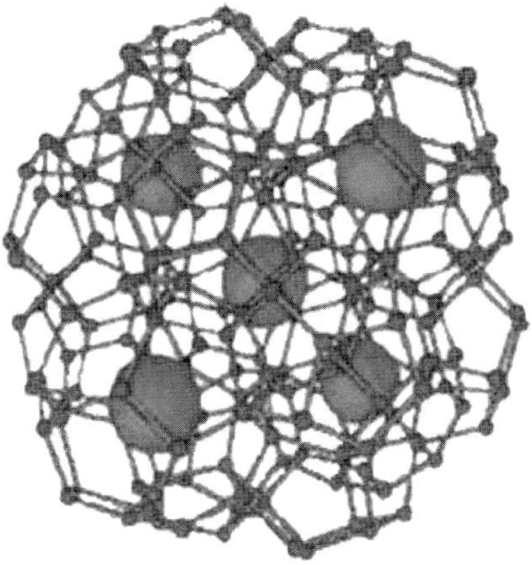

Fig. 5.10. Structure of a gas hydrate (clathrate) dual structure. The encaged molecules are CO_2 in the case of carbon dioxide clathrate (Simplified from Mahajan *et al.*, 2007).

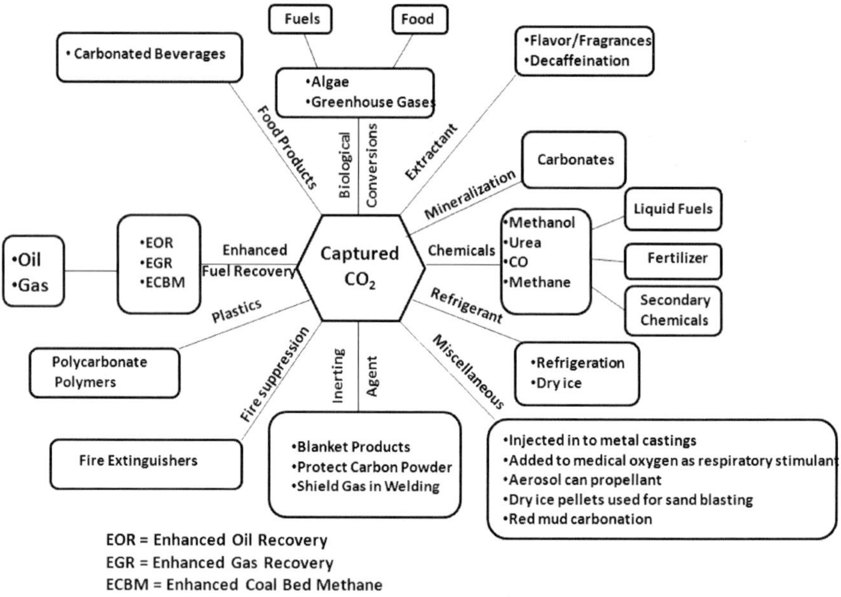

Fig. 5.11. Schematic illustration of various applications of CO_2 (US-DOE, 2014b).

formation. State of Illinois has no ocean access and the ocean storage option is not suitable for the state.

5.5. CO_2 Utilization

There are several applications of CO_2 that can be used to reduce atmospheric CO_2, the major contributor of global warming (see Fig. 5.11).

The current industrial applications of CO_2 include EOR (enhanced oil/petroleum recovery), urea fertilizer production, beverage carbonation, food processing, preservation and packaging, coffee decaffeination, pharmaceuticals, horticulture, fire suppression, winemaking, pulp and paper processing, water treatment, inerting, steel manufacture, metal working, electronics and pneumatics (D'Alessandro *et al.*, 2010; PB and GCCSI, 2011; Agyarko and Mansoori, 2013).

Beside these current applications, there are also emerging industrial applications of CO_2. These include enhanced coal bed methane recovery, algal bio-fixation and biofuel production, enhanced geothermal systems (using CO_2 as a working fluid), bauxite residue processing, carbonate mineralization (aggregate production), polymer processing, CO_2 concrete curing, methanol and formic acid production and more (PB and GCCSI, 2011).

However, the US-DOE efforts are currently focused on applications (use and reuse) of CO_2 that will convert the gas into useable products and fuels to reduce CO_2 emissions at locations where geologic storage may not provide the best option (US-DOE, 2014b). These include enhanced oil/gas recovery, CO_2 as feedstock, nongeologic storage of CO_2, indirect storage of CO_2, beneficial use of produced water and possible new breakthrough concepts (US-DOE, 2014b). Of these, the four main CO_2 utilization research focus areas supported by DOE's Carbon Storage Program include EOR, mineralization, cement production and production of polycarbonate plastics (US-DOE/NETL, 2014). Below, we provide a brief description of some of these applications of CO_2.

5.5.1. *Application of CO_2 in EOR*

The biggest application of CO_2 in the United States in the past few decades has been its utilization in EOR. The EOR technology involves transportation of CO_2 through pipelines to proper oil production fields that are depleted of associated gas (i.e., natural gas found in association with petroleum oil within the reservoir). Through pipelines, the CO_2 is injected into underground oil reservoirs to raise the reservoir pressure and bring the mixture of reservoir fluid (petroleum) and CO_2 to miscibility conditions (Benmekki and Mansoori, 1986; Mansoori and Savidge, 1989). The CO_2 is transported in supercritical state (Kwak and Mansoori, 1986; Mansoori and Ely, 1985) through pipelines at a pressure of 7.58 MPa (1,100 psi) to save on pumping power. The miscibility condition requires a long time (months or years depending on the size and nature of the reservoir). This process requires large quantity of CO_2 to be injected into underground petroleum reservoirs at high pressure. Currently, there are about 6000 km (3730 mile) of CO_2 pipelines installed in the U.S. for this purpose (Watt, 2010; Horanszky, 2012).

It is worth mentioning that when the concept of CO_2 for EOR began in the 1970s, it was planned for CO_2 production and transportation from many CO_2 wells at natural gas storage sites (in the Rocky Mountain areas) to places like West Texas and Southern Illinois gas-depleted oil reservoirs as discussed in Chapter 4. Initially, the CO_2 wells were drilled in the hope of producing natural gas, but many of such wells ended up containing mostly CO_2 (and little or no methane), which were considered as sub-quality natural gas reservoirs. Therefore, to divert these pipelines for transportation of CO_2 captured from electricity generating plants, it has become necessary to install new sets of pipelines. In January 2013, Denbury Resources completed the construction of a 232-mile long pipeline in the Rocky Mountain region, its second major pipeline network delivering CO_2 to a depleted oil field in Montana for EOR operations (Denbury Resources, 2013). But, Denbury's plan to build a third major CO_2 pipeline network in the Midwest including Illinois has been postponed indefinitely. The postponement became necessary following the passage of a measure by Indiana lawmakers in late 2013. The measure blocked the construction of a gasification plant in Indiana that could have been the only source of CO_2 supply for the project, due to the uncertainties surrounding two other potential sources of CO_2 supply from Illinois and Kentucky. However, considering the fact that the Midwest is an industrial hub, Denbury intends to continue with the project in future, should there be any reliable supply of CO_2 from Illinois and Kentucky to justify the investment. It is estimated that at least $\sim$ 14.2 million cubic meters (500 million cubic feet) of CO_2 will be needed per day to justify this investment (Havercroft *et al.*, 2011).

5.5.2. *Mineralization*

As mentioned in Sec. 5.4.1, CO_2 can be converted into stable solid material carbonates through chemical reactions with naturally occurring alkaline and alkaline-earth oxides. The produced solid material can either be a low-cost solid material to be stored away or a useful product with economic value, such as building materials (see Fig. 5.12).

Currently, there is no known company in Illinois involved in such technology development. However, with DOE funding of ~\$20

Fig. 5.12. Top: CO_2 product solids and bricks, Albany Research Center, US-DOE (Walters, 2005); Bottom: Images of full size concrete blocks (left) and 1 ft × 2 ft concrete panels in a test chamber (right) developed using the carbonation systems (US-DOE/NETL, 2013b).

million, the Calera Corporation, in Los Gatos, California is developing a process for direct mineralization of CO_2 in flue gas to carbonates. The carbonates may be used as construction materials or partial feedstock at cement production facilities (US-DOE, 2014b).

5.5.3. *Cement production*

In addition to carbonate mineralization, CO_2 has the potential for improving the mechanical properties of cement. The new cement technology will require use of a substantial amount of CO_2 as feedstock and consequently, will replace the traditional energy intensive steam curing technology for making Portland cement (US-DOE/NETL, 2014). According to DOE, the transition between demonstration and commercial scale for the process and technology should be rapid, since the technology and the process are expected to require limited changes to the existing curing process for making cement. As a result, the research focus of this new technology is to make improvements to the curing rates and CO_2 yield to increase efficiency of CO_2 use. Besides, the technology should have curing processes based on carbonation chemistry rather than hydration chemistry, in order to reduce energy requirements and CO_2 emissions. Moreover, development of the new cement must meet ASTM standards (US-DOE/NETL, 2014). But, Calera Corporation in Los Gatos, California claims to have developed a type of cement known as Calera (see Fig. 5.13).

Calera is apparently a blend of calcium and magnesium carbonates and is produced when brackish water, sea-water, brine or saline water is contacted with waste heat from power plant flue gas. It is claimed that the CO_2 is absorbed in the saline water, resulting in precipitation of the carbonate minerals (IEA-WBCSD, 2009). The basic chemistry for the Calera process is given by the chemical equations (Zaelke, 2011).

$$Ca^{++} + 2OH^{-} + CO_2(g) \rightarrow CaCO_3(s) + H_2O(l) \quad (5.4)$$

$$Mg^{++} + 2OH^{-} + CO_2(g) \rightarrow MgCO_3(s) + H_2O(l) \quad (5.5)$$

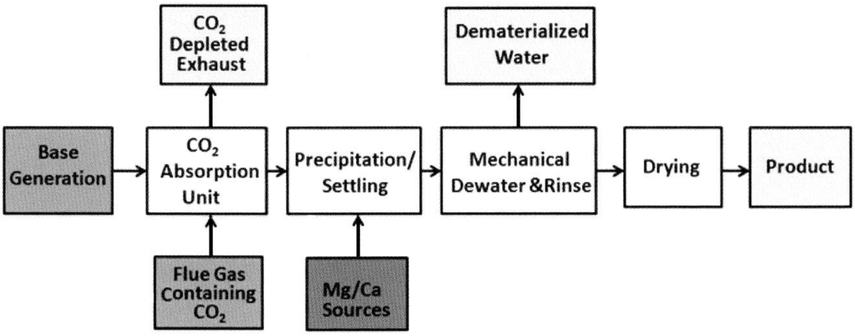

Fig. 5.13. Claimed Celera process diagram: Mineralization via aqueous precipitation (modified from Zaelke, 2011).

In some other countries, firms such as Calix (Australia), Novacem (London), C-Fix (Holland) and TecEco (Australia) are new companies that are focused on producing low-carbon cement (i.e., a type of cement made by using technologies with fewer emissions of CO_2) by eliminating or reducing the carbon emissions (PB and GCCSI, 2011).

5.5.4. *CO_2 for concrete curing*

Concrete curing is a technique used for strengthening and hardening of freshly poured concrete over a period of time. The traditional approach for concrete curing used in producing precast concrete products occurs in a moist, controlled environment. The process is energy-intensive and use of CO_2 in the curing process has been demonstrated to be technically effective (PB and GCCSI, 2011). For example, in 2010 a pilot trial of the technology by Canada-based Carbon Sense Solutions Inc. in partnership with Shaw Brick, Air Liquide and Dalhousie University in Halifax, Nova Scotia, Canada, claimed to have achieved 15% stronger concrete compared to the traditional method. This new technology resulted in a 10% reduction in cement usage, a 38% reduction in energy requirement and improved productivity (CSS, 2014). Additionally, use of the CO_2 sourced from flue gas of concrete plants will help the economic feasibility through lower cost in CO_2 capture and transportation. However, not everyone is convinced by the economic feasibility of CO_2 concrete curing, as remarked by the director of construction and manufacturing technologies at the Portland Cement Association (PCA), in Skokie, Illinois in 2008 in the following quote:

> *"The idea of concrete carbonation has been around for decades but has never been economical as a way to strengthen or improve the finished product. In the late 1990s, researchers showed how carbon dioxide could be turned into a supercritical fluid and injected into concrete to make it stronger, but the required high pressures made the process too energy intensive. Carbon Sense Solutions claims to achieve the same goal but under atmospheric pressure and without the need for special curing chambers. "I'd be really skeptical," "But if someone has a revolutionary process, we'd love to see it"* (Halmiton, 2008).

Nevertheless, since the North American cement and concrete market (including that of Illinois) is made up of about 10 to 15% of precast concrete products, success of this technology will bring some positive impact on the cost of construction projects in the region (Halmiton, 2008). But considering the fact that it took the PCA 25 years to have the cement standards modified to allow 5% limestone in the Portland cement mix, it may take years before these new cement and concrete curing technologies become accepted on the market (Biello, 2008).

5.5.5. *CO_2 as feedstock for polycarbonate plastics*

Synthetic polymers (i.e., polyethylene, polypropylene) are manufactured from derivatives of petroleum such as ethylene or propylene. However, CO_2 can also be used as feedstock for production of chemicals such as polymers and fuels, a potential application that was first discovered in the 1960s by researchers using copolymerization of CO_2 and epoxide over a zinc catalyst (PB and GCCSI, 2011). The general chemical equation for the coupling of CO_2 and epoxide is shown in Fig. 5.14 (Maio *et al.*, 2008).

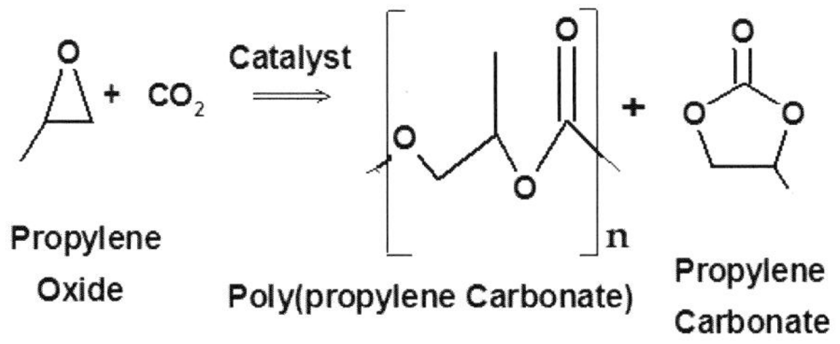

(5.6)

Fig. 5.14. Coupling of CO_2 and epoxides to obtain polycarbonates and/or cyclic carbonates (Redrawn from Miao *et al.*, 2008).

Production of CO_2-based polymers has traditionally been very expensive on commercial scale, due to significant amount of energy needed to split the very stable CO_2 molecule. But in May 2014, Novomer Inc. announced the commercial-scale introduction of Converge® polypropylene carbonate (PPC) polyols (Novomer, 2014). It is claimed the Novomer's technology combines CO_2 with traditional monomers, such as ethylene and propylene, using proprietary zinc-based catalyst system (PB and GCCSI, 2011). According to the company, the CO_2-based polymer can be used in polyurethane formulations targeted at adhesive, coatings, sealants, elastomers and rigid and flexible foams. The company's initial products are claimed to be 1000 and 2000 molecular weight grades of polymers that are manufactured at a multi-thousand-ton commercial-scale "toll processing plant" in Houston, Texas (Novomer, 2014). A "toll processing plant" is referred to a specialized plant or equipment owned by a company that agrees to process raw materials or semi-finished goods for other companies for a fee.

The claimed benefit of this process is significant reduction of the energy requirements, due to direct copolymerization of CO_2 with other monomers, and without the need for initial conversion of CO_2 into carbon monoxide (CO) or some other reactive species (US-DOE/NETL, 2014). Also, since the CO_2-based polymer products contain over 40% by weight CO_2, they have extremely low carbon footprint. Furthermore, it is claimed that the products have favorable manufacturing costs at full commercial-scale production, compared to conventional polyols created from petroleum-based raw materials. Other benefits of the products is low calorific content (40–50% lower than conventional polyether, polyester and polycarbonate polyols) which makes them less flammable; a characteristic that is especially important in polyurethane applications. They also have a unique polycarbonate backbone, which enhances the strength and durability of polyurethane products. Therefore, adding these new polyols into existing formulations produces foams with higher tensile and shear strength, as well as increased load bearing capacity. Additionally, the new polymer is claimed to improve adhesion, cohesive strength, and weatherability of adhesives and coatings, as well as the tensile and flexural strength of elastomers.

Current research work on this technology is focused on utilization of waste energy or alternative energy sources to convert CO_2 into polymers. For successful commercial launch and economic viability of CO_2 polymerization technology, development of novel catalysts to reduce the energy requirements, as well as stabilizers to inhibit degradation of plastics is needed (US-DOE, 2014c).

Another approach to CCS is taken by LanzaTech in Illinois, which is involved in the development of carbon-based industrial waste processes into useful chemicals and fuels. LanzaTech is a start-up company founded in New Zealand, but recently moved to Illinois to benefit from the state's growing market in renewable energy (Scott, 2013). The company claims that it will ultimately establish its global headquarters at the new Science and Technology Park in Skokie, Illinois, a $500 million, 23-acre bioscience campus. The new site will also house the company's R&D center that has recently attracted new investments of over $17 million (Lane, 2014). It is estimated that 30 new jobs will be created in Illinois by the company and the Illinois Department of Commerce and Economic Opportunity (IDCEO) has approved tax credits estimated at $1.1 million over the next 10 years for the company. These credits will help lower the company's state income tax liability and are available under the Illinois

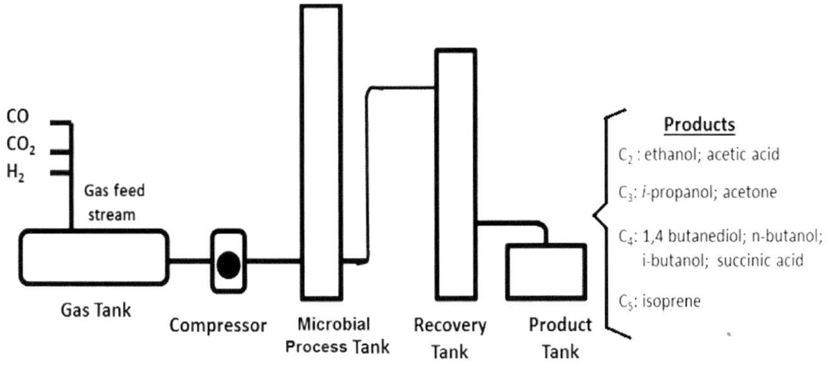

Fig. 5.15. Proposed flowchart for the LanzaTech process (Modified from Wolf, 2013).

Economic Development for a Growing Economy program (EDGE) (Schill, 2014).

In essence, the LanzaTech technology is claimed to capture carbon-rich waste gases from large industrial operations (such as steel plants) and use microbes to convert the waste gas into fuels and chemicals (see Fig. 5.15).

The products from this process is claimed to include acetic acid and 1,3-butadiene, a key chemical intermediate that can be used for production of nylon, synthetic rubber and other plastics. The claimed advantage of this process is that it provides a capital-efficient and cost-effective route for CO_2 storage, compared to traditional CCS technologies discussed earlier in Sec. 5.4. This is because waste gas streams captured from industrial factories can directly be converted on-site to low-carbon liquid fuels and chemicals, by integrating the LanzaTech facility with the source of the pollutant (Reuters, 2014). The company anticipates commercialization of the process within the next one to two years, and has recently closed a $60 million fourth round of venture capital funding, led by Mitsui & Co., Ltd. of Japan, along with Siemens Venture Capital and China International Capital Corporation Growth Capital Fund (Reuters, 2014; Scott, 2013). The technology has apparently attracted the interest of Russia and India, as a means to limit pollution in the two countries.

5.5.6. *Indirect storage of CO_2*

As we discuss in Chapter 7, CO_2 can be stored indirectly by removing the CO_2 gas from air through enhanced photosynthesis or enhancement of carbon intake in terrestrial vegetation and soils (US-DOE/NETL, 2014). Photosynthesis utilizes the sun's energy (light) to convert the carbon in CO_2 to complex carbon-based molecules to store energy. These molecules are often fuels, fuel precursors, food or high value chemicals. In essence, the process of photosynthesis involves a series of enzyme-mediated steps that combines light energy with CO_2 and water to build energy-rich carbohydrates, in accordance with the following general equation (Milne *et al.*, 2009).

$$6H_2O + 6CO_2 \xrightarrow{Light} \text{Carbohydrate}(C_6H_{12}O_6) + 6O_2 \qquad (5.7)$$

A research study sponsored by the Illinois DOE and Natural Resources in Springfield, Illinois and the US-DOE determined the feasibility of alga *Botryococcus braunii* as a biocatalyst for the photosynthetic conversion of flue gas CO_2 to hydrocarbons (Akin *et al.*, 1993). Using photosynthesis, alga *botryococcus braunii* converts CO_2 to cell components and hydrocarbons such as linear alkenes (C_{27}, C_{29} and C_{31} dienes), branched alkenes *(botryococcenes)* and C_{30}-C_{37} isoprenoids, with a general chemical formula C_nH_{2n-10}.

In this study, the estimated cost of algal oil produced from the CO_2-fed algae ranged from \$45–\$75 per barrel. This cost estimate was based on an assumed hydrocarbon yield of 50% of dry weight biomass, as well as the availability of \$60 credit per ton of carbon removed. The oil production in immobilized cell systems ranged from 5–47% of cell biomass dry weight, at an average of 19% of cell biomass dry weight (Akin *et al.*, 1993). Currently, the major problem with this technology is the scale-up. Some studies have suggested that a 1 MW power plant emitting 8,323 metric tons of CO_2 per year would need 64 hectare (158 acre) algal bioreactor facility to yield 20 g of algal oil per meter square of dry weight biomass per day (20g/m^2/day) (Milne *et al.*, 2009; Darzins, 2010). Also, for the Algae technology to be economically viable, it must capture as much CO_2 as projected for the current CCS technologies.

5.5.7. *Conversion of CO_2 into fuel for transportation industry*

There are options to convert CO_2 to liquid fuel for internal combustion engines (see for example Centi and Perathoner, 2009; Jiang *et al.*, 2010). While these options are possible, there are many technological and economic barriers for its practical application. In Fig. 5.16, we report one such possibility.

According to Fig. 5.16, CO_2 produced in, for example, fossil fuel power plants can be converted into a mixture of hydrogen, CO and oxygen through electrolysis with water using the electricity generated by a renewable energy source such as wind or solar. In order to use the mixture of hydrogen and CO in a Fischer–Tropsch synthesis (FTS) process to produce hydrocarbons (such as iso-octane), oxygen must be separated from the $CO + H_2$ gas. Collectively, all stages

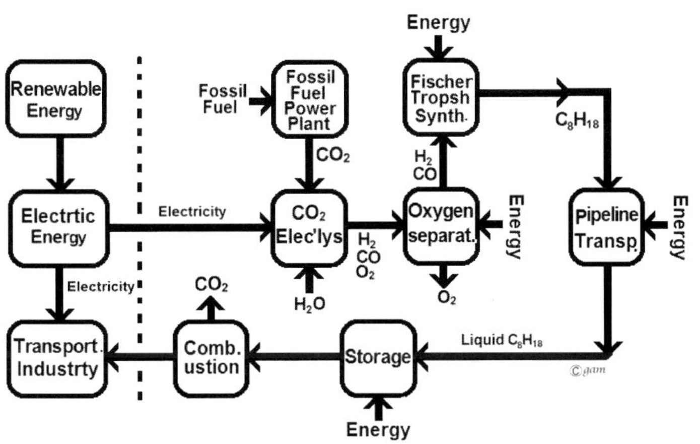

Fig. 5.16. The right side of the dashed line is an option for conversion of CO_2 collected from fossil fuel power plants (like coal plants) to fuel for internal combustion engine using electricity generated by a renewable energy source. The left side of the dashed line is direct use of the same electricity for transportation industry.

of CO_2 electrolysis, Fischer-Tropsch (FT) process, oxygen separation, pipeline transportation and storage are highly energy intensive, in addition to the fact that oxygen separation from the mixture of $CO+H_2$ is quite difficult and highly hazardous due to the explosive nature of hydrogen and oxygen mixture. Considering the fact that automobile's thermal efficiency in converting gasoline internal energy to mechanical energy is about 20%, it is very unlikely that such a process could ever compete with direct use of electricity produced through renewable energy for transportation industry, which has an efficiency of about 80% and also prevents release of CO_2 into the atmosphere.

5.5.8. *Breakthrough concepts*

Besides the abovementioned applications, there is an interest in the development of novel approaches using microbes to consume CO_2 to produce useful products or fuels and other novel applications of CO_2 that would limit its emissions into the atmosphere (US-DOE/NETL, 2014). Currently, researchers are studying the conversion of CO_2 into liquid fuels using a modified cyanobacterium that consumes CO_2 in the presence of sunlight to generate liquid fuels, such as isobutanol, a potential gasoline alternative or diesel fuels (Li *et al.*, 2012). However, without any doubt, direct use of solar and other renewable energies for electricity production as energy input in electric cars will be more efficient and economical than the use of renewable energies to produce gasoline alternatives from CO_2 collected from, for example, fossil-fuel burning power plants stack gases, and using the resulting gasoline alternative to run internal-combustion-engine cars.

5.6. Concluding Remarks

In order to limit the impacts of the greenhouse gasses, the international community has set a goal of limiting increase in average global temperature to 3.6°F (2°C) above pre-industrial levels. To achieve this target, global emissions must drop 40–70% below 2010 levels by 2050 and almost zero by 2100 (IPCC, 2014).

Release of CO_2 to the atmosphere from industrial activities (such as burning of fossil fuels for power production) is a major global concern, mainly because of GHG effects on climate change and fear of environmental degradation. The various major emission sources of CO_2 in Illinois are shown in Table 5.1. According to the table, the largest source of CO_2 emissions is from electricity generation plants, representing roughly 40.5% of the total CO_2 emissions, followed by the transportation sector, accounting for about 28% of the total CO_2 emissions in the state. Therefore, efforts for significant reductions of CO_2 emissions in Illinois must be directed toward these two sectors, which account for more than 68% of the total CO_2 emissions in the state (Agyarko and Mansoori, 2013).

As mentioned in Chapter 2, to meet its environmental obligations and become energy independent, Illinois is transitioning to the use of clean-coal-technologies, The two proposed options that are suitable for CO_2 sequestration are carbonate mineralization and geological storage discussed in Secs. 5.4.1 and 5.4.2, respectively. There are several problems in connecting a conventional process with CO_2 capture technology. But the main problem with deployment of the 1st generation oxy-combustion technology (CO_2 capture device) is the high cost of capturing equipment (US-DOE/NETL, 2013a). It is argued that reduction of this high cost would require development of the 2nd generation and transformational technologies, which could be more efficient and relatively cheaper (US-DOE/NETL, 2013a). Besides the high cost, other problems with carbon capture technologies are their large footprint, operational problems and the high energy penalty for CO_2 removal from flue gases in absorption columns due to regeneration of absorbing solvent.

Obviously, the effective approach to minimize CO_2 emissions from electricity generation sources will be more use of nuclear and renewable energies like wind and solar energies, as they are presented in the later chapters of this book. CO_2 emissions from the transportation sector can be reduced through the use of vehicles with more efficient engine or electric motors. The US-EPA and the US-National Highway Traffic Safety Administration (US-NHTSA) have embarked on a coordinated effort to help produce new generation of clean vehicles, such as electric and possibly fuel cell vehicles. These new generation vehicles together with the enacted and proposed emission standards are expected to save more than 6 billion barrels of oil in the U.S. through the year 2025, while reducing more than 3,100 million metric tons of CO_2 emissions in the nation (US-EPA, 2014b). With the use of more electric vehicles, the load on the refineries, which is the next category of industries contributing to about 5% of the total CO_2 emissions, will be reduced. Furthermore, CO_2 emissions from cement factories can be reduced by integrating mineral carbonation plants to these facilities. As mentioned in Sec. 5.4.1, CO_2 can be converted into a more stable product that can be used as a feedstock in cement plants or it can be converted into cement using the Celera process described in Sec. 5.5.3. Similarly, CO_2 emissions from chemical, iron and steel manufacturing industries, as well as other sources can be reduced by using technologies such as the one offered by LanzaTech at the site of these plants. As mentioned in Section 5.5.5, this technology can convert CO_2 and carbon monoxide (CO) from these sources into fuels and useful chemicals that can serve as intermediate products for polymer manufacturing.

5.6.1. *Carbon tax: an incentive for CCS*

An issue facing fossil fuel power generation is carbon tax, which is a form of explicit carbon pricing. The tax is related to the level of CO_2 emissions and is expressed as a value per ton of CO_2. Since the tax put a price on the amount of GHG emissions, it creates incentives for the emitters to resort to less polluting production methods. İşlegen and Reichelstein (2009) conducted a study to evaluate economics of the currently known CCS technology of competing power generating technologies. Based on their study, $26 per ton of CO_2 emitted is the

Table 5.7. Comparison of cost and CO_2 emissions of selected fossil fuel technologies (İşlegen and Reichelstein, 2009).

	CCS			No CCS		
	Fixed* cost (¢/kWh)	VOC** (¢/kWh)	CO_2 emission (tons/MWh)	Fixed cost (¢/kWh)	VOC (¢/kWh)	CO_2 emission (tons/MWh)
PC plant	33.79	3.39	0.108	19.5	1.93	0.774
IGCC plant	26.27	2.79	0.089	20.39	2.06	0.769
NGCC plant	15.96	5.53	0.042	7.56	4.67	0.361

(PC) — Pulverized Coal.
(IGCC) — Integrated Gasification Combined Cycle coal-fired.
(NGCC) — Natural Gas Combined Cycle.
*Fixed cost represents the capacity investment expenditure required per unit of output.
** Variable operating cost, includes fuel and maintenance costs.

break-even price for the adoption of CCS capabilities by coal-fired power plants. İşlegen and Reichelstein also evaluated construction and operating cost of a few fossil fuel power generating technologies with and without CCS ability. Summary of their results are shown in Table 5.7. Based on these results, application of carbon capture technology will result in higher total cost, especially for pulverized coal (PC) and Natural Gas Combined Cycle (NGCC) technologies.

To investigate greenhouse gas emissions from coal-fired plants, Koornneef *et al.* (2008) conducted Life Cycle Analysis (LCA) to study environmental impacts of three coal-fired electricity generators with and without CCS. Based on their study, application of carbon capture technology reduced total emission of CO_2 by 70% when compared with the state-of-the-art supercritical pulverized coal (SCPC)-fired power plant.

A study by Hamilton *et al.* (2009) showed similar results and confirmed that addition of CCS technology to SCPC lowered CO_2 emission by 86% (based on kg/kWh), but at the expense of 61% increase in total plant cost ($/kWe).

Table 5.8. Estimated LCOE* for power generating plants entering service in 2019 (US-EIA, 2014b).

	Capacity Factor (%)	Levelized Capital Cost ($/MWh)	Fixed O&M ($/MWh)	Variable O&M (including fuel) ($/MWh)	Total system LCOE ($/MWh)
IGCC	85	76.1	6.9	31.7	115.9
IGCC with CCS	85	97.8	9.8	38.6	147.4
Natural gas (Conventional Combined Cycle)	87	14.3	1.7	49.1	66.3
NG (Advanced CC with CCS)	87	30.3	4.2	55.6	91.3

(*) All costs were based on the value of USD in 2012.

Although reduction in CO_2 emissions as a result of employment of CCS is significant, addition of the technology will add greatly to the final cost of the utility. To evaluate the overall competitiveness of different power generating technologies, levelized cost of electricity (LCOE) analysis is often used, which represents the cost of building and operating a power generating plant over an assumed financial life and duty cycle. Table 5.8 shows LCOE from coal and natural gas power generating technologies. Based on these results, addition of CCS technology to coal-fired power plants will add 27% to the total levelized cost.

In the U.S., efforts are underway to decrease emissions of greenhouse gases. In June 2014, the US-EPA proposed the *Clean Power Plan* to reduce power plant carbon pollution by an average of 20% by the year 2025 and 30% around 2030 (Schneider and Boggs, 2014).

A standard, which was reproposed by the EPA in September 2013, calls for limiting emission of CO_2 to no more than 1,100 pounds per MWh of electric generation from new coal-fired power plants and between 1,000 and 1,100 pounds per MWh for new natural gas plants (Folger, 2014). The EPA also identifies CCS as the essential technology required for new coal-fired power plants in the U.S. (Federal Register, 2014).

But, depending on the progress made by individual states since 2005, the EPA is allowing each state a target based on what the agency deems attainable by the state. The reason for the variability in the anticipated emission reductions from state to state is because states with economies less dependent on fossil fuels emit fewer CO_2 to the environment. Therefore, they are allowed to have lower CO_2 emission reductions (19.7% over the next 15 years) to meet the national emission target. On the contrary, states with economies largely dependent on fossil fuels are obliged to have higher CO_2 emissions reductions (35.3% over the next 15 years).

Although the US-EPA believes that under its preferred plan the utilities would spend an estimated $8.8 billion a year to comply with the new proposed rule, it also believes that the overall economic benefits to the nation would far outweigh the cost. These benefits include significant savings for the nation on public health, including consequences of the impact of climate change that could range from $55 billion to $93 billion in 2030. Also, according to US-EPA, an estimated 2,700 to 6,600 premature deaths and 140,000 to 150,000 asthma attacks in children would be avoided in 2030 (US-EPA, 2014a).

This new proposal has been viewed by many as an aggressive target, symbolizing the first ever attempt to curb such a pollutant. However, business groups, including U.S. Chamber of Commerce believe that implementation of this plan would cost the nation's economy about $50 billion a year (Harder, 2014).

Bibliography

Agyarko, L. (2005). Theoretical and experimental studies of the effect of CO_2 on H_2S scavengers performance. MSc thesis, Univ. of Illinois, Chicago.

Agyarko, L. B. and Mansoori, G. A. (2013). Non-renewable Energy Options in Illinois, *Int. J. Oil, Gas and Coal Technol.*, **6**(3): 288–347.

Akin, C., Maka, A., Patel, S. and Conrad, J. (1993). Removal of CO_2 from flue gases by algae. Final Tech. Report, Illinois Dept. of Energy and Natural Resources, US-DOE, Report No. DOE/PC/92521-T97.

Benmekki, E. H. and Mansoori, G. A. (1986). Accurate vaporizing gas derive minimum miscibility pressure prediction. In *Proceedings of* 61st *Ann. Tech. Conf. and Exhib. of the Soc. of Petrol. Engs*, SPE Paper # 15667. New Orleans, 5–8 October, LA.

Biello, D. (2008). Cement from CO_2: A concrete cure for global warming? News section, Scientific American. *scientificamerican.com*

Centi, G. and Perathoner, S. (2009). Opportunities and Prospects in the Chemical Recycling of Carbon Dioxide to Fuels, *Catalysis Today*, **148**(3–4): 191–205.

Cerimele, G. L. and McMillian, M. H. (2011). Mountaineer commercial scale carbon capture and storage project. Topical Report: Preliminary Public Design Report, Am. Electric Power (AEP) Service Corp. and US-DOE Award No. DE-FE0002673.

Changnon, S. A., Kunkel, K. E. and Reinke, B. C. (1996). Impacts and Responses to the 1995 Heat Wave: A Call to Action, *Bulletin of the Am. Meteorological Soc.*, **77**(7): 1497–1506.

Chen, W.-Y and Seiner, J. (2011). *Handbook of Climate Change Mitigation*. Springer.

[CSS] Carbon Sense Solutions, Inc. (2014). Curing concrete of its carbon habit. Halifax, Nova Scotia, Canada. *carbonsensesolutions.com.*

[CSLF] Carbon Sequestration Leadership Forum (2013). CSLF technol. Roadmap. Carbon Sequestration Leadership Forum, US-DOE, Washington DC.

D'Alessandro, D. M., Smit, B. and Long, J. R. (2010). Carbon Dioxide Capture: Prospects for New Materials, *Angewandte Chemie*, **49**(35): 6058–6082.

Denbury Resources, Inc. (2013). Denbury Resources Rocky Mountain activity update. Presented at the Univ. of Wyoming, D. R., Inc., Plano, Texas.

Darzins, A., Pienkos, P. and Edye, L. (2010). Current status and potential for algal biofuels production. Report No. T39-T2 of IEA Bioenergy Task 39, BioIndustry Partners, Nat'l Renewable Energy Lab., Golden, CO.

Fauth, D. J., Soong, Y. and White, C. M. (2002). Carbon Sequestration Utilizing Industrial Solid Residues, *Am. Chem. Soc. Symp. Series*, Div. Fuel Chem., **47**(1): 37–38.

Fed. Register, (2014). Standards of performance for greenhouse gas emissions from new stationary sources: Electric utility generating units. U.S. National Archives and Records Adm., College Park, MD.

Feely, R. A., Sabine, C. L. and Fabry, V. J. (2006). Carbon dioxide and our ocean legacy. PMEL/NOAA. Sci. Brief. pmel.noaa.gov.

Finley, R. J., Gustison, S. R. and Leetaru, H. E. (2005). Geological consideration for CO_2 sequestration in a cratonic basin: An assessment of options in the Illinois basin. In *Proceedings of* 7th *Int'l Conf. on Greenhouse Gas Control Technol.*, Vol. 2(1): 1339–1344. Vancouver, 5 September, BC, Canada.

Finley, R. J., Greenberg, S. E. and Frailey, S. M. (2010). *Carbon Sequestration Atlas of the U.S. and Canada*, DOE/NETL, 3rd Edn. (Atlas III), pp. 48–59. Midwest geological sequestration consortium.

Folger, P. (2013). Carbon capture and sequestration: A primer. CRS Report No. R42532, Congressional Research Service, U.S. Cong., Washington, DC.

Folger, P. (2014). Carbon capture and sequestration: Research, development, and demonstration at the U.S. Dept. of energy. CRS Report No. R42496, Congressional Research Service, U.S. Cong., Washington, DC.

Frailey, S. M. and Finley, R. J. (2006). Site screening and assessment for testing geological sequestration in the Illinois Basin: An assessment of options in the Illinois Basin: A DOE phase II Rregional partnership program. In *Proceedings of* CO_2 *Symp., Lawrence-Berkeley Nat'l Lab.*, Berkeley, CA.

Goldberg, P., Chen, Z.-Y., O'Connor, W., Walters, R. and Ziock, H. (2001). CO_2 Mineral Sequestration Studies in US, *J. Energy and Environ. Research*, **1**(1):117–126.

Hamilton, T. (2008). A concrete fix to global Warming: A new process stores carbon dioxide in precast concrete. MIT Tech. Review, Cambridge, MA.

Hamilton, M. R., Herzog, H. J. and Parsons, J. E. (2009). Cost and U.S. Public Policy for New Coal Power Plants with Carbon Capture and Sequestration, *Energy Procedia*, **1**(1): 4487–4494.

Harder, A. (2014). EPA sets draft rule to cut carbon emissions by 30% by 2030: Framework seeks balance between ambitious target and flexibility. US section, *Wall St. J.*, New York.

Havercroft, I., Macrory, R. B. and Stewar, R. B. (eds.) (2011). *Carbon Capture and Storage: Emerging Legal and Regulatory Issues.* Hart Publishing Co., Portland, OR.

Horanszky, B. (2012). Environmental Issues of CO_2 Pipeline, *Geosci and Eng.*, Univ. of Miskolc, **1**(2): 81–86.

[IDCEO] Illinois Dept. of Commerce and Economic Opportunity (2010). The Illinois coal industry. Report of the IDCEO-Office of Coal Development, Springfield, IL.

[IDCEO] Illinois Dept. of Commerce and Economic Opportunity (2013). Economic impacts of FutureGen 2.0 on Illinois and local economies. Report No. DCEO 14-007 80 7/13, IOCI 14-0043, IDCEO, UIUC, Champaign, IL.

[IEA-WBCSD] Int. Energy Agency-World Business Council for Sustainable Dev. (2009). Cement technol. roadmap 2009: Carbon emissions reductions up to 2050. IEA-WBCSD, Paris, France.

[IPCC] Intergovernmental Panel on Climate Change (2014). Climate change 2014: Mitigation of climate change. 5th Assessment Report of the Intergovernmental Panel on Climate Change, Edenhofer *et al.*, (eds.), Cambridge Univ. Press, Cambridge, UK.

İşlegen, ö. and Reichelstein, S. (2009). Carbon capture by fossil fuel power plants: An economic analysis. Graduate School of Business, Stanford Univ., Stanford, CA.

Jiang, Z., Xiao, T., Kuznetsov, V. L. and Edwards, P. P. (2010). Turning carbon dioxide into fuel, *Phil. Trans. R. Soc.* A., **368**: 3343–3364.

Johnson, D. W., Reddy, S. and Brown, J. H. (2009). Commercially available CO_2 capture. *Power Magazine*, Houston, TX.

Kolstad, C. and Young, D. (2010). Cost analysis of carbon capture and storage for the Latrobe Valley. Bren School of Environ. Sci. and Management, Univ. of Calif., Santa Barbara, CA.

Koornneef, J., van Keulen, T., Faaij, A. and Turkenburg, W. (2008). Life Cycle Assessment of a Pulverized Coal Power Plant with Post-combustion Capture, Transport and Storage of CO_2, *Int. J. Greenh. Gas Control*, **2**(4): 448–467.

Kuipers, D. (2011). Illinois sequestration project is first in U.S. for man-made CO_2. Greenspace, *Los Angeles Times*, Los Angeles, CA.

Kwak, T. Y.and Mansoori, G. A. (1986). Van Der Waals Mixing Rules for Cubic Equations of State, Applications for Supercritical Fluid Extraction Modeling, *Chem Eng Sci.*, **41**(5): 1303–1309.

Lane, I. (2014). LanzaTech moves HQ to Illinois Science and Technol. Park. *Biofuel Digest*, Miami, FL.

Li, H., Opgenorth, P. H., Wernick, D. G., Rogers, S., Wu, T.-Y., Higashide, W., Malati, P., Huo, Y-X., Cho, K. M. and Liao, J. C. (2012). Integrated Electromicrobial Conversion of CO_2 to Higher Alcohols, Science, **335**(6076): 1596.

Mahajan, D., Taylor, C. E. and Mansoori, G. A. (2007). An Introduction to Natural Gas Hydrate/Clathrate: The Major Organic Carbon Reserve of the Earth, *J. Petrol. Sci. & Eng.*, **56**(1–3): 1–8.

Mansoori, G. A. and Savidge, J. L. (1989). Predicting retrograde phenomena and miscibility using equation of state. SPE Paper No. 19809, SPE Ann. Tech. Conf. and Exhib., 8–11 October, San Antonio, TX.

Mansoori, G. A. and Ely, J. F. (1985). Density Expansion (DEX) Mixing Rules: Thermodynamic Modeling of Supercritical Extraction, *J. Chem. Phys.*, **82**(1): 406–413.

Marshall, C. and Climate Wire (2011). Large carbon sequestration project ramps up at corn ethanol facility. Sci. Am. *scientificamerican.com.*

McDonald, D. K., Moorman, S., Darde, A. and de Limon, S. (2011). Oxy-combustion: A promising Tech. for coal-fired plants. *Power Magazine*, Houston, TX.

McDonald, S. (2011). Q&A with Scott McDonald, 2011 Corporate Responsibility Report No. 2011CR, Archer Daniels Midland Co., Decatur, IL.

McDonald, T. M., Mason, J. A., Kong, X., *et al.* (2015). Cooperative Insertion of CO_2 in Diamine-appended Metal-organic Frameworks, *Nature*.

McKinney, J. and Finley, R. J. (2008). ADM, MGSC and ISGS announce carbon sequestration project. News Release, Archer Daniels Midland Company, Decatur, IL.

Miao, C.-X., Wang, J.-Q. and He, L.-N. (2008). Catalytic Processes for Chemical Conversion of Carbon Dioxide into Cyclic Carbonates and Polycarbonates, *The Open Organic Chem. J.*, **2**: 68–82.

Milne, J. L., Cameron, J. C., Page, L. E., Benson, S. M. and Pakrasi, H. B. (2009). Report from Workshop on Biological Capture and Utilization of CO_2. Charles F. Knight Center, Washington Univ. St. Louis, St. Louis, MO.

Mitsubishi Heavy Industries (2009). KM-CDR post-combustion CO_2 capture with KS-1™advanced solvent. In *Proceedings of* 8th *Ann. Conf. on Carbon Capture and Sequestration*. Pittsburg, PA.

Novomer, Inc. (2014). Novomer announces commercial launch of Converge® polyols for broad use in polyurethanes industry. Press Release, Novomer, Inc, Waltham, MA.

Oelkers, E. H., Gislason, S. R. and Matter, J. (2008). Mineral Carbonation of CO_2, *Elements*, **4**(5): 333–337.

[PB and GCCSI] Parsons Brinckerhoff and Global Carbon Capture and Sequestration Inst. (2011). Accelerating the uptake of CCS: Industrial use of captured carbon dioxide. Report by Parsons Brinckerhoff and the Global CCS Inst., Docklands VIC, Australia.

Perlman, D. (2008). Antarctic glaciers melting more quickly. News section, *SFGate*, San Francisco, CA.

Phillips, J. and Maxson, A. (2011). Research and development for future coal generation. *Power Magazine*, Houston, TX.

Pidcock, R. (2013). Scientists shed light on Greenland glacier melt — and what it means for sea level rise. *Carbon Brief*, London, UK.

Rackley, S. A. (2009). Carbon Capture and Storage. Butterworth-Heinemann, Elsevier, Burlington.

Raich, J. W. and Schlesinger, W. H. (1992). The Global Carbon Dioxide Flux in Soil Respiration and its Relationship to Vegetation and Climate, *Tellus B*, **44**(2): 81–99.

Reuters (2014). Mitsui leads investment in LanzaTech's $60M series D round. *Reuters*, New York, NY.

Robine, J. M., Cheung, S. L. K., Le Roy, S., Van Oyen, H., Griffiths, C., Michel, J. P. and Herrmann, F. R. (2008). Death Toll Exceeded 70,000 in Europe During the Summer of 2003, *Comptes Rendus Biologies*, **331**(2): 171–178.

Rochelle, C. A., Camps, A. P., Long, D., Milodowski, A., Bateman, K., Gunn, D., Jackson, P. and Rees, J. (2009). Can CO_2 Hydrate Assist in the Underground Storage of Carbon Dioxide? *Geol. Soc., London*, Special Pub., **319**(1): 171–183.

Rodvelt, G. D. and Oestreich, R. G. (2005). Case history: First commercial Illinois coal bed methane commences via a structured resource evaluation plan. SPE-97720-MS, SPE Eastern Regional Meet., Morgantown, WV.

Schill, S. R. (2014). LanzaTech to move headquarter; Soperton retrofit ongoing. News section., *Biomass Magazine*, Grand Forks, ND.

Schneider, J. and Boggs, J. (2014). America's dirtiest power plants- polluters on a global scale. Environment America Research & Policy Center, Boston, MA.

Scott, A. (2013). Carbon dioxide-to-chemical processes poised for commercialization. *Chem. & Eng. News*, **91**(44): 20–21

Tans, P. (2014). Atmospheric CO_2. Mauna Loa Observatory, NOAA/ESRL, U.S. Dept. of Commerce, Washington, DC.

[US-DOE] U.S. Dept. of Energy (2010). FutureGen industrial alliance announces carbon storage site selection process for FutureGen 2.0. Office of Fossil Energy, Washington, DC.

[US-DOE] U.S. Dept. of Energy (2014a). Pre-combustion carbon capture research. Energy.gov, Office of Fossil Energy, DOE, Washington, DC.

[US-DOE] U.S. Dept. of Energy (2014b). CO_2 Utilization. Energy.gov, Office of Fossil Energy, U.S. DOE, Washington, DC.

[US-DOE] U.S. Dept. of Energy (2014c). Innovative concepts for beneficial reuse of carbon dioxide: Recovery Act. Energy.gov, Office of Fossil Energy, U.S. DOE, Washington, DC.

[US-DOE/NETL] U.S. Dept. of Energy (2006). Sequestration of carbon dioxide emissions in the ocean. Fact sheet, Natl. Energy Technol. Lab., U.S. DOE, Washington, DC.

[US-DOE/NETL] U.S. Dept. of Energy, Natl. Energy Tech. Lab. (2010). DOE/NETL Carbon dioxide capture and storage RD&D roadmap. National Energy Tech. Lab., Golden, CO.

[US-DOE/NETL] U.S. Dept. of Energy (2011). CO_2 capture from biofuels production and sequestration into the Mt. Simon Sandstone. Final Environmental Assessment, ICCS, Area 1 Project, Doc. No. DOE/EA-1828, NETL, DOE, Washington, DC.

[US-DOE/NETL] U.S. Dept. of Energy (2011a). Large-scale industrial carbon capture, storage plant begins construction. National Energy Tech. Lab., U.S. DOE, Washington, DC.

[US-DOE/NETL] U.S. Dept. of Energy and Natl. Energy Tech. Lab. (2012). Carbon utilization and storage atlas, 4th edn. National Energy Tech. Lab., Golden, CO.

[US-DOE/NETL] U.S. Dept. of Energy and Natl. Energy Tech. Lab. (2013a). Carbon capture tech. program plan. CCRP, National Energy Tech. Lab., U.S. DOE, Washington, DC.

[US-DOE/NETL] U.S. Dept. of Energy and Natl. Energy Tech. Lab. (2013b). Beneficial use of CO_2 in precast concrete products. Award No. DE-FE0004285, National Energy Tech. Lab., U.S. DOE, Washington, DC.

[US-DOE/NETL] U.S. Dept. of Energy/Natl. Energy Tech. Lab (2014). CO_2 utilization focus area. National Energy Tech. Lab., U.S. DOE, Washington, DC.

[US-EIA] U.S. Energy Information Administration (2014a). State-Level Energy-Related Carbon Dioxide Emissions, 2000–2011. Environment, Projections and Analysis, U.S. EIA, Washington, DC.

[US-EIA] U.S. Energy Information Administration (2014b). Annual Energy Outlook 2014-with projections to 2040. Doc. No. DOE/EIA-0383(2014), U.S. EIA, Washington, DC.

[US-EPA] U.S. Environ. Protection Agency. (2014a). EPA fact sheet: Clean power plan — By the numbers — cutting carbon pollution by power plants. U.S. EPA, Washington, DC.

[US-EPA] U.S. Environ. Protection Agency (2014b). Regulatory initiatives. Climate Change, U.S. EPA, Washington, DC.

[US-EPA] U.S. Environ. Protection Agency. (2015). Carbon dioxide capture and sequestration. U.S. EPA, Washington, DC. *epa.gov/climatechange/ccs.*

Walters, R. (2005). Sequestration via CO_2 mineralization. Presented at the 4th Ann. Conf. on Carbon Capture and Sequestration, Albany Research Center, U.S. DOE, Washington, DC.

Watt, J. (2010). Carbon Dioxide Transport Infrastructure: Key Learning and Critical Issues, *J Pipeline Eng*, **9**(4): 213–223.

Wolf, C. (2013). Sustainable solutions to global energy challenges. Presented at the Biomass 2013, DOE, Washington, DC.

World Coal Inst. (2009). Securing the future-financing carbon capture & storage in a post-2012 world. London, UK.

Yamin, F. (2005). *Climate Change and Carbon Markets: A Handbook of Emissions Reduction Mechanisms*. Routledge.

Zaelke, D., Young, O. and Andersen, S.O. (2011). Scientific synthesis of Calera carbon sequestration and carbonaceous by-product applications. Donald Bren School of Environ. Sci. and Management, UC, Santa Barbara, CA.

6

Nuclear Energy

6.1. Introduction

The primary peaceful application of nuclear energy, which is the result of splitting the atoms of certain chemical elements, is for electricity generation. Nuclear energy can also be used in water desalination, radioisotopes production for medical purposes (notably diagnostic procedures and cancer treatment), agricultural and industrial uses, food processing, process and district heating, hydrogen production, and other heat applications such as space heating for small towns and villages near nuclear facilities, as well as fishery farms and agricultural greenhouses (Murray and Holbert, 2014).

According to the Nuclear Energy Institute (NEI, 2014a) nuclear energy provided 10.9% of the world's total electricity production in 2012 and comprised 19% of the total electricity production in the U.S., 28% in South Korea, 43% in Sweden and 73% in France. Illinois has been using nuclear energy to meet its base-load power demands which represents about 49% of the electricity produced in the state (ANS, 2003; OECD, 2003; Wernau, 2011a).

The principle behind nuclear energy generation is based on the fact that matter (mass) can be converted into energy based on the equation, $E = mc^2$, as proposed by Albert Einstein in 1905 (Einstein, 1905). In this equation, E is energy, m is the mass and c stands for

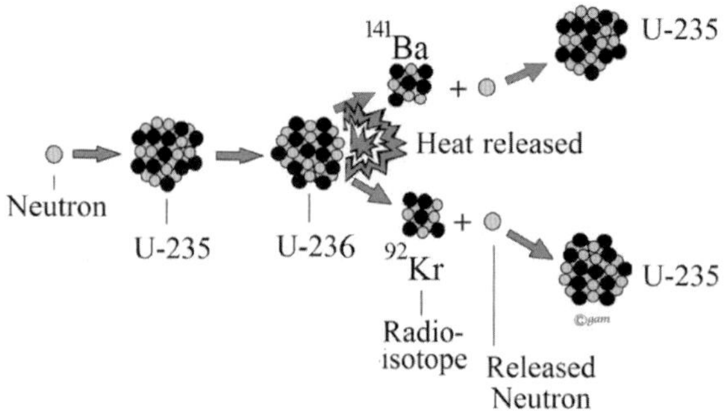

Fig. 6.1. Chain reaction during which a neutron collides with a large U-235 atom, releasing heat and producing radio isotopes plus neutrons. The neutrons then collide with other U-235 atoms and chain reaction continues.

the speed of light. When nucleus of few atoms of matter is split apart by some external force, a tremendous amount of thermal energy is released, which can be controlled and harnessed as a heat source. This energy can be used to evaporate water and run turbines to generate electricity. A nuclear chain reaction (Fermi, 1946) occurs when one nuclear reaction causes more nuclear reactions and subsequently to a self-propagating number of reactions (see Fig. 6.1).

The nuclear reaction is called fission when it involves splitting of a nucleus to lighter nuclei and heat generation. Fission involves use of a heavy molecule, like ^{235}U (an isotope of uranium). The nuclear reaction is called fusion when two nuclei of a light atom (e.g., ^{2}H and ^{3}H, either isotopes of hydrogen) combine to form a heavier nucleus. The energy generated per unit mass of fuel by fusion reaction is more than the one generated by fission reaction. The nuclear chain reaction is unique, since it releases several million times more energy per reaction than a typical exothermic chemical reaction, like combustion of coal or hydrocarbons.

The use of nuclear energy began in Chicago, Illinois, on December 2, 1942 at the University of Chicago, when a team of scientists under the direction of Enrico Fermi initiated the first controlled nuclear chain reaction. Several years later, on July 17, 1955, the town of Arco in Idaho was the first city in the world to be lit by an experimental nuclear reactor (INL, 2014). However, it was in 1958 when the first commercial nuclear power plant in the U.S. was opened in Shippingport, Pennsylvania (Shirk, 2009).

As of December 2014, 438 nuclear power reactors for electricity generation are operating in 30 countries around the world with the total capacity of 375,000 MWe and 71 new nuclear power plants are under construction in 14 countries. Currently, five new nuclear power plants with total net capacity of 5,618 MWe are under construction in the U.S. and are planned to be operational before 2020 (IAEA, 2014; NEI, 2014a; WNA, 2014). Presently, Illinois ranks first among the states in the U.S. in nuclear electricity generation, representing over 12% of the nation's nuclear power.

In this chapter, we present the nuclear reactor technologies and the type of fuels that can be used in nuclear reactors. We then report the nuclear power technology in the State of Illinois, which is an excellent model state for nuclear energy use around the world. Discussion of the state-of-the-art in nuclear fusion energy utilization will be presented in the later part of this chapter.

6.2. Nuclear Fission Reactor Technologies

As mentioned in the Introduction (Sec. 6.1), nuclear fission occurs when a neutron collides with nuclei of certain heavy chemical elements. The collision produces low kinetic energy thermal neutrons and fast neutrons. In addition, the collision releases energy, most of which is in the form of heat. Nuclear reactors are used to create and control fission reactions. The heat from these reactions is used to boil water and generate steam to drive turbines and eventually generate electricity. The advantage of nuclear energy, as compared with other energy sources such fossil fuels, is its high energy density. For example, energy released by fission of one kg of uranium is almost 14,000 times more than the energy released from one kg of liquefied natural gas (LNG) (OECD, 2003).

The basic components of any nuclear fission reactor are: (i) fuel, (ii) moderator, (iii) coolant and (iv) control rods. Structural materials play very important roles in nuclear reactors. For example, the outer layer of fuel rods (fuel cladding) has to withstand heat flux of $\sim$1 MW/m^2 ($\sim$1% of heat flux at the surface of the sun). In most nuclear reactors, zirconium alloys are used as fuel cladding due to their high thermal conductivity and ease of manufacturing (Zinkle and Busby, 2009). Fuel of a nuclear reactor (i.e., uranium) is a fissionable material that is required to start the nuclear chain reaction. After the initial collision, both slow and fast neutrons are produced, but fast neutrons (speed of $\sim$20,000 km/sec) are not efficient in causing fission reactions and in order to lower their energy to the level of slow neutrons (speed of $\sim$2.2 km/sec), a moderator is used (OSPE, 2014). A moderator slows down kinetic energy of fast neutrons, but does not capture them. Use of a moderator in a nuclear reactor increases efficiency of the fuel. Graphite, with 99% purity, is a commonly used fast-neutrons moderator in nuclear reactors. The heat released by a fission reaction is then absorbed and removed by a coolant (water, gas, liquid metals) to maintain the reactor's temperature at a certain range and more importantly, transfer the heat to generate steam and ultimately drive a turbine. Control rods (made of highly neutron-absorbing material like Boron B-10 enriched in steel as alloy for mechanical stability) are used to absorb neutrons to control or stop the fission reactions. It should be born in mind that not all the heat generated by fission reaction can be converted into electricity. Capacity of a nuclear reactor is often reported as thermal (MWt), which indicates the amount of heat that can be generated by the reactor. Capacity of a nuclear reactor is also reported as net electrical (MWe), which indicates the electrical power generated by a generator and is available to be sent to the grid. Between 2002 and 2012, nuclear power plants in the U.S. had an average efficiency, the ratio of MWe/MWt, of 32–33% (US-EIA, 2014a).

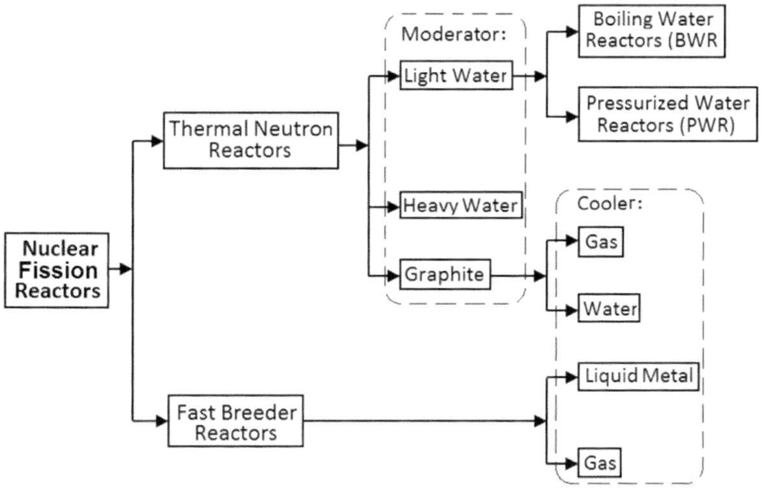

Fig. 6.2. Different types of nuclear fission reactors (Modified from UCDavis-ChemWiki, 2014).

A variety of nuclear reactors are operating around the world as shown in Fig. 6.2. These nuclear reactors can be categorized based on the size or type.

According to the International Atomic Energy Agency (IAEA), nuclear reactors with capacity less than 300 MWe are considered "small", while those with capacity greater than 300 MWe but less than 700 MWe are considered "medium" and reactors with capacity more than 700 MWe are considered "large" (WNA, 2013c). Other categorization is based on the type of moderator and coolant used. 62% of the operating nuclear reactors in the world are pressurized water reactor (PWR) and 19% are boiling water reactors (BWR). The power output of these two types of reactors represents more than 88% of the total power output of nuclear reactors worldwide (ENS, 2014).

In the U.S., as of 2014, 35 out of the 100 operating nuclear power reactors were BWR, with the total licensed capacity of 108,831 MWt. The rest were PWR type reactors, with the total licensed capacity of

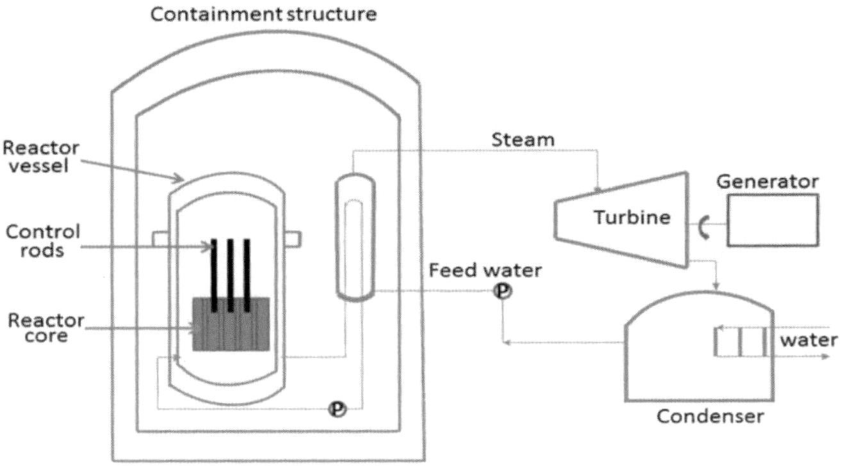

Fig. 6.3. PWR (Modified from OECD, 2003).

199,355 MWt. In the following section, each of these types of reactors is briefly presented.

6.2.1. *Pressurized Water Reactor (PWR)*

The PWR shown in Fig. 6.3 is currently the most commonly used nuclear fission reactor around the world (WNA, 2013c). The PWR is mainly used for electrical power generation. The PWR uses enriched uranium dioxide (UO_2) as fuel and ordinary water as both coolant and moderator. Typically, a PWR uses fuel assemblies containing about 200–300 fuel rods each. The fuel assemblies are arranged vertically in a core, with large reactors having about 150–250 fuel assemblies and containing about 80–100 metric tons of uranium. The reactor has a primary cooling circuit containing water (moderator and coolant) which flows through the core of the reactor at a very high pressure. The reactor has a secondary circuit in which steam is generated through heat exchange with the primary circuit. The steam exits the secondary circuit to drive a turbine (or turbines) to produce electricity before condensing and returning to the heat exchanger to make contact with the primary circuit. The water temperature in the

PWR's core can be as high as 325°C (617°F), making it necessary to maintain the primary circuit pressure at about 15–16 MPa (~2200–2300 psi) to avoid boiling (WNA, 2013c). The system pressure is maintained by steam in a pressurizer. But since ordinary water serves as both coolant and moderator in this reactor, should a portion of the water in the primary cooling circuit turn to steam, it could slow down the fission reaction. This phenomenon is called "a negative feedback effect" and is one of the safety features built into the system. Similarly, addition of boron to the primary circuit can absorb the neutrons to slow down the fission reaction. Thus, both chemistry and control rods are used to control reactivity in this type of reactor (WNA, 2013c). The PWR has high operating reactor pressure and high operation and maintenance costs when compared with a BWR system, which is presented next.

6.2.2. *Boiling Water Reactor (BWR)*

The schematic of the BWR shown in Fig. 6.4 is similar to the PWR, except that the BWR has no secondary circuit. Also, the BWR has lower operating reactor pressure and lower operation and maintenance costs compared to PWR systems. However, the BWR, like

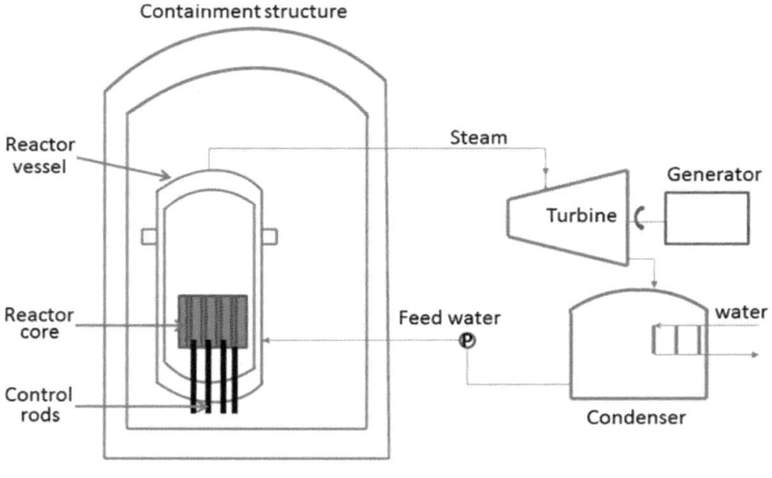

Fig. 6.4. BWR (Modified from OECD, 2003).

PWR, uses enriched uranium oxide (enriched UO_2) as fuel and ordinary water as both coolant and moderator. Typically, a BWR fuel assembly has 90–100 fuel rods, and there are not less than 750 assemblies in a reactor core containing up to 140 metric tons of uranium. The single circuit in the reactor contains ordinary water which boils in the reactor core at about 285°C (545°F) and at 7.6 MPa (1100 psi) (WNA, 2012b). The reactor operates with steam forming about 12–15% of the water at the top of the reactor core, a design which has less moderating effect and hence lower efficiency than in PWR systems. The steam is dried in separators above the core, before exiting the reactor to drive a turbine (or turbines) to generate electricity. Because a steam dryer is installed above the core of a BWR, the control rods are inserted from underneath the core. Also, since the reactor has a single circuit, the water around the core is always contaminated with traces of radionuclide. Therefore during maintenance, the turbine must be shielded in addition to other radiological precautions being taken. The reactor power can be controlled by inserting or withdrawing the control rods or by changing the water flow through the reactor core such that more steam at the top part of the reactor core can reduce the moderation effect (WNA, 2012b).

6.2.3. *Other existing and future potential nuclear fission power generation systems*

The early designs of the nuclear reactors are classified as Generation I, while almost all commercial nuclear reactors operating around the world are designated as Generation II (Nordhaus *et al.*, 2014). The new types of nuclear reactors, designated as Generation III, are the ones with improved fuels, thermal efficiency and safety systems. Generation III nuclear reactors have been constructed in Japan, Russia, China and India, but in the U.S. the focus was switched to Generation III^+ nuclear reactors with advanced passive safety system that does not require human or mechanical intervention to cool or heat the reactor in an emergency. In addition, the Generation III^+ reactors are claimed to have a life span of 60 years as opposed to 30–40 years life span of Generation II reactors (Nordhaus *et al.*, 2014). The more advanced technology, known as Generation IV, is at research and development stage. Their unifying characteristics are that they use different fuel and coolant, have advanced passive cooling systems, use fast neutrons and may have different geometry than the current reactors.

An example of Generation III^+ reactors is the CANDU reactor which runs with natural uranium as fuel and is used in Canada and India (Chetal *et al.*, 2006; WNA, 2012b; EarthlyIssues.com, 2013), advanced gas-cooled reactor used mostly in England (WNA, 2013c; Nonbøl, 1996) and light water graphite-moderated reactor used in Russia (IAEA, 1992; WNA, 2012b).

An interesting nuclear fission reactor, known as fast-breeder reactor (also known as fast neutron-reactor), was proposed during World War II in the U.S. by scientists in the Manhattan Project. Programs for research and development on fast-breeder reactor were established in the U.S. and several other countries. Theory and conceptual designs indicated that fast breeder reactors would be more effective in fissioning all the transuranic isotopes. France was one of the first countries to invest heavily in building (Phénix reactor) and large-scale commercialization (Super-Phénix reactor) of fast breeder reactors (Buck, 2003). But, due to huge technological and safety problems linked with the project after six decades of working, in addition to the expenditure totaling tens of billions of dollars, the commercialization ideas of breeder reactors was abandoned in most countries in the world (Cochran *et al.*, 2010; WNA, 2012b; WNA, 2013b). However, a fast breeder reactor, the 800 MWe, sodium-cooled BN-800 in Russia, planned its commercial operation for early 2015 (Schneider *et al.*, 2014; Pillai and Ramana, 2014).

The USDOE supports research on the next generation reactors that include: the thermal neutron, gas cooled very high temperature reactor (VHTR); the supercritical water cooled reactor (SCWR); the gas-cooled fast neutron reactor (GFR); the lead-cooled fast neutron reactor (LFR) and the sodium-cooled fast neutron reactor [WNA, 2013a].

Many scientists and engineers in Illinois universities and at the Argonne National Laboratory in Illinois, including the senior author of this book (G. A. Mansoori), were involved in various aspects of fast breeder reactor (for example see Baker *et al.*, 1975; Gabor *et al.*, 1977; Mansoori *et al.*, 1981) (with the primary cooling circuit containing liquid metals) research and development.

6.3. Fuels for Nuclear Fission Energy

The two naturally occurring heavy elements that can be used as basic materials for fuels in nuclear fission energy utilization are uranium and thorium. Of the two fuels, uranium is the most commonly used in nuclear reactors because its technology is more advanced compared to thorium. The recoverable resources (i.e., reasonably assured resources in 2011) of uranium in the U.S. are estimated at 207,400 metric tons (WNA, 2014b; WNA, 2014d). But despite this substantial amount of uranium, the current domestic mining represents only 5% of the nuclear fuel used in commercial nuclear reactors in the USA. This is because most of the uranium used in these reactors is imported from overseas, with about 50% obtained from Russian weapons-grade uranium that has been down blended to low enriched uranium (LEU) in Russia (WNA, 2014b). But, this trend of uranium importation will change, as the U.S. begins to increase its nuclear fuel production capability. Currently, one new uranium enrichment plant (National Enrichment Facility in the State of New Mexico, U.S.) is in operation and two more are scheduled to be online before the year 2020 (WNA, 2014b).

6.3.1. *Uranium*

Natural uranium as found in nature contains about 0.7% uranium-235 (U-235), 99.3% uranium-238 (U-238), and a trace (less than 0.01%) of U-234, all being in the form of oxide U_3O_8. Nuclear power plants use elemental U-235 as fuel. The fuel is converted from oxide form to elemental form through an isotopic enrichment process with uranium concentration ranging from 0.7% to about 3.5–5%.

The overall process for making nuclear fuel from the natural uranium ore is as follows:

i. The solid uranium oxide U_2O_8 ore is first crushed through a milling process.
ii. The crushed mineral is suspended in water, forming slurry.
iii. By adding sulfuric acid (H_2SO_4) to the uranium slurry, all the oxides are leached to the liquid phase and are dissolved in water, making the oxides ready for the separation process.

The uranium oxides are then separated by ion exchange, precipitated from the solution, filtered and dried to produce a U_3O_8 concentrate, which has a bright yellow color known as "yellowcake". When it is dried at high temperatures, the color turns to khaki. The U_3O_8 is combined with fluorine to form uranium hexafluoride (UF_6) commonly known as "hex" in the nuclear industry. At ambient temperature and pressure, UF_6 is a white solid. However, the compound converts into vapor or gas at a moderate temperature of 56°C (133°F) at normal atmospheric pressure (FAS, 2013). The gaseous UF_6 is enriched using either a gaseous diffusion process or a gas centrifuge process. Finally, the enriched uranium is converted into pellets and loaded into a nuclear fuel assembly.

The Honeywell Uranium Hexafluoride Processing Facility located in northwest of Metropolis, in southern Illinois, has a nameplate capacity of 15,000 tU (tonne Uranium) of uranium oxide per year. The plant was built in the 1950's under U.S. federal government contract, with the goal of meeting military conversion needs in the nation. However, in late 1960s, the facility was switched to making UF_6 for civilian applications. Since then, the plant's capacity has expanded from 9,000 tU (as UF_6) per year to 18,000 tU (as UF_6) and is expected to increase to 23,000 tU (as UF_6) by 2020

(WNA, 2014b). Among various products manufactured in the plant, about 80% of the work involves refining of raw uranium ore into UF_6. ConverDyn is the exclusive agent responsible for conversion sales from the Metropolis Works. ConverDyn is a general partnership between affiliates of Honeywell and General Atomics, a U.S. Military defense contractor headquartered in San Diego, California specializing in nuclear physics.

The UF_6 from the Metropolis plant was enriched at the Paducah Gaseous Diffusion Plant in Paducah, Kentucky, across the Ohio River near Illinois. However, some of the UF_6 was sold to customers abroad (WNA, 2014b). The Paducah plant was owned by the US-DOE, but was leased and operated by the United States Enrichment Corporation (USEC). In May 2013, USEC terminated all uranium enrichment at the Paducah plant after more than 60 years operation (WNA, 2014). According to a statement released by USEC senior Vice President and Chief Operating Officer Robert Van Namen (Business Wire, 2013): "*While we have pursued possible opportunities for continuing enrichment, DOE has concluded that there were not sufficient benefits to the taxpayers to extend enrichment. I am extremely disappointed to say we must now begin to take steps to cease enrichment.*"

UF_6 is a highly toxic compound which reacts with water in air to produce uranyl fluoride (UO_2F_2) and hydrogen fluoride (highly toxic and corrosive compounds). In 2003, a small release of UF_6 at the Honeywell Metropolis, Illinois, plant prompted evacuation of the personnel within a radius of one mile around the facility. There was no indication of any malevolent activity linked to the incident and initial reports indicated that there were no injuries at the plant. However, three individuals offsite were admitted at the hospital. Of the three individuals, two were released and the third was further evaluated prior to release. The incident lasted for 45 minutes and 25 people offsite were temporarily evacuated, while 75 people remained sheltered for a period of time in their homes (US-NRC, 2003).

Enriched uranium, when going through nuclear chain reactions, produces tremendous amount of heat at a very high desired controlled temperature. For example, fission of one gram of uranium per day generates about 1 MW of energy. This energy is equivalent to burning three tons of coal or 600 gallons of fuels (Lawrence Berkeley National Laboratory, 2014). Fuel cost represents only 15% of the operating costs of a nuclear plant, while fuel cost of a natural gas-fired power plant is about 60% of the operating costs (Rencheck, 2011). As mentioned earlier in the Introduction of this chapter (Sec. 6.1), the thermal energy released from the chain reaction of nuclear fuel is used to produce steam to run a turbine, which drives an electric generator to produce electricity. Nuclear power plants have generally zero greenhouse gas emissions, and they could be cheaper alternative to fossil fuels to produce thermal energy for electricity generation, excluding their rather expensive control and monitoring system for safety requirements.

Enriched uranium is the fuel used in all the U.S. nuclear reactors, including those in Illinois. Natural uranium is a non-fossil energy source, but it is currently not categorized as a renewable source of energy. Uranium mines are in abundance in the States of Arizona, Colorado, Florida, New Mexico, Texas, Utah and Wyoming. Illinois has no commercially viable proven deposits of uranium mineral.

The availability of uranium to meet the world's energy demand for indefinite generations looks very promising because of its abundance and its potential for technology improvements. The World Nuclear Association reported estimated global recoverable uranium reserves as 5.9 million tons, with an annual rate of consumption at 66,000 tons (WNA, 2014d). Besides, uranium from unconventional sources, such as rock phosphate or phosphorite is estimated at 9 to 22 million tons (WNA, 2014c). Also, beginning January 2009, an inventory of the world's uranium revealed that there is an estimated 129,000 tons of uranium for utilities, 10,000 tons of uranium for producers and 15,000 tons of uranium for fuel cycle participants, bringing it to a total of 154,000 tons of uranium worldwide (WNA, 2014d).

However, civilian stockpiles kept by governments and utilities are difficult to quantify due to lack of information. It is predicted that these reserves will increase steadily to provide energy security for utilities and governments. The world's known uranium resources increased 15% in 2005–2006, due to increased mineral exploration (WNA, 2014d). Most of these uranium reserves are located in Australia and Canada. Also, rock phosphate deposits contain several million tons of uranium that may be extracted as a by-product from fertilizers production. The U.S. rock phosphate reserve is about 1,400 million tons, which contains some 170,000 tons of uranium. It is estimated that some 2,300–2,680 tons of uranium per year can be produced from an annual production of 9.6 million tons of diphosphorus pentoxide (P_2O_5) (WNA, 2014c).

Figure 6.5 shows the average price ($/MWh) of uranium (the common fuel for the ongoing nuclear reactors) from 1995–2013. According to this data, uranium is the cheapest fuel for electricity production as compared with fossil fuels.

Figure 6.6 depicts electric power consumption of nuclear energy in Illinois. The reason for the sudden drop of nuclear energy consumption in Illinois in 1997, and its sudden increase in 2000, was due to construction boom of natural gas-fired power plants. Around 1997, investments in nuclear- and coal-fired power plants

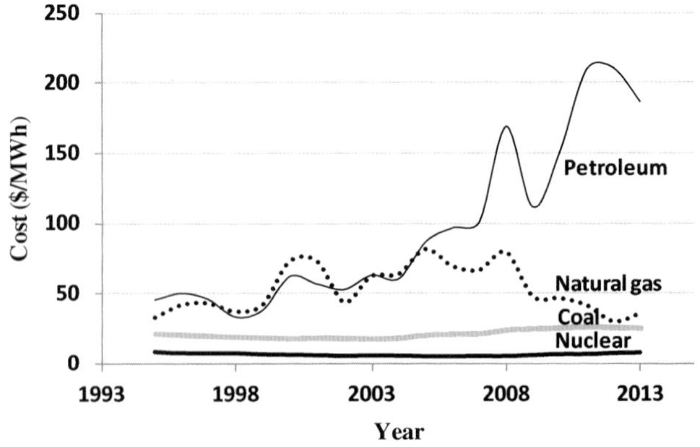

Fig. 6.5. Average fuel cost of U.S. electricity production 1995–2013 (based on the value of USD in 2013) (NEI, 2014b).

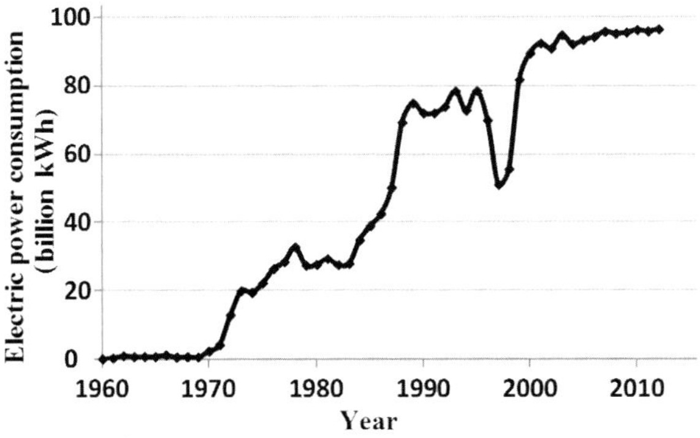

Fig. 6.6. Illinois electric power sector consumption of nuclear energy (US-EIA, 2014b).

technologies nearly disappeared, resulting in increased demand and price of natural gas. Such a price increase and unsustainable demand in natural gas caused an increase in electricity production by nuclear power plants.

The advantages of nuclear power include its relatively cheap price of energy compared to energy produced from coal and natural gas. Also according to EIA, the nuclear energy facilities in Illinois prevent emission of tens of thousands of tons of air pollutants every year, by replacing fossil fuel-fired power plants. The 2010 emissions prevention included 269,840 tons of sulfur dioxide, 82,743 tons of nitrogen oxide and 82,440,000 tons of carbon dioxide (NEI, 2010). Besides, the nuclear facilities employ more than 4,900 highly skilled employees with annual payroll of $400 million. Also, the facilities pay more than $117 million in state and local taxes, with more than $932 million of materials, services and fuel for the nuclear energy industry purchased annually from more than 6,500 Illinois companies (NEI, 2010).

On the negative side, nuclear power plants are too expensive to build due to high cost of their control and monitoring systems for safety purposes. In 2011, the cost of nuclear power plant was estimated at $5,283 per kW, which was more than five times the cost (per kW) of building a natural gas plant. As a result, nuclear power plants are constructed only when they cost lower than alternative technologies, or are backed by federal financing and loan guarantees (Bezdek 2009; NEI, 2011).

6.3.1.1. *Uranium fuel enrichment*

As mentioned earlier in this section, the purpose of enrichment is to increase the percentage of uranium-235 isotope in relation to U-238 in natural uranium. Worldwide, several enrichment processes have been demonstrated on the laboratory scale. These include the gaseous diffusion process, centrifuge process, aerodynamic process, electromagnetic process and the laser process. Among these processes, only gaseous diffusion and centrifuge processes are currently deployed on commercial scale, with both processes using UF_6 gas as feedstock. Both processes are based on physical separation of uranium isotopes in the UF_6 gas, since U-235 atoms are about 1% lighter than U-238 atoms (WNA, 2014a). The performance of an enrichment facility is measured in terms of separative work unit (SWU) which has a unit of kgU/yr. The SWU is a function of the quantity of the uranium processed, as well as the level of the concentration of the U-235 isotope in the UF_6 gas, compared with that of the U-238 isotope (Global Security, 2011).

a. *Gaseous diffusion process*

The gaseous diffusion process was the first process of uranium isotopic enrichment to operate on industrial scale (FAS, 2013). During World War II, the diffusion process was used for the Manhattan Project to produce the first atomic bomb (FAS, 2013). Subsequently, the US-DOE built two of these plants in Piketon, Ohio and in Paducah, Kentucky that were used for commercial purposes. Both plants were leased to the USEC, but the Ohio plant closed in 2001 and the Kentucky plant was closed in 2013, as mentioned earlier in this section (US-NRC, 2014a). In the diffusion process, pressurized UF_6 gas, mostly composed of U-235 and U-238, is forced through a series of porous membranes (about a billion pores per cm^2). Since the $^{235}UF_6$ molecules are relatively lighter than the $^{238}UF_6$ molecules, they cross the membrane barrier a little faster. The gaseous UF_6 that cannot cross the membrane becomes enriched with $^{238}UF_6$. In a typical diffusion process, the UF_6 gas crosses about 1400 stages of packed membranes, known as a cascade, in order to obtain a product with a concentration of 3% to 4% $^{235}UF_6$. Each stage is equipped with a compressor, a diffuser and a heat exchanger for heat removal due to compression of the gas (WNA, 2014a). The cost of an enrichment process is a function of the electrical energy consumption by the process. Therefore, despite the lower capital cost of this process compared to the centrifuge process, the large amount of electrical power needed to operate the compressors of this process is a major setback. For example, the specific energy consumption (i.e., the energy consumption per SWU) for a gaseous diffusion process is 2300–3000 kWh/SWU, compared with 100–300 kWh/SWU for a gas centrifuge process (Global Security, 2011; US-EIA, 2013; US-EIA, 2014c).

b. *Gas centrifuge process*

The gas centrifuge process is currently the most efficient enrichment technology of choice, due its relatively low-energy consumption, short equilibrium time and modular design features (FAS, 2013). The technology has been used for nearly 35 years in Europe (US-NRC, 2014a). In the last decade, the US-NRC has issued licenses for three of such plants to be constructed in the U.S. The first of these licenses was issued in 2004 to Centrus Energy Corp. (formerly known as USEC Inc.) for the construction and operation of a demonstration and test facility (Lead Cascade) at the Piketon site in Ohio, the same site where the now closed USEC's large Portsmouth diffusion plant was once located (US-NRC, 2014a). This prototype facility went online in September 2007, and had the purpose of facilitating the approval of a $2 billion loan guarantee from the US-DOE for completion of the full-scale plant known as the American Centrifuge Plant (Podvig, 2010). Also in 2006, NRC issued a license to Louisiana

Energy Services (also known as Urenco U.S.) to construct and operate a National Enrichment Facility located 5 miles east of Eunice, in New Mexico (US-NRC, 2014a). The first and second phases of the project have been completed, while the Phase 3 was scheduled to start operation in 2015, bringing the capacity to 4.7 SWU/yr and subsequently to 5.7 million SWU/yr (WNA, 2014b). Also in 2011, the NRC issued a license to Areva Enrichment Services LLC (AES) to construct and operate a gas centrifuge uranium enrichment plant (Eagle Rock plant) in Bonneville County, Idaho (US-NRC, 2014a). Work on the $3 billion Eagle Rock enrichment plant was originally scheduled to begin in 2011. However, the date has been postponed three times, due to unresolved issues over financing (WISE, 2014). The gas centrifuge process operates by feeding UF_6 gas into a series of vacuum tubes, with each tube containing a rotor of 3 to 5 meters high and 20 cm in diameter. Spinning of the rotors at a speed of about 50,000 to 70,000 rpm concentrates the heavier $^{238}UF_6$ molecules towards the outer edge of the tubes, while the concentration of $^{235}UF_6$ molecules increases near the center. A thermal gradient creates a countercurrent flow to enable the enriched product to be removed axially (WNA, 2014a). Thus, the lighter $^{235}UF_6$ molecules are removed at one end, while the heavier $^{238}UF_6$ molecules are removed at the other end. The enriched gas becomes part of the feed for the next stages, while the depleted UF_6 gas is recycled to the previous stage. A stage of a typical centrifuge process normally consists of a large number of centrifuges arranged in parallel to form a cascade similar to those of the diffusion process. However, in a centrifuge process the number of stages may range from 10 to 20, rather than 1000 or more, as is the case for the diffusion process (WNA, 2014a). Despite the relatively low electric power consumption of gas centrifuge compared to gaseous diffusion process, there is possibility of converting a gas centrifuge facility intended for civilian enrichment of uranium into a facility that can produce highly enriched uranium. This situation has become a dilemma or a major source of concern from the point of view of nuclear non-proliferation (Glaser, 2008).

c. *Laser enrichment methods*

Uranium enrichment processes discussed above are basically separation of U-235 based on its atomic size. The new developed laser technology uses photo excitation properties of isotopes as a means of separation. There are three types of laser separation techniques, (i) AVLIS (atomic vapor laser isotope separation), (ii) MLIS (molecular laser isotope separation) and (iii) SILEX (separation of isotopes by laser excitation).

The physics of AVLIS process is based on ionization of atoms of a single element. In AVLIS technique, the laser is illuminated on uranium vapor and absorbed by U-235 but not U-238. The excited U-235 ejects an electron and becomes positively charged, which can be deflected by an electromagnetic field and later be collected (Whitaker, 2005; Rousseau and Lepingwell, 2010). AVLIS separates U-235 isotopes, but no other uranium isotopes are separated by this technique (Eerkens and Kim, 2010). The physics of MLIS is based on excitation of molecules of a single element. In MLIS technique, infrared laser is used to selectively excite uranium hexafluoride gas ($^{235}UF_6$) causing $^{235}UF_6$ molecules to dissociate and yield uranium pentafluoride ($^{235}UF_5$). The $^{235}UF_5$ precipitates from the gas in the form of powder, which is filtered from the gas stream and is collected for electric power generation (Whitaker, 2005; Rousseau and Lepingwell, 2010).

Technical details of the SILEX have not been disclosed, but the physics is similar to MLIS and the process uses UF_6 in a carrier gas, where U-235 atoms are selectively excited with a pulsed CO_2 laser and then drawn to a collector (Parvin *et al.*, 2004).

The advantages of laser enrichment are higher separation factor compared with both gas diffusion and gas centrifuge processes. The throughput of laser technology is higher than gas centrifuge, but lower than gas diffusion technology. The cost of laser separation technology is estimated to be lower than both gas diffusion and gas centrifuge techniques (Whitaker, 2005).

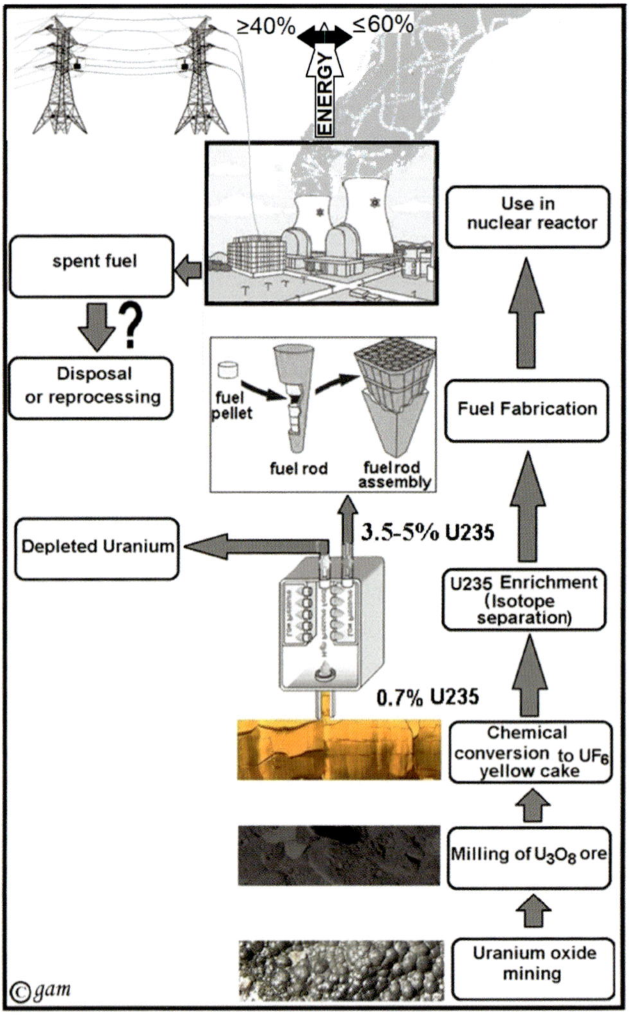

Fig. 6.7. Present status of nuclear technology from uranium oxide mineral to electricity generation and spent fuel accumulation in Illinois and possibly the rest of the U.S.

In Fig. 6.7, we report schematic of the present status of nuclear technology, starting with uranium oxide mineral to electricity generation and spent fuel accumulation as practiced in Illinois and possibly the rest of the states in the U.S.

6.3.2. *Thorium*

The most common source of thorium is the rare earth phosphate mineral (monazite), with chemical formula (Ce, La, Nd, Th)PO_4, containing about 7.5–9% of thorium phosphate (Van Gosen *et al.*, 2009). Monazite is found in igneous and other rocks, but the richest concentrations are in placer deposits. There exist significant thorium deposits in the U.S. estimated at 595,000 metric tons (compared with 207,400 metric tons of known recoverable resources of uranium), mostly in epigenetic veins peripheral to alkaline intrusions, alkaline igneous complexes or carbonatites, as well as alluvial stream and beach deposits (placers) derived from the erosion of alkaline igneous terranes (Van Gosen *et al.*, 2009; WNA, 2012a; WNA, 2014d). In the U.S., the Lemhi Pass district along the border between southwestern Montana and Idaho, as well as the Wet Mountains area of south-central Colorado are the two thorium-vein districts with the majority of the known high-grade thorium resources. Large-volume, low-grade deposits of thorium occur in carbonatite stocks and some alkaline intrusions in the U.S. These include the Hicks Dome in Hardin County of southern most Illinois, measuring about 14.5 km in diameter. The dome-shaped structure in Illinois was formed through the displacement of sedimentary rocks at least 1,200 m upward above an alkaline intrusion at depth. However, the sheer depth to this thorium deposit limits its resource potential. Furthermore, it would require much more surface and subsurface exploration to evaluate the extent and grade of this thorium deposit in Illinois (Van Gosen *et al.*, 2009).

The extraction of thorium from monazite is achieved through a complex multi-stage process as follows (Schulz, 2014):

1. First, the finely ground monazite sand is digested at 310–445°F (155–230°C) with highly concentrated (93%) H_2SO_4. This process converts both the phosphate and the metal content of the monazite to water-soluble species.
2. The solution produced is contacted with aqueous ammonia, first precipitating the hydrated thorium phosphate as a gelatinous

mass and then metathesizing the thorium phosphate to thorium hydroxide.
3. The crude thorium hydroxide is dissolved in nitric acid to yield a thorium nitrate containing feed solution suitable for final purification by solvent extraction.
4. To purify the thorium from residual rare earths and other contaminants present in the nitric acid feed solutions, the crude thorium nitrate concentrate is usually contacted with a tributyl phosphate solution diluted with a suitable hydrocarbon.
5. To remove contaminating rare earths and other metallic impurities to acceptable levels, the resulting organic extract, containing the thorium (and any uranium that may be present), is then contacted countercurrently with a small volume of nitric acid solution.
6. Thermal concentration of the purified thorium nitrate solution produces a product suitable for fabrication of gas mantles. Alternatively, the nitrate can also be calcined to thorium dioxide (ThO_2), which is incorporated into ceramic fuel elements for nuclear reactors or is converted to thorium metal.

Contrary to natural uranium which contains ~0.7% 'fissile' uranium-235 isotope (with the balance 99.3% being mostly the 'fertile' uranium-238 isotope), natural thorium consists of only 'fertile' thorium-232 isotope and does not contain any 'fissile' material (IAEA, 2005). However, similar to uranium-238 isotope, thorium-232 will support a nuclear chain reaction when transmuted into uranium-233, a "fissile" fuel material like plutonium-239 obtained from uranium-238 isotope. This is achieved by neutron irradiation of thorium-232 in a nuclear reactor (such as fast molten salt and other Gen IV nuclear reactors), using a fissile material such as uranium-233, uranium-235, or plutonium-239 as a 'driver'. The transmutation occurs according to the reaction (David *et al.*, 2007).

$$^{232}_{90}\mathrm{Th} + \mathrm{n} \rightarrow {}^{233}_{90}\mathrm{Th} \rightarrow {}^{233}_{91}\mathrm{Pa} + \mathrm{e}^- \rightarrow {}^{233}_{92}\mathrm{U} + 2\mathrm{e}^- \qquad (6.1)$$

In the above reaction, thorium-232 isotope first transmutes into thorium-233 isotope after capturing a neutron and subsequently into protactinium-233 isotope, prior to conversion into uranium-233 isotope. The protactinium-233 isotope has a relatively long half-life (27 days), as compared to thorium-233 (22.3 min), and large neutron absorption cross-section (IAEA, 2005). These characteristics allow a large portion of the protactinium-233 isotope to be converted into non-fissile isotopes (such as uranium-234), instead of decaying rapidly into a uranium-233 isotope, a fissile fuel (IAEA, 2005). Such absorption of neutrons is referred to as "neutron poison" and is known to degrade the reactor efficiency. To circumvent this problem, a heterogeneous fuel arrangement design is frequently used for thorium fuel systems. The design allows the high fissile fuel zone known as the seed region of the reactor to be physically separated from the fertile thorium part of the fuel (frequently referred to as the blanket). An arrangement like this is used to supply surplus neutrons to the thorium nuclei, which convert into fissile uranium-233 isotope. However, since thorium-232 isotope is fissionable to only one tenth of that of uranium-238 (using fast neutrons of over 1 MeV of energy), it is unsuitable for fast neutron reactors. Instead, molten salt reactors (MSRs) are more suitable for thorium fuel, since they avoid the more difficult normal fuel fabrication process, due to the high melting point of thorium oxide (ThO_2). It should be mentioned that the potential for thorium as a nuclear fuel was pursued extensively during the 1950s and 60s in the U.S. In the 1960s, the first privately financed nuclear power plant (Dresden Unit 1) built in Illinois was fueled with "thorium oxide corner rods". The fuel was manufactured by General Electric and later sent to the Savannah River Site in South Carolina, for reprocessing (US-EIA, 1995). Also, the federal government built an 8 MWt experimental molten salt reactor (MSRE) at Oak Ridge National Laboratory (ORNL). The reactor was part of breeder reactor development program in the nation, and is the first and the only thorium-based MSR in the world (IAEA, 2005). Additionally, the U.S. federal government established research and development projects in Pennsylvania, Tennessee and Minnesota to demonstrate the viability of uranium 233 breeder reactors (Alvarez, 2014). In 1962, the Indian Point Unit I PWR near New York city became the first commercial nuclear plant to operate on thorium fuel. However, the first fuel load did not live up to expectations, and subsequently, the operators switched to uranium oxide fuel. But in October 1974, the plant was shut down because of problems with its emergency core cooling systems, which did not meet regulatory requirements. Moreover, the price of fuel oil for the oil-fired system heaters made continued operation of the Indian Point Unit I plant no longer cost effective (NEI, 2004). In 1973, President Nixon terminated all thorium-related nuclear research work in the U.S. (MacPherson, 1985). The reasons given were that uranium-breeder reactors research was more proven and efficient, compared to thorium fuel research. Additionally, unlike thorium spent fuel which is highly contaminated with uranium-232, uranium spent fuel can easily be used for nuclear weapons (IAEA, 2005). By 1977, the federal government had abandoned the thorium fuel cycle in favor of plutonium-fueled breeders. Similarly, by late 1980, the U.S. nuclear power industry had abandoned any future plans for thorium fuel, after failing in several attempts to operate it commercially. It is claimed that the last commercial reactor (the Fort St. Vrain plant in Colorado) to operate on thorium fuel in the U.S. was closed in 1989. This closure was the result of hundreds of equipment failures, fuel failures and leaks, following a decade of operation of plant (Alvarez, 2014; Kazimi, 2003).

But, nearly four decades after the shut-down of all thorium-based nuclear research and development programs in the U.S., there is now growing interest in this fuel source. This rekindled interest stems from the limitations and safety concerns related to the use of uranium-fueled nuclear reactors, including sustainability issues associated with the use of fossil fuel energy sources. Indeed, the importance of thorium as a nuclear fuel is being emphasized through series of scientific conferences around the world. In May 2014, the 6th Annual

Thorium Energy Alliance (TEA) Conference was held in Chicago by Thorium Energy Alliance (TEA), a non-governmental, non-profit and educational organization headquartered in Harvard, Illinois. A conference like this is used to stress on the apparent advantages of thorium as a fuel source over uranium. These include abundance of thorium in the earth crust estimated at 3–4 times more than uranium; fewer radioactive elements contained in thorium-based spent fuels that are smaller in mass and volume compared to conventional uranium-based nuclear wastes; and the possibility of breeding fissile material in slow neutron systems with thorium-based fuels, a unique feature not possible with uranium fuels (WNA, 2012a). Furthermore, proponents of thorium fuel claim that while the spent fuel of thorium is unsuitable for manufacturing of nuclear weapons, it contains uranium-232, a highly radioactive isotope which is difficult to extract, and may serve as high radiation barrier to discourage theft and proliferation of spent fuel (IAEA, 2005). Besides, many thorium deposits contain other valuable mineral resources, such as rare earth elements (cerium, yttrium and lanthanum) (Van Gosen *et al.*, 2009). This group of elements is used for making electronic components, such as permanent magnets for military weapons, electric vehicles, and wind turbines. The other advantage of thorium is the highly stable nature of its oxide (ThO_2), compared with UO_2. This stability of thorium can reduce the likelihood of the fuel pellet reacting chemically with the cooling water or with the metal cladding around them, in case of any malfunction of the protective cladding (Kazimi, 2003). Additionally, the high thermal conductivity of thorium dioxide (about 10–15% more than that of uranium dioxide) can facilitate heat flow from the fuel rods located inside of the reactor. Moreover, the high melting point of thorium dioxide (about 500°C higher than that of uranium dioxide) can serve as additional margin of safety in the event of a transient power surge or loss of coolant (Kazimi, 2003).

Indeed, these positive features of thorium as nuclear fuel have spurred on countries such as Japan, Russia, China, France and the U.S. to develop thorium-fuelled MSRs projects in recent years. India has already made utilization of thorium for large-scale energy production a cornerstone of its nuclear power program, largely due to its indigenous supply of thorium (estimated at 840,000 metric tons), but with scanty uranium deposits. Furthermore, the MSR is one of the six 'Generation IV' reactor designs that have been selected for further development (WNA, 2012a). In essence, the 'Generation IV' reactors (Gen IV) are a set of nuclear reactor designs (mostly theoretical) currently being researched around the world. In the last decade the US-DOE has become interested in fostering research activities in thorium-based fuels (Kazimi, 2003). In one research program, nuclear engineers at Brookhaven and Purdue University examined the use of plutonium-primed thorium as fuel for BWRs. In addition, another research program convened investigators in nuclear power generation from Brookhaven National Laboratory, as well as the Center for Advanced Nuclear Energy Systems (CANES) at MIT, to look for practical methods of simplifying the design of the separated seed and blanket units used in thorium-based nuclear reactors. The third program involved a consortium consisting of two national laboratories (Argonne National Laboratory and the Idaho National Engineering and Environmental Laboratory), three universities (the University of Florida, Purdue University and the Massachusetts Institute of Technology) and two private companies in the business of nuclear fuels fabrication (Framatome Technologies and Westinghouse). The consortium was tasked to produce a plan for using thorium in nuclear reactors, while avoiding the complicated task of dealing with separate types of fuel arrays (from the seed and blanket units), as is the case for the Radkowsky's design (Kazimi, 2003). In the Radkowsky's design, a recycling process is eliminated by using a complex fuel core in which uranium "seeds" with about 20% enrichment of uranium-235 are maintained separate from a surrounding thorium-uranium blanket. In this case, the uranium-235 produces the neutrons to sustain the chain reaction, while slowly producing uranium-233 in the blanket. But as burnup continues, the newly created uranium-233 increasingly takes up more of the fission load (Friedman, 1997).

Despite the attractive features of thorium, its commercialization as a fuel for nuclear reactors encounters some significant challenges. These include testing, analysis and licensing, as well as the qualification work necessary for thorium fuel to enter into service (WNA, 2012a). Furthermore, the availability of large stockpile of depleted uranium that can be used to fuel fast neutron reactors makes any commercial use of thorium as nuclear fuel less likely anytime soon. Nevertheless, the thorium fuel cycle has the advantage of providing the much needed long-term energy security in future due to its potential for being a self-sustaining fuel, which eliminates the need for fast neutron reactors.

6.4. Nuclear Reactor Technology in Illinois

Currently, Illinois has six operating nuclear power plants (see Fig. 6.8) with a total of eleven reactors. These are the Braidwood Generation Station (Units 1 and 2) in Will County, Byron Generating Station (Units 1 and 2) in Ogle County, Clinton Power Station (Unit 1) in DeWitt County, Dresden Generating Station (Units 2 and 3) in Grundy County, LaSalle Generating Station (Units 1 and 2) in LaSalle County and Quad Cities Generating Station (Units 1 and 2) in Rock Island County (US-NRC, 2011).

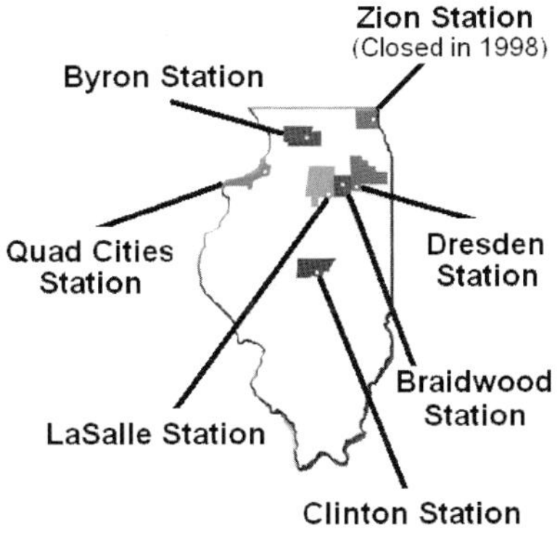

Fig. 6.8. Names and locations of the six operating nuclear reactor stations in Illinois and the closed Zion Station.

Out of the 11 operating nuclear reactors, only four units at the Byron and Braidwood Generating Station are PWRs, the remaining are BWRs. These two types of reactors are described in Sec. 6.2 of this chapter.

Two BWRs in the Quad Cities near Moline and another two of the same type at Dresden in northern Illinois have the same design as the Fukushima Daiichi Unit 1 reactor in Japan, which was heavily damaged by a tsunami on March 11, 2011. Two commercial reactors, Dresden (Unit 1) in Grundy County (see Fig. 6.9) and Zion (Units 1 and 2) in Lake County (see Figs. 6.10, 6.11 and 6.12) have been shut down permanently.

On March 9, 1998 the nuclear fuel was permanently removed from the Zion's Unit 2 reactor vessel and placed in the plant's on-site spent fuel pool, as shown in the picture (Fig. 6.12). The plant's operating license was to expire on November 14, 2013 (Richards, 2012). Fuel transfer to the locations of Independent Spent Fuel Storage

Fig. 6.9. Exelon Dresden Unit 1 which was shut down permanently (Lamar, 2010).

Fig. 6.10. A rusted chain secures a gate to the shuttered Zion Nuclear Power Station along the shore of Lake Michigan, March 11, 2009 in Zion, Illinois — Photo by S. Olson/Getty Images (Daly, 2011).

Fig. 6.11. Zion nuclear power plant in Lake County, IL on the shores of Lake Michigan which was shut down permanently (www.nukeworker.com/pictures/displayimage-135-2652.html, accessed on January 15, 2015).

Installations (ISFSI) began in 2014. But based on a request from Zion Solutions on October 2014, the US-NRC was considering issuing an exemption to the spent fuel storage requirements applicable to Zion Solutions. According to Zion Solutions, a relief from the requirements

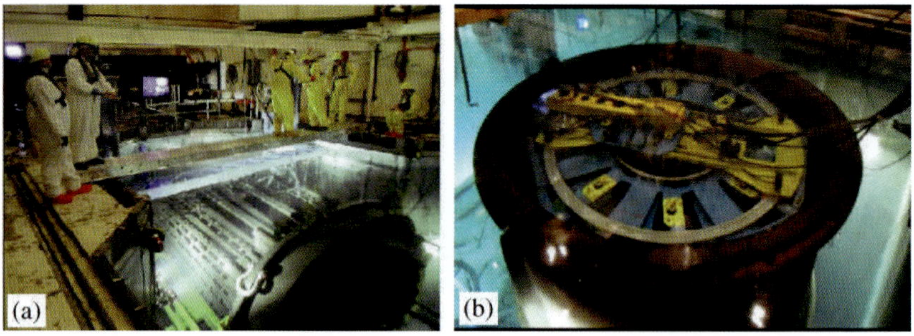

Fig. 6.12. On March 9, 1998, the nuclear fuel was permanently removed from the Unit 2 reactor vessel in Zion and placed in the plant's on-site spent fuel pool, as shown in these pictures (US NRC, 2013a).

will provide flexibility in operations, minimize equipment runtime and repair, as well as minimize personnel dose (Fed. Register, 2014). Preparations for decontamination and dismantlement have begun. Final site survey and license reduction to the ISFSI is currently planned for 2019 — 2020. Submittal of the License Termination Plan (LTP) was scheduled for 2015, with license termination in 2020 (US-NRC, 2013a).

Dresden 1 was the first privately-built commercial nuclear power plant in the U.S., which received its operating permit on September 28, 1959. However, Dresden's Unit 1 reactor was retired on October 1978, after operating for 18 years. Prior to its retirement, the unit had been experiencing steam leaks and corrosion in steam piping in the early and mid-1960s. Additionally, the unit had a history of fuel failures, which resulted in redistribution of radionuclide from the fuel to other parts of the primary system. Most of the problems associated with Unit 1 stemmed from the fact that several of its components, including the heat exchanger surface and the main condenser, were made from admiralty brass (about 70% copper and 30% zinc alloy). Consequently, during the sixth partial refueling, the condenser tubes were changed from admiralty brass to 304L stainless steel. However, the use of copper and zinc alloy surfaces contributed to translocation, as well as deposition of corrosion products in the entire operating systems. Additionally, the use of carbon steel in the secondary feed-water system is suspected to have caused high corrosion radionuclide levels. As a result of these events, the unit underwent renovation in October 1978, followed by chemical decontamination of major piping systems to meet new federal regulations. During the renovation period, more regulations were issued by the US-NRC, following the March 1979 incident at the Three-Mile-Island nuclear power plant (Osif *et al.*, 2004) in Dauphin County, Pennsylvania. The cost for bringing Dresden Unit 1 into compliance with these regulations was estimated to be more than $300 million (US-NRC, 2011). Thus, considering the age and size of the plant, coupled with its lack of benefits, ComEd Company, the operating company of the unit decided to shut it down. In 1984, 753 Curies of cobalt-60 and 12.4 Curies of cesium-137 were removed when chemical decontamination was performed on the primary system of the unit. Shortly after the decontamination procedure, the facility was readied for decommissioning in 1998 because the NRC revised the Dresden Unit 1 license to possess-but-not-operate status. As a result, the periodic surveillance of equipment important to the safety operation of the reactor was no longer required. Since that time, the status of the unit has been at Defueled Safety Analysis Report (DSAR) format. The demolition of Unit 1, including restoration of the site, is scheduled for 2035 and 2036 (US-NRC, 2011).

The Zion Nuclear Power Station is located in the city of Zion in Lake County, Illinois, on a 1.04 square-kilometer (0.4 square miles) of Lake Michigan shoreline, 67.5 kilometers (42 miles) from Chicago (Figs. 6.10, 6.11, 6.12). It was a dual-reactor nuclear power plant (Units 1 and 2) owned by Commonwealth Edison (ComEd) Company network which provided electric power to Chicago and the northern quarter of Illinois. The plant was constructed in 1973 and its Unit 1 began generating power in December 1973, while the Unit 2 was subsequently brought online in September 1974. However, the Zion

Nuclear Power Station was retired on February 13, 1998 and has not been operating since February 1997, due to a couple of incidents which violated NRC regulations.

One incident happened on February 21, 1997, when a reactor operator inserted control rods into the Unit 1 reactor core during a plant shutdown. The operator was transferring control rods into the reactor core to bring the reactor to a very low level of thermal power, but without complete shutdown of the plant. To accomplish this goal, the control rods should have been inserted in the reactor core incrementally, while monitoring a decrease in power level. Inadvertently, the operator inserted two batches of control rods into the reactor continuously, causing a complete shutdown of the reactor. The error was recognized when the operator consulted with the nuclear engineer in the control room, prompting partial withdrawal of the control rods. However, it was too late to reverse any potential damage this error could have caused, as the plant remained completely shutdown. The impact of the incident to the safety of the plant was determined to be insignificant, but the NRC imposed a fine on the company. This was because the NRC determined that there was a breakdown in the management oversight and the control of operational activities, a situation that could have been avoided (US-NRC, 1997a).

The other incident happened on March 7, 1997, when a plant staff observed that there was accumulation of gases at the top of Unit 2 reactor. The unit had already been shutdown in September of the previous year for refueling. This accumulated gas formed a bubble that lowered the cooling water level in the reactor, but nonetheless, it was still well above the top of the reactor fuel. In both cases, the NRC determined that they were not major incidents; however, they could have been prevented if the plant procedures had been followed and therefore the company had to pay a fine (US-NRC, 1997b).

Following these incidents, the operating company of the reactor (ComEd) decided to shut down the unit, instead of paying for new steam generators at the cost of $435 million, an amount which could not be recovered before the plant's operating license expired in 2013. Consequently, on March 9, 1998 the nuclear fuel was permanently removed from the Unit 2 reactor vessel and placed in the plant's on-site spent fuel pool (see Fig. 6.12(a)) for the operating license to expire on November 14, 2013. On August 23, 2010 the NRC authorized the transfer of Exelon's (ComEd's parent company) license to EnergySolutions Company headquartered in Salt Lake City, Utah. The company is contracted to begin a 10-year process of dismantling the site at an estimated cost from $723–$969 million, an amount which Exelon will seek to recover from the Department of Energy (Wernau, 2011b). The dismantled components of the retired Zion nuclear power plant will be transported to the property of Energy Solutions Company in Utah, after which Exelon will take over the responsibility of transferring the spent fuel from its present pool into concrete casks. The estimated date for the closure of the plant is December 31, 2026.

Although Illinois nuclear plants are aging, the state has no current plans to build new plants or halt its reliance on nuclear energy (Daniels, 2009; Bezdek, 2009). Further complicating the attempt to build new nuclear plants is the long search for a national depository for storing radioactive nuclear waste. Without any concrete solution to the nuclear waste problem, it is unlikely to construct new nuclear plants (Bezdek, 2009). Instead, the Chicago-based Exelon Corporation (operator of the Illinois nuclear facilities) is spending $3.6 billion to increase power supply from its existing facilities (Wernau, 2011a). But, opponents of nuclear energy claim that even boosting the energy output from existing nuclear plants could result in a repeat of an incident which occurred in 2002. In the summer of 2002, the Quad City nuclear plant in Illinois was subjected to excessive vibrations, when its power output was raised by approximately 18% (Hughlett & Manor, 2006). This incident, coupled with subsequent safety problems at the Dresden nuclear plant in Chicago has raised some concerns about nuclear safety in the state.

Despite the opposition, in March 2010, the Illinois State Senate voted overwhelmingly to remove a 23-year-old moratorium on the construction of new nuclear power plants in the state, in the wake of escalating cost of crude oil. But following the March 2011 Japanese nuclear disaster, the state reassessed all the safety measures at its nuclear facilities, including its vulnerability to earthquake. According to experts, the 2011 Japanese nuclear disaster was the result of a tsunami, a situation unlikely to occur in Illinois, due to its geographical location and the fact that its nuclear reactors are built to withstand earthquakes. Also considering that Illinois is not known to be prone to earthquakes, the earthquake-related safety of nuclear power plants is high for Illinois (WNA, 2014f; Taira *et al.*, 2014). Nonetheless, consequences of a nuclear incident can be devastating as were in Fukushima and Chernobyl (Wittneben, 2012; Burns *et al.*, 2012).

Faced with competition from natural gas and wind energy, Exelon publicly declared its intention to close three of its six nuclear power plants in Illinois, unless the state passes a legislation that could adequately value Illinois nuclear fleet as a source of reliable, carbon-free generation (Tomich, 2015). These include Byron, Clinton and Quad Cities plants, with a combined capacity of over 5,000 MW. The reason given was that the continuous operation of these plants was no longer profitable, due to weak wholesale power prices. Consequently, on March 4, 2015, a bipartisan coalition of the Illinois Congressional Delegation sent a letter to the Illinois Governor, including the leaders of the Illinois Senate and House "urging them to consider legislative solution that would allow Illinois' nuclear plants to continue to operate". Under the policy supported by Exelon, residential customers would pay about $2 more per month (Sharp, 2015). However, opponents of the policy were considering it as a bailout.

In Sec. 6.4, we presented the concepts of the existing nuclear power generation in Illinois. These include the PWRs and BWRs. In Sec. 6.5, we discuss the containment and management of nuclear spent fuels in Illinois, including the prospects of finding a long-term geologic storage for nuclear waste in the U.S. In Sec. 6.6, we discuss the long term storage of spent nuclear fuel in Illinois that has become a major obstacle to any future expansion of nuclear facilities in the state. Subsequently, Sec. 6.7 discusses the current available nuclear waste reprocessing technologies, which may help the state to reduce its stockpile of spent nuclear fuel in the future. Section 6.8 is about the future of nuclear power generation using fusion reactors, which appears to be a more sustainable form of power generation compared to fission reactors.

6.5. Containment and Management of Nuclear Spent Fuels

According to Nuclear Energy Institute, a policy organization of the nuclear energy and technologies industry, as of 2013, of the roughly 71,780 metric tons of spent uranium fuel stored at the nation's nuclear plants, Illinois has the most spent fuel in storage amounting to almost 9,290 tons (NEI, 2014c). The spent fuel, which has half-life of thousands of years, is the result of over 50 years of operation of Illinois nuclear facilities. Currently, there is no permanent storage location for nuclear spent fuels, including those in Illinois. Therefore, over 9,200 tons of spent fuel produced in Illinois are kept in temporary storage pools of ultra-clean chilled water, which provides radiation shielding and cooling to allow the spent fuel to decay slowly and safely. These storage pools are located at the sites of the seven commercial nuclear power stations which include Clinton, Dresden, La Salle, Braidwood, Byron, Quad Cities and Zion. Additionally, there is one offsite independent spent nuclear fuel storage installation owned by General Electric-Morris, located in Grundy County, Illinois (IEMA, 2011). The GE-Morris installation is the only offsite nuclear spent fuel storage facility in the U.S. Originally constructed in 1964 for spent fuel processing, the GE-Morris facility was never used for its intended purpose, and was contractually tasked to receive some spent fuel. In 2004, the NRC extended the license for General Electric Co. to operate the offsite nuclear spent fuel storage facility through 2022 (US-NRC, 2004). Although spent fuel is normally required to be kept in storage pools for only a few years (1–5 years), at certain stations the fuel has remained in the pools for over four decades (IEMA, 2011). According to the Atomic Energy Act of 1954, federal government is responsible for providing permanent disposal of high-level radioactive waste and spent nuclear fuel to protect the environment and public health (Landis, 2011). To date, there is no permanent solution for a storage location for the spent fuel, as power plant companies continue to run out of space in the storage pools. As a result, the spent fuels are removed from the wet storage and temporarily placed in concrete and metal dry casks filled with inert gas located above the ground (Fig. 6.13).

Since 1982, utility rate payers in the U.S. using nuclear energy have paid over $20 billion into a special fund for the removal of spent fuel from cooling pools for permanent storage. By 2013, Illinois had contributed about two billion dollars into the special fund, representing about 11% of the national contribution (NEI, 2014c). However, due to the delay in the search for uranium spent fuel storage location in the U.S., nuclear power plants owners want part of the special fund to be used to cover the cost for the temporal storage of spent fuels in dry casks (Wernau, 2011b). By 2011, the federal government had paid $776 million in claims to owners of nuclear power plants. These claims were the result of an unfulfilled promise by the federal government to remove the spent fuels around these nuclear facilities for permanent storage, as the facilities are running out of space. Therefore, with any further delay, it is expected that the government's liability will even increase to about $11 billion by the year 2020, if a permanent location is not found sooner to accommodate the spent fuel (Wernau 2011b). In Table 6.1, we present Illinois nuclear power stations and nuclear waste storage.

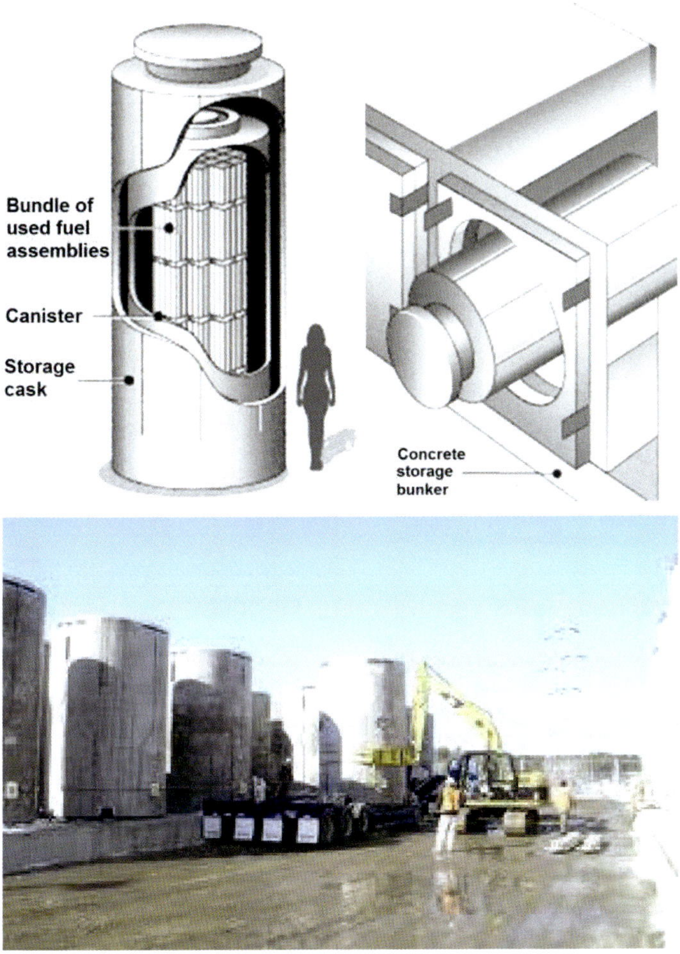

Fig. 6.13. Spent nuclear fuels are removed from the wet storage and temporarily placed in concrete and metal dry casks filled with inert gas located above the ground (US-NRC, 2013b).

The recommended procedures for handling nuclear spent fuel are long-term geologic storage or reprocessing of the fuel for reuse. In what follows, we present the details about these procedures and their status for Illinois.

Table 6.1. Illinois nuclear power stations and nuclear waste storage capacity in 2011 (in metric tons).

Generation Station and/or Spent Fuel Storage Areas(1,2)	Reactors(1,2)	Status(1)	Power Output(2) (MW)	Pool Storage(1) (MetricTons)	Pool Storage(1) (Cores)	Dry Cask Storage(1) (Metric Tons)
Braidwood	Unit 1, Unit 2	Operating	2,330	1,050	13	—
Byron	Station Unit 1, Unit 2	Operating	2,300	1,150	14	—
Clinton	Unit 1	Operating	1,065	450	4	—
Dresden	Unit 1	Shut down	—	—	—	~150
	Units 2 and 3	Operating	1,734	1,000	7	~400
LaSalle	Units 1 and 2	Operating	2,238	1,100	8	—
Quad Cities	Units 1 and 2	Operating	1,774	1,050	8	~350
Zion	Units 1 and 2	Shut down	—	1,020	12	—
GE Morris ISFSI	—	—	—	674	—	—
Total	# of Reactors = 11 (# of Plants = 6)	—	11,441	7,500	—	~900

(1) Union of Concerned Scientists, (2012); (2) US-EIA, (2012).
ISFSI = Independent spent fuel storage installation.

6.6. Long-term Geologic Storage of Nuclear Waste

The first evidence of the feasibility of long-term geologic storage of nuclear waste occurred about 2 billion years ago, due to the formation of natural nuclear reactors in the Franceville Basin in the Western African country of Gabon (Mervine, 2011). The interesting thing about these natural nuclear reactors was that they were stable and operating spontaneously. Besides, most of the radioactive products of the nuclear fission have been safely contained for two billion years (Mervine, 2011).

In October 2011, an international group of scientists researching into the geological disposal of high level radioactive waste concluded that the process is an effective isolation barrier for tens of thousands to millions of years (Nuthall, 2011). But in spite of these findings, there is no country in the world that stores nuclear spent fuel in

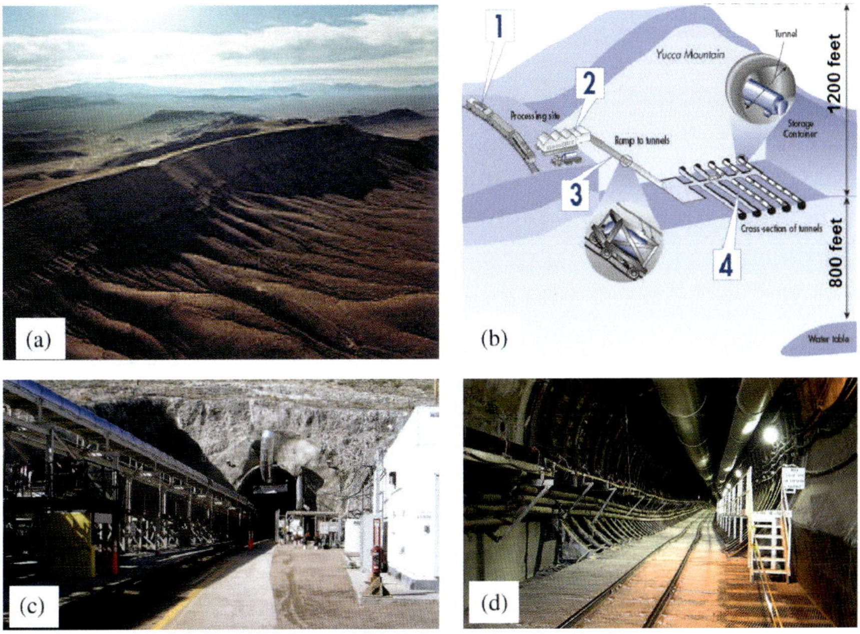

Fig. 6.14. Yucca mountain disposal plan: (a) Overview of the Yucca mountain, (b) Conceptual design of the plan: 1. Canisters of waste, sealed in special casks, are shipped to the site by truck or train. 2. Shipping casks are removed, and the inner tube with the waste is placed in a steel, multilayered storage container. 3. An automated system sends storage containers underground to the tunnels. 4. Containers are stored along the tunnels, on their side. (c) Entrance of the mountain, (d) Inside of the mountain (*Sources*: US-NRC, 2012; Wikipedia; Farnham, 2013; Hargreaves, 2011).

a geologic formation. But after 20 years of intensive research for a suitable geologic formation for the storage of U.S nuclear spent fuels, in 2002, the US-DOE with the U.S. Congress approval finally designated the Yucca Mountain site in Nevada as a permanent storage location (Fig. 6.14).

As of 2014, approximately $10 billion has been spent on developing the Yucca Mountain site. These include testing of the site and construction of concrete tunnels underneath the mountain for storage of the spent fuel. But since its inception, the Yucca Mountain spent fuel storage project has faced fierce criticism at the local, state and national levels. Of particular concern to the residents of Nevada is the fear that the water table underneath the facility will be contaminated by radiation in case of leakage. However, experts are rather of the opinion that the water table is at a level well below that of the maximum projected run off and would not be contaminated (Macfarlane and Ewing, 2006). Furthermore, there are claims that because Nevadans do not depend on nuclear energy, there is no good reason for them to keep watch of the nuclear waste from other states. This latter reasoning makes many believe that Illinois is the appropriate location for nuclear spent fuel storage, since after all, the state is the number one producer of nuclear waste in the nation. The exact storage location of the spent fuel has been identified as a region approximately 16 km (10 miles) northeast of Metropolis, a southern Illinois city bordering with the State of Kentucky (Nikitin *et al.*, 2008). Metropolis and its neighboring cities have a number of nuclear industry facilities. As mentioned in the introduction of this chapter (Sec. 6.1), the Honeywell Uranium Hexafluoride Processing Facility, a uranium conversion facility, is located 3 km (~1.9 miles) northwest of Metropolis, Illinois. In addition, the now closed Paducah Gaseous Diffusion Plant, that was used to enrich UF_6 into various levels for nuclear power or nuclear weapon purposes, was located in Paducah, Kentucky, which is 21 km (13 miles) South of Metropolis. Metropolis area has a very scanty population of approximately 10,000 people within a radius of 32 km (20 miles). However, whether Metropolis, Illinois will embrace this idea of a long-term geologic storage of nuclear waste in their vicinity remains to be seen.

6.7. Reprocessing of Nuclear Waste

The main reason for reprocessing uranium spent fuel is to recover unused uranium in the spent fuel and gain 25–30% more energy from original uranium (WNA, 2014e). Reprocessing is another method of handling nuclear spent fuel which is used in countries such as China, France, India and Russia. France also developed nuclear fuel

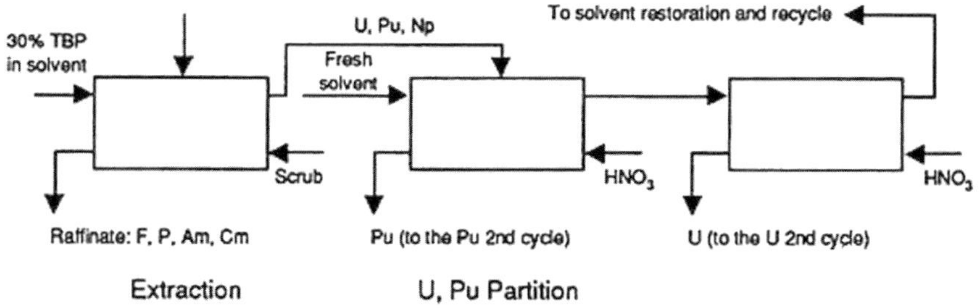

Fig. 6.15. Flow sheet for the first purification cycle in the PUREX process (Rydberg *et al.*, 1992).

reprocessing for their fast-breeder reactors, Phenix and large Super-Phenix reactor, which are now abandoned. Currently, approximately 30% of the spent fuel from the world's LWRs is reprocessed using PUREX (Plutonium and Uranium Recovery by Extraction) technology (Adamantiadesa and Kessidesb, 2009). The process recovers the plutonium and uranium from spent fuel using liquid-liquid extraction (see Fig. 6.15).

As of 2014, there were no operating civil nuclear reprocessing facilities in the U.S., despite construction of three plants for this purpose (WNA, 2014e). The first is a 300 ton per year plant at West Valley in New York, which was operated successfully from 1966–1972. Due to increasing regulation and the high cost involved in its modification, operation of the plant was no longer economically viable and it was shut down. The second reprocessing plant was a 300 ton per year plant built in Morris, Illinois, which incorporated new technology and was operated successfully on a pilot-scale until 1972. However, that technology failed to work in the production plant form which was declared inoperable in 1974. Another reprocessing plant was a 1,500 ton per year plant at Barnwell, in South Carolina. However, this plant was abandoned in 1977, due to a change in government policy that ruled out all U.S. civilian spent fuel reprocessing as one facet of the U.S. non-proliferation policy (WNA, 2014e). In 1992, President G. H. Bush banned nuclear weapons reprocessing in a policy statement on nuclear non-proliferation by declaring: "*I have set forth today a set of principles to guide our non-proliferation efforts in the years ahead and directed a number of steps to supplement our existing efforts. These steps include a decision not to produce plutonium and highly enriched uranium for nuclear explosive purposes....*" (Andrews, 2008).

The emergence of interest in CANDU nuclear reactors in the U.S., in which natural uranium is used as the fuel and does not require the tedious and expensive uranium enrichment, is causing a resurgence of the reprocessing technology in the U.S (WNA, 2014e). As mentioned in Sec. 6.2.3, at least in Canada and India the majority of their nuclear reactor power plants are of CANDU type. These nuclear reactors can be initially fueled with fissile materials extracted from spent nuclear fuels, thus reducing the volume of high level waste to about one fifth, while gaining about 25% more energy from the original uranium. Currently, elsewhere in the world a significant amount of plutonium extracted from spent fuel is recycled into mixed oxide fuel (MOX fuel), while a small quantity of uranium is recovered for recycling (WNA, 2014e). Additionally, new reprocessing technologies that are presently in the research stage could be in deployment in the near future. Using the new technologies, in conjunction with the fast-breeder reactors, may burn all the long-lived actinides (WNA, 2014e). In 2011, a uranium conversion company in Illinois, ConverDyn/Metropolis Works, together with a uranium enrichment company in New Mexico, Urenco U.S., announced the formation of a new partnership called Competitive American Tails Upgrade Partnership (CATUP) (WNN, 2011). This partnership will enable the two companies to meet any potential demand from the US-DOE in managing and upgrading stocks of depleted uranium.

6.7.1. *Advantages of nuclear fission power plant technology*

The argument in favor of the existing nuclear fission power plants has been that they prevent emissions of tens of thousands of tons of air pollutants each year, through the replacement of fossil fuel-fired power plants (NEI, 2013). In 2013, Illinois prevented emissions of 79 million metric tons of CO_2, 67,000 metric tons of NO_x and approximately 160,000 metric tons of SO_2 by using nuclear energy (NEI, 2013). As a result, it is argued that environmentally, nuclear energy has zero emission levels. Unlike fossil energy sources, nuclear energy pays for its waste disposal and decommissioning cost by including them in the actual cost of electricity. Besides, it is easily transportable because of its relatively small quantity. According to the World Nuclear Transport Institute, over several decades of transportation of nuclear materials, there has never been an in-transit accident with serious human health, economic or environmental consequences, due to the radioactive nature of the goods (WNTI, 2011). The reason is that the international transport of radioactive materials is governed by stringent regulatory guidelines, which include standards, codes and regulations that have been subjected to continuous revisions and updates over the past four decades (WNTI, 2011). However, in 1979 a core melt down of Unit 2 nuclear reactor in Three Mile Island in the U.S., caused the release of approximately 2.5 million curies of radioactive gases and approximately 15 curies of iodine-131. Also in 1986, the Chernobyl Nuclear Power Plant explosion in Ukraine, in the former Soviet Union, not only caused uncontrolled release of radioactive materials to the surrounding area, but released high levels of radiation in Europe. Furthermore, the 2011 nuclear incident in Japan caused by a tsunami undermines the sustainability of nuclear power. Economically, nuclear fuel is a highly concentrated energy source for power generation and has the advantage of being economically more attractive than other energy conversion technologies.

6.8. Nuclear Fusion Reactor Technology

Nuclear power can be produced not only from fission reactions using nuclear fuels, as described above, but also from fusion reaction using, for example, hydrogen isotopes as fuel. While nuclear fusion power generation appears to be the long term solution to the world's energy problems because of its sustainability, the technology is still in the research stage. The International Thermonuclear Experimental Reactor (ITER) project currently under construction in Cadarache, France, is envisioned to be the world's first prototype of a nuclear fusion reactor if successful (ITERP, 2013; Wenisch *et al.*, 2007). The nuclei of iron and nickel have the largest binding energies per nucleon, so fusion of two nuclei lighter that iron or nickel releases energy, while fusion of nuclei larger than those two absorbs energy. Thus, in principle any element lighter than iron can be used as a fuel for fusion reaction, but the ITER project in France will use deuterium and tritium as fuel (ITERP, 2013; Wenisch *et al.*, 2007).

The basic concept underlying fusion reactors is that at plasma state the nuclei from two atoms of lighter elements, such as atomic deuterium (D) and atomic tritium (T) can fuse together to produce high energy neutrons (n) and helium (He) (a relatively heavier nucleus) according to the reaction in Eq. (6.1) (WNA, 2012c).

$$ {}^{2}_{1}\mathrm{D} + {}^{3}_{1}\mathrm{T} \rightarrow {}^{4}_{2}\mathrm{He} + {}^{1}_{0}\mathrm{n} + 17.6\,\mathrm{MeV} \qquad (6.2) $$

In the above reaction, the neutrons, which are uncharged particles, can escape the plasma field and be captured by lithium-6 isotope contained in a blanket (about 1 meter in thickness) surrounding the reactor core (WNA, 2012c). After absorbing the neutrons, the lithium-6 isotope transforms into tritium (a nuclear fusion fuel) and helium-4 isotope, according to the reaction in Eq. (6.3) (WNA 2012c).

$$ {}^{1}_{0}\mathrm{n} + {}^{6}_{3}\mathrm{Li} \rightarrow {}^{3}_{1}\mathrm{T} + {}^{4}_{2}\mathrm{He} + 4.784\,\mathrm{MeV} \qquad (6.3) $$

The generated heat energy in the blanket is due to kinetic energy possessed by the neutrons hitting the lithium blanket. The heat is removed using a coolant such as water, helium, or Li-Pb eutectic flowing through a circuit in the blanket, and it may be used to generate steam to produce electricity. Similar to fission reactions, the total mass of the starting fuels for a fusion reaction is slightly higher than the total mass of the products formed after the reaction. This loss in mass is due to conversion of mass into energy as represented by Eq. (6.4) or Einstein's equation (Atzeni and Meyer-ter-Vehn, 2004).

$$Q = \left(\sum_i m_i - \sum_f m_f\right) C^2 \tag{6.4}$$

In Eq. (6.4), Q is the thermal energy (heat) produced, m_i and m_f are the initial and final masses of the reactor fuel, respectively, and C is the speed of light. Unlike neutrons, the helium nucleus produced by the fusion reaction cannot escape the plasma field. Therefore, the energy possessed by the helium nucleus simply contributes to increase in temperature of the plasma (Wenisch *et al.*, 2007).

A major challenge with the fusion process is the high temperature needed to sustain the reaction, especially for the reaction involving two deuterium atoms which is the long-term goal. Another problem is the relatively low power density obtained from fusion reactions (i.e., about 70 times lower compared with fission reaction). This low power density requires construction of a more expensive fusion reactor to obtain the desired reactor power output.

Currently, two major experimental methods are being investigated to improve the power output of fusion reactors. These are the magnetic confinement known as tokamak (see Fig. 6.16) and the inertial confinement fusion (ICF) methods using laser (see Fig. 6.17).

The word tokamak is a transliteration of the Russian word *mokamak*, which stands for "toroidal chamber with (magnetic coils) or (axial magnetic field)". In tokamak, for the magnetic confinement, strong magnetic fields are used to contain the hot plasma required for the fusion to occur.

The ICF method (Fig. 6.17) however, involves the use of strong lasers or particle beams to compress a small pellet containing fusion fuel to extremely high densities and temperatures for fusion to occur (US-DOE, 2014; WNA, 2012c).

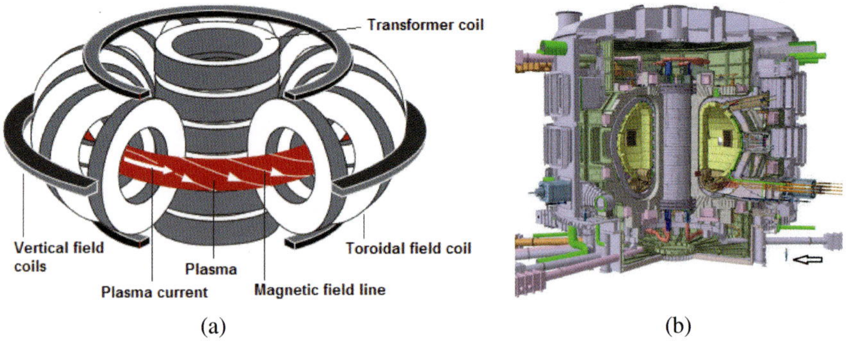

Fig. 6.16. (a) Basic schematic of a tokamak fusion reactor; (b) Diagram of a tokamak fusion reactor to scale with person standing (shown on the arrow) (Pogrebnyak, 2010; Techworldholiday.com).

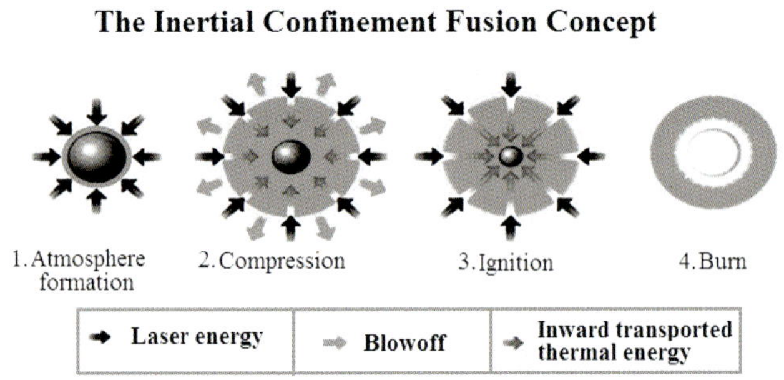

Fig. 6.17. The inertial confinement concept (Modified from US-DOE, 2014).

An alternative approach is the integration of a fusion reactor with a fission reactor to obtain a hybrid system (WNA 2012c). However, such a system will carry existing problems of fission reactors. For a fusion power plant to be feasible, it must produce more energy than they consume. However, this fact has eluded many physicists for several years. But, scientists working at the National Ignition Facility (NIF) are studying various ways to achieve this objective (Clery, 2013).

NIF is a major, laser-based ICF research facility located at the Lawrence Livermore National Laboratory in Livermore, California, USA. NIF possesses powerful lasers to heat and compress small amounts of fusion fuels, such as hydrogen and its isotopes, so that nuclear fusion reactions can occur. Experimental work at NIF has showed some success and for the first time, a fuel capsule gave off more energy than was applied to it. Despite the fact that fusion fuel gain of more than unity was reported by Hurricane *et al.* (2014), there is a long way toward implementing fusion as an energy source. To speed up the fusion energy research in the U.S. and Europe, in January 2014, the U.S. Congress passed a new spending bill. In essence, the bill would increase funding for fusion energy research by more than $100 million over that of the previous year, totaling just over a half-billion dollars. Approximately, $200 million of this funding will be sent to the ITER project based in France (ITERP, 2013), while $305 million will be allocated to domestic research programs to restore the budget cuts incurred over the past two years (Neuhauser, 2014). In Illinois, various fusion energy research activities are underway in universities and other organizations.

6.9. Sustainability of Nuclear Power

As mentioned throughout this chapter, there are major issues surrounding application of nuclear fission power generating technology, namely, safety of operation, disposal of nuclear waste and availability of fuel (i.e., uranium). The issue of fuel availability for nuclear power plants may not be critical at present, but supply of uranium is not unlimited and therefore, using alternative fuels or reactor designs are already under intense investigation. One such an alternative is fast breeder reactor, which because of its huge technological and safety problems has been abandoned in the U.S. and most other countries, after six decades of very expensive R&D work (Cochran *et al.*, 2010; WNA, 2012b; WNA, 2013b).

Another concept of nuclear power technology that is gaining attention is nuclear fusion, brief discussion of which was presented in Sec. 6.8. Fusion reactions have generated energy in the sun and other stars for millions of years, so it would be logical to consider fusion process as a viable source for sustainable energy conversion. The most promising pathway for fusion reaction is the deuterium–tritium reaction. The source of deuterium is water and it is estimated that there are 0.033 grams of deuterium in every liter of water (F4E, 2014). Tritium supply is limited, but it can be generated from lithium via the following reaction (Bradshaw *et al.*, 2011).

$$^{6}_{3}\mathrm{Li} + {}^{1}_{0}\mathrm{n} \rightarrow {}^{4}_{2}\mathrm{He} + \mathrm{T} + 4.8\,\mathrm{MeV} \qquad (6.5)$$

The most recent estimate from the U.S. Geological Survey (Jaskula, 2014) indicated the total world lithium reserves at 13 million metric tons and total world lithium resources at 39.5 million tons. Based on a high estimate by Bradshaw *et al.* (2011), 10 million metric tons of lithium will be required for non-fusion application by the year 2050. The same study estimated that if by the year 2050 fusion technology provides 24,000 TWh of energy, 10,050 tons of lithium will be required annually. Thus, the current reserve of 13 million tons (and considering non-fusion use of lithium), would last almost 300 years. If lithium resources are considered, then there is sufficient lithium for 4,000 years.

Fusion reactors generate a fraction of nuclear waste as compared to fission reactors and radioactive materials from fusion have shorter half-life than their counterparts from fission reactors. Despite the fact that nuclear fusion technology is still at its infancy stage, many believe that the future prospects of this technology are very promising. Certainly there is a long road ahead for development of a controllable fusion technology, but if the process is successful, the high prize may make this journey worthwhile.

6.10. Concluding Remarks

At the national level, about 20% of the electricity used in the U.S. comes from nuclear energy, while in the State of Illinois about 49% of the electricity used comes from its six operating nuclear electric power plants.

Despite the benefits obtained from air pollution prevention of nuclear energy use, the permanent and safe storage of the spent nuclear fuels has become a major obstacle to any new construction of nuclear power plants. As mentioned in Sec. 6.5, in 2013, the U.S. spent nuclear fuels was estimated at about 71,780 metric tons, while that of Illinois was estimated at 9,290 tons, about 13% of the spent nuclear fuel stored at the nation's nuclear plants (NEI, 2014c). Also, as mentioned in this chapter, two nuclear reactors in Illinois, Dresden (Unit 1) in Grundy County and the whole Zion Nuclear Plant (both Units 1 and 2) in Lake County have been shut down permanently, due to persistent problems which were in violation of NRC regulations (US-NRC, 1997a; US-NRC, 1997b; US-NRC, 2011).

The U.S. is the world's largest producer of nuclear power, accounting for more than 30% of worldwide nuclear generation of electricity. The country's 100+ nuclear reactors produced 822 billion kWh in 2013, over 19% of the total electrical output. As of end of 2014, there were 100 nuclear reactors operating and five under construction (WNA, 2014).

Considering the recent awareness in energy conservation, sustainability and the inclination towards the use of renewable energy, it is expected that the public demand for electricity through nuclear energy will reduce in Illinois and across the U.S. (Cooper, 2014). To meet future electricity demands while reducing the use of nuclear and fossil fuels, the U.S. in general, and the State of Illinois in particular, will need to embrace a broad portfolio of renewable energy solutions. Nuclear fusion energy would be part of that energy mix, if this technology becomes commercially available. Indeed, the development of nuclear fusion power plants will change the way nuclear energy is produced and utilized. In essence, nuclear fusion technology, if successfully developed, is capable of meeting the comprehensive goals of energy sustainability. Additionally, the technology has the potential of leading any state or country to energy independence and cleaner environment.

Bibliography

Adamantiadesa, A. and Kessidesb, I. (2009). Nuclear Power for Sustainable Development: Current Status and Future Prospects, *Energy Policy*, **37**(12): 5149–5166.

Alvarez, R. (2014). Thorium: The Wonder Fuel that Wasn't. *Bulletin of the Atomic Scientists*, Hopedale, IL. *thebulletin.org*.

Andrews, A. (2008). Nuclear fuel reprocessing: U.S. policy development. Congressional Research Service, Order Code RS22542, U.S. Congress, Washington D.C.

[ANS] American Nuclear Soc. (2003). Use of nuclear energy for the production of process heat. Position statement No.14, Commu. & Outreach Dept., ANS, La Grange Park, IL.

Atzeni, S., and Meyer-Ter-Vehn, J. (2004). *The Physics of Inertial Fusion*. Oxford University Press, UK.

Baker, L., Mansoori, G. A., Gabor, J. D. and Bingle, J. D. (1975). Investigation of Thermite-type Chemical Reactions to Generate Molten Core Materials for Reactor Safety Experiments, *Am. Nuclear Soc. Transactions*, **22**: 443–447.

Bezdek, R. H. (2009). Nuclear Prospects in the USA: The Continuing Problem of the Waste Issue, *Energy and Environment*, **20**(3): 375–385.

Bradshaw, A. M., Hamacher, T. and Fischer, U. (2011). Is Nuclear Fusion a Sustainable Energy Form?, *Fusion Engineering and Design*, **86**(9): 2770–2773.

Buck, J. -Y. (2003). Le management des connaissances et des compétences en pratique" Deuxième edition, ISBN: 2-7081–2870-1.

Burns, P. C., Ewing, R. C. and Navrotsky, A. (2012). Nuclear Fuel in a Reactor Accident, *Science*, **335**(6073): 1184–1188.

Business Wire. (2013). USEC to Cease Enrichment at Paducah Plant Operations for Inventory: Management and Site Transition to Continue, USEC Inc.

Chetal, S. C., Balasubramaniyan, V., Chellapandi, P., Mohanakrishnan, P., Puthiyavinayagam, P., Pillai, C. P., Raghupathy, S., Shanmugham, T. K. and Pillai, C. S. (2006). The Design of the Prototype Fast Breeder Reactor. *Nuclear Design and Eng.*, **236**(7–8): 852–860.

Clery, D. (2013). Fusion breakthrough at NIF? Uh, not really. News section, *Sci. Insider*, Washington, DC.

Cochran, T. B., Feiveson, F. A., Patterson, W., Pshakin, G., Ramana, M. V., Schneider, M., Suzuki, T., von Hippel, F. (2010). Fast breeder reactor programs: History and status, Research Report 8, The International Panel on Fissile Materials (IPFM), Princeton Univ., Princeton, NJ.

Cooper, M. (2014). The economic failure of nuclear power and the development of a low carbon electricity future. Institute for Energy and the Environment, Vermont Law School.

Daly, M. (2011). Nuclear regulatory commission spent fuel pools. Huff Post Green, *Huffington Post*, Washington, DC.

Daniels, S. (2009). Springfield takes aim at nuke ban. News section, *Crain's Chicago Business*, Chicago, IL.

David, S., Huffer, E. and Nifenecker, H. (2007). Revisiting the Thorium-uranium Nuclear Fuel Cycle, *Europhysics News*, **38**(2): 24–27.

EarthlyIssues.com (2013). 'Nuclear power plants' (HYPERLINK www.earthlyissues.com/nuclearplants.htm viewed on December 23, 2014).

Eerkens, J. W. and Kim, J. (2010). Isotope Separation by Selective Laser-assisted Repression of Condensation in Supersonic Free Jets. *AIChE J.*, **56**(9): 2331–2337.

Einstein, A. (1905) "Ist die Trägheit eines Körpers von seinem Energiegehalt abhängig?" *Annalen der Physik,* **18**: 639.

[ENS] European Nuclear Soc. (2014). Nuclear power plants, world-wide, reactor types. ENS, Brussels, Belgium.

Farnham, A. (2013). Nuclear waste: Yucca Mountain gets reprieve as storage site. World News, abcNews, New York, NY.

[FAS] Federation of Am. Scientists (2013). Uranium Production — History and Usage of Uranium. FAS, Washington, DC.

Fed. Register (2014). Zion Solutions, LLC; Zion Nuclear Power Station, Units 1 and 2; Independent Spent Fuel Storage Installation, Doc. Citation: 79 FR 6221: pp. 62211–62213. U.S. National Archives and Records Adm., College Park, MD.

Fermi, E. (1946). The future of atomic energy. Technical Info. Div., Atomic Energy Commission, Oak Ridge, Tenn.

[F4E] Fusion for Energy (2014). The merits of fusion, Understanding Fusion, Eur. Joint Undertak. for ITER and Devel. of Fusion Energy. Barcelona, Spain. *fusionforenergy.europa.eu*.

Friedman, J. S. (1997). More power to thorium? *The Bulletin of the Atomic Scientist*, **53**(5): 19–22.

Gabor, J. D., Baker, L., Cassulo, J. C. and Mansoori, G. A. (1977). Heat Transfer from Heat Generating Boiling Pools, *Chem. Eng. Progress Series*, **73**(164): 78–85.

Glaser, A. (2008). *Science & Global Security*. Taylor & Francis Pub. Co., New York.

Global Security (2011). Gas centrifuge uranium enrichment. Alexandria, VA. *globalsecurity.org*.

Hargreaves, S. (2011). Nuclear waste: Back to Yucca Mountain? *CNN Money*, CNN, Atlanta, GA.

Hughlett, M. and Manor, R. (2006). Aging nuclear plants pushed to limit: Increased power output raises safety concerns. News section, *Chicago Tribune*, Chicago, IL.

Hurricane, O. A., Callahan, D. A., Casey, D. T., Celliers, P. M., Cerjan, C., Dewald, E. L., Dittrich, T. R., Döppner, T., Hinkel, D. E., Berzak Hopkins, L. F., Kline, J. L., Le Pape, S., Ma, T., MacPhee, A. G., Milovich, J. L., Pak, A., Park, H.-S., Patel, P. K., Remington, B. A., Salmonson, J. D., Springer, P. T. and Tommasini, R. (2014). Fuel Gain Exceeding Unity in an Inertially Confined Fusion Implosion, *Nature*, **506**: 343–348.

[IAEA] Int'l Atomic Energy Agency (1992). The Chernobyl accident: Updating of INSAG-1, Safety Series No. 75-INSAG-7. Int'l Nuclear Safety Advisory Group, IAEA, Vienna, Austria.

[IAEA] Int'l Atomic Energy Agency (2005). Thorium fuel cycle-potential benefits and challenges, Doc. No. IAEA-TECDOC-1450. Nuclear Fuel Cycle and Materials section, IAEA, Vienna, Austria.

[IAEA] Int'l Atomic Energy Agency, (2014). The database on nuclear power reactors. Power Reactor Info System, IAEA, Vienna, Austria.

[IEMA] Illinois Emergency Management Agency (2011). What is spent nuclear fuel? Spent fuel brochure. IEMA, Springfield, IL.

[INL] Idaho Nat'l Lab. (2014). A first-hand account of the lighting of Arco, Idaho. I.N.L., Idaho Falls, ID.

[ITERP] International Thermonuclear Experimental Reactor Project (2013). Assembling the machine, ITER Project. St. Paul-lez-Durance, France.

Jaskula, B. W. (2014). Lithium statistics and information, *Mineral Commodity Summaries*, USGS, Baltimore, MD.

Kazimi, M. S. (2003). Thorium Fuel for Nuclear Energy, *Am. Scientist*, **91**(5): 408–415.

Lamar, T. (ed.) (2010). Exelon nuclear marks 50 years of nuclear power during Dresden unit 1 celebration. *Nuclear News*, Waynesboro, VA.

Landis, T. (2011). Clinton nuclear plant running out of space for spent fuel. The State J. Register, Springfield, IL.

Lawrence Berkeley Natl Lab. (2014). Nuclear fission energy. LBL, Berkeley, CA.

Macfarlane, A. and Ewing, R. C. (2006). Uncertainty Underground: Yucca Mountain and the Nation's High-level Nuclear Waste. MIT Press.

MacPherson, H. G. (1985). The Molten Salt Reactor Adventure, *Nuclear Sci. and Eng*, **90**: 374–380.

Mansoori, G. A., Jedrzejek, J., Shah, N. H. and Blander, M. (1981). Statistical thermodynamic variational equation of state of liquid metals, Chemical Metallurgy — a tribute to Carl Wagner, N.A. Gokcen (ed.), *The Metallurgical Soc. of AIME.*

Mervine, E. (2011). Natures nuclear reactors: The 2-billion-year old natural fission reactor in West Africa, Gabon. *Sci. American.*

Murray, R. and Holbert, K. E. (2014). Nuclear Energy: An Introduction to the Concepts, Systems, and Applications of Nuclear Processes. Butterworth- Heinemann.

[NEI] Nuclear Energy Inst. (2004). Economic benefits of Indian Point Energy Center: An economic impact study by Nuclear Energy Inst, N.E.I., Washington, DC.

[NEI] Nuclear Energy Inst. (2010). Nuclear Energy in Illinois, A fact sheet detailing information about nuclear power in Illinois, 2 pages, N.E.I, Washington, DC.

[NEI] Nuclear Energy Inst. (2011). The cost of new generating capacity in perspective, White paper, N.E.I., Washington, DC.

[NEI] Nuclear Energy Inst. (2013). Nuclear Energy in Illinois-Reliable, Clean and Safe Nuclear Energy, N.E.I., Washington, DC.

[NEI] Nuclear Energy Inst. (2014a). World statistics — nuclear energy around the world, N.E.I., Washington, DC.

[NEI] Nuclear Energy Inst. (2014b). U.S. electricity production costs and components, N.E.I., Washington, DC.

[NEI] Nuclear Energy Inst. (2014c). U.S. state-by-state commercial nuclear used fuel and payments of the nuclear waste fund, N.E.I., Washington, DC.

Neuhauser, A. (2014). Congress supercharges funding for fusion energy technology. News section, *U.S. News & World Report*, Washington, DC.

Nikitin, M. B. D., Parillo, J. M. and Squassoni, S. A. (2008). Managing the nuclear fuel cycle: Policy implications of expanding global access to nuclear power. *Nova Sci.*

Nonbøl, E. (1996). Description of the advanced gas cooled type of reactor (AGR). Doc. No. NKS/RAK-2(96)TR-C2, Risø Nat'l Lab., Roskilde, Denmark.

Nordhaus, T., Lovering, J. and Shellenberger, M. (2014). How to make nuclear cheap: Safety, readiness, modularity and efficiency. The Breakthrough Inst., Oakland, CA. *thebreakthrough.org.*

Nuthall, K. (2011). Nuclear fuel structures OK for disposal. *World Nuclear News*, London, UK.

[OECD] Org. for Economic Coop. and Devel. (2003). Nuclear energy today. Nuclear Energy Agency, Doc. No. 53079 2003, OECD, Paris, France.

Osif, B. A., Baratta, A. J and Conkling, T. W. (2004). TMI 25 Years Later: The Three Mile Island Nuclear Power Plant Accident and its Impact. Penn. State Univ. Press, Univ. Park, PA.

[OSPE] Ontario Soc. of Prof. Engs. (2014). Productive use of nuclear spent fuel, Joint OSPE — PEO Chapter Energy Policy Presentation, OSPE, Ontario, Canada.

Parvin, P., Sajad, B., Silakhori, K., *et al.* (2004). Molecular Laser Isotope Separation Versus Atomic Vapor Laser Isotope Separation, *Progress in Nuclear Energy*, **44**(4): 331–345.

Pillai, S. R. and Ramana, M. V. (2014). Breeder reactors: A possible connection between metal corrosion and sodium leaks. *Bulletin of the Atomic Scientists*, **70**(3): 49–55.

Podvig, P. (2010). USEC began operations of a lead cascade at Am. Centrifuge Plant. News section, *Int'l Panel on Fissile Materials*, Princeton Univ., Princeton, NJ.

Pogrebnyak, I. (2010). The Energy source of tomorrow: Benefits of nuclear fusion power, *PIT J.*, Cycle 1, A pub. of Univ. of N. Carolina, Chapel Hill, NC.

Rencheck, M. (2011). Advancing America's nuclear infrastructure. *Power Magazine,* Houston, TX.

Richards, R. (2012). Size reducing Zion unit-2 for waste disposal, Nuclear Decommissioning Report, Sandrik & Wade Publishing, Ouray, CO.

Rousseau, D. and Lepingwell, J. (2010). Isotopic separation by laser technologies: Safeguards related aspects, Paper No. IAEA-CN-184/262, IAEA, Vienna, Austria.

Rydberg, J., Musikas, C. and Choppin, G. R. (1992). *Principles and Practices of Solvent Extraction.* Marcel Dekker.

Schneider, M., Froggatt, A., Ayukawa, Y., Burnie, S., Piria, R., Thomas, S. and Hazemann, J. (2014). The world nuclear industry status report 2014. *worldnuclearreport.org.*

Schulz, W. W. (2014). Thorium processing. Encyclopedia Britannica.

Sharp, A. (2015). Environmentally-friendly proposal could come at the cost of energy customers' bills. WQAD Broadcasting Station, Moline, IL. *wqad.com*

Shirk Jr. W. L. (2009, spring). Atoms for Peace in Pennsylvania, *Pennsylvania Heritage Magazine*, **35**(2).

Taira, W., Nohara, C., Hiyama, A. and Otaki, J. M. (2014). Fukushima's Biological Impacts: The Case of the Pale Grass Blue Butterfly, *J. Heredity*, **105**(5): 710–722.

Tomich, J.(2015). Bill to support Illinois nuclear plants expected by March. EnergyWire, *Midwest Energy News. midwestenergynews.com*

UCDavisChemWiki (2014) (http://chemwiki.ucdavis.edu/Development_Details/Approaches/ChemCases/Nuclear_Chemistry_and_the_Community/10._Nuclear_Reactors viewed on December 21, 2014).

Union of Concerned Scientists (2012). Nuclear power safety in Illinois. Special report, UCSUSA, Cambridge, MA.

[US-DOE] U.S. Dep't of Energy (2014). Inertial confinement fusion, Nat'l Nuclear Security Admin., U.S. DOE, Washington DC.

[US-EIA] U.S. Energy Info. Adm. (1995). Spent nuclear fuel discharges from U.S. reactors 1993, U.S. EIA, Washington, DC.

[US-EIA] U.S. Energy Info. Adm. (2012). State nuclear profiles, 2010, U.S. EIA, Washington, DC.

[US-EIA] U.S. Energy Info. Adm. (2013). Updated capital cost estimates for utility scale electricity generating plants. U.S. EIA, Washington, DC.

[US-EIA] U.S. Energy Info. Adm. (2014a). What is the efficiency of different types of power plants. U.S. EIA, Washington, DC.

[US-EIA] U.S. Energy Info. Adm. (2014b). State energy consumption estimates, 1960 through 2012. U.S. EIA, Washington, DC.

[US-EIA] U.S. Energy Info. Adm. (2014c). Annual Energy Outlook 2014: Levelized cost and levelized avoided cost of new generation resources in the Annual Energy Outlook 2014. U.S. EIA, Washington, DC.

[US-NRC] U.S. Nuclear Regulatory Commis. (1997a). NRC Special team inspection to review control rod incident at Commonwealth Edison's Zion nuclear power station, Doc. No. RIII-97-20, U.S. NRC Office of Public Affairs, Region III, Lisle, IL.

[US-NRC] U.S. Nuclear Regulatory Commis. (1997b). NRC staff proposes $330,000 fine against Commonwealth Edison for violations of NRC requirements at Zion nuclear station. News Anon. No. RIII-97-79, U.S. NRC Office of Public Affairs, Region III, Lisle, IL.

[US-NRC] U.S. Nuclear Regulatory Commission.(2003) Site area emergency declared due to UF_6 release, 7 Pages Event Notification Report, issues on December 23, 2003, U.S. NRC, Washington, DC.

[US-NRC] U.S. Nuclear Regulatory Commis. (2004). NRC renews license for interim spent fuel storage installation at G.E Morris facility in Illinois. News Anon. No. 04-166, U.S. NRC, Washington, DC.

[US-NRC] U.S. Nuclear Regulatory Commis. (2011). Dresden Unit 1, Docket No. 50-0010, U.S. NRC, Washington, DC.

[US-NRC] U.S. Nuclear Regulatory Commis. (2012). Conceptual design of Yucca Mountain disposal plan. U.S. NRC, Washington, DC.

[US-NRC] U.S. Nuclear Regulatory Commis. (2013a). Zion Units 1 & 2, Docket No. 50-295& 50-304. U.S. NRC, Washington, DC.

[US-NRC] U.S. Nuclear Regulatory Commis. (2013b). Typical dry cask storage system. U.S. NRC, Washington, DC.

[US-NRC] U.S. Nuclear Regulatory Commis. (2014a). Backgrounder on Uranium Enrichment. U.S. NRC, Washington, DC.

Van Gosen, B. S., Gillerman, V. S., and Armbrustmacher, T. J. (2009). Thorium deposits of the United States-energy resources for the future? USGS Circular 1336.

Wenisch, A., Kromp, R. and Reinberger, D. (2007). Fiction or science: Is there a future for nuclear? Austrian Inst. of Ecology, Vienna, Austria.

Wernau, J. (2011a). Illinois options for replacing coal-fired power plants. News section, *Chicago Tribune*, Chicago, IL.

Wernau, J. (2011b). Exelon chief isn't wasting energy worrying about new nuclear regulations. News section, *Chicago Tribune*, Chicago, IL.

Whitaker, J. M. (2005). Uranium enrichment plant characteristics and training manual for the IAEA. Nuclear Sci. & Tech. Div., Oak Ridge Natl Lab., Oak Ridge.

[WISE] World Info. Service on Energy. (2014). Areva Eagle Rock uranium enrichment plant project (USA), NRC Docket No. 07007015. WISE Uranium Proj., Arnsdorf, Germany.

Wittneben, B.B.F. (2012). The Impact of the Fukushima Nuclear Accident on European Energy Policy, *Environmental Science & Policy*, **15**(1): 1–3.

[WNA] World Nuclear Assoc. (2012a). Thorium. WNA, London, UK.

[WNA] World Nuclear Assoc. (2012b). Nuclear power reactors. WNA, London, UK.

[WNA] World Nuclear Assoc. (2012c). Nuclear fusion reactors. WNA, London, UK.

[WNA]World Nuclear Assoc. (2013a). US Nuclear power policy. WNA, London, UK.

[WNA] World Nuclear Assoc. (2013b). Fast Neutron Reactors. WNA, London, UK.

[WNA] World Nuclear Assoc. (2013c). Small nuclear power reactors. WNA, London, UK.

[WNA] World Nuclear Assoc. (2014). Nuclear power in the USA. WNA, London, UK.

[WNA] World Nuclear Assoc. (2014a). Uranium enrichment. WNA, London, UK.

[WNA] World Nuclear Assoc. (2014b). U.S. nuclear fuel cycle. WNA, London, UK.

[WNA] World Nuclear Assoc. (2014c). Uranium from Phosphates. WNA, London, UK.

[WNA] World Nuclear Assoc. (2014d). Supply of uranium. WNA, London, UK.

[WNA] World Nuclear Assoc. (2014e). Processing of used nuclear fuel. WNA, London, UK.

[WNA] World Nuclear Assoc. (2014f). Fukushima accident. WNA, London, UK.

[WNN] World Nuclear News (2011). Heads up on new tails partnership. WNN, London UK.

[WNTI] World Nuclear Transport Inst. (2011). What is transported and how? Nuclear Transport Facts, WNTI, London, UK.

Zinkle, S. J. and Busby, J. T. (2009). Structural Materials for Fission & Fusion Energy, *Mater Today*, **12**(11): 12–19.

7

Biofuels

"We need never fear the exhaustion of our present fuel supplies, so long as we can produce an annual crop of alcohol to any extent desired" **Alexander Graham Bell**, 1917, National Geographic Magazine.

7.1. Introduction

Biofuels are liquid, solid, or gaseous fuels made from renewable biological materials and are one of the more promising forms of energy for the replacement of fossil fuels (Lee *et al.*, 2014; Speight, 2011). Global production of biofuels has been growing steadily over the last decade from 16 billion liters in 2000 to around 110 billion liters in 2013. Today, biofuels provide roughly 3.5% of total road transport fuel globally (IEA, 2014). Biofuels development relies heavily on country-specific programs or mandates and outlooks for transportation fuels. Energy Independence and Security Act of 2007 (EISA) defines three classes of biofuels; first generation (conventional), second generation (cellulosic) and third generation (advanced).

The first generation biofuels are derived from feedstock that are unsustainable in long-term. These include ethyl alcohol (ethanol) produced from fermentation of carbohydrates (from corn or other grains) and sugars using enzymes and microorganisms (yeast), biodiesel, which is a product of transesterification of vegetable oils or fats and green diesel produced from canola oil using fractional distillation (Wang and Mansoori, 1994; Mansoori, 1999; Khoury, 2014).

The second generation (or cellulosic) biofuels are fuels made from materials containing cellulose. Cellulose, a complex carbohydrate, along with hemicelluloses and lignin are the organic matter found in plant walls (Sims *et al.*, 2010). Cellulosic feedstock includes agricultural residues (e.g., corn stover), forestry residues (e.g., wood chips), dedicated energy crops (e.g., switchgrass, hybrid poplar) and urban sources of waste (e.g., municipal solid waste (MSW)). Cellulosic biofuels technology is expected to reduce greenhouse gas(GHG) emissions by at least 60%, as compared to gasoline (USDA, 2010).

The third generation biofuels are related to algal biomass but could, to a certain extent, be linked to utilization of CO_2 as a feedstock.

Use of liquid biofuels as energy source in the U.S. dates back to the years when oil lamps and cooking oil stoves and heaters were used as the general means of lighting, cooking and heating. They included all types of vegetable oils such as castor, rapeseed, peanut; animal oils, especially whale oil and tallow from beef or pork; refined turpentine from pine trees; and alcohols, especially wood alcohol such as methanol (methyl alcohol) and grain alcohol, such as ethanol (ethyl alcohol). The more commonly used fuel in the U.S. prior to the use of petroleum fuels was a blend of alcohol and turpentine called "camphene", which was simply referred to as "burning fluid" (Kovarik, 1998). These were early attempts to use biofuels as transportation fuel. In 1826, Samuel Morey invented a prototype of the first American internal combustion (IC) engine powered by a blend of ethyl alcohol and turpentine (Abbe, 2008). But Morey's work was curtailed by lack of funding and the world's interest directed towards steam engine locomotive using coal (Kovarik, 1998). In 1896, Henry Ford built the first automobile (a quadricycle) to run on pure ethanol. But a $2.08 per gallon liquor tax introduced earlier to pay for the American Civil War, which was still in effect at the time of Ford's automobile, was an impediment to the success of his invention. The liquor tax made ethanol more expensive, forcing people to use coal oil (which was known as kerosene) that was taxed only 10 cents per gallon (Kovarik, 1998; Siegel, 2012). Coal oil and kerosene were often assumed to be the same thing; a clear, liquid fuel used for lamps and cooking. In the early years of the oil industry, the two names were often used synonymously.

After the repeal of the liquor tax in 1906, Ford declared ethanol as the fuel of the future, and in 1908 introduced the Model T vehicle to run on ethanol with gasoline as an option (Abebe, 2008; Solomon *et al.*, 2007). But, the euphoria surrounding ethanol as the fuel of the future was only ephemeral. This was because the discovery of new oil fields by independent oil companies introduced cheaper gasoline on the market (Kovarik, 1998). Nevertheless, in 1917, Alexander Graham Bell, while addressing the graduating class of the McKinley Manual Training School in Washington, D.C., on the theme "Prizes for the inventor: some of the problems awaiting solution", mentioned alcohol as one source of fuel supply which may solve our future energy problem. According to A. G. Bell (1917) "alcohol makes a beautiful, clean, and efficient fuel, and where not intended for consumption by human beings, it can be manufactured very cheaply in an indigestible or even poisonous form" (Bell, 1917). Based on this statement, Bell predicted that the world would probably depend on alcohol more and more as time goes on, with opportunities opening up for engineers who will modify our machinery to enable alcohol to be used as the source of power (Bell, 1917). Among the possible feedstock mentioned by Bell for alcohol production included sawdust, corn stalks, waste products from farms, garbage from our cities, and almost any vegetable matter capable of fermentation (Bell, 1917).

In the late 1970s, the U.S. President Jimmy Carter, while responding to the crisis in the Middle East and its ripple effect on global crude oil prices, called for the production of large quantities of alcohol fuel to help solve the U.S. energy problems (Abebe, 2008; Solomon *et al.*, 2007). Indeed, the fear of uncertainties surrounding global crude oil supply, which is unsustainable and the concern for the environment, necessitated introduction of some energy policies by the U.S. Congress, in the hope of leading the nation towards energy independence and cleaner environment.

In 1992, the Energy Policy Act (EPAct92) was passed to reduce the nation's dependence on imported petroleum and natural gas (Public Law 102–486, 102nd Congress, 1992). The law required certain fleets to acquire alternative fuel vehicles (AFVs), capable of operating on non-petroleum fuels. In 1998, the Energy Conservation Reauthorization Act (ECRA) was passed as an amendment to the EPAct, to allow credit for the use of B20 or petroleum diesel fuel blended with 20% of biodiesel (42 U.S.C 13220 U.S.C HCC, 1998). In 2006, the US-EPA introduced the Renewable Fuels Standards (RFS) Program to encourage use of biodiesel and ethanol in order to double the consumption of these fuels by 2012 (Siegel, 2012). Also in 2007, the Energy Independence and Security Act (EISA2007) was passed to enable the U.S. attain energy independence by systematically weaning itself from fossil fuel imports (Public Law 110–140, 110th Congress, 2007).

In essence, the EISA 2007 called for a five-fold increase in ethanol production by 2022, an equivalent of 36 billion gallons of ethanol, and provided tax incentives and grants to ethanol producers (Byrge and Kliesen, 2008). Out of the 36 billion gallons, 15 billion gallons were proposed to be obtained from conventional biofuel sources, such as corn ethanol. Developments in biofuels technology (cellulosic and biomass-based diesel) are expected to produce the remaining 21 billion gallons. The contribution of biomass-based diesel to the 21 billion gallons goal can be no less than 1 billion gallons and will be determined at a later date by rulemaking. Additionally, 4 billion gallons of advanced biofuels (defined as biofuels that can reduce GHG emissions by at least 50%) is mandated by the EISA 2007 by the year 2022 (USDA, 2010).

Under the EISA 2007, which is mentioned above, all the states in the U.S. are divided into five regional groups. Each of these groups is expected to make regional biofuels contribution toward the 21 billion gallons a year of the RFS2 (U.S. EPA's amended Renewable Fuel Standards, the "2"represents the amendment to RFS1) advanced biofuels goal by 2022. Illinois is part of the Central-Eastern region group, which also includes Delaware, Indiana, Iowa, Kansas, Maryland, Missouri, Minnesota, Nebraska, North Dakota, Ohio, Oklahoma, Pennsylvania, South Dakota, Virginia and Wisconsin. In 2022, the annual advanced biofuels contribution to the RFS2 expected from these Central-Eastern regional states is at least 9.09 billion gallons. This volume represents 43.3% of the national target of 500 million barrels (21 billion gallons) per year of advanced biofuels production. The projected contributions anticipated from the remaining regional groups are 49.8% for Southeast region states, 2.0% for Northeast region states, 4.6% for Northwest region states and $< 0.3\%$ for Western region states. Accordingly, the Central-East region of the U.S., including Illinois, is one of the two regions with the most potential for near and long term development of biofuels. Therefore, based on land, infrastructure, feedstock and demand for biofuels, this region is important in the implementation of a successful biofuels market. These states are expected to produce the last 101.19 million barrels (4.25 billion gallons) or 28.3% of the current national cap of 357.14 million barrels (15 billion gallons) of corn-starch ethanol in the RFS2, as defined by the EISA (USDA, 2010).

The US-EPA has projected that in 2022, the Illinois' annual contribution towards the fulfillment of the EISA cellulosic biofuels standard will be 36.31 million barrels (1,525 million gallons), mostly from the Illinois counties listed in Table 7.1, below (US-EPA, 2010). This amount represents about 9.5% of the 380.95 million barrels (16 billion gallons) of the EISA cellulosic biofuels standard.

As of July 2014, there were 13 operating bioethanol plants in Illinois using corn as feedstock, with operating production of 28.98 million barrels of ethanol per year. Also, according to the U.S. Energy Information Administration, there were six biofuel facilities in Illinois producing biodiesel from soybeans, yellow grease, or multi-feedstock (US-EIA, 2014a). In Table 7.2, we report the list

Table 7.1. Projected cellulosic ethanol facilities capacity in Illinois by counties in 2022 (US-EPA, 2010).

Illinois County	(Gallons/yr) ·10^6	(Barrels/yr) ·10^6
Bureau	130	3.10
Carroll	77	1.83
Champaign	89	2.12
Coles	77	1.83
DeWitt	100	2.38
Du Page	128	3.05
Grundy	77	1.83
Iroquois	80	1.90
Knox	89	2.12
Menard	99	2.36
Montgomery	78	1.86
Morgan	67	1.60
Ogle	95	2.26
Richland	81	1.93
Shelby	68	1.62
Tazewell	107	2.55
Washington	85	2.02
Total	**1,527**	**36.36**

of Illinois biofuels companies that are presently active in ethanol and biodiesel production.

Illinois biodiesel production was estimated to be 4.08 million barrels (171.5 million gallons) per year in 2014, with most of the production coming from the six biofuel companies and the rest coming from other individual producers (US-EIA, 2014c). In all, Illinois 1st generation biofuels production (i.e., ethanol and biodiesel production) in 2014 was estimated at 33.06 million barrels per year (US–EIA, 2014b). It is estimated that the state has the capacity to produce 34.4×10^6 barrels (1.44 billion gallons) of ethanol and 4.60×10^6 barrels (193 million gallons) of biodiesel annually (US-EIA, 2014b; Nebraska Energy Office, 2014).

Table 7.2. Illinois biofuels Industry (2014).

Ethanol Production Companies[a,c]	**City**[a,c]	**Feedstock**[a,c,d]	**Production**[a] [bbl/yr].10^6
Abengoa Bioenergy Corp.	Madison	Corn	2.14
Adkins Energy	Lena	Corn	1.07
Archer Daniels Midland	Decatur	Corn	4.12
Archer Daniels Midland	Peoria	Corn	0.64
Aventine Renewable Energy	Canton	Corn	0.00
Aventine Renewable Energy	Pekin	Corn	3.74
Big River Resources Galva	Galva	Corn	2.38
Center Ethanol Company	Sauget	Corn	1.29
Illinois Corn Processing	Pekin	Corn	2.14
Illinois River Energy	Rochelle	Corn	3.17
Lincolnland Agri–Energy	Palestine	Corn	1.14
Marquis Energy	Hennepin	Corn	2.38
One Earth Energy	Gibson City	Corn	2.38
Patriot Renewable Fuels	Annawan	Corn	2.38
TOTAL			**28.97**
Biodiesel Production Companies[b]	**City**[b]	**Feedstock**[b]	**Production**[b] [bbl/yr].10^6
BioVantage Fuels	Belvidere	Soy, used cooking oil; corn oil	0.08
REG Danville	Danville	High/ low FFA	1.07
Incobrasa Industries	Gilman	Soy Oil	0.71
Stepan Co.-Joliet	Joliet	Soy Oil	0.50
REG Seneca	Seneca	High/ low FFA	1.43
Midwest Biodiesel Products	S. Roxana	Multi-Feedstock	0.29
TOTAL	—	—	**4.08**

(a) Nebraska Energy Office (2014); (b) BBI, 2014; (c) RFA, 2012; (d) Varner, 2008.
FFA = free fatty acids.
(*)= indicates that the company's total capacity is divided among its plants.

In general, use of food crops, such as corn grains for biofuels production can result in increased rate of deforestation. This situation can negatively impact the control of greenhouse gas emissions, as forest plants depend on CO_2 for photosynthesis. Also, use of ethanol as a biofuel would add about \$0.75–1.00 per bushel to the price of corn, which is the mainstay for millions of people around the world (Nelson, 2008). Corn and other agricultural products need to be grown, collected, dried and fermented to produce alcohol. Each of these stages requires use of energy to be administered.

The ethanol fuel energy balance (also called "net energy gain"), which is the difference between "the energy released by burning the resulting ethanol fuel" and "the total amount of fossil energy input into the process of growing corn and producing alcohol from corn" is marginal (Sanders, 2008).

Considering this fact, it may not be economical to produce alcohol from corn and use it as transportation fuel. It should also be mentioned that as farmers are driven by market forces to balance the supply and demand of food crops that are grown for human consumption, they inadvertently divert their land for ethanol production and this practice may become one of the agents of deforestation. The problem of deforestation appears to be inevitable, unless there is an alternative approach for ethanol production which is cellulosic ethanol, or advanced (third generation) biofuels.

The cellulosic (second generation) biofuels are the derivatives of sustainable feedstock. These include cellulosic ethanol, biohydrogen, biomethanol, BioDMF (DMF = 2, 5-dimethylfuran), BioDME (DME = dimethylether) and FT fuels. Requirements for sustainability of a feedstock as defined, among others, include continuous availability of the feedstock, zero or minimum impact on greenhouse gas emissions, as well as on its biodiversity and optimum land use (Pandey *et al.*, 2011). Currently, Illinois is encouraging use of cellulosic (second generation) biofuels because they have positive social and environmental impacts (such as greenhouse gas emissions reduction and avoidance of land competition).

In 2007, a study by the Argonne National Laboratory, located in Illinois, found that using cellulosic ethanol can provide greater benefit by reducing GHG emissions by up to 86%, compared to corn-based ethanol, which reduces life cycle GHG emissions by 19%–52%, depending on the source of energy used during ethanol production process (Wang *et al.*, 2007).

Advanced biofuels can improve the energy balance efficiency, $\eta_{\text{Energy Balance}}$, which we define as,

$$\eta_{\text{Energy Balance}} \equiv \frac{E_{\text{Biofuel}}}{E_{\text{fossil}}} - 1 \tag{7.1}$$

In this equation, E_{Biofuel} is the total energy produced by the biofuel, divided by E_{fossil}, which is the fossil energy needed to produce it. Positive energy balance efficiency means more usable energy is produced than is required to produce the fuel. On the other hand, negative energy balance efficiency means the fuel takes more fossil energy to produce than it yields.

Most recent studies have concluded that the net energy balance is approximately 1.4 for corn ethanol production, 8.32–10.22 for sugar cane ethanol and ranges from 2–36 for cellulosic ethanol production, depending on the production method (Kurki *et al.*, 2006; Macedo *et al.*, 2004; Worldwatch Institute, 2007). In general, corn production in the U.S. requires a significant amount of energy and financial investment. It is estimated that the average corn yield of 8,655 kg/ha (ha = hectare) using average production technology requires about 8.1 million kcal of energy, with production costs of about $917/ha. However, production of one liter of ethanol requires 29% more fossil energy than the energy derived from the ethanol fuel, which cost about $1.59 per gallon. The corn feedstock production alone requires nearly 50% of the energy input for ethanol fuel production (Pimentel and Patzek, 2005). In Table 7.3, we report the data for energy balance efficiencies for U.S. liquid fuels.

In what follows, we report on biofuels production and consumption in Illinois, the economic and environmental impacts that

Table 7.3. Data for energy ratios efficiency, $\eta_{\text{Energy Balance}}$, for the U.S. energy production feedstocks.

Liquid fuels	$\eta_{\text{Energy Balance}}$
Corn ethanol[1,4]	0.4
Sugar cane ethanol[3]*	7.32–9.22
Cellulosic ethanol [2]	1–35
Soy biodiesel [1,4]	−0.02 – 2.2
Diesel oil [4]	−0.16
Gasoline [1,4]	−0.19

(1) Larson (2008); (2) Worldwatch Institute (2007); (3) Macedo *et al.*, (2004); (4) Kurki *et al.*, (2006).
* Data obtained from the production in Brazil.

determine the measures needed to improve the biofuel industry in the state and the data on biofuels production and consumption, which may provide guidelines on how the state could meet the EISA biofuels standard. Then, we report on available and potential biomass resources in the state together with quantities and compositions of different types of sustainable biomass feedstock available for the state's future biorefineries. In addition, we present a discussion pertaining to the potential of abandoned agricultural land areas available in the state, which can be used for energy crops cultivation. Furthermore, we discuss use of urban waste or MSW as a potential biomass feedstock for cellulosic ethanol production and discuss composition of cellulosic polymeric biomass, including those processes that may be used for the conversion of biomass into liquid biofuels.

7.2. Ethanol Production and Consumption

Ethanol (CH_3-CH_2OH) is obtained from grain or cellulosic biomass and can be blended with motor vehicles gasoline to reduce the total emissions of CO_x and NO_x gases by 10–20% (Maiorca, 2009).

This is possible because ethanol contains about 35% by weight oxygen, which serves as oxygenate when blended with gasoline. The blending allows more complete combustion of the fuel mixture and results in lower emissions. Complete purification of alcohol from water and its blending with petroleum fractions to achieve a mixture with the desired property requires knowledge about associating/hydrogen-bonding solutions and many-component (continuous) mixtures (Aghamiri *et al.*, 2001; Hamad and Mansoori, 1990; Massih and Mansoori, 1983; Pires *et al.*, 2001). Another reason for blending ethanol with gasoline is to improve the octane rating of gasoline (Prianto *et al.*, 1997). The octane number is an antiknock index, which is a measure of how much temperature and pressure is needed to ignite the air/fuel mixture. The standard compounds against which the gasoline fuel is rated are isooctane (with octane rating of 100) and n-heptane (with octane rating of 0). Fuels with higher octane rating or antiknock characteristic closer to that of isooctane, like ethanol, can prevent premature ignition of the fuel. The premature ignition can create a pinging or knocking noise in the engine, leading to propagation of stress in the engine with potential damage (Szulczyk, 2010). Lighter aromatic hydrocarbons contained in petroleum were useful in improving gasoline octane number, but they were found to be carcinogenic and needed to be de-aromatized before use (Vakili-Nezhaad *et al.*, 2000). Addition of methyl tertiary-butyl ether (MTBE) to gasoline to improve octane number is also found to have environmental hazards (Rosell *et al.*, 2006). In addition to improving octane number rating, ethanol has no environmental hazards associated with it.

In 2009, Illinois consumed 11.22×10^6 barrels of ethanol as shown in Fig. 7.1, representing 4.3% of the U.S. share (US-EIA, 2014a). The sudden drop in ethanol production in mid-1996 was due to late planting caused by wet weather conditions. This resulted in late corn crop development and higher market prices for corn (US-EIA, 1997). In 2009, Illinois produced 30.498×10^6 barrels

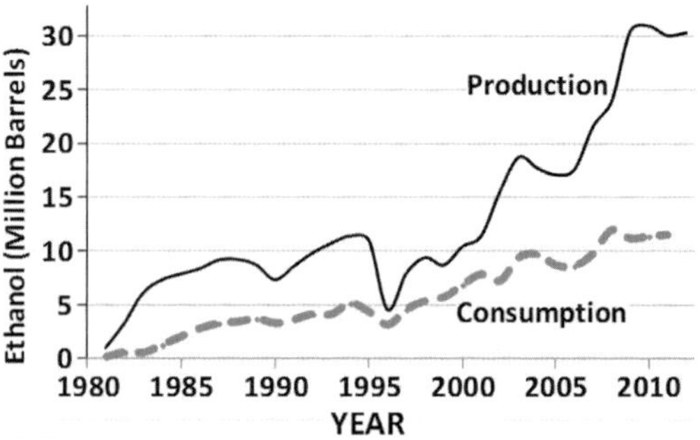

Fig. 7.1. Profile of ethanol production and consumption as automobile fuel in Illinois (Data from US-EIA, 2014a; US-EIA, 2014b).

of ethanol (including denaturants, the material added to ethanol to make it unsuitable for beverage use, but for automotive use) (see Fig. 7.1) (US-EIA, 2014b). Many gasoline stations in Illinois sell a blended fuel called E10, typically made up of 10% denatured ethanol and 90% gasoline, which vehicles can use without modifications to the engine fueling system. Vehicles with modifications to the fueling system called "Flex fuel vehicles" can use E85, or a blend up to 85% denatured ethanol and 15% gasoline.

By the end of 2010, more than 90% of the gasoline used in the U.S. was blended with ethanol (RFA, 2012). In May 2010, Illinois opened the first blender pump in Sullivan, Southern Illinois, to become the first of 20 of such pumps planned for the state in a pilot program approved by the Illinois Department of Agriculture (Zimmermann, 2010). Blender pumps draw fuel (E10 and E85) from two separate storage tanks and dispense pre-programmed blends of the two fuels. These pumps offer FFVs owners a variety of ethanol-blended gasoline products between E15 and E85. The most common blends selected are the E20 (20% ethanol, 80% gasoline) and E30 (30% ethanol, 70% gasoline) (US-DOE, 2013a).

According to the Illinois Institute for Rural Affairs, in 2011, the Illinois ethanol industry contributed $5.3 billion to the state's economy through purchasing of 18.95 million metric tons (746 million bushels) of Illinois corn and through the sales of 39.85 million barrels

(1.674 billion gallons) of ethanol and 6.25 million tons of Dried Distillers Grains with Solubles (DDGS). In the same year, the industry created almost 4000 jobs categorized as 1172 direct employments, 1334 indirect employments and 1493 induced employments (IIRA, 2012). Also, a plan introduced by the governor of Illinois in 2006 expects 50% of Illinois motor fuels to come from local crops and coal by 2017. Additionally, the plan expects every gas station in Illinois to provide E85 (85% ethanol) fuel, which is claimed to create 30,000 new jobs in downstate Illinois (Maiorca, 2009). The high-level ethanol blend (E85) is an alternative fuel used in flexible fuel vehicles (FFVs). A FFV is a vehicle type with IC engine that runs on gasoline or ethanol blend (US–DOE, 2013a). In 2009, the number of FFVs in Illinois reached over 300,000 (RFA, 2012).

Table 7.4 shows the properties of ethanol with those of gasoline (Jessen, 2013). At warm temperatures, the volatile components of gasoline may evaporate into the atmosphere, which would act as greenhouse gases contributing to the depletion of atmospheric ozone layer. This is because lighter hydrocarbons have lower vapor pressures (Edalat *et al.*, 1993, 1997; Hartono *et al.*, 1999). In order to prevent this from happening, the US-EPA prohibits the sale of high vapor pressure gasoline in hot climate regions (Greenhouse, 2011). These standards apply to all consumers during the hot weather season in Illinois effective June 1 to September 15 every year. This includes Cook, Du Page, Kane, Lake, McHenry, Will, and parts of Kendall and Grundy Counties.

In 1990, under the Reformulated Gasoline Program (RFG) currently applied in Illinois and 16 other states and the District of Columbia, the U.S. Congress allowed a "One-Pound Waiver", to accommodate the gasohol industry's practice of "splash blending" 10% ethanol with conventional gasoline. The "One-Pound Waiver" means that fuels containing "gasoline and 10% denatured anhydrous ethanol" can exceed one pound per square inch (6.894 kPa), the set vapor pressure limit (Greenhouse, 2011).

Table 7.4. Fuel properties of gasoline, ethanol and ethanol/gasoline blend.

	Gasoline	Ethanol + Gasoline Blend			Ethanol
Property	**E0**	**E10**	**E20**	**E85**	**E100**
Molecular weight	100–105[1]	~ 104.7	~ 98.18	—	46.07[3]
Oxygen content [% by weight]	~ 0[1]	3.5[6]	7[6]	30[10]	34.7[1,5]
Heat of combustion (HHV), [MJ/kg]	43.6–47.3[1]	42.45[9]	40.67[9]	33.1[11]	29.64[2]
Energy density (HHV) [MJ/L]	34.8[1]	33.18[8]	32[8]	25.6[11]	23.45[1]
Specific gravity at 15°C	0.72–0.78[1]	0.751[8]	0.761[8]	0.78[10]	0.796[3]
Research Octane number	85–100[1,5]	93[8]	94[7,8]	104–106[8]	92–129[1,5,8]
Flash point (closed cup) [°C]	−42.8[1]	−40[8]	−20[8]	−28.9 to −20[12]	12.8[1]
Autoignition temperature [°C]	257.2[1]	260[8]	279[8]	454[13]	422.8[1]
Reid Vapor Pressure [kPa] at 37.6°C (100°F)	48–103.4[1,5]	59.53[9]	54.61[9]	48.3–82.7[10]	15.9[1,4]

(1) US–DOE, Alternative Fuels Data Center (2013b); (2) Wang (2010b); (3) Yacoub *et al.* (1997); (4) Jones (2013); (5) Szulczyk (2010); (6) Naik *et al.* (2010); (7) Ansari (2012); (8) Tangka *et al.* (2011); (9) Gravalos (2011); (10) US-DOE(2013c); (11) Demirel (2012); (12) MDNR (2008); (13) Birath *et al.* (2008).
Energy-volume density = energy obtained from the fuel per unit volume;
HHV = Higher heating value.

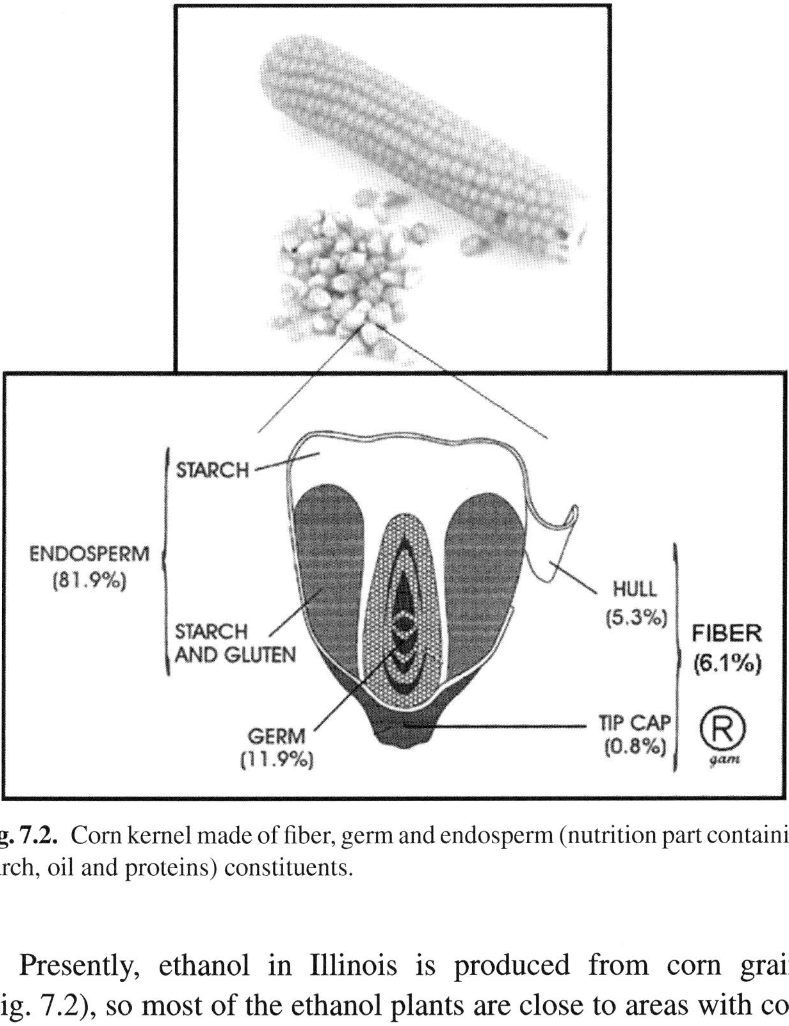

Fig. 7.2. Corn kernel made of fiber, germ and endosperm (nutrition part containing starch, oil and proteins) constituents.

Presently, ethanol in Illinois is produced from corn grains (Fig. 7.2), so most of the ethanol plants are close to areas with consistent supply of grains. Several technological advancements have contributed to the success of ethanol processing methods to improve yield and reduce operating costs. Among those technologies, the most cost-effective is corn fractionation, which is commonly used in wet milling.

The basic principle for corn fractionation (see Fig. 7.3) consists of separating corn kernel into fibers, germ and endosperm (nutrition part containing starch, oil and proteins) as shown in Fig. 7.2.

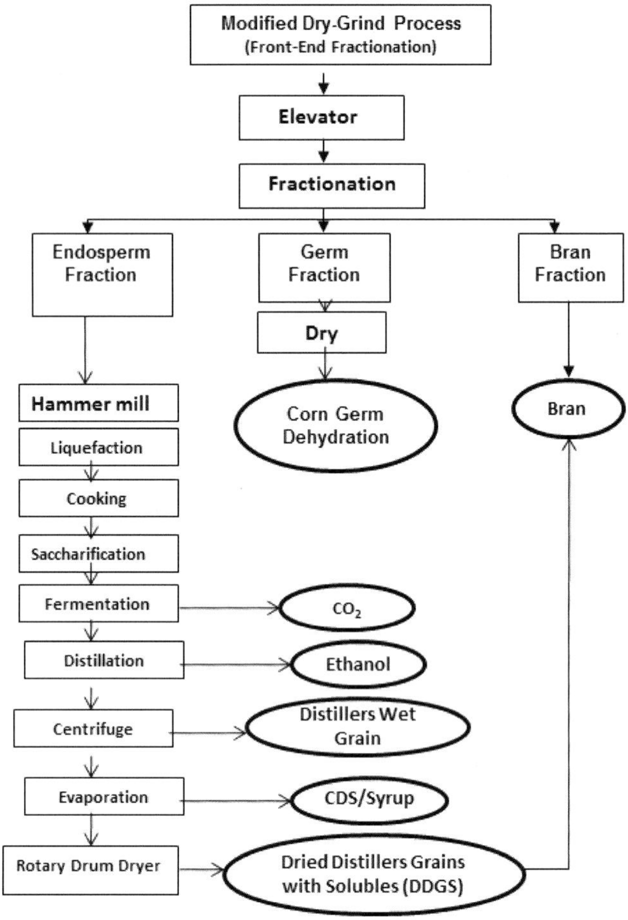

Fig. 7.3. Ethanol production from corn through fractionation technology (Bista *et al.*, 2008).

The method enhances starch availability for ethanol production, including increase in protein concentration of the byproducts. There are also some claims that this method can increase process profitability through higher ethanol yield, including emissions reduction (Bista *et al.*, 2008).

Ethanol plants with production outputs significantly higher than 1.5×10^6 barrels ($>$ 60 million gallons) per year tend to be "major" sources of air pollution, releasing CO_2 in the air (IEPA, 2008). Consequently, building an ethanol plant in an area around a major city in Illinois may face stringent emissions regulations. The plant must be equipped with very stringent control devices, such as the best available control technology for lowest emission rate, and air quality monitoring and acquisition of carbon emission permits (IEPA, 2008). Furthermore, preliminary testing must be performed on the plant to ensure compliance to the emission limitations imposed by the law. This requirement is necessary as a precondition for long-term operation of the plant. The plant owner can apply for operating permit only when the stack test confirms that the plant is in compliance with the emission limits.

Bioethanol is known as “green” energy, but mismanagement of ethanol plants can wreak havoc on the environment. In April 2010, an ethanol production plant in Canton, Illinois was cited by EPA after discovering ash from the facility, including run-off from an adjacent grain processing and storage site that was contaminating a nearby lake leading to the death of fish, turtles and other wildlife (Jellen, 2010).

7.3. Biodiesel Production and Consumption

Biodiesel is a clean-burning renewable diesel fuel produced from biomass (organic compounds made-up of long-chain alkyl groups) normally used in diesel engines. There are also biodiesel blends, which consist of mixtures of petroleum-based diesel fuel and biodiesel (Verry, 2010).

Biodiesel contains 8% less energy per gallon than typical No. 2 diesel in the U.S. (US-DOE, 2006b). As a result, compared to conventional diesel fuel, biodiesel mileage is lower. Biodiesel can reduce carbon dioxide lifecycle emissions by up to 78%, since carbon dioxide is taken up by the annual production of crops, such as soybeans, and is released when vegetable oil-based biodiesel is combusted. Also, compared with regular diesel fuel, biodiesel can reduce carbon monoxide (CO) emission by 48% and particulate matter (PM) emission by 47% from most modern four-stroke compression ignition engines (Maiorca, 2009; Verry, 2010). This is because biodiesel fuel contains up to 11% oxygen by weight, which allows the fuel to burn better, resulting in fewer unburned fuel emissions. For the same reason, air toxics are reduced because they are associated with unburned or partially burned hydrocarbon and PM emissions. However, at lower ignition temperatures, biodiesel has emissions and fuel system problems. NO_x emissions from biodiesel combustion can be reduced at lower ignition temperatures, but particulates emission increases due to unburned carbon (Kotrba, 2005). According to a report by the National Renewable Energy Laboratory (NREL), tertiary butylhydroquinone, an antioxidant, will be an effective NO_x-reducing agent with slight PM increases when biodiesel is used in internal combustion engines (Kotrba, 2005). For on-road diesel application, even low-level blends of biodiesel (1–2%) would enhance the lubricity of diesel fuels. This is particularly useful when using ultra-low level sulfur diesel (ULSD), as these fuels generally have poor lubricating properties. For example, a 2% biodiesel can restore adequate lubricity to kerosene or FT diesel which are "dry fuels" (US–DOE, 2006b). In Table 7.5, we present the EPA's surveyed and averaged emissions studies of regulated and non-regulated emissions reductions for burning of biodiesel in diesel engines intended for on-road use. The negative percentages represent emissions reductions of biodiesel blends compared to that of conventional diesel fuel, and it applies to all the data in the table.

Biomass compounds based on which biodiesel is produced include domestic organic waste and renewable resources like vegetable oil (e.g., soybean oil), used restaurant greases and animal fats. To produce biodiesel, which is chemically known as fatty acid methyl ester (FAME), these fats and oils are esterified with a light alcohol (such as methanol or ethanol) and catalyzed with a base (such as sodium hydroxide or potassium hydroxide) to produce biodiesel and a glycerin co-product as represented by Eqn (7.2) (Holmgren *et al.*,

Table 7.5. EPA's average percentages of biodiesel fuels (B20, B100) emission reductions compared to conventional diesel fuel (B0).

		Biodiesel Type	
	Emission Type	**B100**	**B20**
Regulated	Total Unburned Hydrocarbon(1)	−67%	−20%
	CO(1,2)	−48%	−12%
	Particulate Matter(1,2)	−47%	−12%
	NOx	+10%	−2 − +2%
Non-Regulated	Sulfates (1)	−100%	−20%(*)
	PAH (1)(**)	−80%	−13%
	nPAH (nitrated PAH's)(1)(**)	−90%	−50%(***)
	Ozone Potential of speciated HC (1)	−50%	−10%

(1) Verry (2011); (2) Maiorca (2009).
(*) *Estimated using sulfates emission data from B100.*
(**) *Average reduction across all components measured.*
(***) *2-nirtrofluorine results were within test method variability.*
PAH=Polycyclic Aromatic Hydrocarbons; Npah=Nitrated PAH.
Note: Both PAH and nPAH compounds are identified as potential cancer causing agents.
Regulated emissions=Air pollutants with permissible emissions standard set by EPA.
Non-Regulated emissions=Air pollutants without permissible emissions set by EPA.

2007a; Gerpen, 2005).

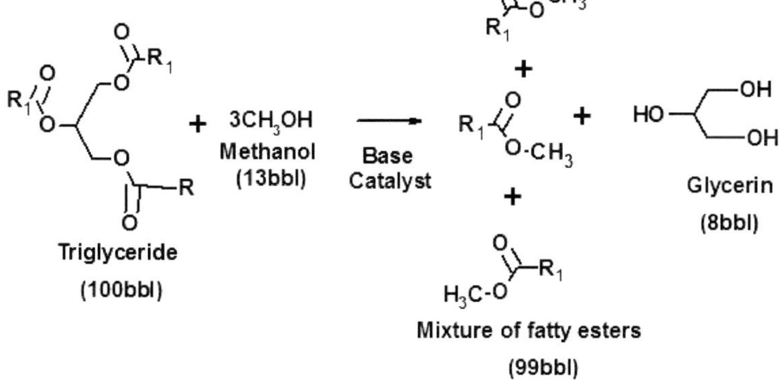

(7.2)

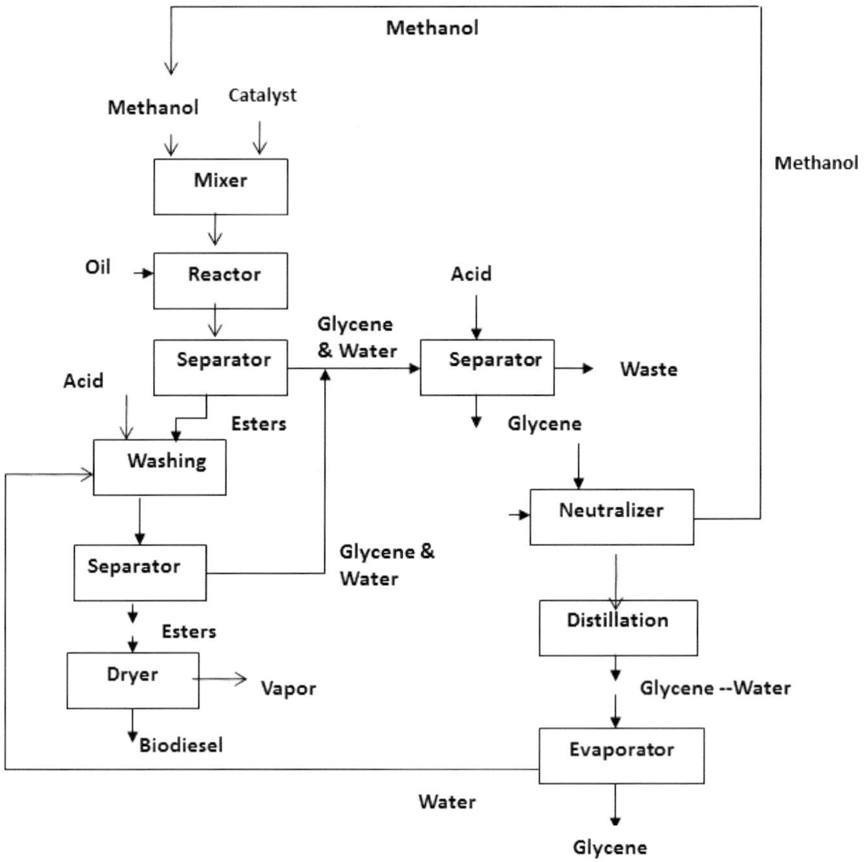

Fig. 7.4. Biodiesel production process from vegetable oil (Kenkel and Holconb, 2008).

The process diagram for biodiesel production from vegetable oils is shown in Fig 7.4 (Kenkel and Holcomb, 2008). While FAME biodiesel has many desirable properties such as high cetane number (the approximate equivalent of octane rating for gasoline which is the rate of fuel combustion quality), there are limitations with its application. These include poor stability and high solvency that can lead to filter plugging problems (Holmgren, *et al.*, 2007a). Consequently, a different process approach (described in Sec. 7.6.7) for producing diesel from vegetable oil has become necessary to convert vegetable oils into a high-quality diesel fuel that is fully compatible with the desired performance of a diesel fuel.

Illinois ranks first in biodiesel consumption in the nation, but lags behind in its production compared to Iowa and Texas. In 2013, the state's annual biodiesel production capacity was 3.98 million barrels (167 million gallons), lagging behind those of Iowa and Texas that were 6.9 million barrels (290 million gallons) and 10.26 million barrels (431 million gallons) per year, respectively (US-EIA, 2013a).

Biodiesel may be used alone (known as B100), or because of its chemical nature, it may be blended with any kind of distillate or diesel fuel in appropriate proportion. This includes light fuels such as kerosene, jet fuel, No.1 diesel, or military fuels (JP8, JP5), as well as normal diesel fuel like No. 2 diesel for diesel engines or gas turbines, and heating oil for boilers in home heating (US–DOE, 2006b). B20 (20% by volume biodiesel and 80% petroleum diesel fuel) is the more commonly used biodiesel blend in the U.S. B20 provides a good balance between material compatibility, cold weather operability, performance, emission benefits and costs. B20 is also the minimum blend level allowed for compliance with the EPAct 1992, which requires the use of renewable fuels and/or AFVs for compression-ignition engines, fuel oil and heating oil boilers and turbines (US-DOE, 2009).

Biodiesel is blended into diesel fuel either at the user end or by a jobber or distribution company and offered for sale as a finished blend (usually up to B20). The blending can also take place at petroleum terminals or racks by a pipeline or terminal company and offered as a finished blend. The product is sold directly to customers or to a petroleum jobber (marketer) or distribution company for further sale to customers. The procedures for biodiesel blending depend on several factors, including the volume of B100 required to make the blend, the volume of blended products being sold, the finished blend level, tank and space availability, operational and equipment costs and customer requirements for blends, both present and in the future.

Biodiesels for blending in the U.S. should meet ASTM D6751 specification before blending (US-DOE, 2009).

In the U.S., including Illinois, the three general techniques used for biodiesel blending with petroleum diesel are the splash blending, in-tank blending and in-line blending (US-DOE, 2006b; Sine Nomine Group, 2007).

In 2003, Illinois passed a sales tax incentive bill (SB46) for the use of biodiesel blends. The bill allowed a partial sales tax exemption of 20% (off the state's 6.25%) for biodiesel blends of 1–10%, and an exemption for the state's entire 6.25% sales tax for blends reaching 11% or higher (Verry, 2010). The immediate impact of this bill was that it helped expand the market for biodiesel in Illinois. The benefits for using biodiesel produced from soybeans include job creation, as well as tax reduction for farmers and biodiesel producers.

Despite the benefits derived from soybeans and corn, there are claims that continuous use of these crops for biofuels production may result in land and food shortages with serious repercussions. Also, as we mentioned in the introduction (Sec. 7.1), corn and other agricultural products need appreciable amount of fossil energy for their cultivation, growing, collection, drying, fermentation and processing in producing alcohol or biofuel with marginal net energy gain.

In the years 2007 and 2008, it was reported that the ethanol industry used 16% of the U.S. corn crop for ethanol production (Sticklen, 2010). Consequently, in June 2011 the U.S. senate voted with broad bipartisan support (73–21) to end tax credits and trade protection that benefited soybean-based biofuel or corn-based ethanol industries (N.V., 2011). In what follows, we describe the available biomass resources in Illinois which may be considered for biofuels production.

7.4. Available and Potential Biomass Resources

This category of biomass resources include low-impact crops for biofuels production, algae as potential biomass feedstock, use of abandoned agriculture lands to grow energy crops and urban waste as feedstock for biofuels production.

7.4.1. *Low-impact crops for biofuels production*

According to the U.S. Department of Energy (DOE) "low-impact" crops are agronomic crops that can sustainably be produced over a broad range of geographic area in the U.S., at a cost that is competitive with other lipid producing crops. Such crops would require a minimum amount of water, fertilizer, pesticides, energy, and land area and are useful for producing high quality biodiesel fuel. Low impact crops would not be consumed by humans and production of such crops would not cause excessive soil erosion, loss of wildlife habitat, decrease in productivity in other rotational crops produced on the same land, or cause other environmental pollution or health hazards while requiring few inputs in providing economic, environmental, and social benefits (Winchester *et al.*, 2013). Lipid is referred to as a broad group of naturally-occurring molecules including fats, fat-soluble vitamins, monoglycerides and diglycerides, phospholipids, sterols, waxes, etc.

In essence, low-impact crops could be valuable resources for producing high quality biodiesel fuels. Since they would require a minimum use of fertilizer, water, pesticides, energy and land, they would be much more economical for biofuels production than crops such as corn and soybean. Use of low-impact crops can mitigate the competition for food and land, while having the potential to achieve higher quantities of biofuel per acre of land than currently being achieved using corn or soybeans. Examples of "low-impact crops" are agricultural residue such as corn stover, as well as wood chips and high yielding herbaceous energy crops, such as perennial grasses.

In Table 7.6, we report the total biomass resources that were available in Illinois in 2005.

In Fig. 7.5, we report the Illinois electric energy sector consumption of biomass during 1990–2012.

Table 7.6. Illinois biomass resources in 2005 (Thousand metric tons/year) (Milbrandt, 2005).

Biomass Resources	1000 Metric Tons/Year
Methane from domestic wastewater	21
Methane from landfills	974
Methane from manure management	76
Forest residue	664
Primary mill Residues	233
Secondary Mill Residues	96
Urban wood Residues	1,337
Switchgrass on CRP Lands	5,290
Crop residues	19,593
Total	**28,284**

CRP = Conservation Reserve Program Lands.

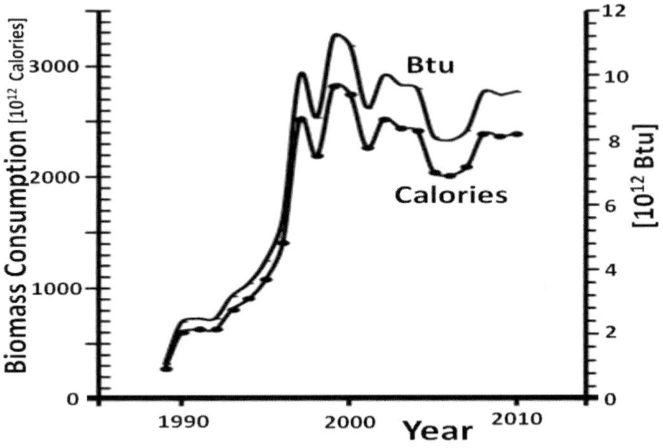

Fig. 7.5. Illinois electric power sector consumption of biomass (Data from US-EIA 2014a) for the period 1990–2012.

7.4.1.1. *Agricultural residue*

Agricultural residues are the leftovers on the field after harvesting of crops. The agricultural residues in Illinois are mainly corn stover, wheat straw, residues of corn, barley, oats, rice sorghum and wheat.

Harvesting of agricultural residues in Illinois, especially corn stover, could provide a substantial amount of feedstock for cellulosic biofuels. However, the amount of residues that can be harvested from the croplands can be reduced by the mechanical efficiency of the harvesting process, as well as the amount of residue needed on the field to maintain the balance of nutrients in the soil (US–EPA, 2010).

a. *Corn stover*

Corn stover is made up of stalks, husk and leaves of maize plants left on the field after harvest. It is a common agricultural biomass found in large quantities in areas of corn production, such as the Illinois corn belt (see Fig 7.6). In 1997, it was estimated that Illinois farmers cultivated 10.7 million acres of corn on 45% of their cropland (English *et al.*, 2000).

The technology for conversion of agricultural cellulosic waste, like corn stover to ethanol, is well developed. The amount of corn stover from these croplands would be enough to feed ethanol plant

Fig. 7.6. Corn stover on the field (from USDOE-NREL).

Table 7.7. Projected ethanol produced to meet EISA in 2022 from corn stover in Illinois (US-EPA, 2010).

Total Harvested Land in 2022 (Million Acres)	**~ 13**
Corn stover yield (dry tons/acre)	5.43
Corn stover available (million tons)	71
Corn stover used (million tons)	15
Corn stover used (%)	21
Ethanol produced (million barrels)	34.38

with an estimated capacity of 291,000 tons per day, when ethanol price is set at $56.7 per barrel (English *et al.*, 2000). Construction of a plant to use 291,000 tons of corn stover per day to produce ethanol will require investment of $2.75 billion. The overall economic impacts in the state will include increased economic activity, creation of 20,208 jobs and moving the state closer to energy-independence (English *et al.*, 2000). By 2022, the US-EPA has projected that corn stover in Illinois can contribute about 34.38 million barrels of cellulosic ethanol per year to meet the U.S. Energy Independence and Security Act (EISA) standard as indicated in Table 7.7 (US-EPA, 2010).

b. *Other agricultural residues*

The other agricultural residues available in Illinois that can be used as feedstock for cellulosic biofuels production are oats, sorghum and wheat. Based on Forestry and Agriculture Sector Optimization Model (FASOM), the US-EPA estimated that these agricultural residues in Illinois will reach 4.0 million wet tons in 2022, as indicated in Table 7.8 (US-EPA, 2010).

7.4.1.2. *Woody biomass*

These types of biomass are categorized as forest residues, primary mill residues, secondary mill residues and urban wood residues (see Fig 7.7). Since woody materials have high energy densities, they can

Table 7.8. FASOM estimated Illinois total agricultural residue feedstock for cellulosic ethanol production possible in 2022 (US-EPA, 2010).

Feedstock	(Million Wet Tons)
Barley	0
Oats	0.1
Rice	0
Sorghum	0.3
Wheat	3.6
Total	**4.0**

Fig. 7.7. Woody biomass (Picture from TheEnvironmentalBlog.org).

produce a large amount of biofuel per ton of feedstock. Below, we describe the different types of woody biomass.

a. *Forest residues*

This category of biomass includes logging residues and other "removals" from forest resources such as unused portions of trees

Table 7.9. Forest residues available for biofuels production in Illinois in 2010 using FASOM model (US-EPA, 2010).

Type of Forest Biomass Residues	Tons Per Year
Logging residue	139,100
Timerland thinnings	115,400
Unused mill residue	18,500
Other removals	117,500
Total	**390,500**

Quantities reported in this table are rounded.

cut, or trees killed by logging and left in the woods. The rest of the removals are considered trees cut or otherwise killed by forestry operations (e.g. pre-commercial thinning, weeding, etc.) or land clearings and forest uses that are indirectly linked with round wood product harvests. However, volume of wood removed from the inventory by reclassification of timberland to productive reserved forestland is excluded.

Illinois available forest residues resources is 664×10^3 dry metric tons per year (Milbrandt, 2005). The U.S. EPA has projected that forest residue in Illinois will contribute 1.38 million barrels (58 million gallons) of cellulosic ethanol per year to meet the EISA standard (US-EPA, 2010) by 2022. Table 7.9 represents the amount of wood residue that would be available in Illinois in 2010, if these residues were recovered.

These projected wood residues are based on Forestry and Agriculture Sector Optimization Model (FASOM) using data from U.S. Forest Service. The U.S. Forest Service data assumes that 50% of logging operations would produce logging residue and 50% would produce forestry thinnings. Also, the U.S. Forest Service data includes unused mill residue, which may not qualify as biofuel feedstock under RFS2 (US-EPA's amended Renewable Fuel Standards), based on the source of the wood. Although this assumption may significantly reduce the amount of wood residue available from current forestry operations, it could not impact the analyses of the data. The reason is because the amount of wood residues used in cellulosic biofuels production, as projected by the FASOM model, is far lower than the total available wood residue in the state (US-EPA, 2010). Also, the data in Table 7.9 suggests that the forests already accessible by roads provide access to low cost forest material from the thinning of timberland. But, getting access to more of the timberland comes with sudden costs increase (US-EPA, 2010).

b. *Primary mill residues*

These types of biomass consist of wood materials (coarse and fine), as well as barks generated at manufacturing plants (primary wood-using mills), when round wood products are processed into primary wood products, such as slabs, veneer clippings, trimmings, edgings, sawdust and pulp screenings. Also included, are mill residues recycled as byproducts, as well as those left unused and disposed of as waste. Illinois available primary mill residues resources is 233×10^3 dry metric tons per year (Milbrandt, 2005).

c. *Secondary mill residues*

The secondary mill residues include wood scraps and sawdust from woodworking shops, wood container and pallet mills, furniture factories and wholesale lumberyards. It also includes wood scraps from wood kitchen cabinet and countertop, non-upholstered wood, household furniture, wood office furniture, wood window and door manufacturers, custom architectural woodwork and millwork. The available secondary mill residues resources in Illinois is 96×10^3 dry metric tons per year (Milbrandt, 2005).

d. *Urban wood residues*

The three main categories of urban wood residues are municipal waste (MSW) wood which includes wood chips, pallets, yard waste, utility tree trimming and/or private tree companies and construction/demolition wood. Illinois available urban wood residues resources is $1,337 \times 10^3$ dry metric tons per year (Milbrandt, 2005). All the above-mentioned woody biomass typically have high lignin contents. This makes them less attractive feedstock for biochemical conversion, but nonetheless they are suitable for thermochemical conversion. In the future integrated biorefinery, the lignin-rich residues from the biochemical process can also be thermochemically converted into bio-oil, coatings and chemical intermediates (US-DOE, 2007). The EPA has identified a number of feedstock pathways (i.e., various feedstock supply chain designs for thermochemical biofuel conversion systems) and imports from countries such as Brazil to meet the US-EPA's amended RFS2. These feedstocks are defined within the context of the EISA of 2007 and include switchgrass, soybean oil, corn oil, crop residues and woody biomass (USDA, 2010). Illinois available annual woody biomass resources totals $2,330 \times 10^3$ dry metric tons, representing 8.23% of the state's total biomass resources of $28,284 \times 10^3$ dry metric tons per year (Milbrandt, 2005). As a potential feedstock for future cellulosic ethanol plants, some of the current and potential applications of woody biomass in Illinois include transformation of the material into pellets or briquettes for home heating, firing of stoves and grills, or for co-firing in power plants.

7.4.1.3. *Dedicated energy crops*

These are renewable energy crops specifically developed and cultivated for biofuels production because of their high energy yields. The energy may be generated through direct combustion or gasification of the crops to create electricity, or through their conversion to liquid fuels such as ethanol for the transportation industry. The geographical location of Illinois makes it one of the most favorable locations for a variety of energy crops cultivation. A number of promising plant species are identified as suitable for liquid fuels which include miscanthus, switchgrass, hybrid poplars, silver maple, reed canarygrass, black locust and sorghum (US-DOE, 2006a). In the continental U.S., there are a number of growing zones with variable factors, such as mean temperature, soil quality and rainfall. However, since there is not a single plant species that is optimal for the entire U.S. growing zones, the use of different species as energy crops is a viable approach to obtain sufficient biomass for large-scale production of liquid biofuels (US-DOE, 2006a). In general, energy crops are classified into two types: (a) Herbaceous energy crops; (b) Short-rotation woody energy crops (Launder, 2002). In the following sections, we describe the characteristics of each of these energy crops.

a. *Herbaceous energy crops*

Herbaceous energy crops are mostly the types of grasses which are harvested like hay. Since these grasses re-grow from their roots, they do not require replanting for long periods of time (15 years or more). They include perennial grasses, such as switchgrass, miscanthus, bluestem, elephant grass and wheatgrass (Launder, 2002). Current research interest is more focused on switchgrass and miscanthus.

a (i). *Switchgrass*

This is a large prairie grass (Fig. 7.8) which grows in the Midwest and can be transformed into cellulosic biofuels. Switchgrass normally attains heights of 3–5 ft; however, it can grow beyond 10 ft in some Southern states in the U.S. The plant is resistant to many pests and diseases and has a deep root system extending several feet below the surface of earth. This allows the plant to tolerate many soil and climate types (US-EPA, 2010).

An evaluation study was conducted on the potential impact of using grass species biomass, like switchgrass, instead of corn as a method of producing ethanol (Ahlberg, 2010; Lynd *et al.*, 1991). The study revealed that growing grass species biomass, such as

Fig. 7.8. Switchgrass harvested, August 2011 (Higgins, 2011).

switchgrass, in the Midwest and in the same cropland area allocated for corn cultivation, could yield an increase in ethanol production, a reduction in nitrogen leaching into the Mississippi River which ends up in the Gulf of Mexico, and a decrease in greenhouse gas emission in the Corn Belt in the Midwest region of the U.S.

Also, it has been estimated that the potential switchgrass production on Conservation Reserve Program (CRP) Lands is 5290×10^3 dry metric tons per year (Milbrandt, 2005). CRP is a voluntary program administered by the USDA Farm Service Agency for agricultural landowners. The program provides financial and technical assistance to eligible ranchers and farmers for the management of water, soil and other related natural resources on their lands (Milbrandt, 2005).

a (ii). *Miscanthus giganteus* (*MG*)

MG is a hybrid and tall perennial grass native to Japan and is currently used in the European Union as a commercial energy crop, mainly as

Fig. 7.9. Miscanthus giganteus (Picture from *insideclimatenews.org*).

a source of heat and electricity, or for conversion into biofuels such as ethanol (see Fig. 7.9).

MG was first planted in Europe in the 1930s as an ornamental introduction from Japan, but it was not until late 1960s that its potential yield for cellulose fiber production was investigated in Denmark (Scurlock, 2009). Subsequent trials for bioenergy production of the plant began in Denmark in 1983, as it spread to Germany in 1987, prior to comprehensive European evaluation. In Illinois and the rest of the U.S., since there is not much experience with this potential energy crop, its performance in the U.S. is still under review. Researchers are studying to determine whether biofuel grasses like miscanthus giganteus could be a viable crop in the U.S. Additionally, the researchers will explore how this viability varies by location (Pyter *et al.*, 2011).

Comparison of the properties of dried herbaceous energy crops (switchgrass and miscanthus giganteus) is shown in Table 7.10.

Table 7.10. Properties of dried herbaceous energy crops.

Property	**Miscantus (*giganteus*)**	**Switchgrass**
Cellulose (%)	40–60[1, 2]	37–38[2, 3]
Hemicellulose (%)	20–40[1, 2]	29–33[2, 3]
Lignin (%)	10–30[1, 2]	17–19[2, 3]
Protein (%)	3[2]	3[2]
—	—	—
Harvestable Biomass (Tons/hectare/year)	8–32[5,6]	9 – 18[5,6]
Ash (%)	2–3.5[1]	2.5–6[2,4,5]
Lower Heating value (MJ/kg DM)*	17.5[5]	17[5]

(1) Brosse *et al.* (2012); (2) Lee *et al.* (2007); (3) Qin *et al.* (2006); (4) Mullen and Boateng (2008); (5) AEBIOM (2008); (6) Heaton *et al.* (2008).
(*) (MJ/kg DM) = Megajoule per kilogram dry mass.

b. *Short-rotation woody energy crops*

These are tree crops grown on short rotations, usually with more intensive management than timber plantations. Short rotation forestry refers to the practice of cultivating fast-growing trees that reach their economically optimum size between 8–20 years (Hardcastle *et al.*, 2006). They include silver maple, cottonwood, black locust, poplar, and willow (Launder, 2002).

b (i). *Black locust*

The black locust tree (*Robina pseudoacacia*) is a nitrogen-fixing legume plant, which is native to southern Illinois and some other parts of the U.S. (McAlister, 1971). It is a potential energy crop for biofuels production. There are more than six ways in which black locust can be utilized as a fiber crop or in producing large amounts of biomass at relatively low energy inputs. These include leaves and young stems for fodder, pulp for paper, leaves and young stems for solid, liquid, or gaseous fuels and extraction of specialty chemicals, such as natural wood preservatives (Hanover, 1993). The initial studies of black locust have shown promising results because the species produced more biomass material than the next closest species by almost three times. Additionally, the species has a rapid growth rate relative to other woody biomass. Based on these initial results, it is anticipated that black locust production may have the potential as energy crop for the Midwest, especially in Northern and Southern Illinois where the land is unsuitable for corn and soybean production (Hanson, 2013a).

b (ii). *Hybrid poplars or willows*

These woody crops have been studied as potential energy crops, due to their high yield rates and adaptation to many growing conditions. It has been estimated that the potential of willow or hybrid poplar production on Illinois CRP lands is ~ 4.5 million dry metric tons per year (Milbrandt, 2005).

Hybrid poplars (*Populus* spp.) are fast growing trees typically grown under intensive control (known as silviculture) and are used as saw and veneer logs, fiber production for the pulp and paper industry, phytoremediation and biofuels feedstock production (White, 2010). Research studies of the hybrid poplars have indicated that the crop can reach a height of 20+ meters (~70 ft) and a diameter of ~30 centimeters (~1foot) within 6 years. Some hybrid poplars yield up to 10 tons of dry biomass per acre per year, which is 5–10 times larger than the yields from natural forests. Because of these attributes, scientists continue to crossbreed hybrids to create trees that will grow faster and are more drought-tolerant and insect resistant (Launder, 2002).

Willows (*Salix* spp.) are fast growing shrubs that can reach between 1–1.5 meters (~3–5 ft) high in the first year of plantation and can be cut to produce multiple stems from the stump the next spring. After the first year, the plant needs to grow for another 3–4 years to about 5–6 meters (~18 feet) before it can be harvested. The harvesting costs are normally lower than other woody crops (Launder, 2002). Also, willow can be planted in temperate climate areas like Illinois, which is a good growing region. Other characteristics that make shrub willows ideal feedstock for biofuels, bioproducts and biopower are ease of propagation from dormant hardwood cuttings, ease of breeding for several characteristics, a broad underutilized genetic base, good chemical composition and energy content, as well as the ability to re-sprout after multiple harvests (Volk, *et al.*, 2010).

Table 7.11 shows a comparison of the properties of short rotation woody energy crops (hybrid poplar, willow and black locust) presented in the above sections.

c. *Pennycress*

Pennycress is an oil seed crop and is a member of the mustard family of plants that can produce seeds containing 20–36% of oil, with 38.1% erucic ($CH_3(CH_2)_7$-CH=CH-$(CH_2)_{11}$-CH_3) as the major fatty acid (WIU, 2014). In Fig. 7.10, field pennycress and perfoliate pennycress are shown.

Table 7.11. Properties of dried short rotation woody energy crops.

Property	Hybrid Poplar	Willow	Black Locust
Cellulose (%)	42–49(1)	43–45(4)	~16.5(5)
Hemicellulose (%)	16–23(1)	17–19(4)	~17.5(5)
Lignin (%)	21–29(1)	25(4)	~7(5)
Protein (%)	2.2–5.3(3)	10–13(5)	22–24(5)
Yield (tDM/(ha/year))	9–16(6)	8–15(6)	5–10(6)
Ash (% dry weight)	1.43–1.5(1,6)	2–2.9(2, 6)	~9(5)
Lower Heating value *(MJ/kg DM)	18.7–19.4(1, 6)	18.5–20(2,6)	19.5(6)

(1) Sannigrahi *et al.* (2010); (2) Lötjönen and Laitinen (2009); (3) Sannigrahi and Ragauskas (2011); (4) Cobas *et al.* (2013); (5) Baertsche *et al.* (1986); (6) AEBIOM (2008).
(*) [MJ/kg DM] Megajoule per kilogram dry mass.

Fig. 7.10. Field pennycress (left) and perfoliate pennycress (middle) (www. illinoiswildflowers.info); pennycress seeds (right) (USDA/ARS)(Hojilla-Evangelist and Evangelista, 2012).

Table 7.12. Composition and physical properties of pennycress and its oil.

Chemical Composition(1)	**Seed**	**Press Cake***
Moisture (%)	8.8	8.4
Oil (wt %)	27.5	5.3
Crude Protein (%)	19.6	25.8
Physical Properties (2)	**Oil**	**Methyl Ethers+**
Pour point (°C)	−18	−15
Cloud Point [°C]	−10	−10
Flash point [°C]	234	136

(1) Selling *et al.* (2013); (2) Isbell (2009).
(*)*A press cake or oil cake is the solid left behind after pennycress seed is pressed to extract the liquids.*
(+)*Methyl ether is a light volatile ether (CH3OCH3) obtained by etherification of methyl alcohol.*

The chemical constituents of pennycress oil (Table 7.12) make it ideally suitable for conversion into biodiesel or green renewable jet fuels.

The cloud point of pennycress biodiesel is −10°C and the pour point is −15°C, compared to soy-based biodiesel, which are −9°C (for cloud point) and −12 to −16°C (for pour point) (Hammond *et al.*, 2005; Isbell, 2009). Pennycress biodiesel is better than soy-based biodiesel in that it has less saturates, making it more stable from oxidation (WIU, 2014). Also, on the same acre of land, pennycress seeds are known to yield two times more oil than soybeans.

The plant can produce between 1.5–2.0 tons of seed per acre of land to yield approximately 1.79–2.38 barrels of biodiesel (Reed, 2011). Pennycress is an oil seed crop that has the potential to produce between 75 to 115 gallons of biodiesel per acre (Reed, 2011; WIU, 2014) (701 to 1075 liter per ha). Assuming that all the marginal lands in Illinois, which according to a research study by Stanford University is estimated at 4 million acres (US-EPA, 2010), could be used for pennycress cultivation, the minimum Illinois annual contribution to the U.S. liquid transportation biofuels would be about 7.1×10^6 barrels.

According to USDA, additional biofuels can be obtained from the press cake of pennycress using pyrolysis process, which increases the yield to about 3.57 barrels per ton of seeds (WIU, 2014). Therefore, USDA has indicated that pennycress has the potential to produce nearly 142.9 million barrels per year of the U.S. liquid transportation biofuels, which will require approximately 40 million acres of land annually (Winchester *et al.*, 2013).

The pennycress plant is grown as a non-food crop in the winter, when Midwest corn belt land is normally left uncultivated, that is between the fall corn harvests and before spring soybean and corn planting seasons (WIU, 2014). Once planted in the winter, the pennycress seed remains dormant till late in the fall before germinating. The reason is insufficient light in the winter to stimulate germination (Bond *et al.*, 2007). In the winter, the leaves of the plant start growing very low on the ground, which prevents soil erosion. By late April and May in spring, the plant begins to flower, reaching a height of about 0.75 meter (∼30 inches) and is ready for harvesting in early June (WIU, 2014).

To transform pennycress into an advanced energy crop similar to soybeans, a start-up company in Peoria, Biofuels Manufacturers of Illinois (BMI LLC), claimed to be collaborating with government researchers, farmers and universities. The company is apparently planning to build a $40 million plant in Mapleton, Illinois, which is claimed to produce 1.5×10^6 barrels (63 million gallons) of biodiesel a year from pennycress. The company is benefiting from a $500,000 Federal Appropriation grant to produce biofuels from pennycress. BMI's claimed plan included creating 150 jobs during construction and 30 jobs when fully operational (Schill, 2008; Tarter, 2009; Alexander, 2011). However, as of early 2013, the starting date for the construction of BMI's multiple-feedstock biodiesel plant was still on hold. The company is also claiming to plan for construction of a futuristic $100 million bioenergy hub to be built in three to five years at the Mapleton site. The claimed plant would crush pennycress to extract the oil and process the meal or press cake. The claimed plant would be designed to blend the oil into biodiesel and convert the cake into energy or biofuel. The bio-hub could crush 600 to 700 tons of pennycress seed per day and produce 20 million gallons of biodiesel/aviation jet diesel and 20 million gallons of bio-oil per year (ATI, 2011).

7.4.2. *Algae as a potential biomass feedstock for biofuels production*

The technology for macro- and microalgae cultivation as biomass and its contribution to reduce carbon dioxide in the air is well established (Mata *et al.*, 2010; Naik *et al.*, 2010; Brennan and Owende, 2010). Algae can efficiently convert carbon dioxide into biomass under solar radiation. It is possible that algae can achieve higher production rates of biomass, compared to land-based crops on the basis of land surface area use. Macro- and microalgae can grow in any kind of water, while utilizing photosynthesis to generate biomass

(Pittman *et al.*, 2011). Also, macro- and microalgae can either be heterotrophic or autotrophic. Autotrophic algae need inorganic compounds such as carbon dioxide and a light energy source for growth (Brennan and Owende, 2010). Heterotrophic algae, however, are non-photosynthetic and require an external nutrient for growth. But some photosynthetic algae are mixotrophic, in that they can acquire exogenous organic nutrients and perform photosynthesis as well (Brennan and Owende, 2010).

Macroalgae, also known as seaweeds, are fast growing multicellular plants that can grow to a size of up to 60 meters in length. Based on their pigmentation, these plants are classified into three broad groups. These are the brown seaweed (Phaeophyceae), red seaweed (Rhodophyceae) and green seaweed (Chlorophyceae). Seaweeds are mainly used for production of food and extraction of hydrocolloids (Wellinger, 2009).

Microalgae are microscopic thallophytic plants (i.e. plants without leaves, stems and roots) (Naik, *et al.*, 2010). The three most important classes of microalgae relative to their abundance are the diatoms (*Bacillariophyceae*), the green algae (*Chlorophyceae*) and the golden algae (*Chrysophyceae*). Diatoms, which are the dominant species of phytoplankton, may be the largest group of biomass producers on earth. Green algae are especially abundant in fresh water and the main storage compound of these algae is starch, but they can also store oils. The golden algae are similar to the diatoms and produces oils and carbohydrates (Wellinger, 2009).

Currently, microalgae are being studied as the third generation biofuel feedstock for biodiesel production (as we mentioned in the introductory section, the first generation biofuels are derived from feedstock, such as corn grains that are unsustainable in long-term, while the second generation feedstock are based on sustainable cellulosic feedstock, such as corn stover) because of their ease of cultivation, fast growth cycle (1–10 days) and their adaptation to different environments requiring only sunlight and simple nutrients for growth

(Mata *et al.*, 2010). Furthermore, microalgae have broad bioenergy potential, since they can be used to produce liquid transportation fuels such as biodiesel, bioethanol, heating fuels and biojet fuels. On the basis of land surface area use, microalgae are capable of producing 15–300 times more oil for biodiesel production than traditional crops (Naik *et al.*, 2010). There are several species of algae capable of accumulating large amounts of oils that can be converted into biodiesel through a transesterification process. A collection of about 3000 strains of algae that have been screened for their oil producing potential are available (Sheehan *et al.*, 1998).

Table 7.13 lists constituents of some of the various strains of algae. Currently, most of the work on these strains of algae focuses on *Botryococcus braunii*. These strain of algae have the highest potential of oil production with cells (in dry weight) containing up to 75% oil (Chisti, 2007; Wellinger, 2009; Menetrez, 2012).

The other candidates are *Nannochloropsis sp.* and *Nitzschia sp*, which have oil contents of 68% and 47%, respectively (Wellinger, 2009; Menetrez, 2012). But, storage of algal feedstock is impeded by the presence of polyunsaturated fatty acids (PUFAs) and high moisture content, which causes oxidation reactions. Also, challenges of a cost-effective large-scale production of algae biomass has slowed down the pace of its commercial-scale conversion into energy.

Table 7.14 shows the comparison of the properties of algal bio-oil (green crude), wood bio-oil and petroleum. Similar to other biomass, the advantages of algal bio-oil over petroleum are that it is renewable, biodegradable and non-toxic. In addition, combustion of algal bio-oil produces lower levels of particulates, carbon monoxide, soot, hydrocarbons and SO_x as compared with petroleum fuels (Brennen *et al.*, 2009).

The Ecofining technology developed by Honeywell UOP in Des Plaines, Illinois, is claimed to be capable of converting multiple sources of algal feedstock (wild algae, enhanced algae strains and heterotrophically grown algae) into green fuels (gasoline, diesel

Table 7.13. Constituents of some of the strains of algae (% of dry matter).

Algae Strain	Protein	Carbohydrates	Lipid
Anabaena cylindrica[1,2]	43–56	25–30	4–7
Aphanizomenon flos-aquae[2]	62	23	3
Botryococcus braunii	40[2]	2[2]	25–75[2,4]
Chlamydomonas rheinhardii[1,2]	48	17	21
Chlorella ellipsoidea[2]	5	16	84
Chlorella pyrenoidosa[1,2]	57	26	2
Chlorella vulgaris[2]	41–58	12–17	10–22
Dunaliella bioculata[1]	49	4	8
Dunaliella salina[1,2]	57	32	6
Dunaliella tertiolecta[1]	29	14	11
Euglena gracilis[1,2]	39–61	14–18	14–20
Nannochloropsis sp.[2,3,4]	—	—	31–68
Nitzschia sp.[2,3,4]	—	—	45–47
Porphyridium cruentum[1,2]	28–39	40–57	9–14
Prymnesium parvum[1]	28–45	25–33	22–39
Scenedesmus dimorphus[1]	8–18	21–52	16–40
Scenedesmus obliquus[1,2]	50–56	10–17	12–14
Scenedesmus quadricauda[1]	47	—	1.9
Spirogyra sp[1,2]	6–20	33–64	11–21
Spirulina maxima[1,2]	60–71	13–16	6–7
Spirulina platensis[1,2]	46–63	8–14	4–11
Synechoccus sp[1,2]	63	15	11
Tetraselmis maculate[1]	52	15	3

(1) Dragone (2010); (2) Menetrez (2012); (3) Wellinger (2009); (4) Chisti (2007).

and jet fuel) (Holmgren *et al.*, 2007; Menetrez, 2012). Also, as discussed in Sec. 7.9.2.4, Elevance Renewable Sciences headquartered in Woodridge, Illinois together with Solazyme, Inc. (a San Francisco, California based company) have recently been involved in R&D projects in Illinois, focusing on the conversion of algal biomass into biofuels.

Algal oil production technology has the potential to produce several billion gallons of renewable fuel per year in the U.S. (Lundquist *et al.*, 2010; Quinn, 2010). But despite the vast energy potential of

Table 7.14. Comparison of the properties of moisture-free algal bio-oil, wood bio-oil and petroleum oil.

	Bio-oils		
Property	**Wood**	**Microalgae**	**Petroleum**
C[(1)]	56.4	62.07	83.0–87.0
H	5.2–7.2[(2)]	8.76[(1)]	10.0–14.0[(1)]
O[(1)]	37.3	11.24	0.05–1.5
N[(1)]	0.1	9.74	0.01–0.7
S	0.0–0.05[(2)]	0[(4)]	0.05–6.0[(4)]
Density (kg/L)[(1)]	1.2	1.06	0.75–1.0
Water content %	15–30[(2)]	26.6[(3)]	—
Viscosity (Pa.s)[(1)]	0.04–0.20 (at 40°C)	0.10 (at 40°C)	2–1000
pH	2.8–3.8[(2)]	4.3[(3)]	—
HHV (MJ/kg)[(1)]	21	29–45.9	42

(1) Brennan and Owende, (2010); (2) Naik *et al.*, (2010); (3) Trinh *et al.*, (2012); (4) Schlagermann *et al.*, (2012).
HHV = High heating value is the amount of heat produced by the complete combustion of a unit quantity of fuel and includes the latent heat of evaporation.

microalgae lipid, data on algal biofuels production costs is limited to biodiesel, while the current production cost of algal biofuels does not favor economics of scale (Lundquist *et al.*, 2010; Menetrez, 2012). According to a 2010 report by Energy Biosciences Institute (EBI) in California, the capital cost for a 250-acre algal biofuel production system for oil production was about $21 million, and with operating cost of about $1.5 million per year. This could produce about 12,300 barrels of oil, offering a break-even price of $330 per barrel of oil (based on an 8% capital charge). But with 1,000 acres algae biofuel production system, the break-even price was reduced to about $240 per barrel. Moreover, about 20% costs reduction was realized when wastewater treatment credits were applied. Although wastewater treatment is considered as the only important co-product for (algal) biofuels production, it is claimed that its availability would significantly reduce production of algae at the national level (Lundquist *et al.*, 2010; Quinn, 2010). It is worth mentioning that Sapphire Energy Inc. recently claimed to have developed a breakthrough technology for producing fungible, drop-in transportation fuel, including gasoline with octane number 91, diesel with cetane number 89, and jet fuel by using algae, sunlight and CO_2 (Saphire Energy, 2009).

7.4.3. *Use of abandoned agriculture lands to grow energy crops*

Adaptation to marginal lands is one of several benefits of using energy crops (perennial species) for biofuels production, especially with the increasing demand for biomass energy. Energy crops can be cultivated on pasture lands, underutilized cropland and land currently used for traditional crops. However, using abandoned cropland or pasture lands for energy crops cultivation could be a more viable option than using agriculture lands or the conversion of forested areas (US– EPA, 2010). This is because these lands require some maintenance to restore their balance of nutrients. Therefore, growing energy crops on these lands can restore some of these nutrients by increasing the soil organic carbon concentration and nitrogen. According to a study by the US-EPA, the potential marginal lands in Illinois suitable for energy crops cultivation are estimated at 3.95 million acres, which is very close to the value obtained by Stanford researchers. Table 7.15 shows Illinois abandoned agricultural lands and their potential ethanol production.

According to Table 7.15, these lands can produce an estimated 124.26 million liters (32.83 million gallons) of ethanol per year, which is about 7% of the present Illinois consumption of ethanol for fuel additive. These estimated marginal lands exclude those lands that are enrolled in the CRP, but nonetheless could be an additional source of land available for energy crops (US-EPA, 2010).

Table 7.15. Abandoned agriculture land and potential ethanol production in Illinois (US-EPA, 2010).

Land Area (Mha)	$\mathbf{1.6 \times 10^6}$
Biomass production rate (tons /year)	11.4×10^6
Ethanol production rate (liters/ton)	10.9
Annual ethanol production (Liters)*	124.26×10^6

(*) Assuming a conservative 302.8 liters/ton (80 gal/ton) conversion rate.

7.4.4. *Urban waste as feedstock for biofuels production*

Cellulosic MSW and construction and demolition (C&D) wood waste are the other potential feedstock for cellulosic ethanol production. Typically, C&D debris contain 20–30% by volume of wood waste, making them potential source for biofuels production. It is estimated that in 2022, urban waste in Illinois will provide about 4.71 million barrels of cellulosic ethanol per year to meet the EISA cellulosic biofuels standard (US-EPA, 2010).

The Illinois Environmental Protection Act of 1997 defined MSW as general household, garbage, institutional, and commercial waste, as well as landscape waste and construction or demolition debris (Fullerton and Miller, 2010). It is estimated that more than half of MSW is biogenic (i.e., waste made from materials that are produced by living organisms or biological processes) (US-EIA, 2010). In 2007, Illinois households, businesses and institutions produced 18.9 million tons of garbage or 3.66 kg (8.06 pounds) per person per day, which was about 19% higher than the national average on the basis of per capita (see Table 7.16).

The average Illinois resident produces an estimated 0.947 metric tons (2,088 pounds) of garbage per year, in contrast to the average American resident who produces 0.794 metric tons (1,751 pounds) of garbage per year. These averages, however, do not include hazardous

Table 7.16. Statewide MSW generation in 2007 in Illinois (IDCEO, 2009).

Components of MSW	(Tons)	%
C&D	4,247,060	22.5
Organics	3,560,120	18.8
Inorganics	877,570	4.6
Paper	5,354,230	28.3
Plastics	2,120,170	11.2
Textiles	1,085,650	5.7
Metals	887,890	4.7
Beverage containers	34,460	0.2
Glass	554,150	2.9
Household hazardous waste	184,790	1.0
Total	**18,906,090**	**100.0**

household waste, construction and demolition waste, flat glass and some other categories (Fullerton and Miller, 2010). Two demographic factors responsible for the increase in waste generation in Illinois include median income and level of urbanization. In 2005, the statewide median income in Illinois was 9% higher than the national average. In addition, 87% of Illinois population lived in urban areas, compared to 79% at the national level (IDCEO, 2009). The Illinois annual MSW generation by counties is shown in Fig. 7.11. These counties are divided into seven regions and the Chicago metropolitan area is in region 2. According to Fig. 7.11, the Region 2 MSW generation is the highest among all the seven regions in the state (estimated at 1–10 million tons per year). The US-EPA has estimated that based on dedicated cellulosic MSW and C&D wood waste generation, the Chicago metropolitan area contribution in meeting the EISA cellulosic biofuels standard in 2022 will be 1.88 million barrels (79 million gallons) (US-EPA, 2010).

Table 7.17 shows a summary of the projected cellulosic ethanol production volumes from potential biofuels feedstock in Illinois. It

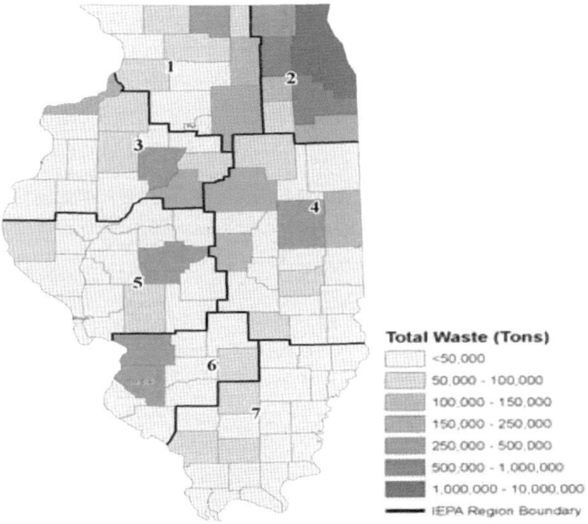

Fig. 7.11. Illinois county annual MSW generation (IDCEO, 2009).

Table 7.17. EPA's projection for cellulosic ethanol production volumes in Illinois in 2022 (US-EPA, 2010).

Component	**Volume (gallons/Yr).10^6**	**Volume (barrels/Yr).10^6**
Agriculture residues	1,270	30.20
Forest residues	58	1.38
Energy crops	0	0.00
Urban waste	198	4.71
Total	**1526**	**36.30**

is expected that at these production capacity, the state will meet the cellulosic biofuels mandate of 380.95 million barrels (16 billion gallons) in 2022 (US-EPA, 2010). The highest cellulosic ethanol projected volume (30.2 million barrels) will be derived from agricultural residues, mainly due to the large quantity of corn stover in the state. However, the state is not anticipating cellulosic ethanol production from energy crops in 2022, as indicated in Table 7.17.

7.5. Composition of Cellulosic Biomass

As the most abundant organic matter on earth, cellulosic polymeric biomass conversion into biofuels may sustainably help Illinois toward solving its energy problems. However, conversion of lignocellulosic biomass to liquid fuels is hindered by the structural and chemical complexity of the biomass. Cellulosic polymeric biomass is contained in the matrix of two other polymers (hemicellulose and lignin), a complex and heterogeneous material, as shown in Fig. 7.12.

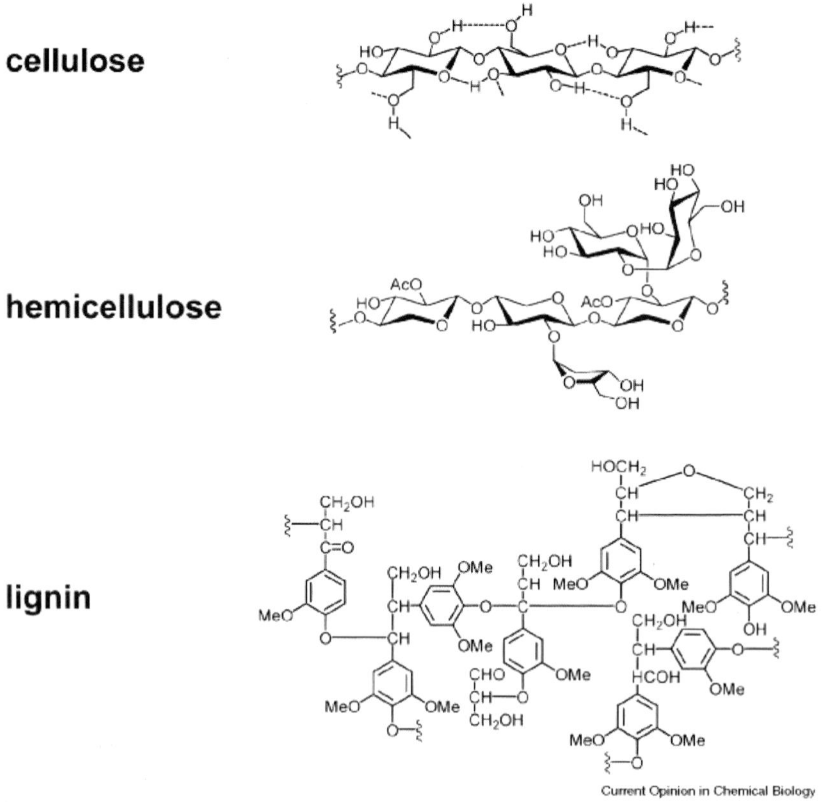

Fig. 7.12. Chemical structures of cellulose, hemicellulose and lignin components of cellulosic polymeric biomass (Chang, 2007).

Table 7.18. Chemical composition of cellulosic polymeric biomass (on % dry weight basis).

Biomass Source	Cellulose	Hemicellulose	Lignin	Protein
Crop residues[(a)]	30–45	16–27	3–13	3.6–7.2
Wood residues[(a)]	45–56	10–25	18–30	—
Cotton[(a)]	89.0	5.0	0.0	1.3
Switchgrass	31–51[(b)]	24–50[(b)]	13–20[(b)]	3[(d)]
Miscanthus	43–44[(b)]	24[(b)]	17–19[(b)]	3[(d)]
Corn Stover	31–38[(b)]	19–44[(b)]	13–21[(b)]	4.0[(c)]
Corn Fiber Hulls[(c)]	16	40	4.0	11
Soybeans Hulls[(c)]	46	18.0	2.0	12.0

(a) ePlantScience.com (an online Botanical Encyclopedia);
(b) Schnepf (2010); (c) van Ee (2010); (d) Lee *et al.*, (2007).

In Table 7.18, we report the ranges of compositions of cellulose, hemicellulose and lignin in various cellulosic polymeric biomasses.

Cellulose is a molecule of glucose with one molecule of water absent ($C_6H_{10}O_5$). It is the most abundant polysaccharide with ß-glycosidic linkages, but nonetheless, a difficult material to hydrolyze because of its high crystallinity. Hemicellulose, on the other hand, is a relatively weak random and amorphous structure that is easily hydrolyzed using dilute acid or base or hemicellulase enzymes. Lignin is a relatively strong complex chemical compound occupying the spaces between cellulose and hemicelluloses in the biomass. Lignin is an excellent solid source of fuel with higher energy density than cellulose, but it is not amenable to enzymatic dissociation because of its heterogeneous composition.

7.6. Processes for Biofuels Production

The rapid expansion of the Illinois biofuels industry following the introduction of the U.S. EISA of 2007 has brought new technologies into the biofuels market. It is expected that these technologies will improve the yield of lignocellulosic biomass derived biofuels, while creating new markets for waste materials and high energy crops. In the ensuing sections, we describe some of these technologies that can be categorized into thermochemical conversion processes, which include gasification, pyrolysis and liquefaction; biological conversion processes using anaerobic digestion, fermentation and enzymes; chemical conversion processes such as hydrolysis or combined gasification and biodegradation (US-EPA, 2010; Speight, 2014). Among these processes, the most technically proven one is the gasification method via the FTS (NSF, 2008).

7.6.1. *Gasification process for biomass conversion to biofuels via FTS*

This process converts cellulosic polymeric biomass into syngas, often referred to as biogas. In 1923, two German scientists, Franz Fischer and Hans Tropsch, invented a process for the production of liquid hydrocarbons and oxygenates from syngas conversion over iron catalysts. Synthesis gas is a mixture comprising of hydrogen (H_2), carbon monoxide and very frequently, carbon dioxide. After success of their invention, several variations on the FTS pathway, such as selective production of methanol, mixed alcohols and isosynthesis products soon followed. Another outgrowth of the FTS was hydroformylation (oxo synthesis) of olefins, discovered in 1938 (Spath and Dayton, 2003). The FTS process involves incomplete combustion of the biomass in a gasifier at high temperatures (over 700°C), using controlled amount of oxygen together with, or without, steam. Typically, the synthesis gas from a biomass gasifier contains 1–1.5% tar, 0.2–0.4% ammonia, and 100–500 ppm H_2S (NSF, 2008). Selection of suitable operating conditions in the gasifier could reduce tar and ammonia formation. But, the preferred method for ammonia removal is to decompose it to N_2 and H_2 using traditional catalysts such as supported Ru, Ni or Fe (NSF, 2008).

The resulting syngas is cleaned up of sulfur compounds and particulate matter, before being burned directly or converted into other

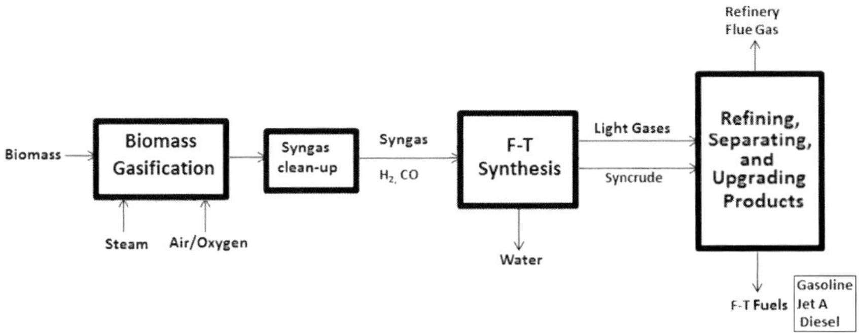

Fig. 7.13. Biomass conversion to liquid fuel via the FTS (NPC, 2007).

heavier hydrocarbon fuels. Figure 7.13 shows biomass conversion into liquid fuels via the FT thermochemical process.

In the FTS, H_2 reacts with carbon monoxide to form mainly aliphatic straight-chain hydrocarbons (C_xH_y), according to the chemical Eq. (7.3).

$$CO + 2H_2 \rightarrow -(CH_2) - +H_2O \tag{7.3}$$

This exothermic reaction releases 20% of the thermal energy involved in the process (Boerrigter, 2003). Since the FTS consumes hydrogen and carbon monoxide in a 2:1 ratio, this feed ratio is adjusted using a water-gas shift process according to the reaction (Hu *et al.*, 2012).

$$CO + H_2O \leftrightharpoons CO_2 + H_2 \tag{7.4}$$

Table 7.19 shows a summary of the FTS conditions as well as the catalyst used.

While ruthenium (Ru) is identified as one of the more effective catalysts for the CO hydrogenation reaction, it has not found any important commercial application as a FT catalyst. One drawback of using Ru catalyst is its relatively higher cost compared to iron (Fe) or cobalt (Co) that are commonly used as FT catalysts (Datye and Schwank, 1985).

Haldor Topsoe Inc. is converting woody biomass to green gasoline in Des Plaines, Illinois, using a full integration and optimization of a multi-step gasification process (US-DOE, 2012b).

Table 7.19. Summary of FTS conversion processes and related conditions (Spath and Dayton, 2003).

	Process Conditions					
Catalyst	**Temperature (°C)**	**Pressure (bar)**	**H_2:Co**	**% Conversion (CO basis)**	**Product**	**% Selectivity**
Iron (Fe)	300–350	10–40	1.7:1	50–90 With recycle	α-olefins gasoline	ASF—48 (max) 15–40 actual
Cobalt (Co)	200–240	7–12	2.15:1	50–90 With recycle	Waxes diesel	ASF— 40 (max)
Ruthenium (Ru)	—	—	—		Waxes	—

ASF = Anderson-Schulz-Flory (ASF) distribution.
% Conversion = amount of feedstock consumed or converted to product(s).
Selectivity = amount of feedstock converted to the desired product(s).
α-olefins (or alkanes) = with chemical formula C_xH_{2x} with a double bond at the primary or alpha (α) position.
Waxes = long-chain hydrocarbons.

Also in December 2012, Paradigm BioAviation (PBA) Company announced construction of a $120 million facility in Bloomington-Normal, Illinois. The plant is claimed to transform municipal solid waste into green electrical power and alternative liquid fuels for the transportation and aviation industries. Construction of the plant was scheduled for 2014 and would be in two phases (Lane, 2012). The first phase includes construction of materials recovery facility for recyclables processing, construction of a gasification plant for syngas production for electrical power generation and construction of a pilot gas-to-liquid (GTL) plant capable of processing 15 tons of garbage per day. The second phase will be construction of a full-scale GTL plant, expected to come online in 2017. Power production was scheduled to begin in 2015 and the plant is expected to generate 5 MWh of green electrical power and 8 million gallons of alternative liquid fuel. Additionally, it is claimed that over 700 direct, indirect and induced jobs will be created by the project. Illinois State University (ISU) will partner with PBA to develop this project (Lane, 2012).

The other partners of the project include Hensel Phelps Construction Company in Orlando, Florida and Southern Research Institute, in Durham, North Carolina. Hensel Phelps Construction Company will be responsible for the design, engineering and construction of the facility (Lane, 2012).

7.6.2. *Fast pyrolysis for biomass conversion to bio-oil*

Fast pyrolysis is a thermochemical process which involves thermal depolymerization of biomass to bio-oil at high temperatures and in absence of oxygen (NSF, 2008). The fast pyrolysis process maximizes bio-oil products, typically at atmospheric pressure and at temperatures ranging from 400 – 500°C (Lehmann, 2007). The products from fast pyrolysis process include bio-oil, gas and char. The products slate from biomass pyrolysis depends on the process pressure (which can be adjusted), temperature and residence time of the pyrolysis vapors.

At short residence times (1–2 s) and high heating rates (i.e., 500°C/s), fast pyrolysis process produces high yields of bio-oils containing up to 70% of the original energy in the biomass feedstock. A sample of products taken from a fast pyrolysis reactor designed to maximize bio-oil liquid contains 65 wt% organics, 10 wt% water (resulting from the reaction, but not entirely, as biomass itself has moisture content), 13 wt% gas and 12 wt% char (NSF, 2008). However, the relative ratios of these fractions are highly dependent on reactor design, reaction conditions and biomass alkali content. The chemical species in the bio-oil leave the pyrolysis reactor either in the form of vapor, as a free radical precursor, or in the form of aerosol.

The product bio-oil is made up of a high oxygen content water-soluble fraction and a low oxygen content water-insoluble fraction. The water-soluble fraction is a derivative of the cellulose and hemicellulose portion of the biomass, while the water-insoluble fraction is derived mainly from the lignin part of the biomass. Since the bio-oil produced from fast pyrolysis process contains hundreds of

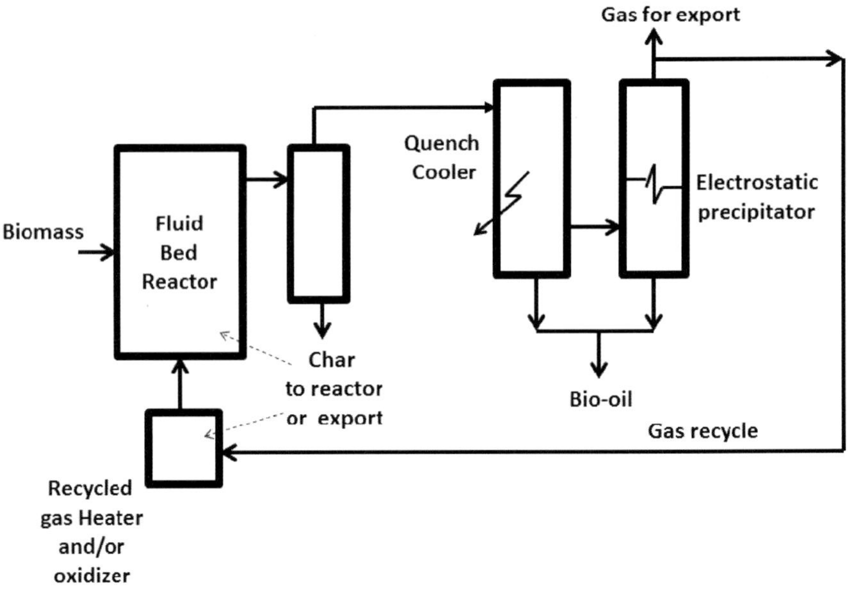

Fig. 7.14. A diagram of the key process components for a representative fast pyrolysis process utilizing a fluidized bed reactor (NSF, 2008).

chemical species, the full characterization of the oil poses some challenges. Fast pyrolysis plants (see Fig. 7.14) can economically be deployed as small scale units using 50–100 tons of biomass/day, which can be mobile and transported close to the biomass source (NSF, 2008). As distributed energy resource, small scale fast pyrolysis plants can create more jobs, compared with other large-scale biofuels technologies. The distributed technology has lower capital cost investment, but a higher operating cost than other biofuels technologies. As will be mentioned in Sec. 7.9.2.2, the Gas Technology Institute (GTI) in Des Plaines, Illinois is conducting research on development of a novel process called "integrated hydropyrolysis and hydroconversion (IH^2)" for economic conversion of wood, agricultural waste and algae biomass into fungible gasoline and diesel.

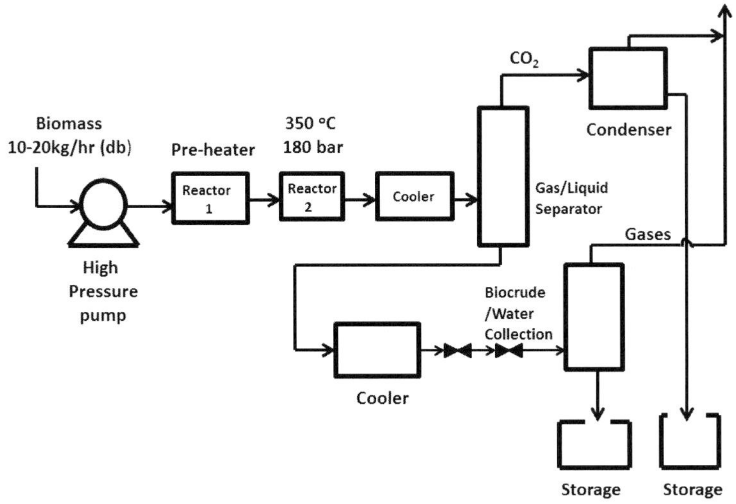

Fig. 7.15. Process diagram for an aqueous liquefaction pilot plant constructed by Biofuel B.V. in the Netherlands (NSF, 2008).

7.6.3. *Hydrothermal liquefaction process for biomass conversion into bio-oil*

Hydrothermal liquefaction process is another thermochemical process for synthesis of crude bio-oil (biofuel) through the application of heat and pressure on biomass.

The hydrothermal liquefaction process shown in Fig. 7.15, allows crude bio-oil production with high efficiency (NSF, 2008).

In this process, the high moisture biomass is subjected to mild temperature (250–350°C) and pressure (10–20 MPa), which break down and reform the chemical building blocks in the biomass into bio-crude oil. At these temperatures and pressures, the water in biomass becomes a highly reactive medium, which accelerates the breakdown and cleavage of the chemical bonds, leading to reformation of new molecules. This conversion technology can be used for a broad range of feedstock and is similar to the naturally occurring geological processes postulated to have led to the formation of fossil fuel reserves. Feedstock that are used with this technology include, swine manure, low lipid algae, sawdust, garbage and even sewage sludge (UIUC, 2013). The products exiting the liquefaction process consist of gas, an aqueous phase, water-insoluble bio-oil and a water soluble product. A sample of product from the liquefaction process relative to the initial biomass feedstock consists of 45 wt% bio-oil, 25 wt% gas (with > 90wt% of gas being CO_2) and 30 wt.% aqueous phase (with the aqueous phase containing 66 wt.% water and 33 wt.% soluble organics) (NSF, 2008).

Recently, marine microalgae *Nannochloropsis sp.* was converted into crude bio-oil and gaseous products using hydrothermal liquefaction (Brown *et al.*, 2010). The highest bio-oil yield from this experiment was reported as 43% by weight at a moderate temperature of 623 K (350°C). The estimated heating value of the bio-oil was 39 MJ/kg, which was found to be comparable to that of a petroleum crude oil (Brown *et al.*, 2010).

Also, hog waste was successfully converted into bio-crude oil using hydrothermal liquefaction route (He *et al.*, 2000; Lovgren, 2004; Vardon, 2011).

A small-scale continuous hydrothermal process reactor system was developed in Illinois in 2006, which is claimed to convert up to 70% of swine manure volatile solids into bio-oil. The system reduced the manure chemical oxygen demand by up to 75% (Ocfemia *et al.*, 2006). The system had the capacity to process up to 48 kg of manure slurry per day and was successfully operated continuously for up to 16 hours per test. Oil yields ranging from 62.0–70.4% were obtained. The heating value of the oil product ranged from 25,176 kJ/kg to 31,095 kJ/kg. The highest temperature was 305°C, while the highest pressure was 10.3 MPa, with a residence time of 80 minutes. Using batch experiments, the energy consumption of the process was found to be less than one third of the energy in the oil product (Ocfemia *et al.*, 2006). Besides production of oil, it is claimed that this process could finally resolve the pollution problem caused by the odor from hog farms, where farmers spend large amount of money to get rid of hog waste (He *et al.*, 2000).

Table 7.20. Comparison of properties of bio-oils from different biomass and heavy fuel oil.

Property	Fast Pyrolysis Bio-oil(1)	Liquefaction Bio-oil*	Small-scale(3) Continuous Hydrothermal Bio-oil	Heavy Fuel Oil(1)
Carbon (wt.%)	54–58	66.6–71.2(2)	65.9–69.0	85
Oxygen (wt.%)	35–40	14.9–18.9(2)	—	1.0
Hydrogen (wt.%)	5.5–7.0	8.9–9.5(2)	9.2–9.6	11
Sulfur [wt.%]	< 1	< 1(2)	0.3–0.4	0.05–6.0
Nitrogen [wt.%]	0–0.2	4.3–6.5(2)	3.7–4.1	0.3
Moisture content [wt.%]	15–30	5.1(1)	—	0.1
Specific gravity	1.2	1.1(1)	—	0.94
Viscosity [Pa.s]	0.040–0.1 (50°C)	15(61°C)(1)	—	0.18 (50°C)
Higher heating value [MJ/kg]	16–19	30.0–34.7(2)	25.2–31.1	40

(1) NSF (2008); (2) Vardon *et al.* (2011); (3) Ocfemia *et al.* (2006).

Pyrolysis bio-oil is produced by direct liquefaction of woody biomass in the presence of water (and may be some catalysts), typically at atmospheric pressure and temperatures <400–500°C.

*Liquefaction bio-oil is produced by direct liquefaction of Spirulina algae, swine manure and digested anaerobic sludge at high temperatures and high pressures.

Comparison of the properties of fast pyrolysis bio-oil, liquefaction bio-oil and petroleum heavy oil and oil produced from a small-scale continuous hydrothermal process are shown in Table 7.20.

The relatively high oxygen content of "fast pyrolysis bio-oil" compared to "liquefaction bio-oil" makes pyrolysis bio-oil less preferable as fuel, at least in the context of heating value. Also, the high oxygen content of fast pyrolysis bio-oil compared to heavy oil explains the instability and low energy density (heating value) of fast pyrolysis bio-oil, which is about 50% of that for conventional fuel oils (NSF, 2008). The relatively low oxygen content of liquefaction bio-oil is due to more decarboxylation reaction taking place in the liquefaction process. Oxygen removal through decarboxylation leads to a more favorable C/H ratio in the bio-oil product from the liquefaction process, making the product water insoluble and with high heating value (NSF, 2008). However, the high viscosity of liquefaction bio-oil limits its applications without some degree of upgrading (NSF, 2008).

7.6.4. *Biodegradation process for biomass conversion to ethanol*

Biomass biodegradation conversion process combines chemical and biological conversion processes to produce chemicals using biological pathways and pretreatment methods. These processes are designed to make cellulose and hemicelluloses in the biomass more accessible to enzymatic attack for conversion into simple sugars (Detroy and St. Julian, 1982; Enayati and Parulekar, 1995). Lignocellulosic biomass costing about $60 per dry ton is equivalent to a unit energy costs of oil at about $20 a barrel. At low price of oil, processing of lignocellulosic materials for conversion into liquid fuels is a real challenge.

In the traditional conversion of biomass into ethanol, the pretreatment step was one of the most expensive in the processing steps. Recently, new methods have been proposed in order to reduce the pretreatment cost of biomass conversion process. A pretreatment process referred to as "co-solvent enhanced lignocellulosic fractionation" has been developed, which claimed to reduce the amount of enzymes needed to breakdown the lignocellulosic biomass by 90% when compared to prior practice (Nguyen *et al.*, 2015). Another method is by altering the chemical structure of lignin in the biomass crop through genetic engineering. For example, pretreatment enzymes can be produced within the leaves and stems of the biomass crops to lower the

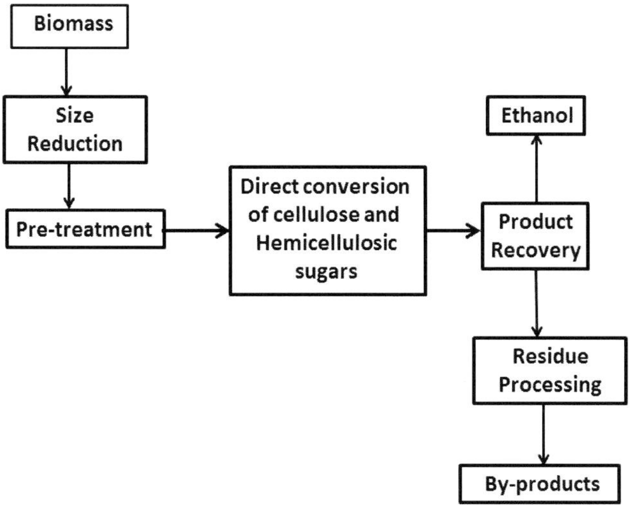

Fig. 7.16. Biorefinery concept incorporating advanced pretreatment and consolidated processing of cellulose to ethanol (US-DOE, 2006a).

amount of lignin as it grows (Sticklen, 2010). Furthermore, recent research and development advances have simplified the traditional biodegradation of biomass process by eliminating and consolidating some of the steps. As a result, there is a direct conversion of cellulose and hemicellulose into ethanol, as shown in Fig. 7.16 (US-DOE, 2006a).

Twenty-one species of short-rotation woody and grassy biomass feedstocks for cellulosic ethanol production were evaluated by the EBI at the University of Illinois. The study has shown that the top five performing species are black locust, northern catalpa, flameleaf sumac, silver maple and sycamore. The top performing grassy plants were three various types of switchgrass and Scout Indiangrass. The study provides an idea of how each crop compares to each other and does not require obtaining data from multiple studies that could have been from different sites, years, weather conditions and management objectives. In addition to cellulosic ethanol producers, the study could be potentially valuable to farmers (Hanson, 2013b).

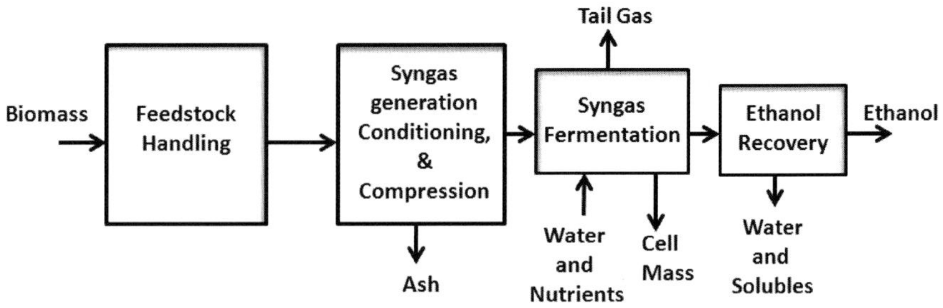

Fig. 7.17. Syngas to ethanol simplified block flow diagram (Spath and Dayton, 2003).

7.6.5. *Combined gasification and biodegradation process for ethanol production*

The combined gasification and biodegradation process converts syngas into pure ethanol (or pure butanol) using microorganisms to ferment the gas, as shown in Fig. 7.17 (Spath and Dayton, 2003).

The process enables lignin in the biomass feedstock, along with hemicelluloses and cellulose, to be converted to ethanol. The bacteria or microorganisms employed for the conversion of syngas mixtures into ethanol are referred to as obligate anaerobes (Spath and Dayton, 2003). These bacteria use a heterofermentative version of the acetyl-CoA (Acetyl group plus Coenzyme A) pathway for acetogenesis. Acetogenesis is a process through which acetate is produced by anaerobic bacteria from a variety of energy and carbon sources. In this pathway, acetyl-CoA is produced from CO or H_2/CO_2 mixtures. The acetyl-CoA intermediate is subsequently converted into either ethanol or acetic acid as a primary metabolic product. The overall reactions (ignoring cell mass production) are given as follows: (Spath and Dayton, 2003).

Acetic acid production:

$$2CO_2 + 4H_2 \rightarrow CH_3COOH + 2H_2O \quad (7.5)$$

$$4CO + 2H_2O \rightarrow CH_3COOH + 2CO_2 \quad (7.6)$$

Ethanol production:

$$6CO + 3H_2O \rightarrow CH_3CH_2OH + 4CO_2 \quad (7.7)$$

$$2CO_2 + 6H_2 \rightarrow CH_3CH_2OH + 3H_2O \quad (7.8)$$

Recent applications of this technology are the BRI Energy, LLC pilot plant located outside Fayetteville in Arkansas (BRI, 2005) and the Coskata Process (Coskata, 2012) patented by Illinois-based start-up company Coskata, Inc. These two processes are described below.

7.6.5.1. *Project 1*

In 2003, Florida-based company BRI Energy, LLC used a pilot plant located in Fayetteville, Arkansas to convert a variety of waste into syngas and subsequently into ethanol (see Fig. 7.18).

After gasification, anaerobic bacteria are used to convert the syngas into ethanol. The heat generated in the gasification process is also used to co-generate excess electricity (BRI, 2005). The organics in the biodegradable feedstock are converted into syngas using

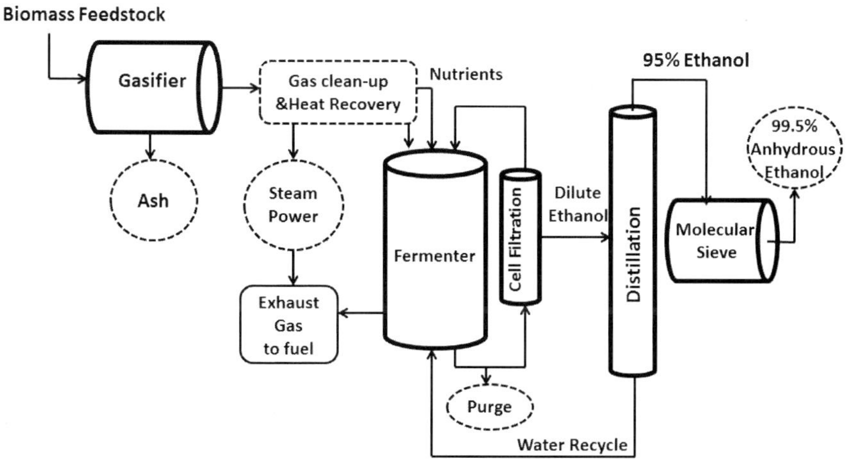

Fig. 7.18. BRI Energy, LLC gasification–fermentation pilot facility in Arkansas (BRI, 2005).

BRI's two-stage thermal gasifier which increases the syngas temperature to above 1, 093°C (2000°F) in the second stage. This high temperature enhances cracking of heavy hydrocarbons to CO and H_2 to maximize ethanol yield. The hot syngas is then cooled to 37.8°C (100°F) in the process generating stream. The syngas enters the fermenter, containing a specialized microbial population and is converted into ethanol and water. Addition of nutrients provides the necessary cell growth and automatic regeneration of the biocatalyst. The aqueous stream containing ethanol from the fermenter is continuously removed via a membrane that retains cells for recycle to maximize reaction rates. Hydrous ethanol (95%) is then produced by conventional distillation (Wang and Mansoori, 1994; Mansoori, 1999), followed by a molecular sieve separation step to further increase the purity of ethanol. The company claims to have produced anhydrous ethanol with 99.5% purity (BRI, 2005). The BRI pilot unit can process a maximum of 1.5 tons per day of various feedstocks at a feed rate of 907 g/min (2 lb/min). The feedstock has 5–10 wt.% moisture content and includes presorted MSW (excluding plastics, metals and glasses) shredded to a size of 5.08 centimeters (2 inches). The entire process claimed by BRI from gasification to distillation is approximately seven minutes.

7.6.5.2. *Project 2*

In 2008, a start-up company, Coskata, Inc., in Warrenville, Illinois announced a successful conversion of corn husk into ethanol, using a combined gasification and biodegradation process (see Fig. 7.19). The process uses the existing gasification technology to gasify corn husk into syngas. But instead of fermenting the syngas or using catalysts to convert it into ethanol, a reactor containing microorganisms converts the resulting syngas into ethanol by consuming CO and H_2 in the gas stream. Once the GTL conversion process is complete, the resulting ethanol is recovered from the solution using established distillation methods.

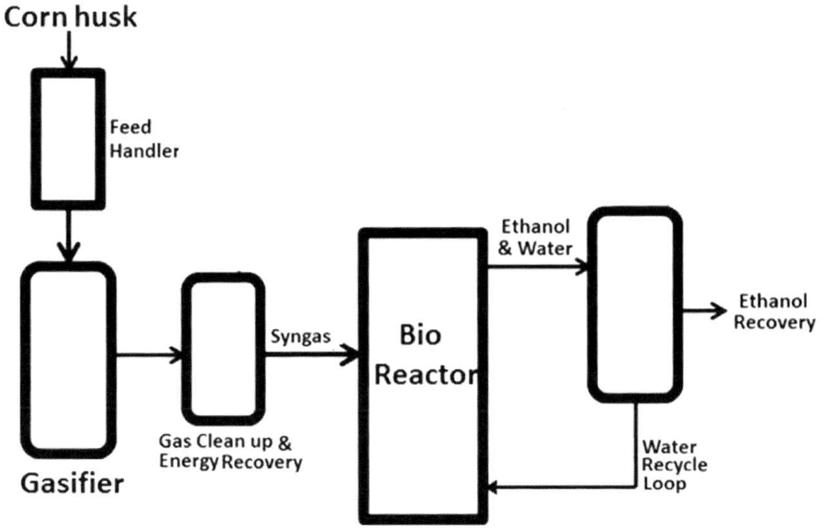

Fig. 7.19. The Coskata Process (Coskata, 2012).

According to the Illinois Renewable Fuels Association (Squatriglia, 2008), this process produces 2.38 barrels of ethanol per ton of dry biomass, compared with 2.14 barrels of ethanol per ton of corn-based feedstock. Additionally, it is claimed that the process requires far less water, heat and pressure (Squatriglia, 2008). According to the company, the ethanol fuel produced from this process will generate 7.7 times more energy than the energy input required for the energy production. This amount of energy is relatively higher than the energy from corn ethanol, which traditionally generates 1.3 times more energy than what is required to produce ethanol (Squatriglia, 2008). Furthermore, the company claims that the process could convert 0.435 metric tons of hay into 18.93 liters (5 gallons) of ethanol for less than $42 per barrel ($1 per gallon or $0.264 per liter), compared to corn-based ethanol which costs $59 per barrel ($1.4 per gallon or $0.37 or liter) to produce (Squatriglia, 2008). Thus, after accounting for the energy needed to produce and transport the feedstock, the Coskata process (Coskata, 2012) claims to be capable of producing

84% less greenhouse gases than fossil fuels, while producing ethanol with 99.7% of purity (Squatriglia, 2008).

Table7.21 shows the quantity and costs of ethanol production in Illinois using various biomass resources. The total cost of ethanol reported in Table 7.21 is based on the wholesale cost at the refinery, including zero return on equity for the year 2007. The feedstock cost for corn ethanol is assumed to be \$5 per bushel of corn for the year 2007. The quantity of corn ethanol is based on an assumed average yield of 145 bushels per acre under a corn-soybean growing rotation, while that of corn stover is based on assumed average yield under a corn-corn growing rotation. The emissions from gasoline are 7.54 kg of CO_2 per energy equivalent of a gallon of ethanol (Khanna, 2008). The co-product credit is the monetary value assigned to high value products (such as corn oil and corn gluten feed), which are obtained as by-products from biofuels processing. These co-products can be sold to defray part of the biofuels processing cost.

7.6.6. *The biorefinery of the future and biofuels production industry*

Biorefinery is an installation for biomass utilization that leads to the production of fuels and other bioproducts. In essence, the term refers to a facility or group of facilities combining production of materials, chemicals, or fuel products with energy production (Naik *et al.*, 2010). A biorefinery of the future will upgrade bio-oils using technologies such as filtration, hydrotreating and hydrocracking to reduce the oxygen content in the bio-oil. These final refining processes convert the bio-oil into hydrocarbon fuel products. A simplified flow diagram of a biorefinery of the future integrating novel technologies is shown in Fig. 7.20. These modifications can lead to a dramatic cost reduction of enzymes, improved fermentation strains and yield, as well as energy requirement for the process. For example, the lignin produced from biomass hydrolysis to ethanol can be burned to generate electricity, as shown in Fig.7.20.

Table 7.21. Quantity and costs of production of ethanol in Illinois using various biomass resources.

			Ethanol Production								CO_2 Emission Reduction Relative to Gasoline kg CO_2/Ga of Ethanol — kg CO_2/L. of Ethanol
			Feedstock Cost		Cost of Conversion		Co-product Credit		Total Cost		
Biomass	Ga*/Acre	L§/ Hectare	$/Ga	$/L	$/Ga	$/L	$/Ga	$/L	$/Ga	$/L	
Corn Ethanol[a]	398.75	3729.5	1.82	0.48	0.78	0.21	0.48	0.127	2.12	0.56	2.79–0.737
Corn Stover[a]	165.04	1543.6	1.27	0.34	1.46	0.39	0.12	0.032	2.61	0.69	7.08–1.87
Switchgrass[a]	214.74	2008.5	3.18	0.84	1.46	0.39	0.12	0.032	4.52	1.19	13.17–3.48
Woody Biomass	125–300[c]	1169–2806	n.a	n.a	n.a	n.a	n.a	n.a	2.25[b]	0.59	n.a
Miscanthus Giganteus[a]	742.45	6944	1.40	0.37	1.46	0.39	0.12	0.032	2.74	0.72	9.83–2.6
Pennycress[d]	75–100	702–935	n.a	n.a	n.a	n.a	n.a	n.a	n.a	n.a	n.a

(a) Khanna (2008); (b) Schnepf (2010); (c) Mandil and Shihab-Eldin (2010); (d) Suszkiw (2008).
* Gallons. § Liter.

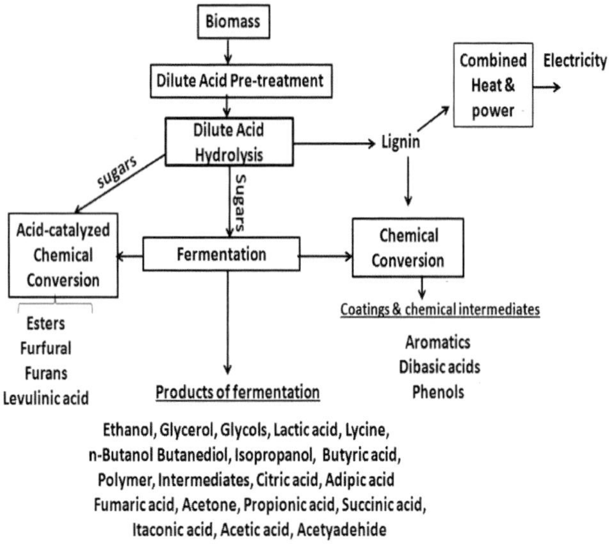

Fig. 7.20. The biorefinery of the future (Blaschek, 2008).

Integrated biorefineries are comparable to conventional refineries in that both technologies can produce a broad range of products that can optimize production economics and use of feedstock (Lee, 2013). However, integrated biorefineries employ novel technologies and different biomass feedstock that require significant investments in research and development, as well as deployment of projects to offset the cost involved. For this reason, in July 2012, the U.S. Secretary of Agriculture announced payments of various grants for 125 advanced biofuel producers across the country, including Illinois. The grants were to support production and expansion of advanced biofuels from a broad range of non-food sources, including waste products. Out of these grants, Illinois companies Archer Daniels Midland, Incobrasa Industries, Ltd. and Midwest Biodiesel Product, LLC received $596,279, $305,038 and $640,572, respectively.

The companies were to use the grants to conduct research in biodiesel transesterification (Freeman, 2012). Two of the current and completed biorefinery research and development (R&D) projects in the State of Illinois are described in Sec. 7.9.

7.6.7. *Renewable liquid fuel production from biomass-derived oils using hydroprocessing*

A growing trend in the biofuels industry is the use of catalytic hydroprocessing of biomass-derived oils (such as vegetable oils or animal fats) for the production of hydrocarbon fuels. These fuels are similar in chemical composition and energy content to standard gasoline, diesel and jet fuel derived from petroleum. The process uses hydrogen and a catalyst (like palladium) to remove oxygen from the triglyceride molecules in the oil feedstock via decarboxylation and hydrodeoxygenation reactions to produce some light petroleum products and water as co-products. Figure 7.21 shows the process flow diagram of the primary steps in renewable diesel production. Renewable diesel can be produced using existing petroleum refineries infrastructure or a stand-alone facility. Prior to entry into the hydrotreatment reactor, the feedstock is pretreated to remove contaminants such as sulfur, nitrogen and trace metals that may poison the catalysts. The

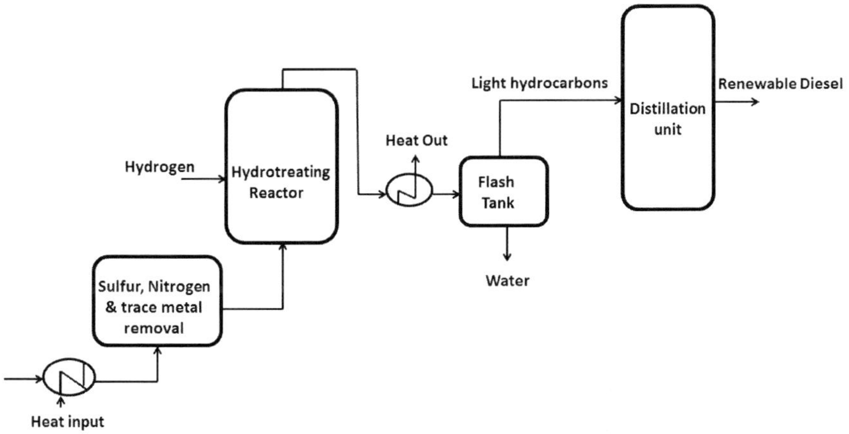

Fig. 7.21. Process flow diagram of primary steps in renewable diesel production (US-EPA, 2010).

process yield is about 90% of renewable diesel, with the rest split between light fuel gas and naphtha (US-EPA, 2010). Major manufacturing companies involved in this Research and Development (R&D) effort are Honeywell UOP in Des Plaines, Illinois, and some other U.S. and international organizations. The USDOE supports the R&D efforts by Honeywell UOP and other U.S. companies which are focused on the development of production technologies (US-EPA, 2010).

Table 7.22 shows comparison of the properties of biodiesel, renewable diesel (green diesel) and ultra-low sulfur diesel (ULSD), also referred to as petroleum diesel. Of the three diesel fuels, green diesel is claimed to be the premium blending component because it has a boiling range comparable to that of conventional diesel, including a significantly higher cetane and lower density (Holmgren *et al.*, 2007a). These properties enable refiners to optimize the amount of lower value refinery streams that can be blended into the refinery diesel pool and still meet all the required diesel specifications (Holmgren *et al.*, 2007a; Holmgren *et al.*, 2007b).

In the U.S., conventional diesel or petrodiesel must have a cetane number of at least 40, but the cetane number for renewable diesel is much higher (70–90). An adequate cetane number is required for good engine performance. Higher cetane numbers ensure good cold start of a diesel engine and minimize formation of white smoke (US-DOE, 2006b).

7.6.8. *MSW for power generation*

Another form of bioenergy that could help Illinois attain energy-independence is conversion of municipal solid waste (MSW) to biofuels using waste-to-energy conversion plants. Such plants include anaerobic digestion of MSW in landfill and small-scale digesters, excluding incinerators. LFG is considered as renewable energy resource by Illinois law (Amendment to SB 878). However, the law specifically excludes incineration (or burning) of tires, garbage,

Table 7.22. Comparison of biodiesel, renewable diesel (green diesel), FT diesel and petrodiesel properties.

Property	Biodiesel (FAME)	Renewable Diesel	FT Diesel	Petrodiesel (ULSD)
Oxygen (%)	11[1]	0[1]	0[3]	0[1]
Sulfur content (ppm)	< 1[1]	< 1[1]	< 1[1]	< 10[1]
Specific gravity	0.88[1]	0.78[1]	0.77[1]	0.81–0.89[9]
Change in NO_x emissions (%)	+10[10]	0–10[10]	−10[3]	Baseline[10]
Higher Heating value (MJ/kg)	40.5[9]	44[1]	44[1]	43.6[1]
Lower Heating value (MJ/kg)	37.6[9]	—	—	40.3[9]
Cloud point (°C)	−3 to +15[6]	−20 to+20[1]	3[3]	−35 to 5[6]
Pour point (°C)	−5 to 10[6]	−10 to −15[7]	0[3]	−35 to −15[6]
Flash point (°C)	100–170[6]	146[8]	72[3]	60–80[6]
Cetane number	48–65[9]	70–90[1]	> 74[3]	40–55[1,9]
Distillation (°C)	340–355[2]	265–320[2]	210–338[3]	200–350[2]
Kinematic Viscosity at 40°C (cSt)	1.9–6.0[4]	3.4[5]	3.57[3]	1.3–4.1[4]
Oxidative stability	Marginal[2]	Good[2]	Good[2]	Good [2]

(1) Kalnes *et al.* (2008); (2) Holmgren *et al.* (2007a); (3) Norton *et al.* (1998); (4) Strong *et al.* (2004); (5) UC (2010);
(6) US-DOE (2009); (7) Liu (2011); (8) Durbin *et al.* (2011); (9) US-DOE Alternative Fuels Data Center (2013b); (10) Holmgren *et al.* (2007b)

general household, institutional and commercial waste. Additionally, the law prohibits incineration of industrial waste, landscape waste other than tree waste, railroad crossties, utility poles, or construction or demolition debris, other than untreated and unadulterated waste wood (Illinois General Assembly, 1997). The reason is that from process engineering point of view, incineration plants are comparable to conventional coal-fired power plants, which emit appreciable amount of pollutants into the environment.

In the 1990s, there was a nationwide fear that the U.S. could soon run out of landfill sites. The situation forced the City of Chicago to allow construction of incinerator plants in the metropolitan areas of the city. These constructions were for combustion of garbage to generate electricity and minimize the use of landfills. At the same time, a Retail Rate Law was passed in the State of Illinois asking utility companies to purchase the electricity, which was produced by thermal energy from the incinerator plants. But as it turned out, the nation would never face such crisis in the future and the Retail Rate Law in Illinois was subsequently repealed by lawmakers. As a result, only a few of the incinerator plants ended up being constructed (Sendzik and Wiewel, 1996). An example of these plants is Geneva Energy Power Plant at Ford Heights, a small village near Chicago, Illinois. This 22 MW capacity power plant is one of the few of such plants in the U.S. The plant uses fuel derived from burning used tires to produce steam for electricity generation. As a result, the company cannot benefit from the Illinois' renewable energy tax credit because according to Illinois Law, energy generation through incineration of tires is not considered as renewable.

In 2009, the Illinois EPA rejected an application submitted by Indiana Recycling and Renewable Fuels, an Indiana-based company, seeking for permits to create bio-oil and bio-char fuels from municipal solid waste in Illinois (IEPA, 2011a). The bio-oil would have been burned in the company's facilities that would have been located in Illinois to produce electricity and create employment. But according to the Illinois EPA, burning of bio-oil is considered as incineration. Since 2000, waste incineration is not allowed as disposal option in Illinois, and even prior to that date, waste incineration was less than 2% of the total waste disposal in Illinois (Fullerton and Miller, 2010). For this reason, most of the municipal solid waste conversion to biofuel energy in Illinois is achieved using LFGE technology. In the following section, we discuss the application of LFGE technology in Illinois.

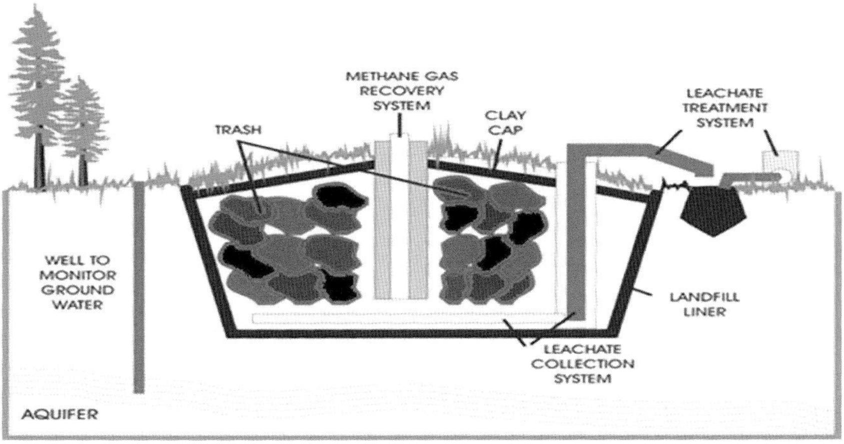

Fig. 7.22. General features of a modern landfill (modified from US-EIA, 2010).

7.6.9. *Power generation using landfill gas-to-energy technology*

The conversion of landfill biomass to energy is a biological process. In the absence of air, anaerobic bacteria in landfills decompose animal manures, organic wastes and crops to produce biogas (see Fig. 7.22).

Biogas contains methane, a major component of natural gas that can be burned to produce heat and electricity. Depending on the origin of the anaerobic digestion process, the composition of biogas can be variable, with a typical landfill gas containing about 40–75% methane by volume (IEA, 2007). Biogas also contains about 45% CO_2 by volume, including a small amount of hydrogen sulfide and ammonia. After purification and upgrading processes, the biogas can be used for various applications similar to natural gas (US-EIA, 2010).

Anaerobic digestion is frequently used in small-size, rural and off-grid applications at domestic and farm-scale. However, LFG can be used directly in various ways, which include IC engines, gas turbines and microturbines. About 85% of LFG in U.S. are used for electricity generation projects through combustion in I.C. engines or turbines.

One million metric tons of landfilled MSW can generate about $(1.7 - 2.5) \times 10^6$ standard cubic meters per year of collectable methane, enough to fuel a gas engine with generating capacity of 0.8–1.25 MW to produce about 6,500–10,000 MWh of electricity per year (Clarke Energy, 2011). In modern landfills (Fig. 7.22), methane production ranges from 50 and 100 kg per metric ton of MSW. Out of this amount, about 50% is recovered and used for power and heat generation. Large-size plants using MSW, agricultural wastes and industrial organic wastes (large-scale co-digestion) require approximately 8,000–9,000 metric tons of MSW per year to feed a power plant of 1 MW capacity (IEA, 2007).

EPA's estimates for preparation of one million ton landfill for energy production requires an initial capital costs of $600,000–$750,000 and operating costs of $40,000–$50,000 a year. These estimates exclude administrative costs associated with permitting, environmental impact studies, legal issues and other costs. However, the capital costs of the process vary, depending on the type of plant used to process the methane as indicated in Table 7.23 (Combs, 2008). Table 7.23 shows the capital, operating and maintenance cost of landfill gas electricity generation projects using various technologies.

Table 7.23. Capital, operation and maintenance (O&M) costs of landfill gas electricity generation projects (US-EPA, 2012b).

Technology	Optimal Project Size/Capacity	Typical Capital Cost ($/kW Capacity)	Typical Annual O&M Cost ($/kW Capacity)
Microturbine	$\leq$ 1MW	5,500	380
Small IC Engine	$\leq$ 1MW	2,300	210
IC Engine	$\geq$ 800kW	1,700	180
Gas Turbine	$\geq$ 3MW	1,400	130

IC — Internal combustion.

In 2007, over 60% of all municipal solid wastes produced in the State of Illinois went to landfills (Fullerton and Miller, 2010). In 2010, 43 active landfills in Illinois accepted more than 35.25 million gate cubic meters (46.1 million gate cubic yards) of municipal waste for disposal (IEPA, 2011b). The "gate cubic yard" refers to the volume of waste as received before compaction in the landfill.

In 1994, the U.S. EPA created a voluntary program known as the Landfill Methane Outreach Program (LMOP). The purpose of this program was to work with companies, utilities and communities to encourage the use of landfill gas for energy and to reduce methane emissions from landfills (US-EPA, 2012a). As of June 2012, there were 33 operational and 25 candidate landfill projects in the State of Illinois, as compared to 36 operational and 38 candidate landfills projects in 1998 (US-EPA, 2012a; US-EPA, 1999). The reason for the reduction is that currently a large portion of the MSW generated in Illinois is recycled.

In 2008, 42.1% of the MSW generated in Illinois was recycled, compared to 37% in 2005 and 21.5% in 1998 (Fullerton and Miller, 2010; IEPA, 2009). Nevertheless, Illinois currently has the third highest number of operational landfill projects in the nation, behind California and Michigan which have 76 and 38 operational landfill projects, respectively. In Table 7.24, we report a comparison of Illinois MSW landfills and their electricity generating potentials in 1998, 2012 and 2014.

The active LFGE projects in Illinois are presented in Table 7.25. The largest project is the Mallard Lake landfill in Hanover Park in DuPage County, Illinois. This landfill has an estimated 24,468,450 tons of waste in place and is capable of producing 23.2 MW of electricity using combined cycle gas turbine.

7.6.10. *Examples of the LFGE projects in Illinois*

Below, we describe three representative examples of the LFGE projects in Illinois based on their technology.

Table 7.24. A comparison of Illinois MSW landfills and their electricity generating potentials in 1998, 2012 and 2014 (US-EPA 1999; US-EPA, 2012a; US-EPA, 2014).

	No. of Landfills			**Capacity (MW)**		
Category	**1998**	**2012**	**2014**	**1998**	**2012**	**2014**
Operational Projects[(a)]	36	33	36	147.9	209	144.5
Candidate Landfills[(b)]	38	25	25	NA	207	NA
Potential Projects[(c)]	NA	31	NA	NA	NA	NA

NA = not available.
(a) Landfill with operational LFGE project or landfill with LFGE under construction. (b) A landfill that is accepting waste or has been closed for five years or less, has at least one million tons of waste and does not have an operational or under construction LFG energy project; or is designated based on actual interest or planning. (c) Landfill that does not meet the candidate definition, whether because of complete or incomplete data. However, the land fill could have LFG energy potential based on site specific needs or if data were complete.

7.6.10.1. *Example 1 LFGE*

In 2003, Antioch Community High School in Lake County, Illinois became the first school in the U.S. to use landfill gas from a small nearby landfill for both heating and electricity generation. The project serves as a model of how a small landfill can be used to produce clean energy. The 360 kW facility consists of 12 capstone microturbines, powered by compressed landfill gas at a relatively small flow rate of 306 standard cubic meters per hour (180 standard cubic feet per minute) (US-EPA 2012b). The landfill gas is delivered from an adjacent landfill (former superfund site) located about half a mile away from the school campus. The energy from the system is used to provide heating and power for the approximately 6 acres (2.42 ha) school building, while the surplus electricity produced is sold to the local electric company (Torresani and Peotter, 2006). The overall cost of the project was estimated at $1.9 million, of which $550,000 was paid with a grant received from the Renewable Energy Resources

Table 7.25. List of operational landfill gas-to-electric-energy projects in Illinois as of June 2014 (US-EPA, 2014).

Landfill Name	City	County	WIP [tons]	Project Type	Elec. [MW]
Advanced Disposal Zion Landfill	Zion	Lake	12,547,170	Reciprocating Engine	5.4
Advanced Disposal Zion Landfill	Zion	Lake	12,547,170	Reciprocating Engine	1.4
Beecher Development Company Landfill	Beecher	Will	5,500,000	Reciprocating Engine	2.12
Brickyard Landfill Units 1 & 2	Danville	Vermilion	2,301,000	Reciprocating Engine	4
CDT LF	Rockdale	Will	1,596,000	Reciprocating Engine	1.6
CID Recycling and Disposal Facility	Chicago	Cook	21,690,122	Gas Turbine	6.6
Clinton Landfill Inc.	Clinton	De Witt	3,800,000	Reciprocating Engine	3.2
Countryside LF	Grayslake	Lake	6,500,000	Reciprocating Engine	7.8
Des Plaines Landfill	Des Plaines	Cook	5,500,000	Reciprocating Engine	3.5
Dixon / GROP Landfill No. 2	Dixon	Lee	1,140,780	Reciprocating Engine	4
Environtech LF	Morris	Grundy	6,387,000	Reciprocating Engine	2.6
Five Oaks RDF	Taylorville	Christian	10,200,000	Cogeneration	3.2
Greene Valley RDF	Naperville	DuPage	18,707,345	Gas Turbine	6.6
H.O.D. Landfill	Antioch	Lake	2,000,000	Cogeneration	0.36
Kankakee County RDF	Chebanse	Kankakee	2,198,000	Reciprocating Engine	1.6
Lake LF	Northbrook	Cook	10,000,000	Gas Turbine	6.6
Land & Lakes #1 and #2	Park Ridge	Cook	11,183,620	Reciprocating Engine	3
Land & Lakes #3	Chicago	Cook	15,000,000	Reciprocating Engine	1
Land & Lakes/Dolton LF	Dolton	Cook	4,202,000	Reciprocating Engine	3
Land & Lakes/Willow Ranch Landfill	Romeoville	Will	2,000,000	Reciprocating Engine	0.4

(*Continued*)

Table 7.25. (*Continued*)

Landfill Name	City	County	WIP [tons]	Project Type	Elec. [MW]
Lansing LF	Lansing	Cook	1,300,000	Reciprocating Engine	2.2
Lee County Landfill	Dixon	Lee	4,030,000	Reciprocating Engine	4
Livingston LF	Pontiac	Livingston	15,132,820	Gas Turbine	15
Mallard Lake LF	Hanover Park	DuPage	24,468,450	Combined Cycle	19
Milam Recycling and Disposal Facility	East St. Louis	St. Clair	16,000,000	Reciprocating Engine	2.4
Prairie View RDF	Wilmington	Will	9,000,000	Reciprocating Engine	4.8
Rochelle Municipal #2	Rochelle	Ogle	4,000,000	Reciprocating Engine	4.8
Roxana LF	Edwardsville	Madison	3,252,000	Reciprocating Engine	2.7
Settler's Hill Recycling & Disposal Facility	Batavia	Kane	13,000,000	Gas Turbine	6.6
South Barrington Landfill	South Barrington	Cook	1,000,000	Reciprocating Engine	1.6
Streator Area LF No. 3	Streator	Livingston	2,500,000	Reciprocating Engine	1
Tazewell RDF	East Peoria	Tazewell	4,049,900	Reciprocating Engine	1.6
Upper Rock Island County LF	East Moline	Rock Island	2,052,000	Reciprocating Engine	2.7
Westchester Landfill	Westchester	Cook	7,000,000	Reciprocating Engine	3.33
Woodland RDF	South Elgin	Kane	7,800,000	Reciprocating Engine	1.6
Woodland RDF	South Elgin	Kane	7,800,000	Reciprocating Engine	3.2
TOTAL					**144.5**

NA = Data not available. WIP = Waste in place.

Program, under Illinois Department of Commerce and Community Affairs. The remaining cost of the project was paid for through the issuance of revenue bonds by school district. The project saves the school about $100,000 annually in energy costs (US-EPA, 2012b). The estimated payback time for the project is 8 years.

7.6.10.2. *Example 2 LFGE*

In December 2011, another LFGE plant in Illinois was opened in Will County, Illinois. The plant is equipped with three 20-cylinder 2,233-horsepower Caterpillar engines and has a generating capacity of 4.8 MW, enough to power 3,000 homes. With future expansion, the facility is expected to power 7,800 homes annually. The 650 square meters (7,000 square feet) plant was partly funded with a $3 million Energy Efficiency and Conservation Block Grant from the US-DOE and $8 million contribution from Waste Management of Illinois which is contracted to operate the plant. The County owns the rights to the landfill gas which is sold to Waste Management of Illinois to generate electricity at the plant. The profit from sale of the electricity is shared between Waste Management of Illinois and the County. Will County is expected to receive a revenue of $441,000 for the first year of operation of the plant, and is expected to reach an average of $1 million per year with future expansion of the plant (AC&C, 2012).

7.6.10.3. *Example 3 LFGE*

In 2009, the Greene Valleys LFGE plant was opened in Naperville in DuPage County, about 45 km Southwest of Chicago, Illinois. This LFGE facility is owned by Forest Preserve District of DuPage County and is operated by Waste Management of Illinois, Inc. The facility collects methane through a network of pipes and wells that are drilled into the wastes. A vacuum system draws the gas from the landfill to a power plant, which fuels two Centaur turbine/generator set to produce electricity. About 113,000 m^3 of landfill gas per day is used by the power plant to produce 6 MW of electrical power (an equivalent of about 135,000 barrels of oil per year) for the surrounding community (Bogner *et al.*, 1997). The electricity is transmitted to ComEd Electric Co. which distributes it to about 5,000 area households and businesses over its electrical grid (Plunkett, 2009).

7.7. Problems Facing 2nd and 3rd Generation (Advanced) Biofuels Production

As Illinois is making a transition to second generation biofuels industry (the use of cellulosic biofuels, etc.), the first generation biofuel industries (conventional ethanol and biodiesel industries) frequently experience a series of problems. These include expiration of federal tax incentive for biodiesel production and high prices for soybeans, which is the main ingredient for conventional biodiesel production. As a result, some conventional biodiesel plants in Illinois have been forced to shut down, while others are looking for alternative feedstock (Dewitte, 2014). Additionally, unfavorable weather conditions such as severe droughts in Illinois could increase soybeans and corn prices, forcing lawmakers to alter the renewable fuel standard (RFS) for corn ethanol. For example, in August 2012, U.S. lawmakers signed a letter asking the EPA for a temporal reduction or suspension of the amount of corn ethanol blended into gasoline under the RFS. The reason for this request was that the widespread drought had devastating effect on corn harvest. This situation was making it impossible for states to meet the 13.2 billion gallons of corn ethanol production target set by the EISA, considering the already tightened corn supplies for livestock feed (Colman, 2012).

In December 2011, the Illinois legislatures passed a tax package which extended the state's biodiesel tax credit for five years (Harker, 2011). In 2010, the federal biodiesel tax incentive was allowed to expire before being extended (Thorstensen, 2010). In December 2011, most of the important biofuels tax incentives approached their expiration date. As a result, several biofuel organizations and companies including Biotechnology Industry Organization, Novozymes and the National Biodiesel Board, submitted a written or verbal testimony to the U.S. Senate Finance Subcommittee on Energy, Natural Resources and Infrastructure, seeking for a hearing titled "Alternative Energy Tax Incentives" (Kotrba, 2011). Consequently, in August 2012, the U.S. Senate Finance Committee released a proposal indicating that the $1 per gallon federal tax credit for biodiesel production would be extended for one year term, with expiration date set on 31 December 2013. This biodiesel tax credit was effective retroactively from 31 December 2011 (Gentile, 2012).

The challenge facing the second generation ethanol and biodiesel production industries is how to optimize their yields to compete with corn seed or soybeans derived biofuels (first generation biofuels) (Dewitte, 2014). Technical problems associated with large-scale use of biological cellulose technology, such as high cost and low specific activity of enzymes and general lack of knowledge regarding enzyme biochemistry and mechanistic fundamentals, have slowed down the pace of cellulosic ethanol fuels production (US-DOE, 2006a). For example, the current conversion efficiency of corn ethanol production is about 399 liters per metric ton of corn, compared with cellulosic ethanol (second generation biofuel) production which ranges from 110–300 liters per dry metric ton of biomass (Mandil and Shihab-Eldin, 2010).

Despite these challenges, the first commercial-scale cellulosic ethanol plant was opened in Emmetsburg, Iowa. Project Liberty, a joint venture of DSM and Poet will convert 285,000 metric tons of biomass per year sourced from within a 45-mile radius of the plant into renewable fuel. Feedstocks will include baled corn cobs, leaves, husk and stalk. As of September 2014, the plant has officially started-up processing its first batch of biomass into cellulosic ethanol and is moving forward towards continuous operation. At full capacity, the plant will convert 770 metric tons of biomass per day to produce ethanol at a rate of 20 million gallons per year. This represents a greenhouse gas reduction of 85–95% over petroleum fuels. The project cost $275 million to develop, with the DOE, Department of Agriculture and the State of Iowa playing important roles in bringing the technology to commercial scale. The DOE awarded a $100 million grant to support the costs of engineering and construction, as well as biomass collection and infrastructure. The State of Iowa contributed $20 million in grants for capital costs and feedstock logistics, while the USDA invested $2.6 million to support the delivery of more than 58,000 dry tonnes of corn crop residue, to help establish the feedstock logistics network (DSM and POET, LLC, 2014). The success of this project could encourage companies to establish similar projects in Illinois because of the state's abundance of corn stover.

Currently, there is no commercial-scale cellulosic ethanol plant operating in Illinois. But as discussed in Sec. 7.9, there are institutions and companies in the state involved in different DOE sponsored research and development projects, working to improve the situations in cellulosic ethanol technology. The high production cost of cellulosic ethanol, including the initial construction costs for unproven technologies and processes on commercial scale involve some investment risk. These problems make cellulosic ethanol commercialization a major challenge to the industry.

According to the US-EIA, despite the expected growth potential for corn ethanol production in the years ahead, the full-scale commercialization of cellulosic ethanol remains a daunting task. The host of factors hindering successful commercialization of cellulosic ethanol include: difficulties in obtaining financing following the U.S. debt crisis, technology scale-up difficulties at startup companies, as well as strategic corporate shifts, due to high production rates of low-cost natural gas (US-EIA, 2013b). The result of these problems is that the RFS set by EPA may not be realized by the year 2022, as projected.

Some of the options being used by advanced biofuels companies to overcome financial constraints during pre-commercial development include alliances with large corporations, venture capital, government grants and loan guarantees (Coyle, 2010). For example, in November 2008, an Illinois-based company, Coskata,

Inc., raised $40 million through venture capital funds for commercialization of multi-feedstock syngas conversion into ethanol. The investors involved in this venture are Blackstone Cleantech Venture Partners, Sumitomo, Arancia, Khosla Ventures, Advanced Technology Ventures, Globespan Capital Partners and TriplePoint Capital. Another example is a joint venture between Danisco and Dupont, which was announced in May 2008. The company, DuPont Danisco Cellulosic Ethanol, is headquartered in Illinois (Curtis, 2010). Although these companies may not necessarily have cellulosic ethanol plants under construction in Illinois, their investments bring economic benefits to the state.

Other problems the Illinois biofuels industry is facing are the regulatory hurdles such as the "blend wall". The blend wall determines the volume and the type of liquid biofuel that can be blended with transportation fuels in accordance with EPA rules (Slating and Kesan, 2011). An example of these regulatory huddles is the reluctance by lawmakers to allow the blending of gasoline with n-butanol manufactured from biomass, also known as biobutanol. Biobutanol is produced using acetone-butanol-ethanol (ABE) fermentation process, a bacterial fermentation process which produces acetone, n-butanol and ethanol from starch. Biobutanol is highly toxic, but it is proven to offer much better advantages as a fuel than ethanol, since it is closer to the average molecular weight of gasoline than ethanol (Sticklen, 2010). Additionally, unlike ethanol, the fuel is non-corrosive and can be transported using existing petroleum pipeline infrastructure.

Another example of a regulatory huddle is a claim of the slow pace adopted by EPA in responding to the incessant requests by the U.S. Corn Growers Association to increase the blend wall for ethanol from 10% to 15%, in order to avoid market saturation of the fuel. To emphasize their demand, in August 2010, a group of farmers and industry leaders across the country, including the Illinois Corn Growers Association, sent a letter to the U.S. President, asking for imposition of an E12 blending policy with immediate effect (Saidak, 2010). They argued that a 12% ethanol blending with gasoline would not damage vehicles and must be an interim step toward EPA's approval of E15. In response to their demand, the EPA finally gave waivers in October 2010 and in January 2011, allowing ethanol blends of up to 15% (E15) in vehicles that were made after the year 2000.

Most of the gasoline consumed in Illinois and in most other states in U.S. contain 10% (by volume) of ethanol (E10), which has been the legal limit for ethanol blends sold for use in vehicles designed to run on gasoline. But, the immediate barriers to the marketing of E15, such as potential misfueling (unsuitable for an engine), associated liabilities, and other related issues may retard the transition to higher blend (US-EIA, 2012b). In addition, since this policy only applies to cars manufactured after the year 2000, owners of gas stations will be unwilling to switch to the E15 blend any time soon, for fear of losing customers whose cars are not adapted to this new blend.

In May 2013, the US-EPA proposed a new rulemaking, which includes several possible modifications to the renewable fuel standard program. The approval of the new proposed rules could allow the US-EPA to categorize ethanol produced from corn fiber and butanol that meets the 50% greenhouse gas emission reduction as advanced biofuels (Jessen, 2013). The proposed rule would make various changes to the E15 misfueling mitigation regulations.

7.8. Sustainability of Biofuels Production

As mentioned in Chapter 1, economics, environmental issues and social acceptance are the three interconnected pillars of sustainability in any renewable energy, including biofuels, which must be taken into account for successful development of that technology. Therefore, to evaluate the overall burden of the biofuels utilization systems, the economic analysis, impacts of biofuels emissions to environment (which includes emissions for producing equipment for biofuels energy systems) and biofuels system water usage and land footprint must be evaluated.

One of the key motivating supports for cellulosic biofuels production is their positive impact on the environment and greenhouse gas emissions. In other words, producing biofuels must have less detrimental effects on the environment when compared with fossil fuels. In order to meet biofuels production goal, large quantities of biofuels crops (cellulosic biomass) must be produced in an environmentally sustainable and economically viable manner. In order to be economically sustainable, biofuels crops must generate an income for a landowner that at least covers the cost of production. The production cost of biofuels crops is highly dependent on location, as the cost of land, planting, chemicals and fertilizer, harvesting, storage, labor, utilities, repair, overhead and price of machinery may vary significantly at different locations.

In theory, biofuels produced from biomass should be carbon neutral, as their combustion only returns the CO_2 to the atmosphere that was withdrawn from the atmosphere by plants during their growth. However, biofuels vary widely in their greenhouse gas balances when compared with petroleum fuels. Depending on the methods used to produce the biofuels feedstock and the conversion process, some crops can generate even more greenhouse gases than fossil fuels. For example, nitrous oxide (N_2O), a greenhouse gas which has a global warming potential roughly 300 times greater than that of CO_2 and the third most important gas in causing global warming after CO_2 and methane, is released from nitrogen fertilizers (FAO, 2008).

The impact of biofuels on climate mitigation (i.e., reduction of greenhouse gas emissions) varies according to type of feedstock, feedstock production location, agricultural practice and conversion technology (FAO, 2008).

Assessment of the net effects of biofuels crops production on greenhouse gas (GHG) emissions must include analysis of emissions throughout the life cycle of the biofuels. This analysis should include planting and harvesting of the crop, processing the feedstock into biofuels, transporting the feedstock and the final fuel, and storing, distributing and retailing the biofuels.

Therefore, while discussing environmental sustainability of biofuels crops production, the effects on greenhouse gas emission, agricultural land, biodiversity, water resources and soil erosion must be addressed. These issues are somewhat interconnected and there is a wealth of information available in the literature, such as the book by Speight and Singh (2014) offering detailed information on this subject, but we present a brief discussion in the following subsections.

7.8.1. *Greenhouse Gas (GHG) emissions*

When sunlight hits the Earth's surface, some of it is reflected back towards space as heat. GHGs are a group of compounds that are able to trap heat in the atmosphere and as a result, enhance global warming. Studies on the subject of GHG emissions of biofuels are controversial, as the arguments are both in favor and against development of biofuels from biomass.

Searchinger *et al.* (2008) argue that while many studies indicate substituting biofuels for gasoline will reduce greenhouse gases, those analyses failed to consider carbon emissions as a result of land-use change (LUC). The land-use change occurs as farmers respond to higher prices and convert forest and grassland to new cropland for biofuels. By using a worldwide agricultural model to estimate emissions from LUC, Searchinger *et al.* found that corn-based ethanol, instead of offering a 20% savings, increases emissions by a factor of two over 30 years. Biofuels from switchgrass, if grown on U.S. corn lands, increase emissions by 50%.

In another study, Howarth and Co-workers (2009) indicated that biofuels grown on cleared forests or grasslands will result in release of GHG from soil and vegetation that exceeds reduction of GHG emission that could result from the use of biofuels. Howarth *et al.*, (2009) study also indicated that release of N_2O, a much more powerful GHG than CO_2, into the atmosphere often increase with enhanced use of nitrogen fertilizer. Another study conducted by Liska *et al.* (2014) showed that removal of corn residue for production of biofuels can decrease soil organic carbon and increase CO_2 emissions because residue (carbon) in biofuels is oxidized to CO_2 at a faster rate than when added to soil.

In contrast, Wang and Co-workers (Wang *et al.*, 2012) showed that GHG emissions of first-generation biofuels is lower than petroleum-based fuels, but higher than second-generation biofuels, which typically offer emission reductions in the order of 88–108%, compared with petroleum gasoline. Baral and Malins (2014) evaluated GHG emission by biofuels using data from the UK Department for Transport biofuels carbon calculator. Their study shows that cellulosic biofuels derived from agricultural residues offer substantial GHG savings exceeding 60% for gasification, pyrolysis and fermentation pathways. The study reports that GHG benefits are even higher for municipal solid waste, since their collection can result in avoided methane emissions.

7.8.2. *Agricultural land use and conversion*

Over the past decade, the rate of growth in demand for biofuels has exceeded the rates of growth in demand for agricultural products and in crop yields. If this trend continues, LUC may become a more important issue with respect to first and second generation biofuels technologies. Increase in demand in agricultural lands limits the use of those lands for biofuels crop(s) production. Estimates of increase in the rate of agricultural land from 1997 to 2050 vary from 0.01%–0.34% per year, which translates to increase in agricultural land between 4 and 137 Mha by the year 2050 (Sala *et al.*, 2009).

A study by Ravindranath *et al.* (2009) estimates that 118–508 Mha of land would be required to meet a target of 10% biofuels in transport fuels globally, depending on the main crops used to meet the targets. The availability of land for biofuels production is subject to debate. Optimistic estimates suggest that globally there could be up to 1,215 Mha of land available, whereas pessimistic estimates are as low as 400 Mha.

There is strong evidence that converting forests for crop production of any kind can release quantities of GHGs that far exceed potential annual savings from biofuels. However, replacing annual crops with perennial feedstocks can improve soil carbon balances (FAO, 2008).

The potential for growing energy crops on land that is currently idle/marginal and not productive for food crops makes energy crops an attractive source of feedstock for biofuels.

United States is the second biggest ethanol producer in the world and has allocated nearly 4 Mha of land to biofuels crops (or 4% of the total cropped area), with an average yields of roughly 3,300 L/ha mostly from maize. Data from various sources indicate that global average ethanol production from 1 ha of land is around 3,500 L (De Fraiture *et al.*, 2008).

A study led by Michigan State University and conducted by Gelfand *et al.* (2013) evaluated the potential of marginal lands in ten Midwestern U.S. states. The study showed that using marginal lands for growing cellulosic biomass crops could provide up to 215 gallons of ethanol per acre (~2000 liter per ha) with substantial reduction in GHG emission. The study also showed that marginal lands represent a huge untapped resource to grow mixed species cellulosic biomass, which could annually produce up to 5.5 billion gallons (130 million barrels) of ethanol in the Midwest alone.

Land use change (LUC) to cultivate biomass feedstock for biofuels production may lead to GHG emissions, if it changes the amount of carbon stored in vegetation and soil. Therefore, both the direct and indirect land-use changes caused by expanded biofuels production need to be considered for a full understanding of potential environmental impacts (Speight and Singh, 2014).

7.8.3. *Biodiversity consequence of biofuels production*

Biodiversity means the degree of variation of plants, animals, or other living things in a particular area. Biodiversity loss is attributed to many factors, namely habitat loss, introduction of non-native species, and overexploitation of natural resources. If not done properly, plantation of biofuels crops can have adverse impacts on biodiversity. The impacts are a function of the biodiversity prior to- and after the land has been converted for biofuels feedstock production and the effects of the biofuels feedstock plantations on the surrounding areas (Campbell and Doswald, 2009).

Assessments have shown that habitat loss will be the major driver of biodiversity loss in the coming decades. The most drastic form of habitat loss occurs when agricultural fertilizers used for plantation of biofuels crops result in runoff of chemicals (i.e., phosphorus and nitrogen) which can impact aquatic ecosystems and water quality, as well as contributing to increased frequency and magnitude of harmful algal blooms (Carpenter *et al.*, 1998). Applying fertilizers as well as pesticides, which include herbicides and insecticides, is a standard agricultural practice. However, these chemicals can wash into bodies of water and affect water quality. For example, excess nitrogen washing into the Mississippi River is known to be a cause of the oxygen-starved "dead zone" in the Gulf of Mexico, in which marine life cannot survive (NAS, 2007).

The impacts of biofuels on biodiversity will depend on the type of crops that are planted and the prior land use. Biofuels can be beneficial to biodiversity when appropriate crops are grown in suitable areas. It is believed that if biofuels contribute to climate change mitigation, they have the potential to be (indirectly) beneficial to biodiversity as well. However, biofuels have already been shown to "negatively impact biodiversity when direct conversion of natural ecosystems or indirect land conversion of non-degraded land occurs" (Campbell and Doswald, 2009). Conversion of existing and new croplands for production of biofuels crops will have environmental effects beyond their impacts on GHG emissions and will likely have detrimental effects on biodiversity.

Tilman *et al.*, (2006) conducted an experimental study to evaluate native perennials as a potential feedstock for biofuels production.

Their data from test plots on degraded and abandoned land showed that low-input high-diversity mixtures of native grassland perennials produce higher net energy gains, greater GHG emission reductions and less agrichemical pollution than maize-ethanol or soybean-biodiesel. In addition, these plants offer a range of ecosystem services, including wildlife habitat, water filtration and carbon sequestration. The authors also found that switchgrass can be highly productive on fertile soils, especially when fertilizer and pesticides are applied, but that its performance on poor soils does not match that of diverse native perennials.

It is therefore necessary that while assessing biofuels crop(s) production, quantitative models to be developed that take into account regional, local and site-specific impacts on, and potential longer-term benefits of biodiversity.

7.8.4. *Water implications*

One of the challenges of biofuels development is their effects on water and related land resources. The questions of how water use and water quality will be affected as agricultural practices shift to include more energy and food crops remain to be thoroughly investigated. Although biofuels crops are mostly rain-fed, they may require irrigation to some extent from groundwater or surface water sources. For crops that grow primarily under rain-fed conditions, water availability is not a constraint, but water pollution associated with the application of fertilizers and agrochemicals, soil erosion and other steps in the ethanol production process are major concerns (Moreira, 2007). The primary concern with regard to water availability is how much irrigation will be required that might compete with water used for other purposes.

Although there are sufficient land and water resources globally, distribution of these resources are not uniform. India and China together have 35% of the world's population and it is estimated that 30–40% of the global energy will be needed by these two countries by 2030. Both countries have already exploited most of their available natural water resources for agriculture and there is little land and water left to expand areas for crop production (De Fraiture *et al.*, 2008; Müller *et al.*, 2008).

Globally, biofuels account for about 100 km^3 (or 1%) of all water evapotranspired by crops worldwide and about 44 km^3 (or 2%) of all irrigation water withdrawals. Even perennial plants such as jatropha and pongamia that can be grown in semi-arid areas on marginal or degraded lands may require some irrigation during hot and dry summers (De Fraiture *et al.*, 2008).

Water use of energy crops is highly dependent upon the geographic location and the type of crop. For example, a liter of ethanol based on irrigated sugarcane in India requires 3,500 liters of irrigation water withdrawals and a liter of ethanol based on irrigated maize in China takes 2,400 liter of irrigation water withdrawals. But in Brazil where sugarcane is largely rain-fed, it only takes 90 liters of water to produce one liter of ethanol. In the U.S., one liter of ethanol obtained from mostly rain-fed maize require 400 liters of irrigation water withdrawals. (De Fraiture and Berndes, 2009; De Fraiture *et al.*, 2008).

King and Webber (2008) conducted a study to estimate water intensity for fuel production. In this context, water intensity is a measure of the water needed for the production and processing of fuel, and therefore a measure of the efficiency of water use. Water intensity was measured in "gallons of water per mile traveled" using a light duty vehicle (LDV). Based on their study, the lowest water consumption rates are for LDVs using non-irrigated biodiesel. Table 7.26 shows the water intensity for fuel production for a variety of fuels.

Use of groundwater for biofuels crops may pose a long-term challenge. Groundwater accounts for almost all of the long-term storage

Table 7.26. Averages of water intensity of selected fuels (King and Weber, 2008).

Energy Source	Fuel	Water Intensity[3,4] (gal/mile)	Water Intensity[3,4] (l/km)
Petroleum	Gasoline	0.07–0.14	0.16–0.33
Petroleum	Diesel fuel	0.05–0.11	0.12–0.26
Oil shale	Gasoline	0.15–0.37	0.35–0.87
Tar sands	Gasoline	0.2–0.46	0.47–1.09
Coal	F-T diesel	0.19–0.58	0.45–1.37
Natural gas	F-T diesel	0.12–0.43	0.28–1.0
Corn grain from irrigated fields[1]	Ethanol	1.3–62	3.07–146.6
Corn grain from non-irrigated fields[1]	Ethanol	0.15–0.35	0.35–0.83
Corn stover from irrigated fields[2]	Ethanol	2.6–46	6.15–108.8
Corn stover from non-irrigated fields[2]	Ethanol	0.25	0.6
Soybean from irrigated fields	Biodiesel	0.6–24	1.4–56.8
Soybean from non-irrigated fields	Biodiesel	0.01–0.02	0.024–0.05

(1) Assume only ethanol is from corn grain.
(2) Assume only ethanol is from corn stover.
(3) The results are for well-to-wheel (or field-to-wheel, as the case may be) analyses.
(4) The vehicle fuel efficiency is 8.7 km/l or 20.5 mpg.
FT diesel = Diesel fuel produced from coal through FT process.

of water and therefore extracting groundwater for irrigation may decrease the water available for future users of the aquifer. In some areas of the U.S., water resources are already significantly stressed. For example, large portions of the Ogallala (or high plains) aquifer, which extends from Texas up into South Dakota and Wyoming, show water table declines of over 100 feet (~30 meters) since the 1940s. Increased biofuels production will likely add pressure to the water management challenges the nation already faces (NAS, 2007).

High water usage of corn ethanol is one of the main reasons that second generation (cellulosic) biofuels are receiving more attention in recent years. Lignocellulosic feedstocks, such as grasses and woody crops, are mainly grown under rain-fed conditions and can support biofuels production with relatively low water requirements. It is estimated that second generation biofuels water requirements for cellulosic ethanol to be 2.5 liter/liter (Czapar, 2008; Lingaraju *et al.*, 2013).

Large-scale biofuels production poses both opportunities and challenges to the water sector. Water consumption for biofuels production depends on the type of feedstock, location of production, productivity, agricultural practices and water management. Multiple approaches in finding creative solutions are needed to ensure sustainable production of biofuels. In general, crops that require less irrigation, less fertilizer and pesticides and provide better year-round erosion protection will likely produce fewer negative water impacts.

7.8.5. *Soil erosion*

Soil erosion is a naturally occurring process that affects all landforms and it occurs when soil particles detach from the soil surface by wind, rain, or flowing water and are then transported by water or wind. Soil erosion involves three distinct actions: soil detachment, movement, and deposition. The effects of soil erosion are beyond just the loss of fertile land. It leads to increased pollution in streams, wetlands, rivers and lakes, clogging the waterways and causing declines in fish and other species and reducing cropland productivity.

Living vegetation or crop residues protect the soil surface from erosion, but when the soil surface is not covered by plant materials, water dislodges soil particles from aggregates. Removal of crop residues beyond a certain threshold substantially increases the rate of soil erosion. In that case, the soil particles bounce around on the soil surface and can plug large pores, which can lead to lower water infiltration (or absorption into the soil), increased water runoff, and in turn, more soil erosion (Cruse and Friedman, 2014). In addition, removal of plant residues can reduce soil nutrient contents and increase GHG emissions through loss of soil carbon.

Land-use change and increase of agricultural production on existing croplands can have significant adverse effect on soils, but these impacts depend on farming techniques. Inappropriate farming practices can reduce soil organic matter and increase soil erosion by removing permanent soil cover.

Growing perennials such as palm, short-rotation coppice, sugar cane or switchgrass instead of annual crops can improve soil quality by increasing soil cover and organic carbon levels. Perennial crops such as eucalyptus, poplar, willow, or grasses require less-intensive management and less energy inputs and can also be grown on poor-quality land, while increasing soil carbon and quality over time (IEA, 2006).

Using perennial grasses and trees as energy crops could help prevent erosion and restore lands degraded by human activities. However, using corn stover could exacerbate erosion if farmers remove too much stover from the ground after harvest (WIE, 2006). When farmers start growing crops for biofuels production, they should consider strategies such as cover cropping and crop rotation to minimize soil erosion. Use of perennials and native grasses, such as switchgrass may reduce soil erosion on marginal lands.

7.8.6. *Sustainability of biofuels crops production in Illinois*

There is a growing interest in biofuels as a renewable alternative to fossil fuels. Replacement of fossil fuels by biofuels requires significant production of biomass that in the long run is economically, environmentally and socially sustainable. First generation production of biofuels from food crops is not sustainable in the long term and therefore, the focus of biofuels production is on the second and third generation technologies.

Among the potential feedstocks for production of biofuels, miscanthus (*Miscanthus* × *giganteus*) was widely studied as an energy crop in Europe (Heaton *et al.*, 2008) and switchgrass (*Panicum virgatum L.*), a perennial grass native to North America was selected by the US-DOE as a model energy crop (Liebig *et al.*, 2008). In general, perennial crops such as miscanthus and switchgrass seem to have fewer negative environmental impacts than annual crops, due to low use of fertilizer and pesticides.

Zeri *et al.* (2011) studied carbon sequestration of miscanthus (*Miscanthus* × *giganteus*), switchgrass (*Panicum virgatum*), a mixture of native prairie plants and maize and soy in Central Illinois, during the years 2008–2011. Their data showed that the net ecosystem carbon balance, which accounts for harvested biomass and grain at the end of 2.5 years of establishment of the grasses, were a sink of carbon, whereas the maize/soybean plot was a net source of carbon to the atmosphere.

In another study, Zeri *et al.* (2013) evaluated water use efficiency of the same crops (maize/soybean, miscanthus, switchgrass and a mix of native prairie species) in Illinois. The results of the study indicated that ecosystem water use efficiency was highest in miscanthus, reaching a maximum value of 12.8 ± 0.3 kg ha^{-1}mm^{-1} in the third year, followed by switchgrass (7.5 ± 0.3 kg ha^{-1}mm^{-1}) and prairie (3.9 ± 0.3 kg ha^{-1}mm^{-1}). It was also revealed that miscanthus had the highest net benefits to the carbon stocks and biofuels production.

Huang and Khanna (2011) studied profitability of biofuels feedstocks such as corn stover, miscanthus, switchgrass, native prairie grasses and poplar under alternative agronomic and economic scenarios in Illinois, Michigan and Oklahoma. Based on their study, the average delivered yield at the farm gate for corn stover is the highest in Illinois at 3.41 metric tons of dry mass per hectare (MTDM/ha), followed by Michigan at 2.69 and Oklahoma at 1.84 MTDM/ha. Huang and Khanna also performed simulations using a crop productivity model, MISCANMOD, and showed that the farm gate biomass yield of miscanthus is high in the Atlantic states and low in the western states due to insufficient soil moisture. They also found that Southern states have higher yields for miscanthus, as compared to Northern states. The average delivered yields and standard deviation (SD) of miscanthus at the farm gate (after accounting for storage losses) were found to be highest in Oklahoma at 29.54 ± 5.59 MTDM/ha followed by Illinois at 25.16 ± 3.88 MTDM/ha and Michigan at 15.45 ± 4.46 MTDM/ha.

Similarly, MISCANMOD simulations showed average delivered yields of switchgrass at the farm gate (after accounting for storage losses) in Illinois, Oklahoma and Michigan to be 10.74 ± 1.23, 9.70 ± 1.53 and 7.53 ± 0.58 MTDM/ha, respectively. These simulated switchgrass yields are within the range reported in literature. DAYCENT simulations model was used by the authors and the results showed the average delivered yield of mixed prairie grasses at the farm gate (after accounting for storage losses) in Illinois to be 7.02 (3.95–10.24) MTDM/ha/yr.

Huang and Khanna also reported that the average delivered yields of hybrid poplar at the farm gate (after accounting for storage losses) was 9.74 ± 1.12 MTDM/ha/yr with a 10-year rotation in Illinois, 9.4 ± 0.69 MTDM/ha/yr with an 8-year rotation in Michigan and 3.15 ± 1.58 MTDM/ha/yr with a 6-year rotation in Oklahoma. Table 7.27 shows the delivered crop yield in Illinois for several biomass feedstocks.

Based on the yield of the crops and considering the cost of machinery and harvest, Huang and Khanna calculated the breakeven farm gate price ($ per ton of dry matter) for each crop. According to their definition, the breakeven cost is the "minimum price per dry metric ton of the bioenergy crop that a cropland owner would need to receive each year to cover all the costs of production over the life of the crop". The results of breakeven cost of biomass production are shown in Table 7.28.

The results show that among the three perennial grass production systems examined, miscanthus has the lowest costs of production when planted on marginal land. In general, the cost of biomass production is lower on marginal land but it will be viable on cropland

Table 7.27. Delivered crop yield in Illinois (Huang and Khanna, 2011).

Crop	Yield (bushel/ha)	
Corn	386	
Soybean	108.5	
Biomass	**Yield (MTDM/ha/yr)**	
Stover (RC, CT)	1.16	
Stover (RC, NT)	1.94	
Stover (MC, CT)	2.05	
Stover (MC, NT)	3.41	
	Low Cost[(a)]	**High Cost[(a)]**
Miscanthus	25.16	18.46
Switchgrass	10.74	9.61
Poplar	9.74	8.63
Prairie grass	7.02	6.28

RC = Rotation corn; MC = Monoculture corn; CT = Conventional tillage; NT = No tillage. MTDM/ha/yr = Metric tons of dry mass per hectare per year.
(a) = $/dry metric ton

only if the price of biomass is sufficiently high and low cost marginal land is unavailable.

7.9. Biofuels Programs in Illinois

7.9.1. *Funded by the state*

In 2011, the Illinois Biomass Working Group (IBWG) was formed. The group aims to study the near-term uses for biomass, as the state transitions to energy-independent economy (Schroeder, 2011). The group consists of academics, government, industry and private sector. Additionally, Illinois Department of Commerce and Economic

Table 7.28. Breakeven cost in (US$/MTDM) of biomass feedstock production in Illinois (Huand and Khanna, 2011).

Biomass	Agricultural Practice	Harvesting Type	Cost Plus Marginal Land Rent at Farm gate*	Cost Plus Crop Land Rent at Farm gate
Corn stover	Rotation corn	CT	NA	60.71
		NT	NA	59.81
	Monoculture corn	CT	NA	96.34
		NT	NA	66.03
Switchgrass	Hi cost	NA	89.97	134.89
	Low cost	NA	59.31	99.35
Miscanthus	Hi cost	NA	72.62	95.89
	Low cost	NA	44.7	61.79
Prairie grass mix	Hi cost	NA	108.59	177.01
	Low cost	NA	83.89	145.15
Hybrid poplar	Hi cost	NA	98.63	148.46
	Low cost	NA	90.06	134.16

Note: CT = Conventional tillage; NT = No tillage.
* For corn stover, the costs of production in this case include no opportunity cost of land.
MTDM = metric tons of dry mass.

Opportunity (DCEO) of Illinois governor's office has several biofuels programs designed to encourage the use of biofuels in the state. These include the Illinois Renewable Fuel Development Program (RFDP), which provides grants of up to $5.5 million per facility for new or expanding biofuels production facilities, Illinois E-85 Clean Energy Infrastructure Development Program (SEP), which provides grants up to $6,000 for conversion of an existing station to E-85 station or up to $30,000 for building a new E-85 station and up to $30,000 for installation of ethanol blender pumps (Mrozowski, 2010).

Other programs are the Illinois Renewable Fuel Research, Development and Demonstration program under DCEO, which provides funding for new technologies to increase ethanol production efficiencies, or commercialize new technologies, or for GHG and water use reduction studies and utilization of new technologies (Mrozowski, 2010). The Illinois Next Generation Biofuels Research, Development and Demonstration grant program (Capital Bill) provides funding for Next Generation Biofuels production in Illinois, including 30% of the cost up to $5 million per project in high priority areas, such as fractionation, corn fiber and stover conversion, cellulosic conversion, water reduction technologies and energy production efficiency (Mrozowski, 2010).

7.9.2. *Funded by the Federal government*

Since 2009 the US-DOE has awarded Illinois companies with various contracts to help them research and develop biomass conversion technologies into biofuels as described below:

7.9.2.1. *Project 1*

With DOE award of approximately $25 million, Archer Daniels Midland (ADM) Company headquartered in Decatur, Illinois is developing and constructing a $36 million facility in Decatur to convert corn stover into biofuels and other products (Lusvardi, 2009). The new ADM Advanced Biorefinery project will use acid to breakdown biomass, which can then be converted into liquid fuels or energy, such as ethanol and ethylacrylate, which is used to make plastics, adhesives, coatings and other materials. Prior to this contract, ADM working in partnership with Deere & Co. and Monsanto Co. developed a technology for converting corn stover into pellets (Lusvardi, 2009). The high transportation cost of corn stover biomass, due to its low density (large volume), poses a major challenge for its transportation to the plant site and conversion into liquid fuels and chemicals. The ADM technology involves pretreatment and pelletizing of the corn stover at the farm to increase its density and facilitate its transportation.

7.9.2.2. *Project 2*

With DOE award of approximately \$2.5 million, the GTI located in Des Plaines, Illinois completed a preliminary engineering design of a novel process for producing green gasoline and diesel from woody biomass, agricultural residues and algae. GTI conducted research and development on a novel process called IH^2 for economic conversion of wood, agricultural waste and algae biomass into fungible gasoline and diesel. GTI determined the optimal catalyst and test conditions for IH^2 and demonstrated the effective use of wood and agricultural waste, including corn stover and algae for gasoline and diesel production. Furthermore, GTI analyzed test data for three types of biomass material and gathered metrics for construction and scale up of a 1 ton per day integrated pilot facility and a commercial-size facility. Additionally, GTI runs a continuous 50kg per day of the IH^2 pilot plant for a prolonged period to obtain data on catalyst life and attrition (US-DOE, 2012a).

7.9.2.3. *Project 3*

With DOE award of approximately \$25 million, Haldor Topsoe Inc, headquartered in Houston, Texas has constructed a biofuel pilot project in Des Plaines, Illinois. The pilot project converted woody biomass to green gasoline using full integration and optimization of a multi-step gasification process (US-DOE, 2012b). The pilot plant resulting from this project is expected to process 21 tons of woody biomass per day. The Haldor Topsoe, Inc.'s pilot project location is within the GTI's Gasification Testing Complex in Des Plaines, Illinois. The objective of this integrated pilot facility is to process approximately 21.7 dry metric tons of woody biomass per day to produce 22.5 barrels of "green" gasoline per day (~345,000 gpy, or 1.3 million liter per year). The company sources the wood feedstock from UPM-Kymmene, a pulp and paper company. The wood feedstock which comes from milling processes is harvested from non-federal lands. The pilot plant start-up was in mid-2012 and was planned to continue operation till December 2014. The pilot plant is estimated to have 92% GHGs reduction compared to emissions from gasoline production plants (refineries) using fossil fuels (crude oil) as feedstock (US-DOE, 2012b).

In October 2013, the GTI in Des Plaines, Illinois working with Haldor Topsoe Inc. of Denmark announced their successful production of 4,000 gallons (15,000 liter) of green gasoline suitable for use as a gasoline blendstock. The green gasoline was used for single-engine emissions testing, in order to demonstrate the effectiveness of renewable gasoline in meeting EPA's standards of gasoline blends for up to 80% (GTI, 2014). In March 2014, the final test campaign of the pilot plant produced sufficient quantities of green gasoline for testing in vehicle engines. This test proved that the green gasoline can be used in existing automobile engines. As a result, about 7,770 gallons (29,400 liter) of the green gasoline was sent to a blending facility in Michigan in order to prepare it for a fleet test at the Transportation Research Center located at East Liberty, Ohio. In this test, four pairs of vehicles were expected to log 75,000 miles (120,000 km) each of comparing performance, using the bio-based gasoline blended with conventional gasoline. The results from this test were expected in September 2014 (Lane, 2014).

7.9.2.4. *Project 4 (R&D)*

Elevance Renewable Sciences headquartered in Woodridge, Illinois, is claimed to be a producer of high-performance renewable specialty chemicals. In 2012, the company completed a DOE project on cellulosic ethanol production in Bolingbrook, Illinois, with a grant worth \$2.5 million. The company claims to have generated specific data for high potential U.S. feedstocks to help in the design of important sections of a biorefinery (Fig. 7.23). The data is intended to help in the conversion of natural oils into fuels and chemicals. Additionally, the company claims to have developed a non-site specific process design, detailed engineering and performed analysis of the sensitivity of the economics of the process using algae oil (US-DOE, 2011b).

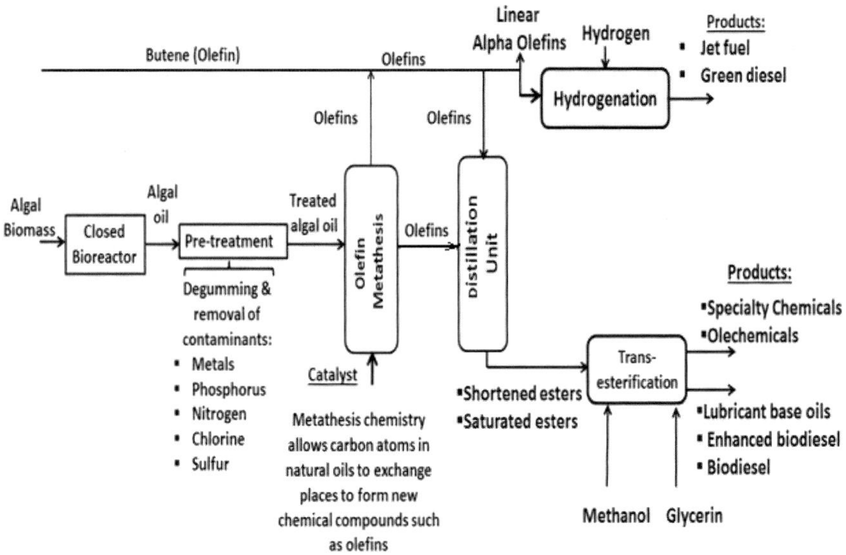

Fig. 7.23. Conceptual project schematic biorefinery (US DOE, 2011b).

A biorefinery resulting from this project is intended to convert renewable feedstock sourced from the U.S. into a wide range of high value products. These high value products include petroleum replacement specialty wax products, olefins, jet fuel, paraffins and specialty chemicals. The conceptual project schematic of the biorefinery is shown in Fig. 7.23. Depending on algae oil content, the GHG reduction of this technology is claimed to be about 67–77% compared with fossil fuels (US-DOE, 2011b).

7.9.2.5. *Project 5* (*R&D*)

In 2010, a San Francisco based company, Solazyme, Inc., received a DOE contract of about $22 million for construction, operation and optimization of a pilot-scale integrated biorefinery in Peoria, Illinois. The company's claimed innovative process harnesses the capacity of some algae to produce as much as 75% of the dry weight of the cells as oil. In this process, algae are efficiently grown in the dark to very high cell densities using industrial fermentation vessels, while being fed with carbon substrates within their growth media (US-DOE, 2011a). The algae metabolize the carbon substrates supplied in the growth media and convert them into triglycerides, almost identical in composition to common vegetable oils. It is claimed that the quantity of oil produced per day per liter of the fermenter volume is extremely high. This high productivity makes the process extremely capital efficient and economically far more attractive than biofuel concepts using photosynthetically grown algae, or many of the other non-ideal fuel substitutes. Based on this technology, the pilot-scale integrated biorefinery in Peoria will be used to validate the projected commercial-scale economics of producing multiple advanced biofuels. The validation would enable Solazyme to gather the necessary data to complete the design of its first commercial-scale facility. This demonstration/commercial-scale plant is claimed to have a capacity of 2 million liters per year of oil. The plant is also claimed to provide a platform for continued work on feedstock flexibility and scaling of new tailored oils into the marketplace. The commercial plant is intended to have a capacity of ~7,500 barrels (300,000 gallons, 1.13 million liters) per year of purified algal oil. The peak capacity feed will be 13 metric tons of dry feedstock per day, comprising of sucrose (from cane), municipal green waste and switchgrass. It is anticipated that the project will create or preserve 88 direct jobs and 256 indirect jobs per year. The greenhouse gas reduction by this technology is claimed to be >90% (US-DOE, 2011a). The company has developed three types of fuels; Soladiesel$_{BD®}$, Soladiesel$_{RD®}$ and Solajet, claiming these fuels meet ASTM standards for diesel fuel oils and aviation turbine fuels (Solazyme, 2015).

7.10. Concluding Remarks

Biofuels are some of the more promising forms of alternative energies that can help Illinois reach sustainability: energy independence with cleaner environment. In all, Illinois 1st generation biofuels production (i.e., ethanol and biodiesel production) in 2011 was estimated at

37.19 million barrels per year (US-EIA, 2012a; US-EIA, 2014b). Also, it is estimated that the state has the capacity to produce 34.4×10^6 barrels (1.44 billion gallons) of ethanol and 4.60×10^6 barrels (193 million gallons) of biodiesel annually (US-EIA, 2014b; Nebraska Energy Office, 2014). In biofuels production, Illinois ranks second only to Iowa in crop residues resources, estimated at 19,593 thousand metric tons per year, enough for sustained production of biofuels (Milbrandt, 2005).

As we mentioned in Sec. 7.4.1, use of low impact crops for biofuels production in Illinois will be a more viable option to reduce the impact of biofuels production on the price of corn and soybeans. These low impact crops include agriculture residues, forest residues, and energy crops, such as miscanthus giganteus, switchgrass, shrub willow and hybrid poplars. Table 7.29 shows comparison of Illinois biomass production, the anticipated cellulosic ethanol production in 2022 and corn ethanol production and consumption in the state.

The US-EPA has projected that in 2022, Illinois biomass will contribute about 9.5% of the 380.95×10^6 barrels (16 billion gallons) of the EISA's cellulosic biofuels standard (US-EPA, 2010). Out of this amount, about 95% will come from corn stovers, which are projected to yield about 34.38×10^6 barrels of cellulosic ethanol per year (US-EPA, 2010). Additionally, Table 7.29 indicates that the estimated

Table 7.29. Projected productions of Illinois biomass and cellulosic ethanol, compared with corn ethanol production and consumption in 2012.

Total biomass resources in 2005 (updated in 2008)[(1)] (dry metric tons/year)	$28,284 \times 10^3$
EPA's projected cellulosic ethanol production volumes in 2022[(2)] (barrels/year)	36.31×10^6
Projected percent cellulosic ethanol contribution to EISA standard in 2022[(2)] (%)	9.5
Estimated corn ethanol produced in 2012[(3)] (barrels)	30.323×10^6
Estimated corn ethanol consumed in 2012[(4)] (barrels)	11.750×10^6

(1) Milbrandt (2005); (2) US-EPA (2010); (3) US-EIA (2014b); (4) US-EIA (2014a).

30.32×10^6 barrels of corn ethanol produced in the state in 2012 is less than the EPA's projected cellulosic ethanol production in 2022. This means that in the long term, there could be more cellulosic ethanol produced in the state than the current production of corn ethanol, if the yield of cellulosic ethanol can be improved sooner.

In Table 7.30, we present the projected Illinois cellulosic ethanol production in 2022, compared with the energy consumption in the state in 2012. According to Table 7.30, EPA does not anticipate any cellulosic ethanol production from energy crops in 2022 in Illinois. So they assigned "zero gallons" to energy crop projected ethanol production in 2022. This is probably because these crops are still under experimentation.

The projected volume of cellulosic ethanol in 2022 is 29.32×10^{15} calories. This represents $\sim$9.9% of the amount of petroleum (294.9×10^{15} calories) consumed in 2012 and just 3% of total amount of fossil energy (9.734×10^{17} calories) consumed in 2012. The amount of petroleum included fuel ethanol (10.2×10^{15} calories) blended into motor gasoline. However, considering the large quantity of fossil fuels consumed in the state (Agyarko and Mansoori, 2013), more cellulosic ethanol needs to be produced in order to realize any appreciable replacement of fossil fuels by cellulosic ethanol in future. The additional cellulosic ethanol can be produced in Illinois from energy crops which projected volume contribution to the EISA's cellulosic biofuel standard in 2022 is zero. Given that in 2022 Illinois' annual contribution toward the fulfillment of the EISA's cellulosic biofuel standard will be 36.31 million barrels (1,525 million gallons) (US-EPA, 2010), the calories equivalent of this amount of energy is

$$1525 \times 10^6 gal\ of\ ethanol \times \frac{76000\ Btu}{1\ gal\ of\ ethanol} \times \frac{251.99\ cal}{1\ Btu} = 29.21 \times 10^{15} \quad (7.9)$$

In the above calculation, the lower heating value equivalent of one gallon of ethanol is 76000Btu, where 1Btu is the equivalent of 251.99 calories.

Table 7.30. Illinois primary energy consumption by source in 2012 and cellulosic ethanol production projected for 2022.

Sources of energy in 2012[(a)]	**[Calorie]**
Nuclear	254.6×10^{15}
Coal	244.2×10^{15}
Natural gas	236.1×10^{15}
Petroleum[(1)]	294.9×10^{15}
Biomass[(2)]	36.3×10^{15}
Other[(3)]	19.7×10^{15}
Net outflow of electricity	112.3×10^{15}
Total energy consumed	973.5×10^{15}
Sources of EPA's cellulosic ethanol projected volumes in 2022[(b)]	**Calories** [(4)]
Agriculture residues	24.4×10^{15}
Energy crops	0
Forest residues	1.11×10^{15}
Urban waste	3.81×10^{15}
Total projected energy of cellulosic ethanol	29.21×10^{15}
End-use energy component replaceable by cellulosic ethanol in 2022[(5)]	**%**
Petroleum	9.9
Total energy consumed	3

(a) US-EIA (2014a); (b) US-EPA (2010).
(1) Petroleum includes fuel ethanol blended into motor gasoline.
(2) Biomass includes wood, wood-derived fuels, and biomass waste.
(3) Other includes geothermal energy and solar/PV energy.
(4) Based on the LHV of ethanol.
(5) Based on the assumption that energy consumption will remain unchanged.

Also, as mentioned in Sec. 7.4.3, the potential marginal lands in Illinois estimated at 3.95×10^6 acres (1.6 million ha) can be used to grow energy crops to produce about 781,657 barrels (124.26×10^6 liters) of cellulosic ethanol per year (US-EPA, 2010). Pennycress is an oil seed crop which has the potential to produce between 75 to 115 gallons of biodiesel per acre (Reed, 2011; WIU, 2014) (701–1075 liter per ha). Assuming that all the marginal lands in Illinois could be used for pennycress cultivation, the minimum Illinois annual contribution to the U.S. liquid transportation biofuels would be about 7.05×10^6 barrels. Furthermore, use of abandoned lands for energy crop cultivation can restore some of the soil nutrients on these lands by increasing the soil organic carbon concentration and nitrogen.

Biofuel technologies potentially suitable for Illinois include gasification process via FTS, fast pyrolysis process for biomass conversion to bio-oil and biodegradation process for biomass conversion to ethanol. Currently, research is focused on how to integrate these technologies with the existing petroleum oil infrastructure for cost reduction.

Also, considering the fact that the average Illinois resident produces an estimated 0.947 metric tons (2,088 pounds) of garbage per year, LFGE projects can help in solving the state's energy problems (Fullerton and Miller, 2010). LFGE projects can accelerate solid waste decomposition by increasing landfill capacity, while reducing the desire to build new landfills or expand the existing ones (US-EPA, 2012b). LFGE projects can also provide the opportunity for cost reduction associated with regulatory compliance in the maintenance of landfills. Furthermore, LFGE source can generate revenue to the state through the direct sales of landfill gas to end users, or through the transfer of gas into existing natural gas pipeline for distribution, or through the sales of electricity generated from the land fill gas (US-EPA, 2012b).

In essence, biofuels projects will have greater economic impact on the state through job creation and market development. However, these benefits can be realized through the concerted efforts of policymakers and private entrepreneurs.

The major problems hurting the biofuels industry in Illinois are regulatory hurdles, such as the percentage of ethanol that can be blended with gasoline (blend wall), expiration of federal tax credits for biodiesel and how to maximize the yield of cellulosic ethanol as discussed earlier in this report in Sec. 7.7.

Biofuels are renewable energy sources derived from biomass which is available in Illinois, as the state is rich in cellulosic polymeric biomass, especially corn stover. Apart from corn stover, there are several other non-food crops in experimentation in Illinois for biofuels production as discussed earlier in Sec. 7.4.1. These include switchgrass, miscanthus and pennycress. It has been estimated that using miscanthus for biofuel production can replace about 30% of Illinois petroleum oil imports, with just 7.7% of Illinois land usage and assuming 50% of biomass energy is lost in conversion. Additionally, over 20% of the state's land would be required to achieve the same with switchgrass (Pyter *et al.*, 2011).

It is estimated that between 600 and 900 million metric tons of biomass (food and non-food crops) could be produced by 2030 in the U.S, at a price of 140 dollars per ton (in 2007 dollars), while still meeting the demand for food, based on current assumptions about yields, production costs and land availability (Ciciora, 2011). The U.S. average levelized cost of electricity for biomass-fired power plants entering service in the year 2019 is estimated at $102.6 per MWh (US-EIA, 2014c). Currently, this cost is relatively higher than coal-, nuclear- and natural gas-generated electricity, however, biomass-fired power plants are expected to be competitive with these energy sources for power generation by 2025 (NEI, 2011).

Production of sustainable cellulosic biofuels would not be possible unless a sustainable farming is practiced. There are educational institutions and private companies that employ complex modeling tools to help farmers make smart economic decisions on their land. For example, scientists have shown that by strategic conversion of as little as 10% of cropped field to perennial prairie, farmers can reduce soil erosion and fertilizer runoff by 85–95% (Martin, 2014). Research has also shown that by using cover crops plants such as crimson clover or annual ryegrass as a soil conservation method, farmers can sustainably remove 1.8 tons more stover per acre than they otherwise would remove (Pratt *et al.*, 2014).

When sustainable production of biofuels from biomass is considered, a complete life cycle analysis (LCA) is required to ensure energy saving of biofuels. This analysis should encompass practice of plantation of energy crop(s), conversion process and delivery of biofuels from biorefinery to distribution center(s). In addition, the analyses should consider the effect of growing crops on GHG emission, land use, water availability, biodiversity and soil erosion.

Environmentally, biofuels emit less pollutant compared to fuels such as petroleum and coal. For example, power generation using biomass emits 0.008 lb NO_x per MMBtu, 195 lb CO_2 per MMBtu and zero emissions of SO_x, compared with coal-fired power plant which emits 0.06 lb NO_x per MMBtu, 206 lb CO_2 per MMBtu and 0.1lb SO_x per MMBtu (US-EIA, 2013c). Although CO_2 emissions by biomass-fired power plants is almost the same as coal-fired power plants, through photosynthesis, biomass plants remove from the atmosphere the same amount of CO_2 as emitted. According to a study in Iowa, in harvesting and converting corn stover into ethanol, about 57% of the stover can be removed from the field without causing soil erosion (NREL, 2001). Consequently, biofuels are sustainable because they are renewable, economically feasible to produce and are environmentally friendly.

Overall, in a CO_2-constrained world, production of biofuels from lignocellulosic materials: (i). Will contribute to resolving energy shortage. (ii). Will improve the economic viability of biofuels manufacturing industry. (iii). Will reduce petroleum-dependence and benefit the environment and society in terms of greenhouse gas emissions reduction and global warming. (iv). Will serve the green transportation industry and (v). Will provide a competitive advantage to the renewable energy industry.

Bibliography

Abebe, M. (2008). History of ethanol, *Journalism & Mass Commu.*, Student Media, Paper 6:24–27, Univ. of Nebraska-Lincoln.

[AC&C] Am. City & County Magazine (2012). Will County's landfill gas-to-energy plant turn garbage to gold. Public Works section, Atlanta, GA. *americancityandcounty.com.*

AEBIOM (2008). New dedicated energy crops for solid biofuels. Europ. Biomass Assoc., Brussels, Belgium.

Aghamiri, S. F., Modarress, H. and Mansoori, G. A. (2001). A New Theoretical Approach to the Hydrogen-Bonded Fluids Based on the Conformal Solution Concept, *J. Physical Chem. B*, **105**: 2820–2825.

Agyarko, L. B. and Mansoori, G. A. (2013). Non-renewable Energy Options in Illinois, *Int. J. Oil, Gas and Coal Technol.*, **6**(3): 288–347.

Ahlberg, L. (2010). Grasses have potential as alternate ethanol crop, Illinois study finds. *News Bur.*, Office of Public Affair, UIUC, Champaign, IL.

Alexander, T. (2011). Pennycress commercialization could add $4B in farm income. News section, *Farm World*, Knightstown, IN.

Ansari, F. T., Verma, A. P. and Choube, A. (2012). Experimental Determination of Suitable Ethanol-gasoline Blend for Spark Ignition Engine, *Intl J. of Eng. Research & Technol.*, **1**(5): 1–10.

[ATI] Arvens Tech., Inc. (2011). Crusher planned for 2011: $100 Million bioenergy hub on the drawing board. News item, Peoria, IL. *arvenstech.com.*

Baertsche, S. R., Yokoyama, M. T. and Hanover, J. W. (1986). Short Rotation, Hardwood Tree Biomass as Potential Ruminant Feed-chemical Composition, Nylon Bag Ruminal Degradation, *J. Anim. Sci.*, **63**: 2028–2043.

Baral, A. and Malins, C. (2014). Assessing the climate mitigation potential of biofuels derived from residues and wastes in the European context, White paper, The Int'l Council on Clean Transportation, Washington, DC. *theicct.org.*

[BBI] (2014). 2014 Biodiesel Industry Directory. BBI International, Grand Forks, ND. *bbiinternational.com.*

Bell, A. G. (1917). Prizes for the inventor: Some of the problems awaiting solution. *National Geog. Magazine*, **31**(2): 131–146.

Bista, B., Hubbs, T., Richert, B. T., Tyner, W. E. and Preckel, P. V. (2008). Economic value of Ethanol byproducts in swine diets: evaluating profitability of corn fractionation techniques, Paper No. 48708. In *Proceedings of the Transition to a Bio Economy Conf., Integration of Agricultural and Energy Systems Conf.* Atlanta, 12–13 February, GA. *ageconsearch.umn.edu.*

Blaschek, H. P. (2008). What are the possibilities for the new bioeconomy? Paper No. 48717. In *Proceedings of the Transition to a Bio Economy Conf., Integration of Agricultural and Energy Systems Conf.* Atlanta, 12–13 February, GA.

Boerrigter, H., den Uil, H. and Calis, H. P. (2003). Green diesel from biomass via Fischer-Tropsch synthesis: New insights in gas cleaning and process design. In *Proceedings of Pyrolysis and Gasification of Biomass and Waste*, Expert meeting. Strasbourg, 30 September–1 October, France.

Bogner, J., Spokas, K., Niemann, M., Niemann, L. and Baker, J. (1997). Emissions of non-methane organic compounds at an Illinois (USA) landfill: Preliminary field measurements. In *Proceedings of. Sardinia '97 Sixth Intl. Landfill Symp.* Argonne Nat. Lab., Argonne, IL.

Bond, W., Davies, G. and Turner, R. (2007). The biology and non-chemical control of field pennycress (*Thlaspi arvense L.*). HDRA, Ryton Organic Gardens, Coventry, UK. *gardenorganic.org.uk*

Brennan, L. and Owende, P. (2010). Biofuels from Microalgae — A Review of Technologies for Production, Processing, and Extractions of Biofuels and Co-products, *Renew Sust. Energ Rev.*, **14**(2): 557–577.

[BRI] BRI Energy, LLC and Bioengineering Resources, Inc. (2005). Gasification-fermentation Pilot Facility (Arkansas). BRI Energy, Fayetteville, AR.

Brosse, N., Dufour, A., Meng, X., Sun, Q. and Ragauskas, A. (2012). Miscanthus: A Fast- growing Crop for Biofuels and Chemicals Production, *Biofuels, Bioprod, Bioref.*, **6**(5): 580–598.

Brown, T. M., Duan, P. and Savage, P. E. (2010). Hydrothermal Liquefaction and Gasification of *Nannochloropsis sp*, *Energ & Fuel*, **24**(6): 3639–3646.

Byrge, J. A. and Kliesen, K. L. (2008). Ethanol: Economic Gain or Drain? Fed. Reserve Bank of St. Louis, St. Louis, MO. *stlouisfed.org.*

Campbell, A. and Doswald, N. (2009). The impacts of biofuel production on biodiversity: A review of the current literature., U.N. Env. Prog. — World Conservation Monit. Centre, Cambridge, UK. *unep-wcmc.org.*

Carpenter, S. R., Caraco, N. F., Correll, D. L., Howarth, R. W., Sharpley, A. N., Smith, V. H. (1998). Non-point Pollution of Surface Waters with Phosphorus and Nitrogen, *Ecol Appl.*, **8**(3): 559–568.

Chang, M. C. Y. (2007). Harnessing Energy from Plant Biomass, *Curr Opin Chem Biol.*, **11**(6): 677–684.

Chisti, Y. (2007). Biodiesel from Microalgae, *Biotechnol Adv.*, **25**(3): 294–306.

Ciciora, P. (2011). A billion tons of biomass a viable goal, but at high price, new research shows. *News Bur.*, Office of Public Affair, UIUC, Champaign, IL.

Clarke Energy (2011). Landfill gases, Liverpool, UK. *clarke-energy.com.*

Cobas, A. C., Felissia, F. E., Monteoliva, S. and Area, M. C. (2013). Optimization of the Properties of Poplar and Willow Chemimechanical Pulps by a Mixture Design of Juvenile and Mature Wood, *Bioresources*, **8**(2): 646–1656.

Colman, Z. (2012). Lawmakers urge EPA to lower ethanol target amid drought. News, Capitol Hill Pub. Corp., Washington DC. *thehill.com.*

Combs, S. (2008). The Energy Report 2008, Chapter 17: Landfill Gas, pp. 247–255, Texas Comptroller of Public Accounts, Austin, TX. *window.state.tx.us/specialrpt/energy.*

Congress, U.S. (2007). "Energy independence and security act of 2007", HR 6, 110 Congress. 1st Session, HR, 6.

Coskata (2012). Process overview. Coskata Inc, Warrenville, IL. *coskata.com/process.*

Coyle, W. T. (2010). Next-generation biofuels: Near-term challenges and implications for agriculture, *Outlook No. BIO-01-01*, Econ. Research Service, U.S. Dep't of Agri., Washington, DC.

Cruse, R. and Friedman, D. (2014). Soil erosion and sustainable biofuel production. America's Research-based Learning Network, Lincoln, NE. *extension.org.*

Curtis, B. (2010). U.S. Biofuels Industry: Mind the Gap — 2008–2009 Review. Concentric Power, Inc., Los Gatos, CA. *concentricenergies.com.*

Czapar, G. (2008). Impacts of Biofuel Production on Water Resources. Univ. of Illinois Extension, Champaign, IL.

Datye, A. K. and Schwank, J. (1985). Fischer-Tropsch Synthesis on Bimetallic Ruthenium-Gold Catalysts, *J Catal.*, **93**: 256–269.

De Fraiture, C., Giordano, M. and Liao, Y. (2008). Biofuels and Implications for Agricultural Water Use: Blue Impacts of Green Energy, *Water Policy*, **10**(1): 67–81.

De Fraiture, C. and Berndes. G. (2009). Biofuels and water, pp. 139–153 in Howarth, R. W. and Bringezu, S. (eds.), Biofuels: Environ. Consequences and Interactions with Changing Land Use. In Proceedings of Sci. Comm. on Prob. of the Env. (SCOPE), Intl Biofuels Proj. Rapid Assessment. Gummersbach Germany, Cornell Univ., 22–25 September 2008, Ithaca, NY.

Demirel, Y. (2012). *Energy: Production, Conversion, Storage, Conservation, and Coupling*. Springer.

Detroy, R. W. and St. Julian, G. (1983). Biomass Conversion Fermentation Chemicals and Fuels, *Crit Rev Microbiol.*, **10**(3): 203–228.

DeWitte, D. (2014). Shuttered biodiesel plant in Crawfordsville to be auctioned. News section, *The Gazette*, Iowa City, IA.

Dragone, G., Fernandes, B., Vicente, A. A. and Teixeira, J. A. (2010). Third generation biofuels from microalgae" Current Research, Tech. and Edu. Topics in App. Microbiol. and Microbial Biotech., Mendez-Vilas A (ed.), Microbiology Series 2, **2**: 1355–1366.

DSM and POET, LLC. (2014). First commercial-scale cellulosic ethanol plant in the U.S. opens for business. Sioux Falls, SD. *dms.com.*

Durbin, T. D., Miller, J. W., Johnson, K., Hajbabaei, M., Kado, N. Y., Kobayashi, R., Liu, X., Vogel, C. F. A., Matsumura, F. and Wong, P. S. (2011). Biodiesel characterization and NO_x mitigation study, Final report, CARB Assessment of the Emissions from the Use of Biodiesel as a Motor Vehicle Fuel in California.

Edalat, M., Mansoori, G. A. and Bozar-Jomehri, R. B. (1997). Vapor pressure of hydrocarbons, generalized equation. Encyclopedia of Chem. Processing and Design — 61, pp. 362–365, Marcell-Dekker.

Edalat, M., Mansoori, G. A. and Bozar-Jomehri, R. B. (1993). Generalized Equation Predicts Vapor Pressure of Hydrocarbons, *Oil & Gas J.*, 39–40.

Enayati, N. and Parulekar, S. J. (1995). Enzymatic Saccharification of Soybean Hull-based Materials, *Biotechnol. Prog.*, **11**(6): 708–711.

English, B., Menard, J. and Ugarte, D. D. L. T. (2000). Using corn stover for ethanol production: a look at the reg. economic impacts for selected Midwestern states, AIM-AG Agri-Industry Modeling Analysis Group Contract No. 4500010956, Univ. of Tennessee, Knoxville, TN.

ePlantScience (2009). Biomass: A renewable source of energy, Biotechnology & Environment, An Online Botanical Encyclopedia. *eplantscience.com.*

[FAO] Food and Agric. Org. of U.N. (2008). The state of food and agriculture. FAO, Rome, Italy.

Freeman, W. (2012). Agriculture Secretary Vilsack announces support for producers to grow renewable feedstocks for advanced biofuels. News release, Office of Commun., U.S. Dep't of Agriculture, Washington, DC.

Fullerton, D. and Miller, S. (2010). Waste and recycling in Illinois: Illinois communities cope with waste in different ways, Inst. of Gov. and Public Affairs, UIUC, Champaign, IL.

Gelfand, I., Sahajpal, R., Zhang, X., Izaurralde, R. C., Gross, K. L. and Robertson, G. P., (2013). Sustainable Bioenergy Production from Marginal Lands in the US Midwest, *Nature*, **493**: 514–517.

Gentile, G. (2012). US Senate Tax Bill Would Extend Biodiesel, Cellulosic Tax Credits, *Platts*, McGraw Hill Financial, New York, NY.

Gerpen, J. V. (2005). Biodiesel Processing and Production, *Fuel Processing Tech.*, **86**(10): 1097–1107.

Gravalos, I., Moshou, D., Gialamas, T., Xyradakis, P., Kateris, D. and Tsiropoulos, Z. (2011). Performance and emission characteristics of spark ignition engine fuelled with ethanol and methanol gasoline blended fuels, Chapter 7, *Alternative Fuel*, Maximino Manzanera (ed.), ebook.

Greenhouse, J. P. (2011). E15: Cracking the RVP Nut. News article, *Ethanol Producer*, Grand Forks, ND.

[GTI] Gas Tech. Inst. (2014). GTI announces production of high octane renewable gasoline. *Biomass Magazine*, Grand Forks, ND.

Hamad, E. Z. and Mansoori, G. A. (1990). A Fluctuation Solution Theory of Activity Coefficients: Phase Equilibria in Associating Molecular Solutions, *J Phys. Chem.*, **94**: 3148–3152.

Hammond, E. G., Johnson, L. A., Su, C., Wang, T. and White, P. J. (2005). Soybean oil, *Bailey's Industrial Oil and Fat Products*, F. Shahidi (ed.), Wiley, 6th Edn., **2**: 277–641.

Hanover, J. W. (1993). Black locust: an excellent fiber crop, *New Crops*, Janick, J. and Simon, J. E. (eds.), Wiley, New York, pp. 432–435.

Hanson, C. (2013a). Black locust trees evaluated in Illinois for biofuel potential. *Biomass Magazine*, Grand Forks, ND.

Hanson, C. (2013b). Study provides comparison of biomass crop growth in the Midwest. *Ethanol Producer*, Grand Forks, ND.

Hardcastle, P. D., Calder, I., Dingwall, C., Garrett, W., McChesney, I., Mathews, J., Savill, P., Copeland, J. and Ray, D. (2006). A review of the potential impacts of short rotation forestry, LTS Int'l Ltd., Near Edinburgh, Scotland.

Harker, J. (2011). Illinois extends its biodiesel tax credit for 5 years. News section, *Brownfield Ag News for America*, Jefferson City, MO.

Hartono, R., Mansoori, G. A. and Suwono, A. (1999). Prediction of Molar Volumes, Vapor Pressures and Supercritical Solubilities of Alkanes by Equations of State, *Chem Eng Commun*, **173**: 23–42.

He, B. J., Zhang, Y., Funk, T. L., Riskowski, G. L. and Yin, Y. (2000). Thermochemical Conversion of Swine Manure: An Alternative Process for Waste Treatment and Renewable Energy Production, *Transactions of the ASAE*, **43**(6): 1827–1833.

Heaton, E. A., Dohleman, F. G. and Long, S. P. (2008). Meeting US Biofuel Goals with Less Land: The Potential of Miscanthus, *Glob. Change Biol.*, **14**: 1–15.

Higgins, R. (2011). Report from the Research Farms-Northern Illinois Agronomy Research Center. *The Bulletin*, Pest Management and Crop Devel. Info. for Illinois, Issue No. 20, UIUC.

Hojilla-Evangelist, M. P. and Evangelista, R. L. (2012). Effects of Oil Extraction on Functional Properties of Protein in Pennycress (*Thlaspi arvense L.*) Seed and Press Cake. Agri. Research Service, U.S. Dep't Agri., Washington, DC.

Holmgren, J., Marinangeli, R., Marker, T., McCall, M., Petri, J., Czernik, S., Elliot, D., Shonnard, D. (2007). Opportunities for Biorenewables. In *Proceedings of the PETROTECH-2007-7th Intl Petrol. Conf. and Exhib*. New Delhi, 15–19 January, India.

Holmgren, J., Gosling, C., Marinangeli, R., Marker, T., Faraci, G. and Perego, C. (2007a). New Developments in Renewable Fuels Offer More Choices: Vegetable Oil-based Diesel Can Offer Better Integration Within Crude-oil Refineries for Fuels Blending, *Hydrocarb Process*, Sep. (67–71).

Holmgren, J., Gosling, C., Couch, K., Kalnes, T., Marker, T., McCall, M. and Marinangeli, R. (2007b). Refining Biofeedstock Innovations, *Petrol. Tech. Quarterly*, **12**(4): 119–125.

Howarth, R. W., Bringezu, S., Bekunda, M., de Fraiture, C., Maene, L., Martinelli, L. and Sala, O. (2009). Rapid Assessment on Biofuels and the Environment: Overview and Key Findings. In *Biofuels: Environ. Consequences and Interactions with Changing Land Use. Proc Sci Committee on Prob of the Env., Intl Biofuels Proj. Rapid Assessment*, Howarth, R. W. and Bringezu, S. (eds.), pp. 1–13. Gummersbach Germany, 22–25 September, Cornell Univ., Ithaca NY.

Hu, J., Yu, F. and Lu, Y. (2012). Application of Fischer–Tropsch Synthesis in Biomass to Liquid Conversion, *Catalysts*, **2**(2): 303–326.

Huang, H. and Khanna M., (2011). The Breakeven Costs of Alternative Feedstocks for Cellulosic Biofuels, *Aspects of Applied Biology*, **112**: 153–162.

[IDCEO] Illinois Dept of Commerce and Economic Opportunity. (2009). Illinois Commodity/Waste Generation and Characterization Study, Illinois Dept of Commerce & Economic Opportunity, Springfield, IL.

[IEA] Intl Energy Agency (2006). World Energy Outlook 2006. IEA, Paris, France.

[IEA] Intl Energy Agency (2007). Biomass for Power Generation and CHP. IEA-Energy Tech. Essentials (ETE03), Paris, France.

[IEPA] Illinois Environ Protection Agency (2008). Building Biofuels Plant in Illinois: A Guide to Permit Requirements, Funding Opportunities, and Other Considerations, Report No. IEPA/DO/08-002, IEPA, Springfield, IL.

[IEPA] Illinois Environ Protection Agency (2009). Non-hazardous solid waste management and landfill capacity in Illinois: 2008. IEPA, Springfield, IL.

[IEPA] Illinois Environ Protection Agency (2011a). Indiana recycling and renewable fuels: Illinois EPA review and public comment. General Public Notices, IEPA, Springfield, IL.

[IEPA] Illinois Environ Protection Agency (2011b). Non-hazardous solid waste management and landfill capacity in Illinois, 2010. IEPA, Springfield, IL.

[IIRA] Illinois Inst. for Rural Affairs (2012). Illinois Ethanol Industry: Annual Economic Impact to the 2011 Illinois economy. Illinois Inst. for Rural Affairs, Western Illinois Univ., Macomb, IL.

Illinois General Assembly (1997). Renewable energy, energy efficiency, and coal resources development Law of 1997. Illinois General Assembly.

Isbell, T. (2009). US effort in the development of new crops (Lesquerella, Pennycress Coriander and Cuphea), *Oilseeds & Fats Crops and Liquids* (*OCL*), **16**(4–6): 205–210.

Jellen, J. (2010). Illinois ethanol plant cited for toxic waste: Mismanagement of biofuel facility in Canton causes caustic waste to kill local wildlife. *Mother Nature Network*, Atlanta, GA.

Jessen, H. (2013). Proposed rule defines corn fiber ethanol, butanol as advanced. *Ethanol Producer Magazine*, Grand Forks, ND.

Jones, D. T. (2013). Biobutanol, Biotechnology, Encyclopedia of Life Support Systems, Vol. 6, EOLSS, Paris, France.

Kalnes, T. N., Marker, T., Shonnard, D. R. and Koers, K.P. (2008). Green Diesel Production by Hydrorefining Renewable Feedstocks, *Biofuels Tech.*, **4**: 7–11.

Kenkel, P. and Holcomb. R. (2008). Feasibility of on-farm or small scale oilseed processing and biodiesel production, In *Proceedings of Transition to a Bio Econ Conf., Integ of Agricultural and Energy Sys Conf.*, Paper No. 48710. Atlanta, 12–13 February, GA.

Khanna, M. (2008). Cellulosic Biofuels: Are They Economically Viable and Environmentally Sustainable? *Choices Magazine*, **23**(3): 16–21.

Khoury, F. M. (2014). Multistage Separation Processes, 4th Edn. CRC Press, Boca Raton, FL.

King, C. W. and Webber, M. E. (2008). Water Intensity of Transportation, *Environ Sci. and Tech.*, **42**(21): 7866–7872.

Kotrba, R. (2005). Knocking out NO_x. *Biodiesel Magazine*, Grand Forks, ND.

Kotrba, R. (2011, Dec. 14) "Senate hearing held on alternative energy tax incentive", Biodiesel Magazine, Grand Forks, ND.

Kovarik, B. (1998). Henry Ford, Charles Kettering and the Fuel of the Future, *Automotive History Review*, **32**: 7–27.

Kurki, A., Hill, A., Morris, M. and Lowe, A. (2006). Biodiesel: The sustainability dimensions, ATTRA-National Sustainable Agric. Info. Service.

Lane, J. (2012). Paradigm BioAviation Announces $120M Aviation Biofuels Project in Illinois. *Biofuels Digest*, Miami, FL.

Lane, J. (2014). GTI and Haldor Topsoe complete pilot scale wood-to-gasoline biorefinery. *Biofuels Digest*, Miami, FL.

Larson, E. D. (2008). Biofuel production technologies: Status, prospects and implications for trade and development, Report No. UNCTAD/DITC/TED/2007/10, UN Conf. on Trade and Development, Geneva, Switzerland.

Launder, K. (2002). Energy crops and their potential development in Michigan, Michigan Biomass Energy Program, Michigan Dept of Consumer and Industry Services, Energy Office, Biomass Energy Program, Lansing, MI.

Lee, D. K., Owens, V. N., Boe, A. and Jeranyama, P. (2007). Composition of herbaceous biomass feedstocks, Sun Grant Initiative-North Central Center, South Dakota State Univ., Report No. SGINC1-07, SDSU, Brookings, SD.

Lee, J. W. (ed.) (2013). Advanced Biofuels and Bioproducts. Springer, New York, NY.

Lee, S., Speight, J. G. and Loyalka, S. K. (eds.) (2014). Handbook of Alternative Fuel Technologies, 2nd Edn. CRC Press.

Lehmann, J. (2007). Bio-energy in the Black, *Front. Ecol. Environ.*, **5**(7): 381–387.

Liebig, M. A., Schmer, M. R., Vogel, K. P. and Mitchell, R. B. (2008). Soil Carbon Storage by Switchgrass Grown for Bioenergy, *Bioenerg. Res.* **1**(2–3): 215–222.

Lingaraju, B. P., Lee, J-Y. and Yang, Y. J. (2013). Process and Utility Water Requirements for Cellulosic Ethanol Production Processes via Fermentation Pathway, *Environ Prog & Sustainable Energy*, **32**(2): 396–405.

Liska, A. J., Yang, H., Milner, M., Goddard, S., Blanco-Canqui, H., Pelton, M. P., Fang, X. X., Zhu, H. and Suyker, A. E. (2014). Biofuels from Crop Residue Can Reduce Soil Carbon and Increase CO_2 Emissions, *Nature Climate Change*, **4**: 398–401.

Liu, Y. (2011). One-step Hydrotreatment Process for Conversion of Vegetable Oils to Renewable Diesel and LPG, *Green Car Congress*.

Lötjönen and Laitinen (2009). Energy from field energy crops-a handbook for energy producers. Jyväskylä Innovation, Finland.

Lovgren, S. (2004). Pig manure converted to crude oil. News article, *Nat'l Geog. Soc.* Washington, DC.

Lusvardi, C. (2009). ADM gets $24.8 million to build biofuel plant in Decatur. News section, *Herald Review*, Decatur, IL.

Lundquist, T. J., Woertz, I. C., Quinn, N. W. T. and Benemann, J. R. (2010). A realistic technology and engineering assessment of algae biofuel production, Energy Biosci. Inst., Berkeley, CA.

Lynd, L. R., Cushman, J. H. and Nichols, R. J. (1991). Fuel Ethanol from Cellulosic Biomass, *Science*, **251**: 1318–1323.

Macedo, I. D. C., Leal, M. R. L. V. and Da Silva, J. E. A. R. (2004) "Assessment of greenhouse gas emissions in the production and use of fuel ethanol in Brazil, Gov. of State of São Paulo, Brazil.

Maiorca, J. (2009). The Illinois labor market review: developing a biofuel industry in Illinois, State of Illinois, Dept of Employment security, Chicago, IL.

Mandil, C and Shihab-Eldin, A. (2010). Assessment of biofuels potential and limitations, *Int'l Energy Forum,* Riyadh, Saudi Arabia.

Mansoori, G. A. (1999). Fundamental thermodynamics for distillation Proc. *Distillation Seminar*, Lecture 1, AIChE-Chicago sect., Chicago, IL.

Martin, J., (2014). 5 Things I learned in Iowa about biofuels, Union of Concern Scientists, UCSUSA, Cambridge, MA.

Massih, A. R. and Mansoori, G. A. (1983). Conformal solution theory of polar fluids, *Fluid Phase Equil.*, **10**: 57–72.

Mata, T. M., Martins, A. A. and Caetano, N. S. (2010). Microalgae for Biodiesel Production and Other Applications: A Review, *Renewable and Sustainable Energy Reviews*, **14**(1): 217–232.

McAlister, R. H. (1971). Black Locust (Robinia pseudoacacia L.), USDA Forest Service, Washington, DC.

[MDNR] Missouri Dept of Natural Resource. (2008). Ethanol and gasoline fuel blend release, Environ Emergency Response/Div. of Public Safety — Div. of Fire Safety, MDNR, Jefferson City, MO.

Menetrez, M. Y. (2012). An Overview of Algae Biofuel Production and Potential Environmental Impact, *Environ. Sci. Technol.*, **46**(13): 7073–7085.

Milbrandt, A. (2005). A geographic perspective on the current biomass resource availability in the United States, Report No. NREL/TP-560-39181, NREL, DOE, Washington, DC.

Moreira, J. R. (2007). Water use and impacts due ethanol production in Brazil, Presented at Int'l Conf. at ICRISAT Campus, Hyderabad, India.

Mrozowski, A. (2010). Status of biofuel programs in Illinois, Presented at the Illinois Dept of Commerce and Economic Opportunity, Springfield, IL.

Mullen, C. A. and Boateng, A. A. (2008). Chemical Composition of Bio-oils Produced by Fast Pyrolysis of Two Energy Crops, *Energy & Fuels*, **22**(3): 2104–2109.

Müller, A., Schmidhuber, J., Hoogeveen, J. and Steduto, P. (2008). Some Insights in the Effect of Growing Bioenergy Demand on Global Food Security and Natural Resources, *Water Policy*, **10**: 83–94.

[NAS] Nat'l Academy of Sciences, (2007). Water implications of biofuels production in the United States, Nat'l Academies Press, Washington, DC.

Naik, S. N., Goud, V. V., Rout, P. K. and Dalai, A. K. (2010). Production of First and Second Generation Biofuels: A Comprehensive Review, *Renewable and Sustainable Energy Reviews*, **14**(2): 578–597.

Nebraska Energy Office (2014). Ethanol facilities capacity by state and plant, Nebraska Energy Office, Lincoln, NE.

[NEI] Nuclear Energy Inst. (2011). The cost of new generating capacity in perspective, White paper, NEI, Washington, DC.

Nelson, S. (2008). Analysis-Ethanol no longer seen as big driver of food price. Thomson Reuters, UK Edn., New York, NY.

Nguyen, T. Y., Cai, C. M., Kumar, R. and Wyman, C. E. (2015). Co-solvent pretreatment reduces costly enzyme requirements for high sugar and ethanol yields from lignocellulosic biomass, *ChemSusChem*.

Norton, P., Vertin, K., Bailey, B., Clark, N. N., Lyons, D. W., Goguen, S. and Eberhardt, J. (1998). Emissions from trucks using Fischer-Tropsch diesel fuel, SAE Tech. Papers Series No. 982526, Int'l Fall Fuels and Lubricants Meeting, 19–22 October, San Francisco, CA.

[NPC] National Petrol. Council (2007). Coal to liquids and gas, Topical Paper No. 18, Working Doc. of the NPC Global Oil & Gas Study, Nat'l Petrol. Council, Washington, DC.

[NREL] Nat'l Renewable Energy Lab. (2001). Corn stover for bioethanol-your new cash crop? Report No. DOE/GO-102001-1273, NREL, Golden, CO.

[NSF] Nat'l Sci. Foundation (2008). Breaking the chemical and engineering barriers to lignocellulosic biofuels: Next generation hydrocarbon biorefineries, Research roadmap report, Huber, G. W. (ed.), 25–26 June Workshop, NSF, Chemical, Bioengineering, Environmental, and Transport Systems Div. Washington, DC.

N. V., (2011, Jun. 24) "The Difference Engine: the beef about corn", 2 pages, Blogs sect., The Economist, London, UK.

Ocfemia, K. S., Zhang, Y. and Funk, T. (2006). Hydrothermal Processing of Wine Manure Into Oil Using a Continuous Reactor System, *Am. Soc. of Agricultural and Biological Engs*, **49**(2): 533–541.

Pandey, A., Larroche, C., Ricke, S. C., Dussap, C-G. and Gnansounou, E. (eds.) (2011). *Biofuels: Alternative Feedstocks and Conversion Processes*, 1st Edn. Academic Press, USA.

Pimentel, D. and Patzek, T. W. (2005). Ethanol Production Using Corn, Switchgrass, and Wood: Biodiesel Production Using Soybean and Sunflower, *Natural Resources Research*, **14**(1): 65–76.

Pires, A. P., Mohamed, R. S. and Mansoori, G. A. (2001). An Equation of State for Property Prediction of Alcohol–hydrocarbon and Water–hydrocarbon Systems, *J. Petrol. Sci. Eng.*, **32**: 103–114.

Pittman, J. K., Dean, A. P. and Osundeko, O. (2011). The Potential of Sustainable Algal Biofuel Production Using Wastewater Resources, *Bioresource Technology*, **102**(1): 17–25.

Plunkett, B. (2009). Closed 13 years ago, landfill generates clean energy for 5,000 DuPage homes, *TribLocal*, *Chicago Tribune*, Chicago, IL.

Pratt, M. R.,Tyner, W. E., Muth Jr., D. J. and Kladivko, E. J. (2014). Synergies Between Cover Crops and Corn Stover Removal, *Agr Syst*, **130**: 67–76.

Prianto, S., Mansoori, G. A., Suwono, S., Samudro, H. M. and Sitompul, J. P. (1997). Preliminary studies of octane number enhancement of non-lead fuel: Solubility prediction of methane in liquid hydrocarbons. In *Proceedings of. Intl Conf. on Fluid and Thermal Energy Conversion '97*, pp. 235–244.

Public Law 102–486, U.S. (1992). Energy Policy Act of 1992, Lib. of Congress, H.R.776.ENR.

Public Law 110–140, U.S. (2007). Energy Independence and Security Act of 2007, Library of Congress, H.R. 6.

Pyter, R., Voigt, T., Heaton, E., Dohleman, F. and Long, S. (2011). Giant Miscanthus: Biomass Crop for Illinois, Reprinted from: Issues in new crops and new uses (2007), Janick, J. and Whipkey, A. (eds.). ASHS Press, Alexandria, VA.

Qin, X., Mohan, T., El-Halwagi, M., Cornforth, G. and McCarl, B. (2006). Switchgrass As An Alternate Feedstock for Power Generation: Integrated Environ, Energy, and Economic Life-cycle Analysis, *Clean Technol Envir*, **8**(4): 233–249.

Quinn, N. (2010). Algae for Biofuels: Moving From Promise to Reality, but How Fast? Energy Biosci. Inst., Berkeley, CA.

Ravindranath, N. H., Manuvie, R., Fargione, J., Canadell, J. G., Berndes, G., Woods, J., Watson, H. and Sathaye, J. (2009). Greenhouse gas implications of land use and land conversion to biofuel crops. In Biofuels: Environ Consequences and Interactions with Changing Land Use, Howarth, R. W. and Bringezu, S. (eds.). In *Proceedings of Scientific Committee on Problems of the Env. (SCOPE) Intl Biofuels Proj. Rapid Assessment*, pp. 111–125. Gummersbach Germany. 22–25 September, Cornell Univ., Ithaca, NY.

Reed, K. C. (2011). Pennycress biodiesel good for America? *Biodiesel Magazine*, Grand Forks, ND.

[RFA] Renewable fuels Assoc. (2012). Accelerating industry innovation: 2012 ethanol industry outlook, Renewable Fuel Assoc., Washington, DC.

Rosell, M., Lacorte, S. and Barceló, D. (2006). Analysis, Occurrence and Fate of MTBE in the Aquatic Environment Over the Past Decade, *Trac-Trend Anal Chem.*, **25**(10): 1016–1029.

Saidak, T. (2010). Illinois Farm Group, biofuel producers petition Obama on E12. *Biofuel Digest*, Miami, FL.

Sala, O. E., Sax, D. and Leslie, H. (2009). Biodiversity consequences of biofuel production. In Biofuels: Environ Consequences and Interactions with Changing Land Use, Howarth, R. W. and Bringezu, S. (eds). In *Proceedings of Scientific Committee on Problems of the Env., Intl Biofuels Proj. Rapid Assessment*, pp. 127–137. Gummersbach Germany, 22–25 September, Cornell Univ., Ithaca, NY.

Sanders, R. (2006). Ethanol can replace gasoline with significant energy savings, comparable impact on greenhouse gases, *UC Berkeley News*, Berkeley, CA.

Sannigrahi, P., Ragauskas, A. J. and Tuskan, G. A. (2010). Poplar As A Feedstock for Biofuels: A Review of Compositional Characteristics, *Biofuels, Bioprod, Bioref.*, **4**(2): 209–226.

Sannigrahi, P and Ragauskas, A. J. (2011). Characterization of Fermentation Residues From the Production of Bio-ethanol from Lignocellulosic Feedstocks, *J Biobased Mater Bio*, **5**: 514–519.

Sapphire Energy (2009). Sapphire Energy statement before the senate committee on environment and public works, Sapphire Energy Inc., San Diego, CA.

Searchinger, T., Heimlich, R., Houghton, R. A., Dong, F., Elobeid, A., Fabiosa, J., Tokgoz, S., Hayes, D., Yu, T-H. (2008). Use of U.S. Croplands for Biofuels Increases Greenhouse Gases Through Emissions from Land-use Change, *Science*, **319**(5867): 1238–1240.

Schill, S. R. (2008). Making pennycress pay off. *Biodiesel Magazine*, Grand Forks, ND.

Schlagermann, P., Göttlicher, G., Dillschneider, R., Rosello-Sastre, R. and Posten, C. (2012). Composition of Algal Oil and its Potential as Biofuel, *J. of Combustion*.

Schnepf, R. (2010). Cellulosic ethanol: Feedstocks, conversion technologies, economics, and policy options.

Schroeder, J. (2011). Illinois Biomass Working Group Formed, Domestic-fuel, Cantonment, FL.

Scurlock, J. M. O. (1999). Miscanthus: A review of European experience with a novel energy crop, Report No. ORNL/TM-13732, Environ. Sci Div., ORNL, Oak Ridge, TN.

Selling, G. W., Hojilla-Evangelista, M. P., Evangelista, R. L., Isbell, T., Price, N. and Doll, K. M. (2013). Extraction of Proteins from Pennycress Seeds and Press Cake, *Ind Crop Prod*, **41**:113–119.

Sendzik, M. E. and Wiewel, W. (1996). Solid waste incineration in the Chicago metropolitan area: The battle over the Illinois retail rate law, Great Cities Inst., College of Urban Planning and Public Affairs, UIC, Chicago, IL.

Sheehan, J., Dunahay, T., Benemann, J. and Roessler, P. (1998). A look back at the U.S. Department of Energy's aquatic species program-biodiesel from algae, Report No. NREL/TP-580-24190, NREL, DOE Office of Fuels Development, Washington, DC.

Siegel, R. P. (2012). A brief history of biofuels: The future of biofuels. *Triple Pundit*, San Francisco, CA.

Sims, R. E. H., Mabee, W., Saddler, J. N. and Taylor, M. (2010). An Overview of Second Generation Biofuel Technologies, *Bioresource Technology*, **101**(6): 1570–1580.

Sine Nomine Group (2007). Assessment of the biodiesel distribution infrastructure in Canada, CANMET energy Tech. Centre, Ottawa, Montreal, Canada.

Slating, T. A. and Kesan, J. P. (2011). Making regulatory innovation keep pace with technological innovation, Wis. Law Rev., Illinois Public Law Research Paper No. 10–34, Illinois Program in Law, Behavior and Social Science Paper No. LBSS11–17.

Solazyme, Inc. (2015). Renewable fuels for the future. Available at solazyme.com/solutions/fuel/?lang=en. Accessed on 15 March 2015.

Solomon, B. D., Barnes, J. R. and Halvorsen, K. E. (2007). Grain and Cellulosic Ethanol: History, Economics, and Energy Policy, *Biomass and Bioenergy*, **31**(6) 416–425.

Spath, P. L. and Dayton, D. C. (2003). Preliminary screening — Technical and economic assessment of synthesis gas to fuels and chemicals with emphasis on the potential for biomass-derived syngas, Report No. NREL/TP-510-34929, NREL, Golden, CO.

Speight, J. G. (ed.) (2011). The Biofuels Handbook. Royal Society of Chemistry.

Speight, J. G. (2014). *Gasification of Unconventional Feedstocks*. Gulf Pub. Co.

Speight, J. G. and Singh, K. (2014). *Environmental Management of Energy from Biofuels and Biofeedstocks*. Wiley-Scrivener.

Squatriglia, C. (2008). Start-up says it can make ethanol for $1 a gallon, and without corn. News section, *Wired*, Boone, IA.

Sticklen, M. (2010). Is large-scale production of biofuel possible, Actionbioscience, Am. Inst. of Biological Sci., Reston, VA.

Strong, C., Erickson, C. and Shukla, D. (2004). Evaluation of biodiesel fuel: Literature review, Report No. FHWA/MT-04-001/8117-20, Western Transportation Institute, Montana State Univ. — Bozeman, MT.

Suszkiw, J. (2008). Experimental plots of pennycress tested for biodiesel potential, Agri. Research Service, U.S. Dep't Agri., Washington, DC.

Szulczyk, K. R. (2010). Which is a Better Transportation Fuel-butanol or Ethanol? *Intl J. of Energy and Env.* (*IJEE*), **1**(1): 1–11.

Tangka J. K., Berinyuy J. E., Tekounegnin and Okale A. N. (2011). Physicochemical Properties of Bio-ethanol/Gasoline Blends and the Qualitative Effect of Different Blends on Gasoline Quality and Engine Performance, *J. of Petrol. Tech. and Alternative Fuels*, **2**(3): 35–44.

Tarter, S. (2009). Biodiesel plant gets $500,000. News section, *Peoria J. Star*, Peoria, IL.

Thorstensen, H. (2010). Biodiesel industry wants Federal tax credit back. *Agri News*, Rochester, MN.

Tilman, D., Hill, J. and Lehman, C. (2006). Carbon-negative Biofuels from Low-input High-Diversity Grassland Biomass, *Science*, **314**(5805): 1598–1600.

Torresani, M. and Peotter, B. (2006). A Case study in the development of a landfill gas-to-energy project for the Antioch, Illinois, Community School District. In *Gas Turbine Handbook: Principles and Practices*, Giampaolo, T., Chapter 17, 3rd Edn. Fairmont Press, Lilburn, GA.

Trinh, N. T., Jensen, P. A., Sørensen, H. R., Dam-Johansen, K. and Hvilsted, S. (2012). Flash pyrolysis properties of algae and lignin residue. In *Proceedings of 20th Europ. Biomass Conf. and Exhib.*, Milan, Italy.

[UC] Univ. of Calif. (2010). California renewable diesel multimedia evaluation, Tier I report, UC-Davis and UC-Berkeley.

[UIUC] Univ. of Illinois at Urbana Champaign (2013). Illini Algae — hydrothermal liquefaction waste biomass to bioenergy conversion, Dept of Agri. and Biolog. Eng, UIUC, Champaign, IL.

[U.S.C HCC] U.S. Congress House Comm, on Commerce. (1998). Energy conservation reauthorization Act of 1998: Report together with additional views (to accompany HR 4017) (including cost estimate of the Congressional Budget Office), **105**(727), USC.

[USDA] U.S. Dept of Agric. (2010). A USDA regional roadmap to meeting the biofuels goals of the renewable fuels standard by 2022, USDA, Washington, DC.

[US-DOE] U.S. Dept of Energy (2006a). Breaking the biological barriers to cellulosic ethanol: a joint research agenda, Report No. DOE/SC-0095, A research roadmap resulting from biomass to biofuels workshop, 7–9 December, Rockville, MD.

[US-DOE] U.S. Dept of Energy (2006b). Biodiesel handling and use guidelines, 3rd Edn., Report No. DOE/GO-102006-2358, NREL, Golden, CO.

[US-DOE] U.S. Dept of Energy (2007). Research advances cellulosic ethanol, NREL leads the way, Report No. NREL/BR-510-40742, NREL, Golden, CO.

[US-DOE] U.S. Dept of Energy (2009). Biodiesel handling and use guide (4th Edn.), Report No. NREL/TP-540-43672, NREL, Golden, CO.

[US-DOE] U.S. Dept of Energy (2011a). Solazyme pilot-scale biorefinery, DOE, Office of Energy Efficiency & Renewable Energy, Washington, DC.

[US-DOE] U.S. Dept of Energy (2011b). Elegance pilot-scale biorefinery, DOE, Office of Energy Efficiency & Renewable Energy, Washington, DC.

[US-DOE] U.S. Dept of Energy (2012a). Gas Technology Institute R&D project: Gasoline and diesel from wood, agricultural waste, and algae R&D, Doc. No. DOE/EE-0819, DOE, Office of Energy Efficiency & Renewable Energy, Washington, DC.

[US-DOE] U.S. Dept of Energy (2012b). Haldor Topsoe, Inc.'s project: Green gasoline from wood pilot biorefinery demonstration project,

Doc. No. DOE/EE-0820, DOE, Office of Energy Efficiency & Renewable Energy, Washington, DC.

[US-DOE] U.S. Dept of Energy (2013a). Ethanol blends, Alternative Fuels Data Center, DOE Clean Cities Program, Washington DC.

[US-DOE] U.S. Dept of Energy (2013b). Alternative fuels data center — fuel properties comparison, Alternative Fuels Data Center, DOE Clean Cities program, Washington DC.

[US-DOE] U.S. Dept of Energy (2013c). Handbook for Handling, Storing, and Dispensing E85 and Other Ethanol-gasoline Blends, Doc. No. DOE/GO-102013-3861, DOE, Washington DC.

[US-EIA] U.S. Energy Info. Admin. (1997). Renewable energy annual 1996, Report No. DOE/EIA-0603(96), Office of Coal, Nuclear, Electric and Alternate Fuels. EIA, Washington, DC.

[US-EIA] U.S. Energy Info. Admin. (2010). EIA energy kids biomass. EIA, Washington, DC.

[US-EIA] U.S. Energy Info. Admin. (2012a). Petroleum and other liquids: Monthly biodiesel production report. EIA, Washington, DC, 10 pages.

[US-EIA] U.S. Energy Info. Admin. (2012b). Annual Energy Outlook 2012 with projections to 2035: Infrastructure hurdles limit near-term growth in consumption of E15 and E85 fuels, Report No. DOE/EIA0383, EIA, Washington, DC 20585, 120 pages.

[US-EIA] U.S. Energy Info. Admin. (2013a). Petroleum and other liquids: Monthly biodiesel production report achieves. EIA, Washington, DC.

[US-EIA] U.S. Energy Info. Admin. (2013b). Cellulosic biofuels begin to flow but in lower volumes than foreseen by statutory targets. EIA, Washington, DC.

[US-EIA] U.S. Energy Info. Admin. (2013c). Updated capital cost estimates for electricity generation plants. EIA, Washington, DC.

[US-EIA] U.S. Energy Info. Admin. (2014a). State energy data system (SED): Table C3, electric power sector consumption estimates, 1960–2012, Illinois, Report No. DOE/EIA-0214. EIA, Washington, DC.

[US-EIA] U.S. Energy Info. Admin. (2014b). State energy production estimates, 1960 through 2012, Table PT1, Energy Production Estimates in Physical Units, Illinois, 1960–2012. EIA, Washington, DC.

[US-EIA] U.S. Energy Info. Admin. (2014c). Annual energy outlook 2014: Levelized cost and levelized avoided cost of new generation resources in the annual energy outlook 2014. EIA, Washington, DC.

[US-EPA] U.S. Environ Protection Agency (1999). Landfill gas-to-energy projects opportunities: Landfill profile for the state of Illinois, Report No. EPA-430-99-009, Nat'l Service Center for Environ Pubs (NSCEP), U.S. EPA/NSCEP, Cincinnati, OH.

[US-EPA] U.S. Environ Protection Agency (2010). Renewable Fuel Standard Program (RFS2) Regulatory Impact Analysis, Report No. EPA-420-R-10-006, EPA, Washington, DC.

[US-EPA] U.S Environ Protection Agency (2012a). Energy projects and candidate landfills, Landfill Methane Outreach Program. EPA, Washington, DC.

[US-EPA]. U.S. Environ Protection Agency (2012b). Local government climate and energy strategy guides: Landfill gas energy, a guide to developing and implementing greenhouse gas reduction programs, Report No. EPA-430-R-09-047. U.S. EPA, Washington, DC.

[US-EPA]. U.S. Environ Protection Agency (2014). Landfill methane outreach program: operational projects. U.S. EPA, Washington, DC. Available at *epa.gov/lmop/projects-candidates/operational.html*

Vakili-Nezhaad, G. R., Modarress, H. and Mansoori, G. A. (2000). Thermodynamic Modeling and Experimental Studies of Dearomatization Process From a Complex Petroleum Fraction, *Separ Sci Technol.*, **35**(5): 743–754.

Van Ee, J. H. (2010). Stover handling, fragmentation and upgrading. In *Proceedings of. Intl Conf. on lignocelluloses Ethanol from Demonstration to Market Deployment*. Copenhagen, 13–15 October, Denmark.

Vardon, D. R., Sharma, B. K., Scott., J., Yu, G, Wang, Z., Schideman, L., Zhang, Y. and Strathmann, T. J. (2011). Chemical Properties of Biocrude Oil from the Hydrothermal Liquefaction of Spirulina Algae, Swine Manure, and Digested Anaerobic Sludge, *Bioresource Technol.*, **102**(17): 8295–8303.

Varner, B. L. (2008). The ethanol industry in Illinois, Illinois Commis. on Gov. Forecasting and Accountability, Springfield, IL.

Verry, T. (2010). Biodiesel mass transit demonstration, Report No. FTA-MO-26-7009-2010.1, Fed. Transit Admin Office of Res, Demo and Innov., U.S. DOT, Washington, DC.

Volk, T. A., Buford, M. A., Berguson, B., Caputo, J., Easton, J., Perdue, J. H., Rials, T. G., Riemenschneider, D., Stanton, B. and Stanturf, J. A. (2010). Woody feedstocks — management and regional differences, Chapter 7, Sustainable Feedstocks for Advanced Biofuels. In *Proceedings of the Sustainable Feedstocks for Advance Biofuels Workshop*. Atlanta, 28–30 September, GA.

Wang, J. L. and Mansoori, G. A. (1994). A Revision of The Distillation Theory, *Scientia Ir.*, **1**(3): 267–287.

Wang, M., Wu, M. and Huo, H. (2007). Life-cycle Energy and Greenhouse Gas Emission Impacts of Different Corn Ethanol Plant Types, *Environ Res Lett.*, **2**(2).

Wang, M. Q. (2010). The greenhouse gases, regulated emissions, and energy use in transportation (GREET) model. Version 1.8, ANL, Argonne, IL.

Wang, M., Han, J., Dunn, J. B., Cai, H. and Elgowainy, A. (2012). Well-to-wheels Energy Use and Greenhouse Gas Emissions of Ethanol from Corn, Sugarcane and Cellulosic Biomass for US Use, *Environ. Res. Lett.*, **7**(4).

[WIE] Wood Inst. for the Environment (2006). The impacts of large-scale biofuel use on water resources. WIE, Stanford Univ., Palo Alto, CA.

Wellinger, A. (2009). Algal biomass: Does it save the world? Europ. Biofuel Tech. Platform, Bioenergy Task 37. IEA, Paris, France.

White, E. M. (2010). Woody biomass for bioenergy and biofuels in the United States: A briefing paper, Report No. PNW-GTR-825, Pacific Northwest Research Station, USDA, Washington DC.

Winchester, N., McConnachie, D., Wollersheim, C. and Waitz, I. A. (2013). Market cost of renewable jet fuel adoption in the United States, Report No. 238, MIT Joint Program on the Sci. and Policy of Global Change, MIT, Cambridge, MA.

[WIU] Western Illinois Univ. (2014). History of pennycress (*Thlaspi arvense* L.). Pennycress Resource Network, Western Illinois Univ., Macomb, IL.

Worldwatch Institute (2007). Biofuels for transportation: Global potential and implications for sustainable energy and agriculture, Deutsche Gesellschaft für Technische Zusammenarbeit (GTZ) GmbH, Germany.

Yacoub, Y., Bata, R., Gautam, M. and Martin, D. (1997). The performance characteristics of C11-C5, alcohol-gasoline blends with matched oxygen content in a single cylinder SI engine, Preprints, Div. of Fuel Chem., Am. Chem. Soc., **42**: 723–727, Washington, D.C.

Zeri, M., Anderson-Teixeira, K., Hickman, G., Masters, M., DeLucia, E., J. and Bernacchi, C. J. (2011). Carbon Exchange by Establishing Biofuel Crops in Central Illinois, *Agric., Ecosystems and Environment*, **144**: 319–329.

Zeri, M., Hussain, M. Z., Anderson-Teixeira, K. J., DeLucia, E. and Bernacchi, C. J. (2013). Water Use Efficiency of Perennial and Annual Bioenergy Crops in Central Illinois, *J Geophys Res-Biogeo*, **118**: 581–589.

Zimmermann, C. (2010). Ethanol pump opening celebrations, Domesticfuel, Cantonment, FL.

8

Wind Energy

"Of all the forces of nature, I should think the wind contains the largest amount of motive power — that is, power to move things. Take any given space of the earth's surface — for instance, Illinois; and all the power exerted by all the men, and beasts, and running-water, and steam, over and upon it, shall not equal the one hundredth part of what is exerted by the blowing of the wind over and upon the same space. And yet it has not, so far in the world's history, become proportionally valuable as a motive power. It is applied extensively, and advantageously, to sail-vessels in navigation. Add to this a few windmills, and pumps, and you have about all. . . . As yet, the wind is an untamed, and unharnessed force; and quite possibly one of the greatest discoveries hereafter to be made, will be the taming, and harnessing of it." **Abraham Lincoln**, Lecture 'Discoveries and Inventions', (1860) in *Discoveries and Inventions* (1915).

8.1. Introduction

The concept of harnessing wind energy goes back to early recorded history when people used wind energy to propel boats along the Nile River as early as 5000BC. By 200BC, windmills in China were used to pump water, while vertical-axis windmills with woven reed sails were used for grinding grain in Persia. By the 11th century, people in the Middle East used windmills extensively for food production.

The European merchants returning from the Middle East brought the idea to Europe and the Dutch adapted it for draining lakes and marshes in the Rhine River Delta. By the late 19th century, the settlers took this technology to the New World and began using windmills to pump water for farms and ranches and later, to generate electricity for homes and industry. American colonists used windmills to grind wheat and corn, to pump water and to cut wood at sawmills. With the development of electric power, wind power found new applications in lighting buildings remotely from electric power plants (WEF, 2014).

Wind energy can be converted into electrical energy or used to produce mechanical power. With advances in materials science, control technology, wind turbines, and electric generator technology of various sizes, we can now utilize wind energy with rather high efficiencies. U.S. Energy Information Administration (US-EIA) has estimated that wind power capacity in the U.S. will reach 87 GW by 2040, a 45% increase in electricity production by wind power which was 60 GW in 2012 (US-EIA, 2014). On average, a 1 MW turbine can power approximately 300 U.S. homes (Baker *et al.*, 2014), so additional 8 million more homes will be powered by wind energy by 2040.

The intermittency of wind energy presents a practical challenge to its functionality and developers choose areas that have high and consistent wind speeds, and then pick their turbines accordingly.

Illinois is endowed with wind patterns that are favorable for wind utilization, especially for electric power generation. Illinois inland wind power potential at 80-meter height above the ground and at gross capacity factor (without losses) of 30% and greater is about 250 GW (US- DOE, 2011). The total windy land area for this potential wind power generation is about 71,000 km^2 (~27,500 square miles). Out of this total windy land area, about 70% (~50,000 km^2) is available for installation of wind turbines (at 80 meter height above the ground and at gross capacity factor of 30%), representing roughly 34% of the State of Illinois total area, which is estimated at 149,998 km^2 (57,918.38 square miles) (US-DOE, 2011; U.S. Census

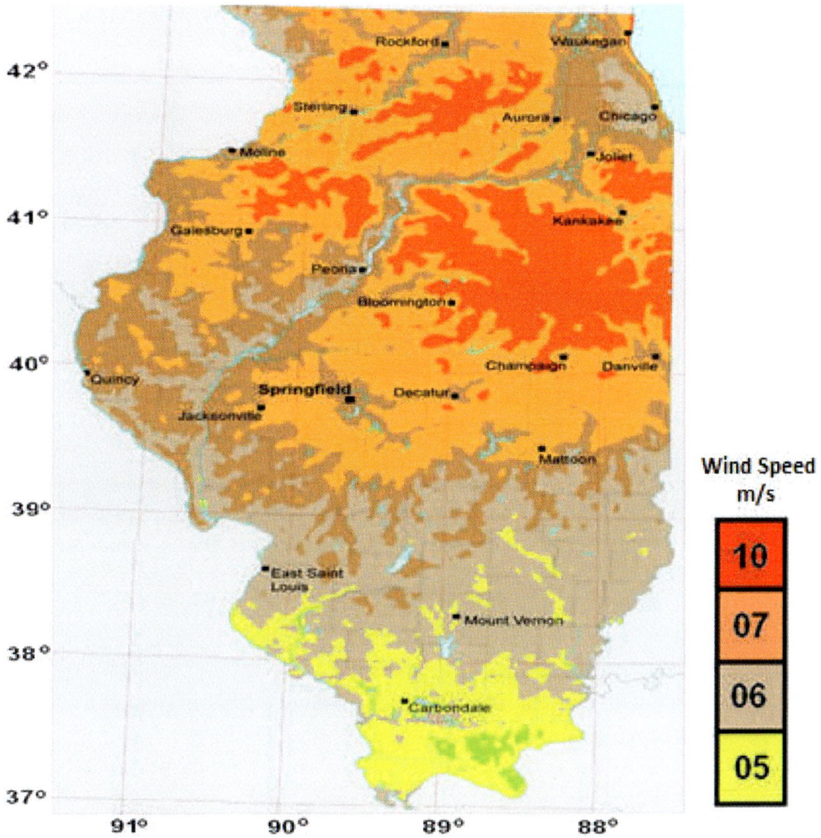

Fig. 8.1. This Illinois wind map shows the wind resource at an 80 m height (US-DOE, 2011).

Bureau, 2004). It should be mentioned that Illinois has the 25th largest total land area among the 50 states in the U.S.

It is worth mentioning that the cumulative installed capacity for land-based wind power in Illinois in the year 2012 was 3.568 GW. Also, the corresponding net electricity generation from land-based wind power in 2012 was 7,727 GWh (US-EIA, 2013b). Figure 8.1 shows the map of Illinois wind resource potential at 80-meter height above the ground and at wind speed ranging from 0.5–10 m/s.

According to Fig. 8.1, Illinois high wind speed areas (∼10 m/s) are close to cities such as Champaign, Kankakee, Bloomington and

Table 8.1. Estimates of windy land area and wind energy potential in Illinois (US-DOE, 2011).

Height Above Ground (m)	Gross Capacity Factor (%)	Total Windy Land Area (10^3 km^2)	Available % State Land	Wind Energy Potential	
				Installable Capacity (GW)	Annual Generation (10^3 GWh)
80	>30	71	34.25	250	764
80	>35	31	16.53	121	392
80	>40	1	0.62	4.5	16
100	>30	100	45.18	330	1,073
100	>35	65.5	32.00	233	799
100	>40	20	10.79	79	287

(a) 'Total windy area' refers to all the windy land areas in the state, including areas that are unlikely to be developed, such as wilderness areas, parks, urban areas and water features.
(b) 'Available windy' area refers to all the windy land areas in the state, excluding protected lands (national parks, wilderness, etc.), urban, airports, wetland, water features and other considerations.
(c) 'Available % State land' represents the percentage of Illinois land area available for wind turbine installation.
(d) Gross capacity factor is the actual amount of energy produced divided by the maximum amount of energy that could be produced (expressed as %), if the wind turbine were to operate at rated power for the entire year.
(e) The installable capacity assumes 5 MW/km^2 of installed nameplate capacity.

Peoria. The high wind speed areas are free from significant obstructions to wind currents. The lowest wind speeds (~0.5m/s) in Illinois are found in the areas around Carbondale. The state's largest inland wind power potential occurs at 100 meters above the ground. The installable wind power at this height is 330 GW at a gross capacity factor of 30% and greater, as shown in Table 8.1. The total windy land area for this potential wind power generation is 99,669 km^2.

Illinois can also take advantage of offshore wind energy from Lake Michigan. Figure 8.2 shows the total estimated technical potential for inland and offshore wind power in the U.S.

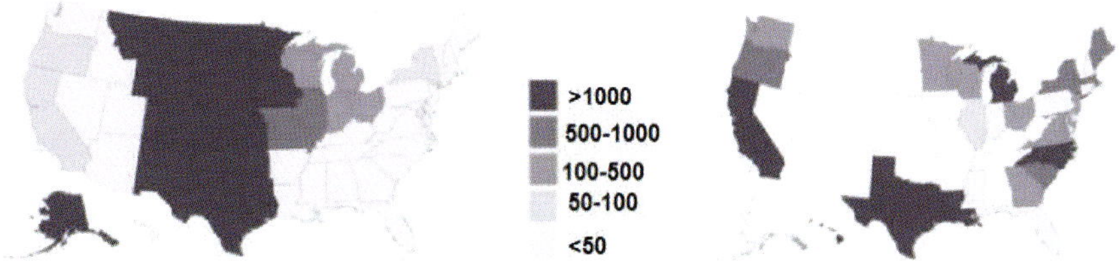

Fig. 8.2. Total estimated technical potential for inland (left) and offshore (right) wind power (in 1000s of GWh) in the U.S. (Lopez *et al.*, 2012).

Fig. 8.3. Mendota Hills Wind Farm, spring 2007 (NIU, 2007) (Courtesy of College of Liberal Arts and Sciences, Northern Illinois University).

According to Fig. 8.2, Illinois' offshore (Lake Michigan) total wind energy potential is in the range of 50,000–100,000 GWh, which is an order of magnitude less than inland wind energy potential of the state. The inland wind energy potential in the state of Illinois is greater than that of California and those of most of the Midwestern states, but less than those of the U.S. Central and Southwestern states, such as Texas.

The first utility-scale wind farm in Illinois was established in 2003 at Mendota Hills, near the village of Paw Paw in Lee County, about 160 km (100 miles) west of Chicago (see Fig. 8.3).

This 53-turbine wind farm is operated by Navistar, which sells the energy to Commonwealth Edison (ComEd) utility company. The Mendota Hills farm has a capacity of 51.66 MW, enough to power 13,000 homes.

In 2013, Illinois provided 4.7% of its total electricity by wind energy. As of first quarter of 2014, Illinois ranked fourth in the nation in wind energy with capacity of 3570 MW (US-IEA, 2013b; NEO, 2014), despite ranking 15th in wind resources nationally (Lydersen,

Table 8.2. Yearly wind power generation percentage increases required to meet the RPS for the electric utilities in Illinois (PJM and EIS, 2014).

Year	2008	2010	2012	2014	2016	2018	2020	2022	2024	2025
(%)	1.5	3.75	5.25	6.75	8.625	10.875	13.125	15.375	17.625	18.75

2012; AWEA, 2014). The Illinois Power Agency Act (IPAA) enacted the Renewable Energy Portfolio Standards (RPS) in 2007 (Public Act 095-0481). The RPS mandates statewide gradual consumption of more renewable energy with the requirement that 25% of the state's power demand be met by clean energy by 2025. RPS required industry-owned utility (IOU) companies with more than 100,000 customers to supply 1.5% of their power from wind energy in 2008–2009 (1 June 2008 to 31 May 2009), reaching a goal of 18.75% (75% of all the renewables) between the years 2025–2026, as shown in Table 8.2.

Also, Illinois House Bill HB-1865 enacted in August 2011, allows multi-jurisdictional utilities with fewer than 100,000 Illinois customers to request a procurement plan from the IPA. These utilities will be subject to the RPS requirements (US-DOE, 2012a). Illinois electric power sector generation by wind energy from 1989–2012 estimates is shown in Fig. 8.4.

As of 2011, there were over 150 wind energy companies in Illinois, with a combined workforce of over 15,000. About 30% of these companies were based in Chicago (Graig *et al.*, 2011). The Illinois wind energy companies include turbine and tower makers; manufacturers of gears, couplings, bearings and fasteners; legal, financial, engineering and consulting firms; and diagnostic software designers. As of 2014, Chicago hosted at least 13 global or U.S. headquarters of major wind energy companies, more than any city in the U.S. (Graig *et al.*, 2011; Chicago Tribune, 2014). These companies include Acciona, a $7 billion Spanish-based multinational corporation that develops and manages clean energy projects and operates a turbine assembly and U.S. Mainstream Renewable Power Company, an Irish

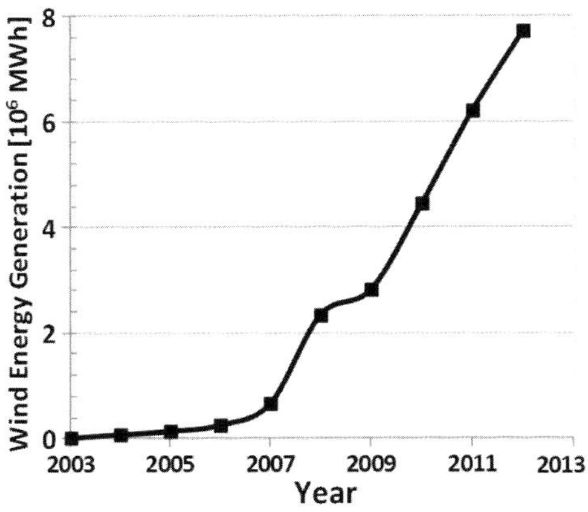

Fig. 8.4. Illinois electric power sector generation by wind energy, 1989–2012 (Data from US-EIA, 2012).

company that claimed to have plans to invest $1.6 billion over the next four years in three Illinois wind farm developments (Yerak, 2009). However, the expiration of Federal renewable energy production tax credit in December 2012 has delayed many U.S. wind energy projects (Wernau, 2012; Lydersen, 2013). Mainstream Renewable Power Company claimed to produce sufficient clean energy from these projects to power 200,000 homes by 2013, but apparently this promise cannot be fulfilled without Federal subsidy (Landis, 2011). Prior to the expiration date of 31 December 2013, the Federal renewable energy tax credit offered wind energy producers 2.3 cents per KWh of electricity produced (Wernau, 2012; DSIRE, 2014). But despite these problems, the Illinois wind energy industry has made significant progress (see Table 8.3).

From 2003 to 2012, the state's wind power capacity increased from 50 to 3,568 MW, a growth of more than 66 times in less than a decade (Loomis *et al.*, 2012; Lydersen 2012). In 2014, the number of online wind farm projects in Illinois increased to 46, compared to just 17 in 2011, with 23 having the capacity greater than 50 MW, representing 99% of wind power generation in the state (Lydersen, 2012; AWEA, 2014). In 2012, the Center for Renewable Energy of Illinois

Table 8.3. Summary of Illinois wind energy generation profile.

% Share of wind in State RPS(3)	18.75
Target year for wind power RPS(3)	2025
2003 Installed MW capacity(2)	50
2011 Installed MW capacity(2)	692
2012 Installed MW-capacity(2)	823.3
2012 Cumulative installed capacity(2)	3,568
No. of wind turbines in 2012(2)	2195
2011 % of Illinois power by wind(2)	3.2
2012 % of Illinois power by wind(2)	3.9
2013 % of Illinois power by wind(2)	4.7
No. of homes powered in 2014(2)	$>10^6$
No. of companies in 2011(1)	~150
Work force No. in 2011(1)	~15,000
No. of jobs supported in 2012(2)	6,001–7000
No. of jobs supported in 2013(2)	3,001–4,000
2012 capital investment(2)	$\$7.2 \times 10^9$
2012 land lease payments(2)	>$10,705,000
2012 water savings(2) [gallons]	$>2 \times 10^9$
2013 CO_2 avoided(2) [metric tons]	5.69×10^6

Sources: (1) Graig *et al.*, 2011; (2) AWEA 2013; (3) Loomis and Pagan 2011. RPS = Renewable Energy Portfolio Standard

State University estimated that the largest 23 wind farm projects created about 20,000 full-time jobs during the construction period, with total payroll estimated at over $1 billion. Additionally, the projects provided about 814 permanent jobs in rural Illinois areas, with total annual payroll of $48 million. The projects support local economies by generating $28.5 million in annual property taxes, and generate $13 million in annual income for Illinois landowners who lease their land to wind farm developers (Lydersen, 2012). The dramatic growth of Illinois wind energy industry is shown in Fig. 8.5.

Encouraging Federal and state policies, as well as increasing energy cost and energy security concerns are among the host of factors that have contributed to the rapid growth of wind power generation in the state. For example, a business establishing a new wind energy

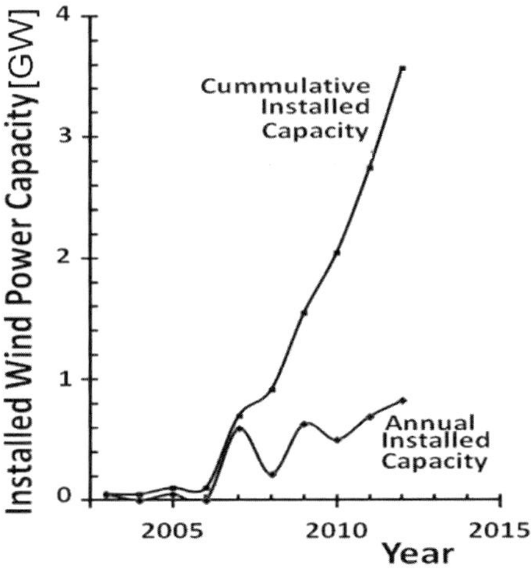

Fig. 8.5. Illinois installed wind power. Data from Illinois Wind Mill, 2012; Gelman, 2008, 2009, 2010, 2011, 2012; AWEA, 2013.

facility in Illinois may be eligible for a designation as 'High Impact Business', which allows state and local tax exemptions for the construction materials (Graig *et al.*, 2011). In addition to the RPS, other policies to encourage use of wind energy in Illinois are sales-tax incentive, state bond program and property tax consistency program. For the state bond program, Illinois Finance Authority can issue credit enhancements and tax-exempt bonds for renewable energy projects that would meet the eligibility criteria and would provide significant benefits to the public in Illinois. In the case of property tax consistency, Illinois provides enhanced tax for wind farms, by keeping uniform property tax assessment of wind energy devices in all counties in the state. Another factor that has contributed to the Illinois wind energy development, especially in Northern and Central Illinois, is the location of the state in the Pennsylvania New Jersey Maryland (PJM) Interconnection market. PJM is currently the world's largest competitive wholesale electricity market. Therefore, since Illinois has better wind resources than many of the PJM states to the East, added to its robust transmission infrastructure, its wind power can be sold on the PJM market to assist other states meet the renewable energy portfolio standards. Also, the strong electricity demand in the Chicago area (surrounded by farmland with good wind potential and many transmission lines) can provide an immediate market for wind energy (Lydersen, 2012).

Southern Illinois is part of the Midwest Independent System Operator (MISO) Regional Transmission Organization (Lydersen, 2012). MISO guarantees consumers with unbiased regional grid management, as well as open access to the transmission facilities under MISO's functional supervision.

Besides inland areas, the US-DOE has estimated that Illinois has the potential to produce ~20 GW from its offshore wind energy resource of Lake Michigan (Kessler, 2011; Loomis and Carter, 2011). Lake Michigan is one of the five Great Lakes in North America. The Lake has a shoreline of 1,640 miles (2640 km), extending 307 miles (494 km) long by 118 miles (190 km) in width. From West to East, the lake is bounded by states of Wisconsin, Illinois, Indiana and Michigan. It is the second largest of the Great Lakes by volume, and the third largest by surface area, after Lake Superior and Lake Huron (NOAA, 2014). Nearly 12 million people live along the shores of Lake Michigan, mostly in the Chicago and Milwaukee metropolitan areas. The estimated surface area of the lake is 22,400 square miles (58,000 km^2), which makes it the largest lake entirely within one country by surface area and the fifth largest lake in the world (Wright, 2007). The lake's water volume measures about 1,180 cubic miles (4,918 km^3) and has average depth of 279 ft (85 meters), with the deepest depth occurring at 923 ft (281 m) (NOAA, 2014). The bathymetric data developed by the National Oceanic and Atmospheric Administration (NOAA) indicates that water depths are shallow enough for the placement of turbines in Lake Michigan (Kessler, 2010). In 2010, Illinois offshore wind resource potential represented 32% of the state's total electric power generation that was 199,475.2 GWh (Musial and Ram, 2010). In August 2011, the governor of Illinois signed into law House Bill 1558 (Public Act 97-0266), allowing creation of Lake Michigan Offshore Wind Energy Advisory Council within the Department of Natural Resources. The council was tasked to study the potential for wind energy projects in Lake Michigan, which could allow counties along the lake to establish wind farm districts. In this study, the council considered wildlife impact, protected habitats and recreational uses of Lake Michigan to identify locations that would be appropriate for offshore wind development (Kessler, 2011; SJR, 2012). The charter of the council was to assist the Illinois Department of Natural Resources by providing recommendations and preparing reports to the Illinois General Assembly and the governor, based on the study of the following topics (IDNR, 2012):

(i) Appropriate criteria to be used by the Department to review applications for offshore wind development of Lake Michigan lakebed leases.
(ii) Criteria for identification of areas that are favorable, acceptable and unacceptable for offshore wind development including, but not limited to, impacts to wildlife, protected habitats, navigation, commercial fisheries and recreational uses of Lake Michigan.
(iii) A recommended process for ensuring public engagement in the Department's process for leasing Lake Michigan lakebed for offshore and wind energy projects.
(iv) Options for how the State shall be compensated for Lake Michigan lakebed leasing.
(v) A summary of the lessons learned from other domestic and international offshore wind development experiences, including, but not limited to, those related to public policy, regulatory and siting concerns for offshore wind development.
(vi) Identification of local, State, and Federal authorities with permitting, siting, or other approval authority for wind power development in Lake Michigan.
(vii) Recommendations for needed State legislation and regulations governing offshore wind farm development.

On June 30, 2012, the council released its final report of the study to the Illinois Governor and the General Assembly (IDNR, 2012). Prior to the passage of House Bill 1558, the City Council of Northern lake-front suburb of Evanston, near Chicago had unanimously voted to move forward on a proposal to place 40 wind turbines on Lake Michigan. The decision could have made the city the first in the U.S. to have an offshore wind farm (Mollmann, 2010). An advantage of offshore wind for power generation in Illinois is proximity to the city of Chicago (a major U.S. industrial city), which can bring a significant reduction in power transmission cost. Also, unlike the U.S. Gulf coast, the Great Lakes are free of natural disasters, such as hurricanes. Additionally, offshore winds tend to blow stronger with more uniformity than land-based winds because of lack of surface infrastructure which can obstruct the path of wind currents. Such wind patterns have been observed as a potential source for more electricity generation, as well as smoother and steadier operation of wind turbines, compared to land-based wind power systems (Musial and Ram, 2010).

According to the American Wind Energy Association (AWEA), Illinois wind power resource is capable of meeting more than 5.2 times the state's current electricity needs (AWEA, 2014). Table 8.4 shows Illinois' offshore and land-based windy areas, as well as the wind energy potential compared with other Great Lakes States and Iowa (a non-Great Lake neighboring state).

8.2. Concepts of Wind Power Generation

The kinetic energy of wind can be converted into mechanical shaft power. This mechanical shaft power can be used for specific tasks (such as grinding grain or pumping water) or can be converted into electricity using a generator. Wind speed is the feature of wind for mechanical shaft energy production and wind energy is proportional to wind speed (Manwell *et al.*, 2010).

There is an empirical mathematical relationship between wind speed and height, which is used to formulate the turbine power as follows: for wind turbine engineering purposes, an exponential variation

Table 8.4. Comparison of Illinois offshore and inland wind energy potential with other Great Lakes states.

	Windy Land Area ≥ 30% Gross Capacity Factor at 80 m Height					Land-Based Potential		Offshore (>7 m/s) Potential Within 50 Nautical Miles at 90 m Height	
State	Total(1) (km^2)	Excluded(1) (km^2)	Available[1] (km^2)	Available % of State(1)	% windy Land Excluded(1)	Installed Capacity(1) (GW)	Annual Generation(1) (GWh)	Estimated Area(2) (km^2)	Estimated Capacity(2) (GW)
Illinois	70,764	20,787	49,976	34.25	29.4	249.882	763,529	4,192	20.960
Indiana	46,255	16,610	29,645	31.63	35.9	148.227	443,912	584	2.919
Iowa	134,900	20,757	114,143	78.32	15.4	570.714	2,026,340	NA	NA
Michigan	19,761	7,953	11,808	7.85	40.2	59.042	169,221	96,642	483.212
Wisconsin	30,229	9,477	20,751	14.29	31.4	103.757	300,136	23,298	116.5

(1) US-DOE (2011); (2) Schwartz *et al.* (2010)
NA = Not available.

in wind speed with height may be defined relative to wind speed measured at a reference height of h_r. The reference height is usually chosen to be 10 meters. Thus, the relationship will be (Zaman and Shakouri, 2011):

$$V_{wind}(\mathrm{h}) = V_{10}\left(\frac{\mathrm{h}}{\mathrm{h}_{10}}\right)^{a} \tag{8.1}$$

where $V_{wind}(\mathrm{h})$ is the velocity of the wind in [m/s] at an arbitrary height of h [m]; V_{10} is velocity of the wind at height, $\mathrm{h}_{10} = 10$ meters; a is known as the Hellman exponent, which is an empirically derived coefficient that varies depending upon stability of the atmosphere. For neutral stability conditions, a is approximately 1/7, or 0.143.

The kinetic energy E (in J) of an air mass m (in kg) moving at a speed V_{wind} (in m/s) is given by:

$$E = \frac{1}{2} m V_{wind}^2 \tag{8.2}$$

Designating ρ [kg/m^3] as the air density, mass flow through an area A is given by:

$$\dot{m} = \rho A V_{wind} \tag{8.3}$$

Thus, an equation for the power (in Watt) through a cross-sectional area A normal to the wind is:

$$P_{wind} = \frac{1}{2}\rho A V_{wind}^3 = \frac{1}{2}\rho A V_{10}^3 (\mathrm{h}/\mathrm{h}_{10})^{3a} \tag{8.4}$$

This equation shows the relationship between wind power and wind speed and the height of the wind mill.

Generally, wind turbines consist of three principal components, the nacelle, rotor and tower. The nacelle compartment contains the generator, gears and controlling mechanisms. The rotor, which consists of wing-shaped blades connected to a central hub, converts the kinetic energy of the wind into rotational energy. The tower, including the supporting foundation, provides the height necessary to access the wind resource and the conduit required to transfer

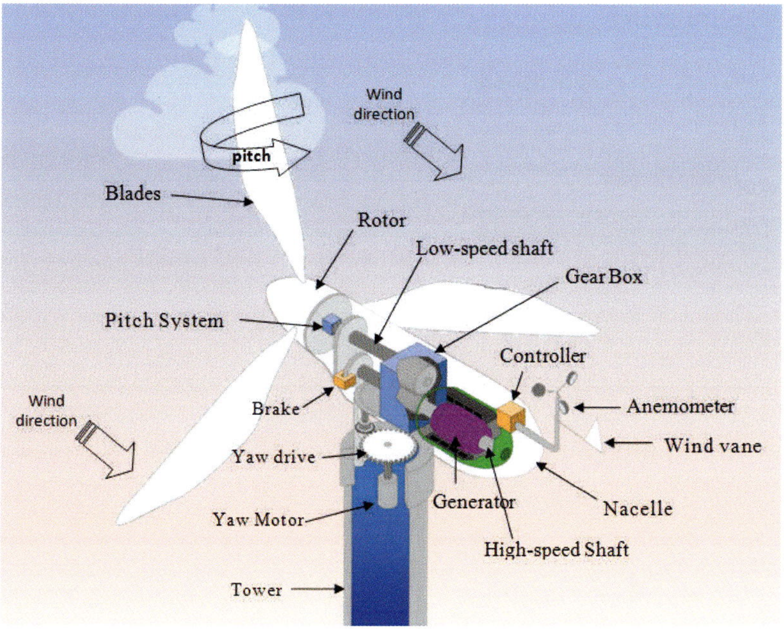

Fig. 8.6. Components of an upwind horizontal-axis wind turbine (modified from www.energy.gov, 2014).

the turbine-generated electricity to the collection system (Wilburn, 2011). Figure 8.6 shows the components of an upwind horizontal-axis wind turbine.

Wind power is classified based on typical wind speeds. These classifications range from Class 1 (the lowest with wind speed less than 5.6 m/s at 50 meters high above the ground) to Class 7 (the highest with wind speed 11.9 m/s at 50 meters high above the ground) (US-DOE, 2011). Most wind turbines start generating electricity at wind speeds of around 3–4 meters per second (m/s), generate maximum (rated) power at around 15 m/s and shut down to prevent storm damage at 25 m/s or above.

8.2.1. *Betz law*

Betz (1966) formulated a law that calculates the maximum power that can be extracted from the wind, independent of the design of a wind turbine in an open flow system. According to the kinetic energy equation, power extracted from the wind by a rotor is equal to the mass of air multiplied by the drop in the wind speed to power two:

$$P = (1/2)m(V_1^2 - V_2^2) \tag{8.5}$$

Considering the surface area swept by the rotor blades, A, the mass of the air flowing through the rotor area per unit time is

$$\dot{m} = \rho A(V_1 + V_2)/2 \tag{8.6}$$

where $\dot{m}$ is the mass flow per second, ρ is the density of air and $[(V_1 + V_2)/2]$ is the average wind speed through the rotor blades area. Substituting $\dot{m}$ into Eq. (8.5), we get the following expression for the power extracted from the wind:

$$P = (\rho/4)(V_1^2 - V_2^2)(V_1 + V_2)A \tag{8.7}$$

Now, let us compare our result with the total power in an undisturbed wind streaming through exactly the same area of Eq. (8.4), with no rotor blade blocking the wind. We call this power P_0:

$$P_0 = (\rho/2)V_1^3 A \tag{8.8}$$

The ratio between the power we extract from the wind and the power in the undisturbed wind is then:

$$(P/P_0) = (1/2)(1 - (V_2/V_1)^2)(1 + (V_2/V_1)) \tag{8.9}$$

Let us now define $\mathrm{Y} \equiv P/P_0$ and $\mathrm{X} \equiv V_2/V_1$. Then this equation will be in the following form:

$$\mathrm{Y} = (1/2)(1 - \mathrm{X}^2)(1 + \mathrm{X}) = (1/2)(1 + \mathrm{X} - \mathrm{X}^2 - \mathrm{X}^3) \tag{8.10}$$

Now, by setting $\mathrm{dX/dY} = 0$ and $\mathrm{dX^2/(dY)^2} < 0$ we will find the maximum of Y with respect to X as follows

$$\text{At X} = 1/3, \quad \mathrm{Y} = 16/27 = 0.593 \tag{8.11}$$

Graphically, we have the curve shown in Fig. 8.7 of P/P_0 *vs.* V_2/V_1.

We can see that the function reaches its maximum for $V_2/V_1 = 1/3$, and that the maximum value for the power extracted from the

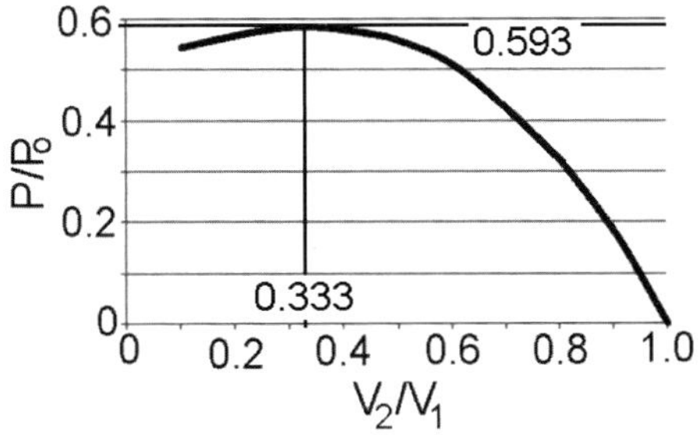

Fig. 8.7. Graph of the ratio of power entering and leaving a wind turbine versus velocity ratios (of wind entering and leaving a turbine), indicating the value of the maximum power ratio (0.593) at the velocity ratio of 1/3 = 0.333.

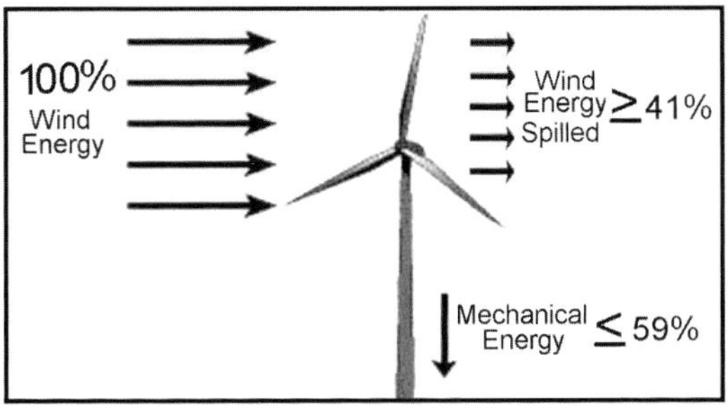

Fig. 8.8. Demonstration of the Betz Law, showing the maximum fraction of the power that a wind turbine can extract from the total wind power entering the turbine area.

wind is 0.59 or 16/27 of the total power in the wind. This result is shown in Fig. 8.8, which indicates that only up to 16/27 = 0.593 fraction (59.3%) of the wind energy can be converted to mechanical energy.

The calculation and the results reported above is known as the Betz law, which basically talks about how a wind turbine cannot extract more than 59.3 % of wind kinetic energy.

8.3. Wind Turbines Classification

Wind turbines can be classified as horizontal-axis or vertical-axis and can either be impulse (drag) or aerodynamic (lift) devices. The most commonly used wind turbines are the horizontal-axis devices

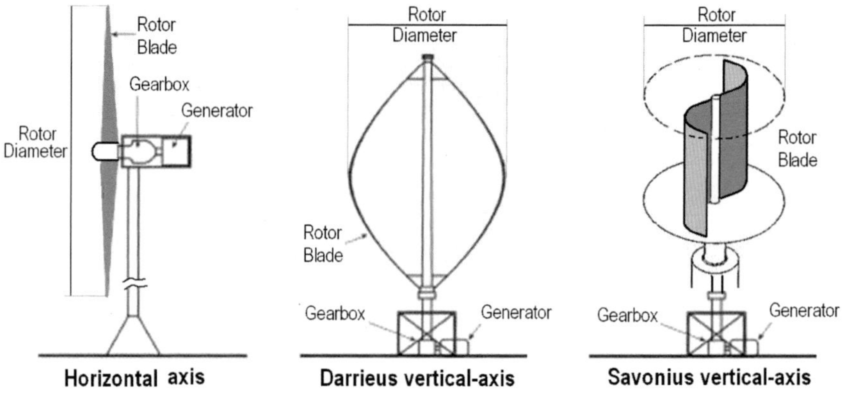

Fig. 8.9. Horizontal-and vertical-axis wind turbines (Scottish Gov., 2006).

because they are much more efficient. However, vertical-axis wind turbines have certain advantages, making them attractive for small scale use and in populated areas (See Fig. 8.9).

Below, we present the descriptions of the different types of wind turbines.

8.3.1. *Horizontal-axis wind turbines (HAWTs)*

HAWTs are the ones which the axis of the rotor's rotation is parallel to the wind stream and the ground and are used for utility and residential power generation. To achieve this orientation, utility- (large-) scale horizontal-axis wind turbines must be equipped with proper sensors and servomotors, while small-scale horizontal-axis wind turbines are equipped with wind vanes (John *et al.*, 2014).

These wind turbines capture the kinetic energy of wind with a propeller type of rotor and convert it into electrical energy. The rotor of a horizontal-axis wind turbine can be equipped with one, two, or three blades. The blades are attached to the rotor shaft, which is mounted on a high tower through a drive train or a gearbox. The wind currents passing over the blades cause rotation of the shaft which converts mechanical energy into electrical energy using a generator (Paraschivoiu, 2009).

The advantages of horizontal-axis wind turbines include access to stronger wind currents, since the blades can be mounted on a tall tower; and high efficiency, since the blade always moves at right angle to the direction of wind currents. Optimal layout of an actual wind farm depends on critical factors such as annual distributions of wind alignment and site-specific wind roses (information about wind speed and frequency of wind blowing from various directions). Studies have found that in order to maintain 90% of the performance of isolated HAWTs, the turbines in a HAWT farm must be spaced 3–5 turbine blade diameters apart in the cross-wind direction and 6–10 diameters apart in the downwind direction (Sørensen, 2004; Hau, 2013). The power density of such wind farms, defined as the power extracted per unit land area, is between 2 and 3 watts per meter square (MacKay, 2009).

The disadvantages of HAWTs include the cost of transportation and installation of the large and tall tower and blades, the additional yaw control mechanism to turn the blades toward the wind and disruption of landscape appearance caused by these turbines (Lack, 2010). Furthermore, there are predictions that the economic development and viable use of horizontal axis wind turbines would be limited by design constraints in the future, due to the high stress load imposed by the large size of the blades (Ragheb, 2014). In most cases, wind power Class 4 (with wind speed of 7.5 m/s at 50 meters high above the ground) or higher can be used for horizontal-axis wind power generation with large turbines. Class 4 and higher are considered good wind resources (US-DOE, 2011).

There are scattered regions of Class 4 wind resource in central and northern Illinois, with not less than five prime wind zones located at south east of Quincy, the Bloomington area, north of Peoria, the Mattoon area, and between Sterling and Aurora, as well as other potential sites, especially in northern Illinois (Daniels, *et al.*, 2004).

Current wind farms using horizontal-axis wind turbines have power densities that can range from 2–3 watts per square meter of land area. However, there are claims that power densities with several order of magnitudes higher than horizontal-axis wind turbines are potentially achievable, using vertical-axis wind turbines arranged in layouts that enable them to tap energy from above the wind farm and from adjacent turbine wakes (Dabiri, 2011; Bullis, 2013). As invisible currents, wakes are similar to water waves in nature in that they flow like water, creating ripple effects or disturbances in the atmosphere. In a wind farm, those disturbances in the atmosphere are created by the wind turbine blades and can potentially damage the wind turbines or decrease their efficiency downstream of the wind currents (Stark, 2011).

Horizontal-axis wind turbines can serve as remote or stand-alone power generating systems and can be used by small businesses to back up their electricity supply from the utility. Illinois companies that presently manufacture horizontal axis wind turbines include Acciona Wind Energy, Vestas Nacelles Americas GP & SC, Broadstar Wind Systems LP and Suzlon Wind Energy Corporation (Graig *et al.*, 2011). As of June 2014, there were 53 completed wind energy projects with the total capacity of 3,570 MW in 31 counties in Illinois. By the same time, permits had been issued for construction of additional 26 projects with the total capacity of 3,299 MW and 45 proposals had been submitted for construction of new wind projects. The summary of these data is presented in Fig. 8.10.

8.3.2. *Vertical-axis wind turbines (VAWTs)*

Vertical-axis wind turbines can catch wind currents irrespective of the direction of the wind. The axis of rotation of their rotor is perpendicular to the wind currents and they normally operate at low wind speeds approximately 6 miles per hour (2.68 m/s) compared to horizontal-axis wind turbines, which have wind speeds of 7.5 m/s (at 50 meters high above the ground) or higher, as mentioned above (EDL and Huntsman, 2010; Müller *et al.*, 2009; US-DOE, 2011; Ragheb, 2014). This allows vertical-axis wind turbines to be installed closer

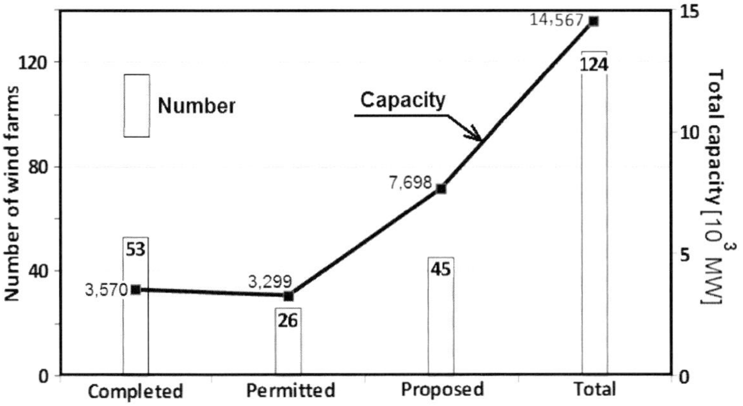

Fig. 8.10. Total number and capacity of completed, permitted and proposed wind farm projects in Illinois as of 2014 (Data from Center for Renewable Energy, ISU, 2014).

to the ground, where the wind speeds tend to be low and inconsistent, due to infrastructure development obstructing the path of the wind currents (Sutherland *et al.*, 2012).

However, there are claims that vertical-axis wind turbines are not as efficient as horizontal-axis wind turbines (Bullis, 2013). The reason is that half of their operating time, the blades of vertical-axis wind turbines move against the wind currents instead of generating the lift required for spinning the generator. As a result, the structures of vertical-axis wind turbines are subject to wear and tear, as the blades alternatively catch and move against the wind currents (Bullis, 2013). Although individual VAWTs often exhibit lower power coefficients than HAWTs, this deficiency can be compensated by the fact that VAWTs can be placed closer together.

Currently, there are few of these wind turbines in Illinois that belong to private companies and individuals. Examples of these turbines are the Aeroturbines, manufactured by Chicago-based Aerotecture International Inc. (Aerotecture International Inc, 2014; Graig, *et al.*, 2011). Another type of vertical-axis wind turbines is the magnetically levitated barrel generators (MagLev), manufactured by Chicago-based U.S. Way Turbines Corporation. The company is planning for the mass production of these turbines, based on the following applications and power outputs: conventional (5 kW), residential (10 kW), commercial (100 kW), industrial (500 kW) and utility (750 kW) (U.S. Way, 2014).

Vertical-axis wind turbines (see Fig. 8.9) are advantageous for small-scale wind energy utilization (for single-family homes) and have the potential for large-scale utilization at wind-farms as well. These wind turbines use straight or curved bladed rotors to capture the kinetic energy of the wind and convert it into electricity or drive a pump.

Vertical-axis wind turbines are more capable of operating in areas such as the inner city locations, where wind currents are inconsistent. For small-scale use, they may be installed on the roof of buildings since they are quiet and operate at lower wind speeds. But they may stall under gusty winds and their blades are prone to fatigue and may crack, leading to much less useful lifespan than horizontal-axis wind turbines. VAWTs require cheaper materials of construction and can easily be maintained, since the blades, generator and gearboxes are placed close to the ground. For large-scale energy systems (wind farms) they can be packed closer together and therefore, need less space (Jha, 2010).

The vertical-axis wind turbines are classified as Darrieus or Savonius rotor type and are described in the following sections.

8.3.2.1. *Darrieus vertical-axis wind turbines*

The Darrieus rotor (see Fig. 8.9) is constructed with high-performance symmetric profile blades. In general, the rotors have between two and three blades, and can be vertical, inclined or 'folded' in a semicircle or a parabola that is aerodynamically (lift) driven (Voneschen, 2012). The generation of the lift is caused by the shape of the turbine blades, which is airfoil (or like the wings of an airplane). A pressure differential is created as the blades of the rotor cut through the wind at an angle, causing a lift force that propels the blade forward. To propel the blades, the net torque caused by the drag forces must be less than the net torque caused by the lift forces (Hughes *et al.*, 2004). The Darrieus rotors are not self-starting and are usually used for turbines that are connected to external source of power, which is a disadvantage. The practical performance coefficient (ratio of the rate of net energy output to the total energy input) of the Darrieus vertical-axis wind turbine is about 40% at a tip-speed ratio (ratio of rotor speed to wind speed) of 5.0 (Lack, 2010). The tip-speed ratio determines whether a wind turbine is impulse (drag) or aerodynamic (lift) device (Müller *et al.*, 2009; Ragheb, 2014).

8.3.2.2. *Savonius vertical-axis wind turbines*

In its primitive form, the Savonius type wind turbine consists of two (see Fig. 8.9) or three cups attached to a central shaft in opposite directions. As each cup catches wind currents, a torque is exerted on the shaft to cause a rotation. This rotation will continue, in as much as there are wind currents. The energy from the rotating shaft can be used to drive a pump or be converted into electricity using a small generator. The modern Savonius vertical-axis wind turbines are equipped with fluted blade devices, which can operate at much higher performance coefficients with less vibration than the cups (Müller *et al.*, 2009; Ragheb, 2014). The practical performance coefficient of the Savonius vertical-axis wind turbine is about 15% at a tip-speed ratio of 1.0 (Lack, 2010).

Wind turbines control systems may be equipped with an anemometer instrument, which continuously monitors the wind speed. At sufficiently high wind speed ('cut-in' wind speed, generally about 4 or 5 m/s) and in order to overcome the frictional force in the drive train or gearbox, the control system allows the turbine to rotate to generate a minimum power. In general, wind turbines have a 'rated wind speed' (normally about 10–12 m/s) which corresponds to the maximum output power. Whenever the wind speed exceeds the rated wind speed, the control system avoids any additional power surge until the 'cut-out' wind speed (about 25 m/s) is reached (Paraschivoiu, 2002).

Other vertical-axis wind turbines emerging on the market are the helical turbines. These turbines are considered to be much safer for urban environments, due to their low rotational speeds, low noise, and ability to capture wind currents from all directions. There are claims that if a vertical-axis wind turbine with a rated power output of 10 MW, with almost the same availability as horizontal-axis wind turbine could be developed, then the economies of scale dictate that it would not make any difference even if the blade efficiency of the vertical-axis wind turbine is reduced from 56% to approximately 19–40%, in as much as the cost per unit of the rated power output is lower than that of horizontal-axis turbines. Therefore, it can be inferred that the future prospects of the vertical-axis wind turbine technology look very promising. It is also claimed that an array of counter-rotating vertical axis wind turbines will allow the use of a configuration that can generate twice as much power per acre of land compared to horizontal-axis wind turbines. The wind farm power flux for a vertical axis wind turbine is estimated to be 18 watts per square meter, which is 6–9 times the power density of modern horizontal-axis wind turbines. In other words, vertical axis wind turbines require less footprint or spacing compared to horizontal axis wind turbines (Müller *et al.*, 2009; Ragheb, 2014). Table 8.5 shows comparison of practical performance coefficients, tip-speed ratios and power densities of the different types of wind turbines.

As we mentioned in Sec. 8.3.2.1, tip-speed ratio determines whether a wind turbine is impulse (drag) or aerodynamic (lift) device, and is defined as the ratio of the rotor speed to the wind speed. The coefficient of performance is the ratio of the actual power captured by the wind turbine rotor to the available power. The power density of a wind turbine is defined as the wind power available per unit area swept by the wind turbine blades.

Table 8.5. Comparison of horizontal and vertical axis wind turbines.

	Vertical Axis		Horizontal Axis
	Darrieus	**Savonius**	**Triple blade**
Tip-speed ratio[(1)]	5	1	7
Performance coefficient[(1)] [%]	40	15	48
Power density[(2)] [MW/m^2]	18[(*)]	18[(*)]	2–3

Sources: (1) Lack (2010); (2) Ragheb (2014). (*) Based on mathematical modeling.

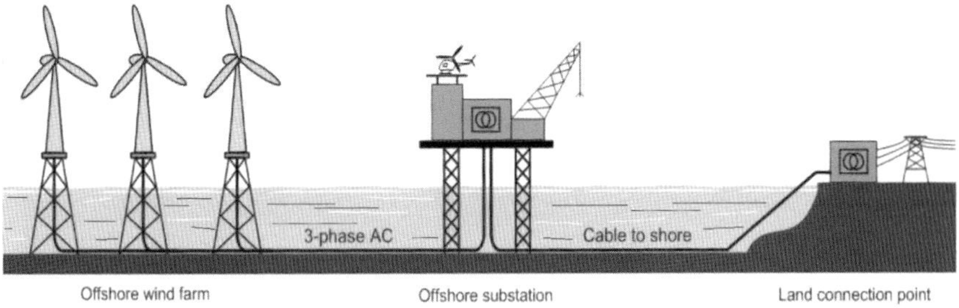

Fig. 8.11. A typical system of an offshore wind farm (modified from AWS Truewind LLC, 2009).

8.4. Offshore Wind Turbine Technologies

Offshore wind turbine is a nascent technology based on upgrade of the conventional land-based wind turbine technology. The principles of operation of offshore wind turbines are the same as onshore turbines; the critical differences being the support structure. These turbines are designed to meet aquatic or marine conditions. The rationale behind such a design is to make use of the relatively high wind speeds that exists offshore. This is possible because economies of scale permit installation of bigger size wind turbines offshore, allowing conversion of that wind resource into more electricity than using inland wind turbines. Figure 8.11 shows a typical offshore wind farm electrical system made up of the individual turbine transformers, a collection system, an offshore substation and transmission line to the mainland and the inland transmission components (AWS Truewind LLC, 2009).

The offshore substation steps up the voltage of the electricity generated from the wind turbine using a transformer, either offshore or inland. The transformer converts the electricity into a high voltage (usually between 33 kV and 132 kV) before connecting to the grid at an inland substation (BWEA, 2004). Compared with land-based wind turbine systems, this technology is relatively new and expensive, but it can be improved to achieve lower cost, expanded access to wind resources and improved performance of the system. Due to the fact that they can be built larger and mounted at higher elevation, offshore wind turbines have higher power capacity than standard land-based turbines, and ranges from 2 MW to 5 MW. The present design of offshore wind turbines typically consists of a three-bladed horizontal-axis system (with a nominal size of 80 meters to 130 meters in diameter,) a yaw controller, an active blade-pitch-to-feather controller and upwind rotors. Also, offshore wind turbines have higher blade tip speed (80 meters per second or greater) than those of land-based turbines. This is because at offshore, there are no aerodynamic noise concerns in the immediate vicinity. However, availability of offshore wind turbines is 80–95% compared to 95–99% for land-based wind turbines (IEA, 2009; Lijuan, 2014). Typically, offshore wind turbines are mounted on tubular towers ranging from 60–105 meters above the surface of sea. The towers are connected to the foundation, often using a transition piece as an interface between the foundation and the tower. The significance of the tower is to allow the turbine to capture winds at heights far above the surface of the water, where there are strong wind resources with less turbulent (AWS Truewind LLC, 2009). The design criteria for offshore wind turbine substructure depend on the soil conditions, the maximum wind speed, wave heights, as well as depth of water and vessel constraints.

Offshore wind projects require large capital investment and therefore necessitate involvement of local, state and federal governments. In September 2014, the US-DOE gave a $150 million loan guarantee for the 468 MW Cape Wind offshore wind farm in Massachusetts. This is the second offshore wind energy project in the U.S., as the 30 MW pilot Block Island offshore wind farm in Rhode Island is set for completion in 2016 (Casey, 2014). Next, we describe the different types of substructure for offshore wind turbines. These are classified as fixed-bottom and floating-platform technologies, as described below.

8.4.1. *Fixed-bottom offshore technology*

Fixed-bottom offshore wind turbines are installed at water depths that allow erection of a tall tower with the foundation touching the water floor. Therefore, they are classified based on the water depth, which includes shallow water and transitional water technologies.

8.4.1.1. *Shallow-water technology*

This technology is used in water depths of up to 30 meters or less. The substructure extension to the sea floor consists of fixed-bottom monopile and gravity-base, with proven turbine technology adapted from land-based systems (Musial and Ram, 2010).

8.4.1.2. *Transitional technology*

Transitional technologies are applied at water depths greater than 30 meters, but less than 60 meters. In shallow waters with soft soils or extreme breaking waves, the transitional technology could be used to withstand overturning forces, while conforming to turbine design requirements for stiffness (Musial and Ram, 2010). This new fixed-bottom technology includes jacket substructures and multi-pile foundations, such as tripods and trusses, which provide stiffer base for turbines used in shallow water (AWS Truewind LLC, 2009).

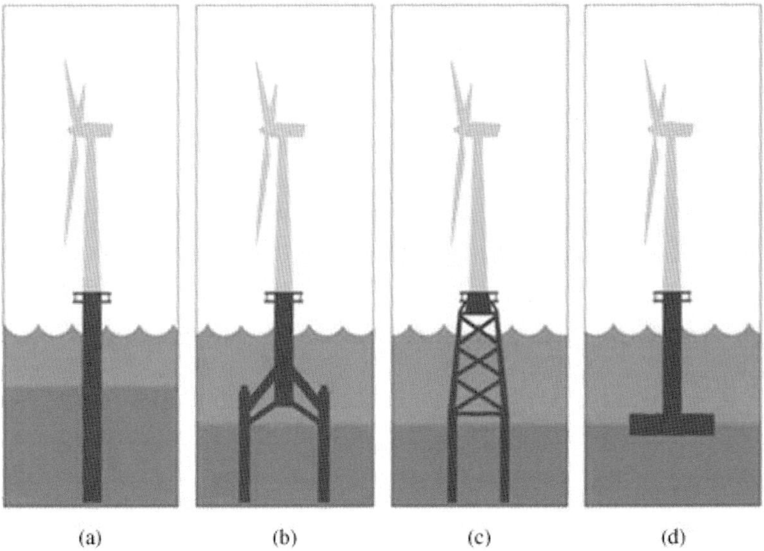

Fig. 8.12. Four main types of offshore wind turbine foundations (a) monopile, (b) tripile, (c) jacket and (d) gravity base (modified from Czyzewski, 2012).

8.4.1.3. *Substructure extensions for fixed-bottom offshore technology*

The four substructure extensions used in fixed-bottom offshore technology are the monopile, tripile, jacket and gravity base (Czyzewski, 2012). Figure 8.12 shows the different types of offshore wind turbine foundations.

a. *Monopiles*

These are the most commonly used substructure extensions in shallow water (less than 20 meters), owing to their design simplicity and relatively low cost. Additionally, they have minimum footprint and can be used on both land and in water, usually with minimal changes in the design. However, since they are flexible and slender, the performance of monopiles can be limited by resonance effects (Musial and Ram, 2010).

b. *Tripile*

The tripile foundations are improved versions of monopile foundations specifically designed for harvesting offshore wind currents. Tripile structure consists of three piles and a transition piece and can be used in water depths ranging from 25–50 meters (Czyzewski, 2012). Similar to monopiles, piles are typically made of cylindrical tubes of about 3 meters in diameter and 50 to 60 meters long. The thickness of the cylindrical tube of the piles can range from 40 mm to 65 mm, depending on water depth, turbine size and seabed conditions (de Vries, 2012). Also, depending on the depth of water and soil conditions, the piles can weigh about 400 metric tons, with each pile measuring about 65–90 meters high (Zaaijer, 2003; Fenech, 2011). For example, the piles and the hub height of the turbines above sea water level of the 400 MW Bard Offshore I North Sea project in Germany (claimed to be the third largest wind farm in the world during its official inauguration in 2013) is 90 meters high (de Vries, 2012; Power Technology, 2014). Depending on soil properties, about 20–30 meters of the pile can remain in the soil, as is the case of the Bard Offshore I North Sea project in Germany (de Vries, 2012). It should be mentioned that construction of the Bard Offshore I project began in March 2010 in Germany. The wind farm covers an area of about 60 km^2 and is located approximately 100 km northwest of Borkum Island in the North Sea. It consists of 80 Bard 5.0 wind turbines, enough to power up to 400,000 homes with annual CO_2 emissions reduction of about 450,000 metric tons (Power Technology, 2014). Above water level, the transition piece of the tripile foundation connects the three piles. The transition piece consists of a flange upon which the wind turbine is mounted, while the three legs remain within the accompanying piles. Also, the transition piece is fitted with work platform and stairs, while the landing for the boat is mounted on one of the piles. Compared with monopiles, the advantages of tripiles include simplicity of their design, improved stiffness and their automated leveling process. The disadvantages include the large amount of steel and intense labor required in their manufacturing (Lorc, 2012; Seidel, 2014).

c. *Gravity bases*

The gravity base foundation is traditionally deployed in shallow waters usually at depths less than 15 meters, but they can be installed at depths up to 29 meters as well (AWS Truewind LLC, 2009). Gravity base substructures are the most commonly used alternatives to the monopiles because of their improved performance in overcoming structural flexibility (Musial and Ram, 2010). They are different from monopiles in that they are not driven into the seabed, but instead they rest on top of the sea floor. However, these foundations may need more site preparation including dredging, filling, leveling and scour protection, depending on the geologic conditions of the site, (AWS Truewind LLC, 2009). Besides, the technology relies on massive weight and wide footprint to overcome the forces exerted on the turbine from the wind and waves, while in general the cost increases rapidly as the water depth increases (Musial and Ram, 2010).

d. *Jacket*

The Jacket foundation is used in water depths ranging from 25 to 120 meters. The construction of jacket frames is similar to an electricity pylon, weighing up to 828 metric tons. The design concept of jackets is borrowed from the oil and gas industry, where they have been used for supporting oil rigs at depths of more than 100 meters. A jacket is made up of three or four main legs, connecting to each other by bracings. But unlike inland lattice structures, which are usually made from angular profiles, all the elements of the jacket are tubular. The bracings and legs are connected up on to tubular joints, a key component of the construction. The advantages of jackets include their light weight, efficient construction, as well as their stability. The disadvantages include many man-hours of welding and complicated transportation to site (Lorc, 2012).

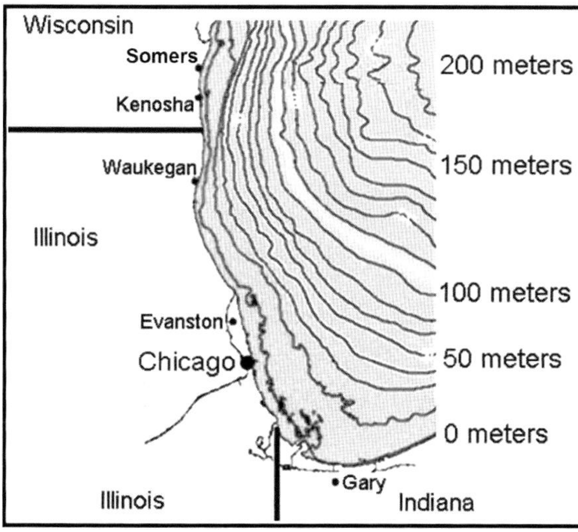

Fig. 8.13. Lake Michigan bathymetric chart around State of Illinois (modified from NOAA, 2014).

The suitability of the above-mentioned foundations in Lake Michigan will depend on the bathymetry map of the Lake (see Fig. 8.13).

Bathymetric charts provide information about lakes and oceans floors terrain or depth contours (isobaths). For example, at lake depths ranging from 0–50 meters (see Fig. 8.13), the selection of monopiles, gravity bases or tripiles foundations will be the appropriate choice because of the relatively shallow depth. But beyond 50 meters of depth, jacket foundations or floating platforms will be needed.

8.4.2. *Floating-platform offshore wind technology (Deep-water technology)*

The floating-platform offshore wind technology shown in Fig. 8.14 is needed in water depths greater than 60 meters, where a tall tower and foundation is not cost effective or economically feasible. For example, at water depths greater than 60 meters, there is no economic

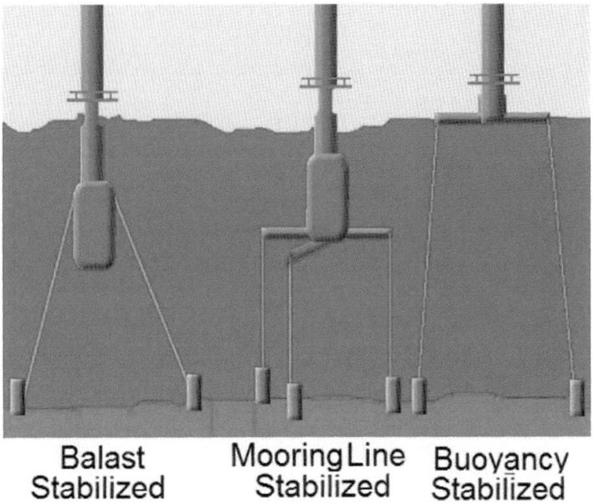

Fig. 8.14. The three main types of floating wind platforms: (Left) the ballast stabilized uses spar buoy platforms with catenary mooring drag-enabled anchors, (Middle) the tension leg platforms or 'mooring line stabilized platforms', is attached to the seabed with suction pile anchors, (Right) the 'buoyancy stabilized' platform employs a 'barge' type device with catenary mooring lines. Modified from (Williams, 2011).

justification for using any of the above-mentioned substructure foundations (monopoles, gravity basis) for an offshore wind turbine. A floating platform may become the technology of choice, however, this technology is still in the research and development stage and may not be available in the U.S. until 2020 (Black and Veatch, 2012). Typical substructures under consideration for deep-water technology include semi-submersibles, spar buoys and tension-leg platforms. A floating platform technology will require a modification of the current offshore wind turbine design specifications to offset the effect of coupled hydrodynamic/aerodynamic forces, including weight and buoyancy stability requirements that may arise (Musial and Ram, 2010). The main advantages of developing floating systems for deep water offshore wind turbines include: (i) access to potentially much greater wind resource, (ii) much greater total energy potential that could increase resource potential past 60 meters in deep water, (iii)

Table 8.6. Comparison of costs and characteristics for offshore and inland wind turbines.

Costs and Characteristics	Inland	Offshore
Investment Cost$^{(2)(*)}$ (\$/MW)	$(1.4-2.6) \times 10^6$	$(3.1-4.7) \times 10^6$
% cost share$^{(3)}$	65–84	30–50
% grid connection cost share$^{(3)}$	9–14	15–30
% construction cost share$^{(3)}$	4–16	15–25
Operation & maintenance cost$^{(2)}$ (\$/MWh)	12–32	21–48
Lifecycle cost of energy$^{(2)(*)}$ (\$/MWh)	70–130	110–131
Installed capacity of turbine$^{(1,3)}$ (MW)	1.5–2.5	2–5
% availability$^{(2)}$	>97	80–95
Blade tip speed$^{(1)}$ (m/s)	~75	>80

(1) Musial and Ram (2010); (2) IEA (2009); (3) IRENA (2012).
$^{(*)}$ Analysis was done based on the value of \$(U.S.) in 2008.

reduced impacts on human activities and environmental ecosystems, (iv) less shipping traffic and potentially out of migration routes and (v) absence of constructional restrictions regarding height and rotor diameter. The disadvantages include: (i) more expensive components, installation and maintenance, (ii) research and development needed for floating supports, (iii) greater cost risk and lack of support infrastructure and (iv) harsh environment to people and equipment (Manwell, 2010). Comparing inland and fixed-bottom offshore systems, there are fewer studies and published sources on floating-platform offshore wind technology systems. The 2020 estimated capital cost for the floating-platform offshore technology is \$4,200/kW + 35%, with cost improvements of 10% assumed through 2030 and capacity factor improvements for lower wind classes assumed until 2030 (Black and Veatch, 2012).

Table 8.6 shows the comparison of costs and characteristics of offshore and land-based wind turbine technologies.

In the U.S., data for fixed-bottom offshore wind turbines based on engineering, procurement and construction (EPC) projects are generally non-existent. But based on reviewed engineering studies and published data from the National Renewable Energy Laboratory (NREL) and European projects, the 2010 capital cost for a fixed-bottom offshore wind project was estimated at \$3,310 per kW + 35%. This estimate also assumes cost improvements of approximately 10% through the year 2030 and capacity factor improvements for lower wind classes through 2030 (Black and Veatch, 2012). Because offshore wind power technology in the U.S. is still in its infancy relative to land-based technology, there is a need to strike a viable balance between technological and economic challenges for it to become successful. Offshore wind power technology can take advantage of economies of scale in the future, as the size of turbines continues to increase. Table 8.7 shows Illinois' wind power resource potential for the various categories of water depths and the status of technology classes for offshore wind turbines.

As indicated in Table 8.7, substantial amount of wind power exists above the deep waters of Lake Michigan in Illinois. But as mentioned above, it will require deep-water technologies such as floating-platforms to tap into such potential power. Currently, there is no company in Illinois manufacturing these platforms. However, according to a technical report published in 2013, there are already certain employments in the various sectors of the state's economy that

Table 8.7. Illinois wind power resource potential (in MW) at 90-meter height and at 7.0–9.0 m/s wind speed above different categories of water depth, at different nautical miles (nmi) from the shore, including the status of technology classes for offshore wind turbines.

Offshore Technology[1,2]	Depth Category[1,2] (m)	Technology Status[1]	0–3 nmi (MW)	3–12 nmi (MW)	12–50 nmi (MW)	Total (MW)
Shallow water	<30	Proven	2502	4219	35	6756
Transitional water	30–60	Demonstration	5	2944	2137	5086
Deep water	>60	Demonstration	0	704	8414	9118

(1) Musial and Ram (2010); (2) Schwartz *et al.* (2010).

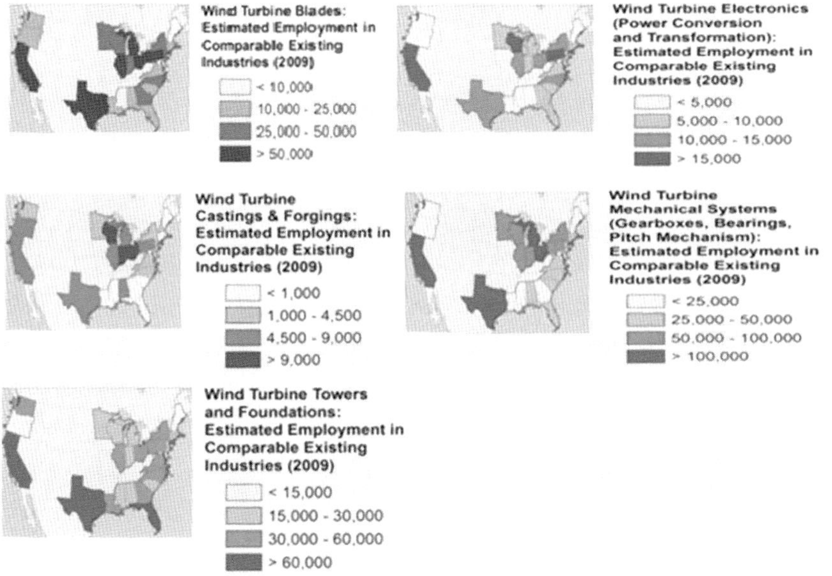

Fig. 8.15. Comparison of Illinois current employment in industrial sectors relevant to offshore wind power generation with states within the continental U.S. (Navigator Inc., 2013).

are important to offshore wind power generation (Navigant Consulting, 2013). These include foundries for casting and forging of wind turbine components, manufacturing of electrical systems and components for wind turbines, manufacturing of wind turbine blades, manufacturing of mechanical systems for wind turbines and manufacturing of wind turbine towers and foundations (see Fig. 8.15).

In 2009, the estimated number of employments in existing industries in Illinois that could manufacture wind turbine blades was greater than 50,000, while that for wind turbine towers and foundations manufacturing ranged from 30,000–60,000 (see Fig. 8.15).

On August of 2013, Illinois governor signed the Lake Michigan Wind Energy Act (Public Act 098-0447), that requires the Illinois Department of Natural Resources (IDNR) to develop a detailed offshore wind energy sitting matrix for the public trust lands of Lake Michigan. The Act authorizes the IDNR to grant offshore wind energy development site assessment permits and leases. The Act also authorizes the IDNR to convert site assessment leases to construction and operation leases and grants other rulemaking powers. Additionally, the Act creates the Offshore Wind Energy Economic Development Policy Task Force. The Task Force is charged with analyzing and evaluating policy and economic options to facilitate development of offshore wind energy and proposing an appropriate Illinois mechanism for purchasing and selling power from offshore wind energy projects (Hahn and Gilaman, 2014).

8.5. Small Scale Wind Energy Utilization

In this section, we present small scale wind energy utilization systems. Small wind systems are commonly defined as those with rated nameplate capacities between 1 and 100 kW; however, some have included wind turbines of up to 500 kW in that category (Corfee *et al.*, 2014). These small scale systems are specifically designed for places such as homes, small businesses, educational institutions, hospitals and farms. In the past, these turbines were generally installed in remote areas far from the grid, or locations where connection to the main grid was too expensive. Now-a-days small scale wind energy technology has become increasingly popular everywhere because individuals can now produce their own power and reduce their energy bills, while protecting the environment. Furthermore, small scale wind energy technology can relieve transmission problems such as power losses, while reducing the amount of electricity that must be transmitted long distances down high-voltage power lines. But, despite these advantages, grid interconnection has been identified by industry groups as the most important barrier to the installation of distributed generation technologies (Forsyth *et al.*, 1999; US-DOE, 2012c). Consequently, in August 2007, Illinois policymakers enacted a legislation (Public Act 095-0420). The purpose of this legislation was to encourage private investment in renewable energy resources, stimulate economic growth and enhance continued diversification of

Illinois' energy resources mix, while protecting the environment (IL General Assembly 2007). The bill required the Illinois Commerce Commission (ICC) to establish standards for interconnection and net metering covering renewable energy systems. However, while the promulgation of the bill was only applicable to interconnection standards for 'eligible renewable generating equipment', the ICC extended the standards to cover all distributed generation systems up to 10 MW of power as well. In August 2008, the ICC adopted the final interconnection standards and in March 2010, large distributed generation facilities over 10 MW were also added to be subject to the new interconnection standards (US-DOE, 2012b). Systems that could benefit from the interconnection standards include wind, solar thermal electric, photovoltaics, landfill gas, biomass and hydroelectric. Also included are geothermal electric, fuel cells, municipal solid waste, combined heat and power (CHP) cogeneration, anaerobic digestion, microturbines and other distributed generation technologies. The applicable sectors included commercial, industrial, residential, non-profit, schools, local and state government, federal government, agricultural and institutional (IL General Assembly, 2007). There are quite a large number of applications for distributed off-grid wind power generation systems in Illinois. However, in recent years, the small scale wind market has been shifting momentum from off-grid systems to more grid-connected systems, largely due to the passage of the legislation (Public Act 095-0420). The legislation allows owners of small wind turbines to take full advantage of the incentives contained in the Illinois' Renewable Energy Portfolio Standards. Below, we describe the configurations of a couple of small-scale wind energy utilizations in Illinois as examples.

In 2003, Chicago's Department of Housing selected Chicago's Mercy Housing Lakefront (formerly known as Lakefront Supportive Housing) to develop a new housing project as part of 'Chicago's 10-Year Plan to End Homelessness'. The building (see Fig. 8.16) was completed in 2007 and is located at 1244 N. Clybourn Avenue

Fig. 8.16. Wind power turbines atop the Mercy Lakefront SRO, a 96-unit housing development in Chicago's Near North neighborhood (Mercy Dream Productions) (Johnson, 2012).

on the fringes of what was once a public-housing project known as Cabrini-Green in Chicago's Near North.

The 96 apartment units are powered by eight rows of twisting wind turbines and flat solar panels hidden from public view. The cost of the building was estimated at just $14 million, with the eight wind turbines adding about $1 million to the final cost. Most of the money for the project was in the form of a grant from the City of Chicago. According to the developers, working together, the wind turbines and solar panels would result in the building's energy consumption being 22% less than a traditional building. Additionally, they claimed that energy savings for the building would pay for the wind turbines in 17 years. Also, compared with a Near North Apartments without wind turbines or solar panels, the developers estimated that this green technology would result in up to $18,000 in annual energy savings (Rafter, 2008).

In 2008, Chipotle Restaurant in Gurnee, IL opened a freestanding store outside Chicago with an on-site, 6 kW wind turbine (see Fig. 8.17). The 60-foot-tall wind turbine supplies up to 10% of the restaurant's electricity needs. During construction, the Gurnee store recycled and diverted 75% of construction waste (Cornish, 2009). The restaurant is equipped with an energy management system (EMS) capable of monitoring all major building systems (heating, ventilation and air-conditioning (HVAC), wind turbine, water heater, make up air, hood and lights) (USGBC, 2010).

In 2010, Clarendon Hills Middle School in Hinsdale, Illinois installed a vertical-axis wind turbine on its campus as students' project (see Fig. 8.18).

The cost of the 3 kW wind turbine was just under $40,000, including the installation. To facilitate education, the turbine is hooked to a computer display unit inside the school to allow students to monitor its daily performance. The anticipated savings from the turbine is between $500 and $1,000 per year. Also, it has been claimed that close to one metric ton (2,152 pounds) of CO_2 was saved in about two years since the installation of the wind turbine (Vandyck, 2012).

Fig. 8.17. Chipotle Restaurant in Gurnee, IL with an on-site, 6 kW wind turbine (Picture from Google maps).

Fig. 8.18. A wind turbine is installed in front of Clarendon Hills Middle School in Hinsdale, Illinois, in September 2010 (from Google maps).

8.5.1. *Configurations for grid-connected small-scale wind turbines*

Small-scale wind turbines connected to the grid usually have capacities/powers ranging from 10–100 kW. However, it is possible to have turbines that are rated as low as 1 kW, which may be connected to the grid as well, despite not necessarily designed for that purpose (Forsyth, *et al.*, 2000). Figure 8.19 shows the flow diagram of three commonly used configurations for grid-connected wind turbines.

For configuration 1 (in Fig. 8.19), the output of the turbine can be from either a wound-field synchronous generator or permanent-magnet (PM) alternator as shown in Fig. 8.20.

The wound field synchronous generator is a device that uses field coils to create magnetism from an electric current. It consists of a magnetic field on the rotor that rotates and a stationary stator containing multiple windings that supplies the generated power. The relative movement of this magnetism through other coils creates electricity. It

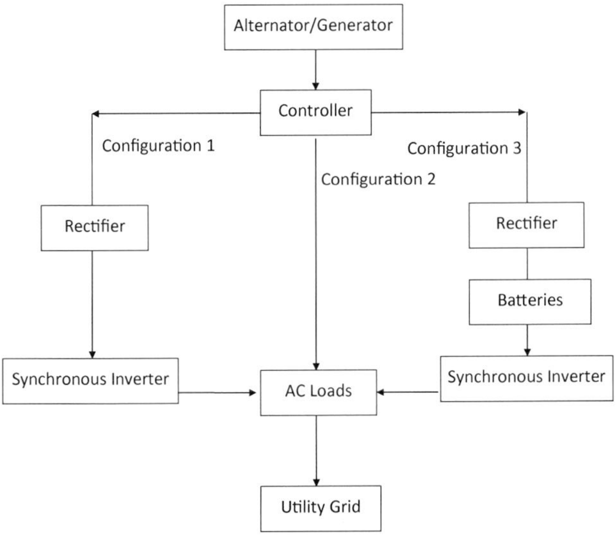

Fig. 8.19. Flow diagram of small wind turbines configurations (modified from Forsyth, *et al.*, 2000).

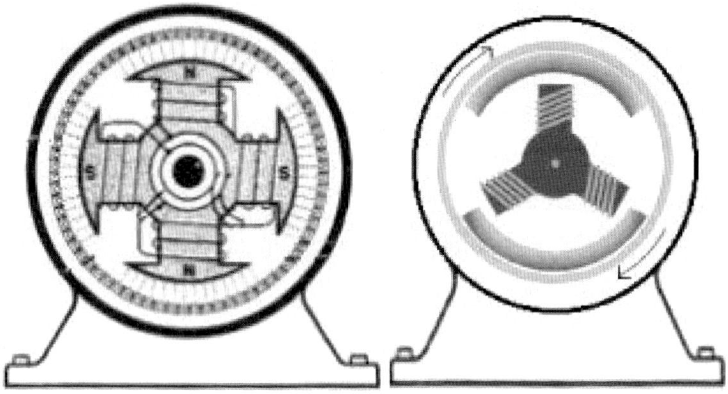

Fig. 8.20. Left: an elementary synchronous generator. Right: a permanent magnet generator (PMG).

is more widely used alternating generator, especially for hydropower and thermal power plants. For synchronous generators, an excitation direct electrical current (or a DC current) flows to the rotor windings and an alternating current (AC) output current is obtained from the stator winding. A PMG, however, uses high strength permanent magnets to create a magnetic field on the rotor (Boone, 2013). Therefore, since there is no current flow to the rotor, PMGs do not require brushes in their design, a situation which improves the reliability of the generator. But, one problem with the use of PMGs is the sensitivity of the stator insulation (mica) to vibration and thermal cycling. Also, for certain rotor designs in permanent magnets, direct-drive generators can impose significant heat-transfer problems which can complicate the design of the cooling system (Boone, 2013).

In any of the three cases of Fig. 8.19, the turbine output will be three-phase AC with variable voltage and frequency. Therefore, a line-commutated inverter will be necessary to condition the power prior to being transmitted to the utility grid. A line-commutated synchronous inverter, however, will stop functioning during power interruption or outage. For configuration 2 in Fig. 8.19, a constant-speed, grid-quality AC power is produced from an induction generator without requiring an inverter. The disadvantage of configuration 2 is that since the output of the induction generator is regulated by the utility power, in case of power interruption, the generator will not produce any output. The third configuration grid connection, as shown in Fig. 8.19, is a modified form of the first configuration, except that the former uses a self-commutated inverter instead of line-commutated inverter (Forsyth *et al.*, 2000). A self-commutated inverter when added to a battery will maintain the power in the inverter, even with utility power interruption.

8.5.2. *A grid-connected small scale wind energy system*

A typical grid-connected distributed small-scale wind energy generation system suitable for Illinois households is shown in Fig. 8.21. The system is equipped with a meter (usually a single, bi-directional meter is required) to enable it to dispatch excess electricity generation to the utility grid. In this case, the property owner can reduce the utility-supplied electricity consumption, or under a situation where the turbine cannot meet the energy needs of the household, the utility can

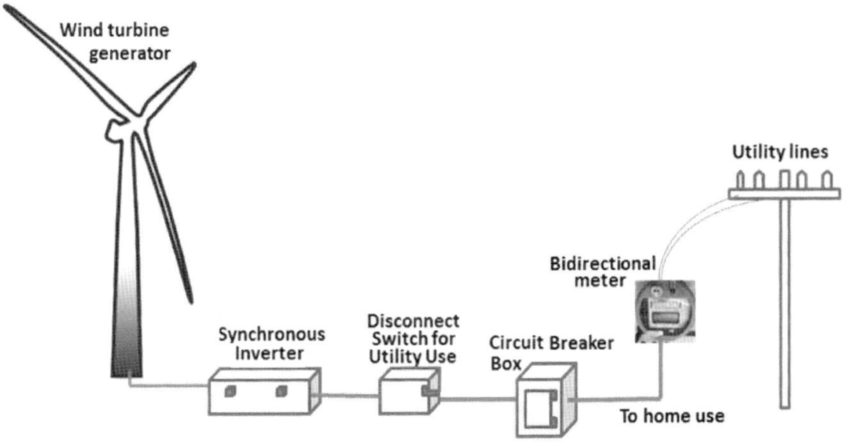

Fig. 8.21. Configuration of a typical on-grid small wind turbine with a permanent magnet alternator and synchronous inverter (modified from Iowa Energy Center, 2012).

make up for the difference (Svejkovsky, 2007). Illinois' program managed through utility companies, allowing renewable energy producers to send or sell excess electricity to the grid is termed 'a net-metering program'. As part of Illinois grid interconnection and net metering standards, net-metering requirements are frequently accompanied by 'feed-in tariffs', which requires the utility to accept customer-produced electricity at a predetermined price for at least 15–20 years period (Conger, 2012). Under the net-metering program, qualifying customers with generators of 40 kW or less will receive a one-to-one retail rate credit for their excess electricity generation. In addition, customers with eligible generators from 40 kW–2 MW will receive credits equal to the utility's avoided cost for their excess electrical energy generation. The utility's 'avoided cost' is the cost the utility would have paid to purchase the electricity from another source, assuming the customer is not providing the energy. Customers with service under time-of-use rates will receive net metering credits at time-of-use rates. Credits are carried over on month to month basis, with the annual period running from May to April, or November to October, normally at the choice of the customer. In Illinois, each utility must offer net metering program up to 1% or more of their previous year's peaks demand (IL-AG, 2013). Normally, the grid-connected wind energy systems become more beneficial when located in areas with average annual wind speed of at least 4.5 m/s (10 mph) or in areas where the utility-supplied electricity is expensive (10–15 cents per kWh). The technology can be attractive in areas where the utility's demands for connecting a system to the grid are not too expensive, and moreover, when there are good incentives for the purchase of wind turbines or for the sale of excess electricity (Svejkovsky, 2007).

8.5.3. *Hybrid wind energy systems*

Small wind turbines can be an integral component of a hybrid power system that depends on more than a single source of energy to provide a reliable power supply. These energy sources could be a cluster of

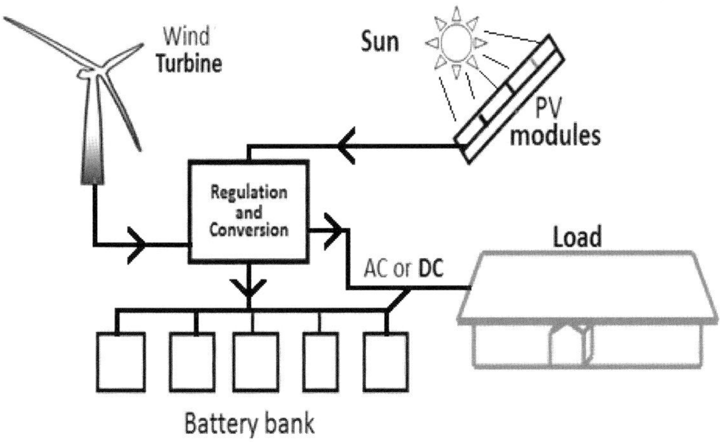

Fig. 8.22. Solar and wind hybrid system (modified from Rolland and Auzan, 2012).

wind energy, solar energy, fuel cell or microturbines using landfill gas as energy source. A hybrid power system (see Fig. 8.22) can provide uninterrupted power supply to remote locations, especially in areas of high population growth, partly due to their flexibility in the delivery of power. These remote locations could be far away from the power transmission lines or could be areas frequently characterized by frequent power outages (Rolland and Auzan, 2012). Because the state of Illinois experiences about 30% of sunny days throughout the year, integration of solar energy into a hybrid power grid could be a viable option in meeting its energy needs. However, the combination of these energy sources will require more controls, since the system design will be more complicated than stand-alone systems. Besides, the design must meet the cost effectiveness and technical reliability criteria. While batteries may be required in the system to accommodate the excess power during periods of high wind, the cost of battery storage system could be high, especially if the system is providing a large amount of power (Rolland and Auzan, 2012).

Wind turbine can be added to a hybrid HVAC system to obtain an optimum combination of convective and radiant heating and cooling system, with in-space thermal energy storage (Kilkis, 1999). Such

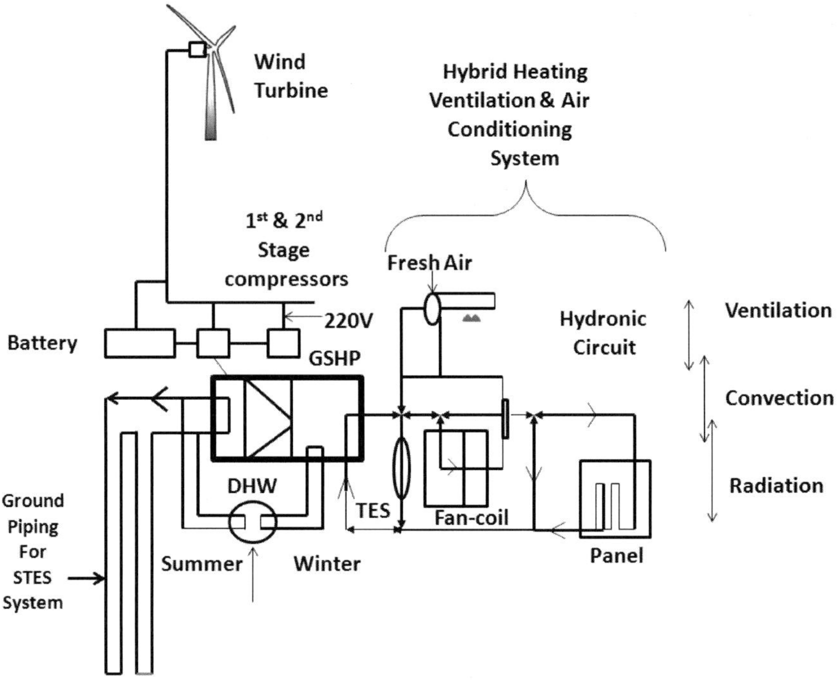

Fig. 8.23. Wind turbine driven GSHP coupling to a hybrid HVAC system (Modified from Kilkis, 1999).

a design (see Fig. 8.22) will be suitable for a single family home in Illinois. Besides, the design can reduce use of the more polluting primary energy source (natural gas) for space heating, cooling and air-conditioning. In Fig. 8.23, a 6 kW wind turbine is used to power a ground source heat pump (GSHP) coupled to a hybrid HVAC system to meet the heating demand of a 1100 square foot (∼100 square meter) home.

The sensible heating and cooling loads are met by the high mass radiant floor which matches the daily peak demand, as well as the available peak wind energy. The latent heating and cooling loads, along with ventilation requirements are met using a forced-air system. The GSHP is claimed to serve both the variable radiant and convective split type of control systems, while meeting the demand for domestic hot water (DHW) (Kilkis, 1999).

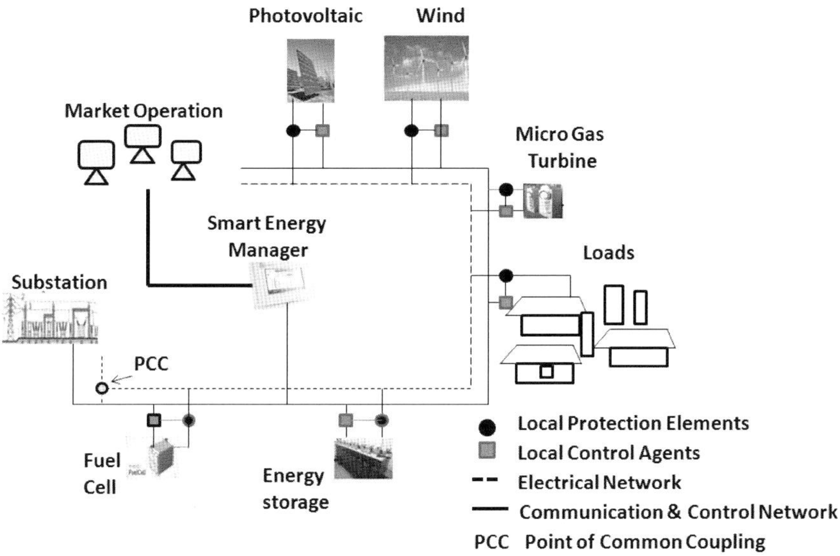

Fig. 8.24. Smart Microgrid (Modified from Zeitgeistlab, 2012).

8.5.4. *Small-scale wind turbine integration with micro-grid technology*

Wind energy can be integrated into smart micro-grid systems to provide cheaper, environmentally friendly, and a more reliable power supply to consumers (see Fig. 8.24).

Microgrid systems are not nascent technologies as they have been in existence since 1882 when the Edison Illuminating Company headed by Thomas Edison constructed the first centralized power plant (the Pearl Street Station) in Manhattan in New York City. Indeed, without a centralized grid system at that time, the Manhattan Pearl Street Station was in essence a microgrid. It is claimed that by 1886, Edison's firm had installed 58 direct current (DC) microgrids (Asmus *et al.*, 2009). But, following the evolution of the electric services industry in the U.S., which culminated into a state-regulated monopoly market, incentives for microgrid developments were removed (Asmus *et al.*, 2009).

The modern microgrid system, however, is a regionally-based energy system of distributed energy resources, which consists of a central controller, loads, and power sources that include wind turbines, photovoltaic systems, fuel cells and energy storage systems (see Fig. 8.24). Microgrids have several advantages, such as the ability to isolate preemptively from the point-of-common-coupling (PCC) to the utility grid, increased system reliability, consumers' ability to interact with the EMS and adjust their energy usage to reduce costs, as well as reduced need for central dispatch from utility (Zeitgeistlab, 2012). Additionally, the system may continuously run in on-grid, off-grid or dual mode by changing the grid connection status. Furthermore, in case of power outage, the system can operate autonomously in off-grid mode by drawing power from its generating sources. The system has the capability to automatically resynchronize with the grid when normal conditions are restored. In the grid-connected mode, the microgrid operator can use the market operations monitoring system to find out the current energy prices on the market and make economic decisions, such as energy sales and purchases, depending on on-site generation capability (Dohn, 2011). Currently, the following major market segments are taking advantage of this innovation: commercial/industrial, community/utility, institutional/campus and remote/off grid. Microgrid projects currently in the U.S. are the 42 MW microgrid at the University of California in San Diego (UC, San Diego), Michigan-based TARDEC (Tank Automotive Research, Development and Engineering Center) microgrid and the Illinois Institute of Technology's (IIT) microgrid or the 'Perfect Power at IIT' (Johnstone, 2013).

In 2009 a statewide Illinois Smart Grid Collaboration consisting of public and private partnership initiative was established. The collaboration aims to speed up the adoption of the Smart Grid technology in Illinois and nationally. The key private and municipal partnerships include the City of Chicago, ComEd, Ameren, the Village of Oak Park, the Galvin Electricity Initiative and more than 50 other companies (IIT, 2009). A city in Illinois that is already benefiting from the smart grid technology is Naperville, a suburb of Chicago. In 2009, the city began a smart grid initiative project with a total budget of $21,988,220, out of which $10,994,110 was Federal grant. The project involved a city-wide deployment of an advanced metering infrastructure (AMI), including the expansion of distribution automation capabilities. These include circuit switches, smart relays and remote fault indicators. The modifications to the power grid allows customers to purchase devices that assist in managing electricity usage and costs, including in-home displays, programmable communicating thermostats, and direct load control devices for participation in load management programs. The benefits of this project are that they will allow participants to view their energy usage through in-home displays, a web portal, or both. Besides, the project allows the city to manage, measure and verify targeted demand reductions during peak load periods. Furthermore, the new AMI and distribution automation technologies will help improve service quality and reliability, through outage management, distribution circuit monitoring, and automated circuit switching (US-DOE, 2012d).

8.5.5. *Interconnection rules for distributed generation systems of size up to 10 MW*

For distributed wind turbine generation systems of size up to 10 MW, the Illinois interconnection rule contains four levels of review for interconnection requests (see Table 8.8).

Thus, to be eligible for a specific level of expedited review, a project must satisfy all the requirements of a given classification. According to the rules, all systems must be equipped with an external disconnect switch directly accessible to the utility. Furthermore, the rule requires that systems larger than 1 MW carry liability insurance, which protects the customer from claims for bodily injury and/or property damage. The minimum coverage of the liability insurance is $2 million per occurrence and $4 million in aggregate. Standardized interconnection agreements are available for all the four tiers. The

Table 8.8. Illinois interconnection standards (Sheaffer *et al.*, 2011).

Tier 1	Includes certified, inverter-based systems with a capacity rating of 10 kW or less.
Tier 2	Includes certified systems with a capacity rating of 2 MW or less, interconnected to a radial distribution network or a spot network serving one customer.
Tier 3	Includes non-exporting certified systems with a capacity rating of 50 kW or less interconnected to an area network, or certified, non-power-exporting systems with a capacity rating of 10 MW or less interconnected to a radial distribution network.
Tier 4	Include systems with a capacity of 10 MW or less not meeting the criteria for inclusion in a lower tier. The required level of review is generally dependent on system capacity, whether system components are certified by a nationally recognized testing laboratory (NRTL) and the type of network connection.

Tier 1 agreement is a simplified version of that used for projects requiring higher levels of review. In general, higher level applications are subject to more intensive screening and prolonged time limits (US-DOE, 2012b).

8.5.6. *Interconnection rules for distributed generation systems of size larger than 10 MW*

For distributed generation systems of size larger than 10 MW that are not subject to the above-mentioned small-scale distributed generation systems interconnection standards, ICC adopted a separate set of rules. These rules require the electric utility company to use relevant technical interconnection standards adopted by the applicable regional transmission organization (RTO) (Sheaffer *et al.*, 2011). This is because interconnections that are covered by the Federal Energy Regulatory Commission (FERC), the Midwest Independent Transmission System Operator or PJM Interconnection are not subject to these new interconnection standards. The new rules provide a standard set of procedures covering the interconnection process, including standard agreements. But, in case the RTO lacks applicable standards, the utility and the interconnection customer can agree on modifications to Institute of Electrical and Electronics Engineers (IEEE) Standard 1547 to address the electric system and distributed generator conditions. In the U.S., the IEEE has developed a series of standards that address interconnection. The base standard (IEEE 1547), 'Standard for Distributed Resources Interconnected with Electric Power Systems' provides requirements related to the performance, testing, operation, safety considerations and maintenance of the grid interconnection (US-DOE, 2012b). In essence, the ICC rule is similar to the Mid-Atlantic Demand Resource Initiative (MADRI) Level 4 review process, at least from a procedural standpoint (Sheaffer *et al.*, 2011). MADRI is a collaborative process of federal and state utility, electric distribution companies, energy and environmental regulators, distributed generation developers and owners, demand response companies, PJM and other interested stakeholders. The MADRI Model Small Generator Interconnection Procedures seek to promote, advance and enable distributed generation, while maintaining system safety and reliability (MADRI, 2005).

8.6. Large Scale Wind Energy Generation Projects

As we mentioned in the introduction (Sec. 8.1), there are currently 42 wind farm projects online in Illinois, accounting for over 3,300 MW of wind power generating capacity in the state or 99% of the state's wind power generating capacity. Below, we describe some of these wind farm projects in the state.

8.6.1. *EcoGrove Wind Farm I*

The EcoGrove Wind Farm I project in Stephenson County, Illinois was brought on line in September 2009. The wind farm is owned and operated by ACCIONA Energy and has sixty-seven 1.5 MW Acciona Windpower turbine generators located on private agricultural land. The farm is capable of generating sufficient clean energy to power over 25,000 U.S. homes and offset 176,000 tons of carbon every year. The EcoGrove Wind Farm I project employed 125 workers during the height of its construction phase and currently employs 10 permanent employees (IWWG, 2012).

8.6.2. *Grand Ridge wind farm*

The Grand Ridge wind farm consists of four phases, which were constructed between 2007 and 2009. The farm is located in La Salle County, Illinois, about 80 miles Southwest of Chicago. The 210 MW wind farm includes 140 turbines, generating enough clean energy to power approximately 65,000 homes each year. The wind farm began its first commercial operations in 2008 (IWWG, 2012). The total land area of this project is 71,300 acres, with an expected area of disturbance of about 175 acres (HDR Engineering, Inc 2008; Rogers and Ragheb, 2010).

8.6.3. *Twin Groves wind farm*

The Twin Groves Wind Farm Phases I and II located in McLean County, Illinois is owned and operated by Horizon Wind Energy. The phase I was brought online in 2007, while the phase II was brought on line in 2008 (IWWG, 2012). The farm is made up of 240 Vestas wind turbines distributed over 22,000 acres of land. The wind turbines have a projected life span of 25–30 years. Currently, the Twin Grove Wind Farm has a capacity of 398 MW and produces enough power for 120,000 homes. For each of the turbines, approximately one acre of land was taken out of production for the gravel roads, switchyard, control building and site location. Landowners were paid royalties ranging from $5,000–6,000 per turbine (Kirchhofer, 2009).

8.6.4. *The Bishop Hill wind farms*

The Bishop Hill I wind farm project is located in Henry County, Illinois, near the towns of Galva and Kewanee, approximately 40 miles Southwest of Moline, Illinois. The wind farm is owned by Invenergy and consists of 133 General Electric wind turbines. The wind farm is located on 22,000 acres of farmland and has a capacity of 211.4 MW, enough to power approximately 65,000 homes with clean energy each year (IWWG, 2012). The Bishop Hill I wind farm project began commercial operation in 2012. Tennessee Valley Authority has signed an agreement to purchase the power generated from the wind farm. During the peak period of its construction, the farm employed 220 workers, but currently, it is operated and maintained by approximately 15 full-time employees. In addition to job creation, the wind farm facilities will economically benefit the host community by providing tax payments to the county and local municipalities, as well as land lease payments to landowners (Dvorak, 2012). The second phase of the project, Bishop Hill II wind farm, began commercial operation in December 2012 in Galva, Illinois, 40 miles Northwest of Peoria, Illinois. The 81 MW wind farm will feature 50 General Electric wind turbines. The second phase of the project was acquired in March 2012 by MidAmerican Energy Holdings Co. Ameren Illinois has signed a 20-year power purchase agreement to buy electricity from the Bishop Hill II wind farm. The third phase of the project, an 83-turbine Bishop Hill III wind farm is currently in the developmental stage at Henry County, Illinois (Weiderman, 2012).

8.6.5. *Proposed wind farm in Sangamon County*

In 2008, Springfield-based American Wind Energy Management (AWEM) Corporation announced the construction of a 200-turbine wind farm project in Sangamon County in West Springfield, Illinois. Each of the wind turbines will have a capacity to generate 2 MW, making a total of 400 MW for the 200 turbines (Yeagle, 2010). Following this announcement, in January 2012, the Sangamon County

Board approved a moratorium on wind turbines to enable the county officials to review zoning rules, gather public input and decide on some needed modifications to the old rules. The reason for the modifications was that while the county's zoning rules allow installation of wind turbines 305 meters (1,000 ft) away from residential homes, some inhabitants still found it too close to their homes. As a result, two public meetings were scheduled in the County in late February 2012. The meeting was to allow the public to have the opportunity to comment on the proposed wind farm project (Associated Press, 2012). Among the issues discussed at the meeting were setback distances of about 243.8 meters (800 ft) to residential structures or up to 2.4 km (1.5 miles) from residential homes for non-participating properties. The other matters of concern were noise from turbine blades, flashing red lights on turbines and lighting regulations, health issues for families and wild life, malfunction of turbine such as blades falling off or fires in turbines, property values and taxes, property rights for non-participating property owners near projects, decommissioning of turbine and liability problems, as well as weather related problems (Associated Press, 2012). AWEM intends to lease about 25,000 acres of land from local landowners for the project, with each famer receiving about $100 per acre per year. Additionally, between $3,500 and $4,000 for each turbine will be paid to the farmers annually, depending on the gross revenue of the project. This amounts to between $12,000 and $14,000 per landowner per year (Yeagle, 2010).

8.6.6. *Wind turbine production facility plant in Elgin*

In 2009, Winergy Drive Systems Corp. and Siemens Drive Technologies officially opened a second wind turbine production facility plant in Elgin, Illinois. The companies were planning to boost production of wind turbine gear drives to meet customers' expectations (Siemens Industry, 2008). The new plant, which represents an investment of over $20 million, is used for assembling and testing of mechanical gear drives, while the existing plant will be maintained to manufacture gears and components that will be assembled and tested in the new plant. The new facility is expected to create approximately 300 new jobs in production phase and 55 new office jobs over the next three to four years. This amounts to more than double the current number of employees in Elgin working for the company. The state of Illinois and the City of Elgin provided a number of incentives to Siemens to help secure the investment, including a Wind Energy Development Grant and Employer Training Investment Program job training funds (Siemens Industry, 2008).

8.7. Sustainability of Wind Energy

Over the past 30 years, wind energy has become one of the major sources of electricity generation in many countries. The future of wind energy technology will depend on the ability of the industry to devise cost-reducing strategies (Hau, 2013). In the span of 20 years, from 1980 to early 2000s, the wind energy technology has achieved significant reductions in capital cost and increase in performance of wind turbines. The combined effect resulted in reduction of the levelized cost of energy (LCOE) for onshore wind energy from \$150/MWh to approximately \$50/MWh (Lantz *et al.*, 2012). The LCOE is an estimate of total electricity cost including payback of initial investment and operating costs over full lifetime of a power plant. It also reflects the price a plant must sell electricity in order to break even. Key inputs to calculate LCOE include capital costs, fuel costs, fixed and variable operations and maintenance (O&M) costs, financing costs, and an assumed utilization rate for each plant type.

The industry experienced an increase in capital cost between 2004 and 2009 in Europe and the U.S., due to raw material, energy price increase and labor cost. During the same period, wind turbine upscaling, which involves sophisticated design and application of advanced control and sensing systems, has been the single largest contributing factor in price increase in the U.S. (Bolinger and Wiser,

2011). However, the industry has witnessed 8% reduction in capital cost from 2009 to 2011 (Wiser and Bolinger, 2011) and continuous improvements in turbine technology are expected to reduce LCOE of wind energy, especially in low and medium turbine power categories (i.e., 6.0 m/s to 8.5 m/s average annual wind speed) (Wiser *et al.*, 2012).

Investors are concerned with the rate of return of any potential investment and similar to all energy sources, wind energy technologies must be economically and environmentally sustainable and socially justifiable.

Energy is required over the whole life cycle of a wind farm, which include equipment manufacturing, construction of turbine and supporting infrastructure (foundations, site operation and maintenance) and in decommissioning the site (CSE, 2011). These energy expenditures do not make wind energy GHG emission-free and concerns are expressed against installation of wind farms. Hundreds of life cycle analyses (LCA) of wind energy have been conducted with variations in results. According to Dolan and Heath (2012), these differences can be attributed to several factors, including specifics of a turbine model, operating conditions of the system studied, assumptions made by those conducting the study, variability in data sources and LCA approach. The approach by Dolan and Heath (2012) was to systematically review previously published wind power LCAs and harmonize their GHG emission estimates by establishing more consistent methods and assumptions. Their study found that GHG emissions of onshore and offshore wind energy had a mean value of 15 gCO_2-eq/kWh and 12 gCO_2-eq/kWh, respectively.

Proponents of natural gas argue that the overall cost of a natural gas plant is less than a wind power plant. Although the levelized cost analyses indicate the accuracy of such claim, GHG emissions of a natural gas power plant is much higher than a wind farm. Skone and James (2012) conducted a LCA to determine greenhouse gas emission of natural gas combined cycle (NGCC). They considered

Table 8.9. Estimated average levelized cost (based on the value of \$U.S. in 2012) of electricity (LCOE) and GHG emissions for onshore and offshore wind technologies as compared with advanced NGCC in U.S. The estimates are for plants entering service in 2019.

	Onshore Wind	Offshore Wind	NGCC
Capacity Factor (%)	35[1]	37[1]	87[1]
Levelized Capital Cost (\$/MWh)	64.1[1]	175.4[1]	15.7[1]
Fixed O&M (\$/MWh)	13[1]	22.8[1]	2[1]
Variable O&M (including fuel) (\$/MWh)	0[1]	0[1]	45.5[1]
Transmission investment (\$/MWh)	3.2[1]	5.8[1]	1.2[1]
Total system LCOE (\$/MWh)	80.3[1]	204[1]	64.4[1]
GHG emissions (gCO_2-eq/kWh) (mean)	15 [2][§]	12 [2][§]	347.5 [3]

(1) US-EIA (2014a); (2) Dolan and Heath (2012); (3) Skone and James (2012).
[§]Numbers shown are after harmonization.

four scenarios in their study and found the minimum and maximum GHG emission from a NGCC power generation facility was 162 and 533 gCO_2-eq/kWh, respectively with the mean value of 347.5 gCO_2-eq/kWh. The study showed that even at minimum value, GHG emissions of a NGCC plant is an order of magnitude higher than a wind power farm, as shown in Table 8.9.

A report prepared for the NREL and conducted by Black and Veatch (2012) estimated the cost of onshore, fixed-bottom offshore and floating-platform offshore wind turbine installations. The results of the study indicated the cost of onshore wind installation to be at 1,980 \$/kW (±25%). Similarly, the cost of fixed-bottom and floating-platform wind installations were estimated at 3,310 \$/kW (±35%) and 4,200 \$/kW (±35%), respectively, as shown in Table 8.10.

Currently, the U.S. land-based market maintains a viable supply chain for towers for the 1–3 MW turbines. Fullenkamp and Holody (2014) prepared a comprehensive report detailing cost breakdown

Table 8.10. Capital cost (in 2009 U.S.-$) breakdown per kW, ($/Kw), of wind power technologies. The number in parentheses indicates the percentage of the total cost (Black & Veatch, 2012).

	Wind Turbine	Distribution	Balance of Plant	EP& CS	Owner's Cost	Total Cost
Onshore	1346 (68)	198 (10)	257 (13)	79 (4)	100 (5)	1980
Offshore Fixed-Bottom	1665 (50.3)	397 (12)	894 (27)	165 (5)	189 (5.7)	3310
Offshore Floating-Platform	1890 (45)	546 (13)	1260 (30)	252 (6)	252 (6)	4200

EP&CS = Engineering, Procurement & Construction Services.

of four major components of a wind turbine namely, tower, blades, PM generator and jacket foundation (for offshore installation). They compared the supplier results from the U.S., Germany and China for manufacturing of a 5 MW turbine. Based on their study, U.S. offered the lowest price for tower and blade, while China offered the lowest price for the generator. For the jacket foundation, the U.S. suppliers had a very slight advantage over their Chinese counterparts. These comparisons were made after inclusion of tariff tax and logistics when applicable.

As shown in Table 8.11 below, the highest cost associated with a wind turbine installation is for offshore jacket foundation. If considering onshore installation of a 5 MW wind turbine and linear proportionality to a 1 MW system, the highest cost will be incured by PM generator.

Fullenkamp and Holody also provided information on transportation of components of a 3 MW and 5 MW wind installations by rail and truck in the Central U.S. In general and when possible, transportation by rail is more economical as compared with transportation by truck, as shown in Table 8.12.

Table 8.11. Cost breakdown of major components of a 5 MW wind turbine installations (excluding logistics) in the U.S. Numbers in the parenthesis indicate the % of total (Fullenkamp and Holody, 2014).

	Tower	**Blade**	**PM Generator****	**Offshore Jacket foundation**
Materials	$380,251 (59.4)	$147,825 (46.4)	$534,500 (55.4)	$344,384 (30.7)
Labor	$51,352 (8)	$31,300 (9.8)	$107,500 (11.1)	$196,746 (17.5)
Burden	$55,354 (8.6)	$64,165 (20.1)	$215,000 (22.3)	$393,492 (35.1)
SGA	$87,723 (13.7)	$36,494 (11.5)	$55,900 (5.8)	$44,431 (4)
Engineering	$20,000 (3.1)	$14,597 (4.6)	$25,800 (2.7)	$44,431 (4)
Profit	$45,291 (7.1)	$24,329 (7.6)	$25,800 (2.7)	$97,749 (8.7)
Total	$639,971	$318,710	$964,500	$1,121,233
Grand total		**$3,044,414**		
% of grand total	21	10.5	31.7	36.8

SGA = sales general and administrative.
(**)Assumed linear proportionality of all costs to 1 MW generator.

As mentioned earlier in the introduction section, the IPA Act of 2007 requires the utilities to use wind energy to satisfy 18.75% of the electricity production in the state by the year 2025. Modeling work by the IPA as corroborated by similar findings in Massachusetts suggests that for 2011, integration of renewable resources into the power grid has lowered Illinois average LMPs by $1.30 per MWh, from $36.40–$35.10 per MWh (IPA, 2012). LMP is a methodology for pricing electricity and managing transmission congestion. Craig *et al.* (2010) conducted a study that indicates 17 new jobs are created from every new MW of power developed in the state of Illinois. However, a simulation study by Tuerck *et al.* (2014) showed that the IPA Act of 2007 would not benefit Illinois residents since it would result in increase in electricity cost and loss of 8000 jobs by 2026.

Various benefits are attributed to wind energy, including the impact on electricity prices, economic development, local economies, and environmental benefits. But critics of wind energy point to factors

Table 8.12. Characteristic and average transportation cost (in Central U.S., not including any load or unload charges) of major components of a wind turbine installation in the Central U.S. (Fullenkamp and Holody, 2014).

Component/ Manufac. time	Characteristics	Wind Energy Installation 3 MW	Wind Energy Installation 5 MW
Tower/1,175 (hrs)	Height (m) (ft)	105 (344)	105 (344)
	Diameter (m) (ft)	4.5 (14.7)	6 (19.7)
	Weight (tons)	400	550
	Truck transp.($/mile)	3–4	3–4
	Rail transp. [$/mile]	1–2	1–2
Blade/493 [hrs]	Height (m) (ft)	49 (160.8)	65 (213)
	Diameter (m) (ft)	NA	3.5 (8.2)
	Weight (tons)	10.9	22.5
	Truck transp. ($/mile).	15–20	18–25
	Rail transp. ($/mile)	2	2
Generator/NA	Height (m) (ft)	1.2 (3.9)	1.4 (4.6)
	Diameter (m) (ft)	4.5 (14.8)	6.5 (21.3)
	Weight (tons)	40	68
	Truck transp. ($/mile)	33	84
	Rail transp. ($/mile)	10.5	NA

NA = Not available.

such as heavy initial investment cost, government subsidies and reduced land values as issues that may offset some of the advantages of wind energy. It is very difficult to accurately and reliably quantify the negative impacts of wind energy compared with other sources of energy. Nevertheless, these impacts must be considered in any policy discussion regarding wind energy industry.

8.7.1. *Advantages and disadvantages associated with wind power generation*

The major benefits of wind energy utilization for electricity production in Illinois include price stability and less dependency on fossil fuels as it has been the case till now (Agyarko and Mansoori, 2013), steady income tax revenue for farmers/local government and creation of permanent employment.

According to a U.S. Census American Community Survey conducted in 2012, a total number of 4,221 non-working individuals in Illinois were identified as having the skills relevant to offshore wind power generation (Navigant Consulting, 2013).

The annual land lease payments to Illinois farmers in 2012 were over $10,705,000. In addition, the annual saving in amount of carbon dioxide (CO_2) emissions by Illinois wind power installations in 2011 was 4.7 million metric tons, compared to similar electricity generation technologies using fossil fuel (Lydersen, 2012). As mentioned in the introduction of this chapter, Illinois inland wind energy resource can produce about 5.2 times of the state's electricity needs. As of 2014, the capital investment in Illinois wind energy reached $7.2 billion, and it is estimated that using wind energy, the state could expect nearly $6 billion of revenue added to its economy over the course of several decades (Lydersen, 2012; AWEA, 2014). Most of these investments would occur in cash-strapped rural areas in Illinois. Furthermore, the cost for land-based wind energy production is declining. For example, the U.S. average levelized cost (2011 $ per MWh) for plants entering service in 2018 is projected to be $86.6 per MWh for inland wind turbines, compared with $100.1 per MWh for conventional coal-fired power plant or $123.0 per MWh for advanced coal power plants (US-EIA, 2013a).

On the other hand, noise from horizontal- and vertical-axis wind turbines remains a nuisance, while wind resource may compete for land in the future. Transmission cost is another problem hurting the Illinois wind energy industry, as a significant number of backlog wind energy projects in Illinois are waiting for approval and financing to be connected to the electricity power grid (Lydersen, 2013). Unquestionably, electric utilities claim that they place a high premium on the safety of their workers, as well as the reliability of their electrical systems. Therefore, confronted with large numbers of distributed generators owned and operated by non-utilities, some members of the utility industry have the perception that distributed generators could be a threat to their priorities. This situation has led some utilities to impose overly conservative restrictions on interconnected systems, creating additional costs that could make an installation of small turbines economically unfeasible (US-DOE, 2012c). These requirements include manual disconnects that are easily accessible to utility personnel, as well as power quality requirements such as limits on the interconnected system's effects on 'flicker' and other types of distortion. Also included, is equipment that prevents power from being fed to the grid when the grid is de-energized. Systems may be required to automatically shut down in the event of electrical failures. This automatic shutdown can be accomplished by using protective schemes at the grid interface, which may include a synchronizing relay, protection against under- and over-voltage, protection against under- and over-frequency, phase and ground over-current relays, ground over-voltage relays and more. Even more restrictive (and expensive) requirements can include an isolation transformer for the system and liability insurance against worst-case scenarios of damage to utility equipment and harm to utility personnel (US-DOE, 2012c). Horizontal-axis wind turbines may be installed at less populated areas with frequent wind currents. This situation could require the energy to be transmitted to consumers elsewhere, which requires transmission lines. Furthermore, since there has never been any offshore wind turbines sited in freshwaters such as Lake Michigan, its deployment could present some engineering challenges. These challenges may include winter ice floes that could impose substantial forces on turbine bases, floats and mooring lines [Musial and Ram, 2010]. Similar to the state of Michigan, some Illinois residents may oppose any proposed offshore wind turbine projects that could have visual and other impacts on the Lake (Vuljevic, 2011). In 2011, a group of Michigan residents expressed their dissatisfaction with Scandia Wind LLC, a company planning to install 50–100 offshore wind turbines in the Lake. This situation forced the wind energy company to move the project to Canada (Vuljevic, 2011). In Table 8.13, we present a

Table 8.13. Advantages and disadvantages of the wind power industry in Illinois.

Advantages	
2012 land lease payments to farmers [1] ($)	10,705,000
Capital investment by 2012 [1] (in $10^9$$)	7.2
Anticipated revenue added to economy in several decades [2] (in $10^9$$)	6
2011 amount of CO_2 avoided [1] (10^6 metric Tons)	5.69
U.S. ALC for inland plants entering service in 2019[4] ($/MWh)	80.3
Non-working individuals with skills relevant to offshore plants [3]	4,221
Disadvantages	
U.S. ALC for offshore plants entering service in 2019[4] ($/MWh)	204.1
Cost of wind power integration to the grid [5,6] ($/MWh)	$<1\sim12$
Insufficient investment, regulatory & infrastructure challenges for offshore projects persist	

(1) AWEA (2013); (2) Lydersen (2012); (3) Navigant Inc (2013); (4) US-EIA (2014a); (5) IPA (2013); (6) IPA (2014).
ALC = average levelized cost.

summary of these advantages and disadvantages of the wind power industry in Illinois.

8.7.2. *The last words on wind energy sustainability*

As mentioned in Chapter 1, economics, environmental issues and social acceptance are the three interconnected pillars of sustainability in any renewable energy, including wind energy, which must be taken into account for successful development of that technology (Welch and Venkateswaran, 2009; Warren and McFadyen, 2010). Therefore, to evaluate the overall burden of wind utilization systems, the economic analysis, impacts of wind energy emissions (which includes emissions for producing equipment for wind energy systems) to environment and wind energy system water usage (which is zero) and land footprint must be evaluated. While the amount of water usage in wind energy utilization is practically zero, its land footprint (1,335 m^2/GWh $\cong$ 14,370 ft^2/GWh) is appreciable, as reported in

Chapter 10 (Fig. 10.24) in which land footprints of different power generating technologies are compared.

Throughout this chapter, we have discussed the economics of various wind energy utilization methods. It goes without saying that economics of wind power technology have improved rapidly in the past few years, thanks to improvements in machine performance and increases in size of turbines, both of which are contributing to reduce costs. These trends are examined and future costs are further assessed by Sathyajith (2006).

Generally, municipalities in U.S. would not allow installation of horizontal-axis windmills in residential areas due to their noise generation, land use and habitat loss. However, vertical-axis windmills are generally acceptable in residential areas. Without any doubt, wind energy utilization has been socially accepted and is one of the oldest energy technologies known to human beings, as mentioned earlier in this chapter.

8.8. Concluding Remarks

Wind is a renewable energy resource since it is inexhaustible and could be exploited indefinitely to provide energy demands of the future generations. The Intergovernmental Panel on Climate Change (IPCC) has estimated the total amount of globally extractable wind power to be 2–4 TWe (US-DOE, 2005).

There is abundance of wind energy available in Illinois, which can be used for production of electricity. In 2012, the Illinois electric power sector consumed about 7,727 GWh of wind power, up from 18 GWh in 2003, representing 4% of the net electricity generation in the state (US-EIA, 2012). While this may not be a significant amount of power to reckon with, it indicates which direction the state is moving in terms of electricity production. It is noteworthy that wind energy investments in the state have been increasing in recent years, which could help the industry to reach its full potential in the foreseeable future if more investment is put into this sector. Furthermore, the

US-DOE has estimated that Illinois has the potential to generate at least 20 GW of electric power from its offshore wind energy resource in Lake Michigan (Loomis & Pagan, 2011).

Economically, the cost of wind-generated electricity has been declining steadily, due to improved technology and increased efficiency in production of wind turbines. It is estimated that the U.S. average levelized cost of electricity for wind power generation to be $80.3 per MWh (onshore) and $204.1 per MWh (offshore) in 2019 (US-EIA, 2014a), which is predicted to be less than that of coal and nuclear and more competitive with natural gas by the year 2019, as the technology improves (EIA, 2014a).

Environmentally, wind energy does not emit pollutants, apart from few wastes created during the construction of turbines. The environmental concerns raised about wind farms are minor, compared with conventional power generation technologies, such as coal, oil, or nuclear plants. The full aspect of ecological effects of wind farms and related accessories on in-land and offshore wildlife habitat are being investigated (see for example Mcnew *et al.*, 2014 and Snyder and Kaiser, 2009).

Furthermore, the use of wind turbines of 100 kW or smaller have become very common over the past few years, as the cost and demand for energy increases. These turbines are suitable for residential, farms, schools or small businesses since their noise generation is negligible and they can reduce energy bills. Consequently, wind power can contribute appreciably to the energy and environmental sustainability.

Bibliography

Aerotecture Int'l Inc. (2014). Pepsico/Quaker Oats — sustainability center, Chicago, IL. Aerotecture Int'l Inc., Park Forest, IL.

Agyarko, L. B. and Mansoori, G. A. (2013). Non-renewable Energy Options in Illinois, *Int. J. Oil, Gas and Coal Technol.*, **6**(3): 288–347.

Asmus, P., Cornelius, A. and Wheelock, C. (2009). Microgrids — Islanded power grids and distributed generation for community, commercial, and institutional applications, Pike Research LLC., Boulder, CO.

Associated Press (2011). Quinn signs bill to study offshore wind energy. News section, *Daily Herald*, Arlington Heights, IL.

[AWEA] Am. Wind Energy Assoc. (2013). State wind energy statistics: Illinois. AWEA, Washington, DC.

[AWEA] Am. Wind Energy Assoc. (2014). Wind energy facts: Illinois. Fact Sheet, AWEA, Washington, DC.

AWS Truewind LLC. (2009). Offshore wind technology overview. NYSERDA PON 995, Task Order No. 2, AWS Truewind, LLC., Albany, NY.

Baker, K., Huang, M., Januzzi, A. and Mansoor, A. (2014). Agency coordination of offshore wind projects in Washington State, Environ Management Certificate Program Keystone Project Report, Univ. Washington, Seattle, WA.

Betz, A. (1966). *Introduction to the Theory of Flow Machines*, (D.G. Randall, Translations), Pergamon Press, Oxford.

Black and Veatch. (2012). Cost report: Cost and performance data for power generation technologies, Black & Veatch Engineering Consultant, Overland Park, KS.

Bolinger, M. and Wiser, R. (2011). Understanding trends in wind turbine prices over the past decade, Doc. No. LBNL-5119E, LBL, Berkeley, CA.

Boone, B. (2013). Design ideas in permanent magnet generators, Short comm., Wind Power Eng & Develop. *windpowerengineering.com.*

Bullis, K. (2013). Will vertical turbines make more of the wind? MIT Tech. Rev., Cambridge, MA.

[BWEA] British Wind Energy Assoc. (2004). Briefing sheet-wind turbine Technology. BWEA, London, UK.

Casey, T. (2014). Floating wind turbines float into US waters (Finally!), CleanTechnica, Laramie, WY.

Chicago Tribune (2014). Wind energy benefit Illinois. Opinion section, *Chicago Tribune*, Chicago, IL.

Conger, H. (2012). Opening the loop: A proposal to allow the installation of distributed renewable generation tech. in downtown Chicago, Presented at Land-Use Seminar, School of Law, Loyola Univ. of Chicago, Chicago, IL.

Corfee, K., Stevens, G. and Goffri, S. (2014). Distributed generation resource assessment for long-term planning study, Ref. No. 171094, Navigant Consulting, Chicago, IL.

Cornish, A. (2009). Gurnee, IL Chipotle among first LEED® certified restaurants in the US, Greenbeanchicago, Chicago, IL.

Craig, A., Learner, H. and Gray, P. (2010). The wind energy supply chain in Illinois, Envtl. Law & Policy Ctr, Chicago, IL.

[CSE] Centre for Sustainable Energy (2011). Common concerns about wind power. CSE, Bristol, UK.

Czyzewski, A. (2012). Wind energy gets serial: Solid foundations: Onshore assembly could enable serial production of offshore wind turbines, *Process Engineering*, Berkhamsted, UK.

Dabiri, J. O. (2011). Potential Order-of-magnitude Enhancement of Wind Farm Power Density via Counter-rotating Vertical-axis Wind Turbine Arrays, *J. of Renewable and Sustainable Energy*, **3**(4): 1–12.

Daniels, L. M., Johnson, S. E. and Slaymaker, W. (2004). Harvest the wind: A wind energy handbook for Illinois, (IVARDC Reports 614), Illinois Inst. for Rural Affairs, Western Illinois Univ., Macomb, IL.

de Vries, E. (2012). Exploring the trends in foundation design. *Windpower-Offshore*, London, UK.

Dohn, R. L. (2011). White paper: The business case for microgrids — the new face of energy modernization, Siemens AG., München, Germany.

Dolan, S. and Heath, G. A. (2012). Life Cycle Greenhouse Gas Emissions of Utility-scale Wind Power, *J Ind Ecol*, **16**(S1): s36–s54.

[DSIRE] North Carolina Solar Center (2014). DSIRE: Database of state incentives for renewables & efficiency, North Carolina State Univ., Raleigh, NC. *dsireusa.org*.

Dvorak, P. (2012). 200-MW Bishop Hill wind farm under construction in Illinois. News section, *Wind Power Eng & Develop.*, Cleveland, OH.

[EDL and Huntsman] Energy Dynamics Lab and Jon M. Huntsman School of Business (2010). Renewable energy: A path forward for Park City, The Energy Dynamics Lab and the Jon M. Huntsman School of Business, Utah State Univ., Logan, UT.

Fenech, L., Sant, T. and Muscat, M. (2011). Design and cost evaluation of a deep water support structure for a wind turbine in central Mediterranean waters. In *Proceedings of the AWEA 2011*, pp. 316–320. Brussels, 14–17 March, Belgium.

Forsyth, T., Tu, P. and Gilbert, J. (1999). Economics of grid-connected small wind turbines in the domestic market, Paper No. NREL/CP-500-26975. In *Proceedings of AWEA Wind Power '99 Conf.* Burlington, 20–23 June,VT.

Fullenkamp, P. H. and Holody, D. S. (2014). U.S. wind energy manufacturing and supply chain: a competitiveness analysis, Doc. No. DE-EE-0006102, Global Wind Network, Cleveland, OH.

Gelman, R. (2008). 2007 Renewable energy data book, NREL, Golden, CO.

Gelman, R. (2009). 2008 Renewable energy data book, Doc. No. DOE/GO-102009-2827, NREL, Golden, CO.

Gelman, R. (2010). 2009 Renewable energy data book, Doc. No. DOE/GO-102010-3074, NREL, Golden, CO.

Gelman, R. (2011). 2010 Renewable energy data book, Doc. No. DOE/GO-102011-3310, NREL, Golden, CO.

Gelman, R. (2012). 2011 Renewable energy data book, Doc. No. DOE/GO-102012-3598, NREL, Golden, CO.

Graig, A., Learner, H., Gray, P and Dobies. J. (2011). The clean energy supply chain in Illinois: Wind, solar, and geothermal, Environ Law and Policy Center, Chicago, IL.

Hahn, M. and Gilman, P. (2014). Offshore wind market and economic analysis, Doc. No. DE-EE0005360, Navigant Consulting, Chicago, IL.

Hau, E. (2013). *Wind Turbines: Fundamentals, Technologies, Application, Economics*. Springer.

HDR Eng'g Inc (2008). Grand Ridge wind farm project Phase I, Phase II and III- storm water pollution prevention plan for construction activities, HDR Eng. Inc. Minneapolis, MN.

Hughes, L., Fung, L., Koksal, M., DeCoste, J., Smith, A., White, D., Berkvens, D. and Crawford, J. (2004). Self-starting Darrieus wind turbine, Dept of Mechanical Eng, Dalhousie Univ., Halifax, N.S., Canada.

[IDNR] Illinois Dept of Natural Resources (2012). Lake Michigan offshore wind energy advisory report, IDNR, Springfield, IL.

[IEA] Int'l Energy Agency (2009). Technology roadmap: Wind energy, Report No. OECD/IEA 2009, IEA, Paris, France.

[IIT] Illinois Inst. of Tech. (2009). Illinois Institute of Technology leads statewide initiative to create jobs and National hub for smart grid. News item, IIT, Chicago, IL.

[IL-AG] Illinois Attorney General (2013). Energy efficiency: Net metering and interconnection. Illinois Attorney General, Springfield, IL.

[IL General Assembly] Illinois General Assembly (2007). Amendment of Public Utilities Act: Net electricity metering, Illinois General Assembly, Public Act 095-0420.

Illinois Windmills (2012). Wind farms. Woodridge, IL. *illinoiswindmills.org*.

Iowa Energy Center (2012). On-grid small wind turbine configuration with a permanent magnet alternator. Wind Energy Manual, Iowa Energy Center, Ames, IA.

[IPA] Illinois Power Agency (2012). Annual Report: The costs and benefits of renewable resource procurement in Illinois under the Illinois Power Agency and Illinois Public Utilities Acts. IPA, Chicago, IL.

[IPA] Illinois Power Agency (2013). Annual Report: The costs and benefits of renewable resource procurement in Illinois under the Illinois Power Agency and Illinois Public Utilities Acts. IPA, Chicago, IL.

[IPA] Illinois Power Agency (2014). Annual Report: The costs and benefits of renewable resource procurement in Illinois under the Illinois Power Agency and Illinois Public Utilities Acts. IPA, Chicago, IL.

[IRENA] Int'l Renewable Agency (2012). Renewable energy technologies: Cost analysis series, **1**(5), Power sector. IRENA, Abu Dhabi, UAE.

[IWWG] Illinois Wind Working Group (2012). Illinois windy city: Prairie power wind farm and growth. Center for Renewable Energy, Ill. State Univ., Normal, IL.

Jha, A. R. (2010). *Wind Turbine Technology*. CRC Press.

John, M., John, R., Syamily P. S and Vyshak P. A. (2014). MagLev Windmill, *Int'l J. of Research and Eng. Technol.*, **3**(5): 50–56.

Johnstone, H. (2013). The rise of the microgrid. *Power Eng.International*, Tulsa, OK.

Johnson, C. (2012). Solar and wind energy Schaumburg's Sustainable Future, A Sustainability Studies Project, Roosevelt Univ., Chicago, IL.

Kessler, R. A. (2010). Chicago mayor open to offshore wind energy develop. *Recharge News*, Oslo, Norway.

Kessler, R. A. (2011). Council to evaluate Illinois offshore wind energy potential. News section, *Recharge News*, Oslo, Norway.

Kilkis, I. B. (1999). Utilization of Wind Energy in Space Heating and Cooling with Hybrid HVAC Systems and Heat Pumps, *Energy Buildings*, **30**(2): 147–153.

Kirchhofer, P. (2009). Wind farming in Illinois. Interbusiness Issues, *Peoria Magazine*, Peoria, IL.

Lack, C. A. (2010). Urban wind turbines, MSc Thesis, Tallinn Univ. of Tech., Barcelona, Spain.

Landis, T. (2011). Follow-up file: Sangamon County wind farm. *The State J. Register*, Springfield, IL.

Lantz, E., Wiser, R. and Hand, M. (2012). IEA wind task 26: The past and future cost of wind energy, Tech. Report No. NREL/TP-6A20-53510, NREL, Golden, CO.

Lijuan, D. (2014). Safe and efficient operation and maintenance of offshore wind farms. Doctoral thesis, ISSN 1503-8181, Norwegian Univ. of Sci. and Technol., Trondheim, Norway.

Loomis, D and Carter, J. (2011). Economic impact: Wind energy Development in Illinois, *Center for Renewable Energy*, Illinois State Univ., Normal, IL.

Loomis, D. and Pagan, A. (2011). The Illinois RPS: Context, structure and history of the policy, Center for Renewable Energy, Illinois State Univ., Normal, IL.

Loomis, D., Noll, S. and Hayden, J. (2012). Economic impact: Wind energy develop. in Illinois, Center for Renewable Energy, Illinois State Univ., Normal, IL.

Lopez, A., Roberts, B., Heimiller, D., Blair, N. and Porro, G. (2012). U.S. renewable energy tech. potentials: A GIS-based analysis, Report No. NREL/TP-6A20-51946, NREL, Golden, CO.

Lorc. (2012). 'Tripiles-three monopiles in one", 2 Pages, Lindoe Offshore Renewables Center, Denmark.

Lydersen, K. (2012). Study foresees $6 billion impact from Illinois wind industry. *Midwest Energy News*, St. Paul, MN.

Lydersen, K. (2013). Wind energy lessons from Illinois: Q&A with AWEA's Rob Gramlich. *Midwest Energy News*, St. Paul, MN.

MacKay, D. J. C. (2009). *Sustainable Energy Without the Hot Air*. UIT Cambridge Ltd., Cambridge, UK.

[MADRI] The Mid-Atlantic Distributed Resource Initiative (2005). MADRI model small generator interconnection procedures. MADRI Working Group. *sites.energetics.com/MADRI*.

Manwell, J. F. (2010). Offshore wind energy in deep water: The next generation. In the *Proceedings of New England Conf. of Public Utilities Commissioners*. U Mass., Amherst, MA.

Manwell, J. F., McGowan, J. G. and Rogers, A. L. (2010). Wind Energy Explained: Theory, Design and Application. Wiley.

Mcnew, L. B., Hunt, L. M., Gregory, A. J., Wisely, S. M. and Sandercock, B. K. (2014). Effects of Wind Energy Develop. on Nesting Ecology of Greater Prairie-chickens in Fragmented Grasslands, *Conserv Biol.*, **28**(4): 1089–1099.

Mollmann, M. (2010). Chicago Suburb Approves Nation's First Offshore Wind Farm. *AOL Real Estate*.

Müller, M., Jentsch, M. F. and Stoddart, E. (2009). Vertical Axis Resistance Type Wind Turbines for Use in Buildings, *Renewable Energy*, **34**(5): 1407–1412.

Musial, W. and Ram, B. (2010). Large-scale offshore wind power in the United States: Assessment of opportunities and barriers, Report No. NREL/TP-500-40745, NREL, Golden, CO.

Navigant Consulting (2013). U.S. offshore wind manufacturing and supply chain development, Report No. DE-EE0005364, Navigant Consulting, Inc., Burlington, MA.

[NEO] (2014). Wind energy facilities installed capacity by state. Nebraska Energy Office, Lincoln, NE.

[NIU] Northern Illinois Univ. (2007). Mendota hills wind farm, spring 2007. College of Liberal Arts and Sci., N.I.U., DeKalb, IL.

[NOAA] Natl Oceanic and Atmospheric Administration (2014). Great Lakes bathymetry. N.O.A.A., Boulder, CO.

Paraschivoiu, I. (2002). Wind turbine design-with emphasis on Darrieus Concept, Polytechnic Int'l Press, Canada.

[PJM and EIS] PJM and Environ Information Services (2014). Comparison of Renewable Portfolio Standards (RPS) Programs in PJM States, Doc. No. 292227v14, KAS-1/6/2014, PJM and EIS, Norristown, PA.

Power Tech. (2014). BARD Offshore I Wind Farm, North Sea, Germany. Industry projects section, *Power Tech.*, John Carpenter House, London, UK.

Rafter, D. (2008). Near North apartments with wind turbines, Chicago, Illinois. Projects section, *Solaripedia.*

Ragheb, M. (2014). Vertical axis wind turbines. *mragheb.com.*

Rogers, K. and Ragheb, M. (2010). Symbiotic coupling of wind power and nuclear power generation, 1st Int'l Nuclear & Renewable Energy Conf. (INREC), 21–24 March, Amman, Jordan.

Rolland, S. and Auzan, B. (2012). The potential of small and medium wind energy for developing countries: A guide for energy sector decision-makers, Alliance for Rural Electrification, Renewable Energy House, Brussels, Belgium.

Sathyajith, M. (2006). *Wind Energy: Fundamentals, Resource Analysis and Economics*. Springer.

Schwartz, M., Heimiller, D., Haymes, S and Musial, W. (2010). Assessment of offshore wind energy resources for the United States, Report No. NREL/TP-500-45889, NREL, Golden, CO.

Scottish Gov. (2006). Planning for micro renewable — annex to PAN 45 renewable energy technologies. The Scottish Government, Edinburgh, Scotland.

Seidel, M. (2014). *Substructures for Offshore Wind Turbines Current Trends and Developments*. Festschrift Peter Schaumann, Hannover, Germany.

Sheaffer, P., Schwartz, L., Basso, T. and Fox, K. (2011). Interconnection of distributed generation to utility systems: Recommendations for technical requirements, procedures and agreements, and emerging issues, RAP Montpelier, VT.

Siemens Industry (2008). New multi-million dollar Siemens plant will increase production of mechanical drives for wind turbine industry, Siemens Industry, Alpharetta, GA.

[SJR] The State Journal Register (2012). Sangamon County officials set meetings to discuss wind turbines. SJR, Springfield, IL.

Skone, T. J. and James, R. (2012). Life cycle analysis: Natural gas combined cycle (NGCC) power plant, Doc. No. DOE/NETL-403-110509, NETL, U.S. DOE, Washington DC.

Snyder, B. and Kaiser, M. J. (2009). Ecological and Economic Cost-benefit Analysis of Offshore Wind Energy, *Renewable Energy*, **34**(6): 1567–1578.

Sørensen, B. (2004). Renewable energy: Its Physics, Engineering, Use, Environmental Impacts, Economy and Planning Aspects. Elsevier.

Stark, A. M. (2011). In the wake of the wind. News Release, LBL, Berkeley, CA.

Sutherland, H. J., Berg, D. E and Ashwill, T. D. (2012). A Retrospective of VAWT Tech., Report No. SAND2012-0304, SNL, Albuquerque, NM.

Svejkovsky, C. (2007). Small-scale wind energy on the farm, National Sustainable Agriculture Information Service, Butte, MT.

Tuerck, D. G., Bachman, P. and Head, M. (2014). The economic impact of the Illinois Renewable Portfolio Standard, The Beacon Hill Inst., Suffolk Univ., Boston, MA.

U.S. Census Bureau (2004). 2000 Census of population and housing, population and housing unit counts PHC-3-1, U.S. Dept of Commerce, Washington, DC.

[US-DOE] U.S. Department of Energy. (2005). 'Basic research needs for solar energy utilization, report on the basic energy science workshop on solar energy utilization', Office of Science, U.S. DOE, Washington, DC.

[US-DOE] U.S. Dept of Energy (2011). Wind powering America: Wind resource potential. U.S. DOE, Washington, DC.

[US-DOE] U.S. Dept of energy (2012a). Illinois incentives/policies for renewables and efficiency: Renewable Energy Portfolio Standard. U.S. DOE, Washington, DC.

[US-DOE] U.S. Dept of Energy (2012b). Interconnection standards. U.S. DOE, Washington, DC.

[US-DOE] U.S. Dept of Energy (2012c). Learning about renewable energy: Grid interconnection. U.S. DOE, Washington, DC.

[US-DOE] U.S. Dept of Energy (2012d). City of Naperville, IL City of Naperville smart grid initiative. U.S. DOE, Washington, DC.

[US-EIA] U.S. Energy Info. Admin. (2012). State energy consumption estimates 1960 through 2012: Table CT8, electric power sector consumption estimates, state profile and energy, Report No. DOE/EIA-0214(2012). U.S. EIA, Washington, DC.

[US-EIA] US Energy Independent Administration (2013a). 'Levelized cost of new generation resources in the Annual energy Outlook 2013', U.S. EIA, Washington, DC.

[US-EIA] Energy Information Administration (2013b). Electric power annual 2012. U.S. EIA, Washington, DC.

[US-EIA] U.S. Energy Information Administration (2014). Annual energy outlook 2014, Report, No. DOE/EIA-0383 (2014). U.S. EIA, Washington, DC.

[US-EIA] U.S. Energy Independent Administration (2014a). Levelized cost of new generation resources in the Annual Energy Outlook 2014. U.S. EIA, Washington, DC.

[USGBC] U.S. Green Building Council (2010). USGBC project profile: Chipotle's vision for sustainability. USGBC, Washington, DC.

U.S. Way (2014). U.S. Way Urban Turbines: Magnetically levitated barrel generator. U.S. Way Corp., Chicago, IL.

Vandyck, T. (2012). With long paybacks, vertical turbines still a tough sell. *Midwest Energy News* , St. Paul, MN.

Voneschen, M. F. (2012). Brief introduction to the Darrieus wind turbines. *amics21.com/laveritat*.

Vuljevic, S. (2011). West Michigan residents voice opposition to offshore wind turbines. *Environment Report*, Ann Arbor, MI.

Warren, C. R. and McFadyen, M. (2010). Does Community Ownership Affect Public Attitudes to Wind Energy? A Case Study from Southwest Scotland, *Land Use Policy*, **27**(2): 204–213

[WEF] Wind Energy Foundation (2015). Wind energy history. W.E.F., Washington, DC.

Weiderman, G. (2012). Ameren to Purchase Power from Bishop Hill II Wind Farm. News section, *St. Louis Business J.*, St. Louis, MO.

Welch, J. B. and Venkateswaran, A. (2009). The Dual Sustainability of Wind Energy, *Renewable and Sustainable Energy Reviews*, **13**(5): 1121–1126.

Wernau, J. (2012). End of tax credit a blow for wind power industry up to 37,000 jobs, many in Illinois, could be lost as projects are halted or abandoned. Business section, Chicago Tribune, Chicago, IL.

Wilburn, D. R. (2011). Wind energy in the United States and materials required for the land-based wind turbine industry from 2010 through 2030, Report No. 2011–5036, USGS, Washington, DC.

Williams, A. (2011). A buoyant future for floating wind turbines? *Renewable Energy World*, Nashua, NH.

Wiser, R. and Bolinger, M. (2011). 2010 wind technologies market report, Report No. DOE/GO-102011-3322, Office of Energy Efficiency and Renewable Energy, U.S. DOE, Washington, DC.

Wiser, R., Lantz, E., Bolinger, M. and Hand, M. (2012). Recent Develop. in the levelized cost of energy from U.S. wind power projects, NREL, Golden, CO.

Wright, J. W. (ed.) (2007). *The New York Times Almanac*. Penguin Books.

Yeagle, P. (2010). Cashing in on wind energy: Sangamon County may get a wind farm. *Illinois Times*, Springfield, IL.

Yerak, B. (2009). Illinois wind farms bought by an Irish energy company: President Barack Obama's stimulus package spurs investment by Dublin firm. Business section, *Chicago Tribune*, Chicago, IL.

Zaaijer, M. B. (2011). Comparison of monopile, tripod, suction bucket and gravity base design for a 6 MW turbine, section Wind Energy, Delft Univ. of Tech., Netherlands.

Zaman, H. and Shakouri, H. G. (2011). A Simple Non-linear Mathematical Model for Wind Turbine Power Maximization with Cost Constraints, *Iraq J. Electrical and Electronic Eng.*, **7**(1): 60–62.

Zeitgeistlab (2012). Unveiling the hidden connections between e-mobility and smart microgrid, Optical Zeitgeist Laboratory, Montréal, QC, Canada.

9

Solar Energy

"We are like tenant farmers chopping down the fence around our house for fuel when we should be using Nature's inexhaustible sources of energy — sun, wind and tide. ... I'd put my money on the sun and solar energy. What a source of power! I hope we don't have to wait until oil and coal run out before we tackle that." **Thomas Alva Edison**: In conversation with Henry Ford and Harvey Firestone (1931); as quoted in Uncommon Friends : Life with Thomas Edison, Henry Ford, Harvey Firestone, Alexis Carrel & Charles Lindbergh (1987) by James Newton, p. 31.

9.1. Introduction

Solar power is energy from the sun that can be harnessed and converted into thermal or electrical energy. Solar energy is considered by many as the most scalable and universal energy resource that can help communities become energy self-sufficient. Solar energy utilization involves both active and passive harnessing processes. Active processes generally involve some sort of energy conversion (e.g., *via* solar panel or thermal radiative devices) to utilize energy of the sun's radiation. Examples of active processes include water heating, space cooling and heating, agricultural crop and vegetable drying, water desalinization and distillation, cooking, design of heat actuated heat pumps, water pumping in arid areas, steam turbines operation for power generation, dissociation of water for production of hydrogen, advanced magnetohydrodynamics and gas turbine processes, solar furnaces, photovoltaics (PV) and thermionic converters and more. In contrast, passive solar energy utilization is mostly through fixed architecture that uses direct, indirect or isolated heat gains from the sun (Lewis and Nocera, 2006; Mansoori, 1976). The early utilization of solar energy was reportedly solely limited to passive solar processing techniques. It is known that the Greek and Persian civilizations, as early as 4th century BC, and the Chinese, around 20th century AD, utilized solar architectures for various urban reasons (Butti and Perlin, 1980; Tchernev, 2001). Perhaps the earliest record of an active solar energy technology dates back to 1860's, when the French mathematician August Mouchet and his assistant Abel Pifre constructed the first solar powered engine. Later in 1905, Albert Einstein published a paper on photoelectric effect, along with a paper on the theory of relativity, to be considered the first work that profoundly studies the conversion of solar power to anything other than thermal energy. Einstein's work (1905) on photoelectric effect has had a significant impact on the development of solar PV technology for electrical power generation to date (US-DOE, 2009). In 1954, the first PV cell for converting solar energy to electricity was developed by the Bell Laboratories in the U.S. This breakthrough eventually led to the production of a lightweight, portable and dependable power source for space travels in the 1960s (Adkins, 1981). Also in 1979, as part of his effort to solve the U.S. energy crisis, President Jimmy Carter ordered about 32 solar panels to be installed on the roof of the White House as a demonstration to the American people to follow. But skeptical as he could be about the policies of his opponents, the president remarked that '*a generation from now, this solar heater can either be a curiosity, a museum piece, an example of a road not taken, or it can be a small part of one of the greatest and most exciting adventures ever undertaken by the American people: harnessing the power of the sun to enrich our lives as we move away from our crippling dependence on foreign oil*' (US Government, 1979).

Table 9.1. Days of sunshine per year in Illinois (currentresults.com, 2014).

City	Sunny Days(1)	Partly Sunny Days(2)	Total Sunny Days(3)	% Sun(4)
Cairo	113	104	217	61
Chicago	84	105	189	54
Moline	101	100	201	55
Peoria	95	97	192	56
Rockford	93	98	191	NA
Springfield	104	94	198	59

(1) 'Sunny Days' includes the days when cloud covers up to 30% of the sky during daylight hours.
(2) 'Partly Sunny Days' have cloud covering from 40–70% of the sky during the daytime.
(3) 'Total Days with Sun' is the sum of the 'Sunny Days' plus the 'Partly Sunny Days'.
(4) '% Sun' measures the percentage of time between sunrise and sunset that sunshine reaches the ground. The rest of the days are mainly overcast, with at least 80% cloud cover.

As a particular case study, we consider use of solar energy in the state of Illinois. The active use of solar energy as a power source for residential and utility-scale electric power generation in Illinois was not very common until 1970's, partly because of the mostly-cloudy weather conditions across the state, which is still the case, and partly due to cheap and abundant fossil fuels and little concern about global warming due to carbon dioxide emissions. Table 9.1 shows the annual averages of sunny days in some of the major cities in Illinois. These numbers place Illinois among the lowest 30th percentile of all states in the U.S., in terms of available sunshine.

Despite the limitations, solar energy utilization has grown reasonably since the 1970's. A survey published by the Illinois Institute of Natural Resources (IINR) in 1980, demonstrated the growth of solar energy utilization within almost a decade. According to this survey, there were more than 200 solar equipment suppliers and installers, 120 solar energy consultants, 16 manufacturers of collectors and component parts, and more than 30 home contractors working on passive solar facilities in Illinois in the late 1970's (Adkins, 1981). According to the same report, one-fourth of the solar system installations in the state in 1978 were for agricultural purposes, such as crop driers and warming of hog buildings. It has also been reported that in 1978, the first 'solar mini-farm' was established in the rural township of Pembroke, near Kankakee, Illinois, a town where the unemployment rate was as high as 35%. The same farm was later turned into Illinois' largest solar greenhouse, which provided soil for other solar greenhouses in the local area, as well as offering starter plants for the local farmers, including laying hens and earthworms (Adkins, 1981).

Today, Illinois' stance on solar energy usage and market is dramatically different from what it used to be decades ago. According to Intelligent Generation LLC, a Chicago-based firm that specializes in networked energy storage modules, the Chicago metropolitan market is currently the largest untapped solar market in the U.S. (Tweed, 2014). Furthermore, a U.S. Department of Energy (DOE) report published in July 2012, ranked Illinois the 3rd place in the U.S. in terms of total technical potential for urban utility-scale solar PV power estimated at 64 GW (Lopez *et al.*, 2012). In Figs. 9.1 and 9.2, the total estimated technical potential for urban, rural and rooftop PV are shown for different states in the U.S.

The same database ranks Illinois in the 9th place in the total estimated technical potential for rural utility-scale solar PV power with 4,969 GW output (Lopez *et al.*, 2012). Additionally, Illinois is in

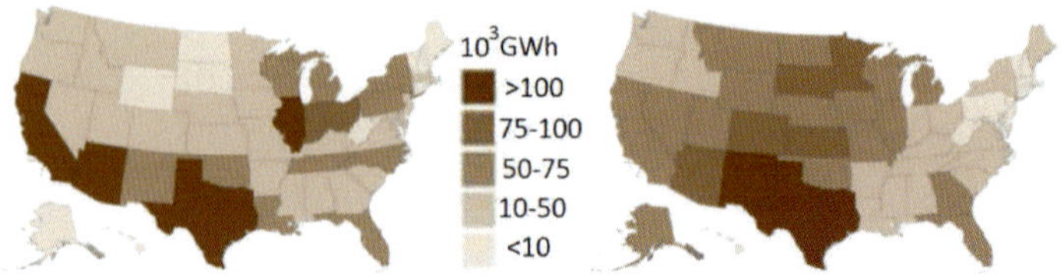

Fig. 9.1. Total estimated technical potential for urban utility-scale PV (left) and rural utility-scale PV (right) in the U.S. (Lopez *et al.*, 2012).

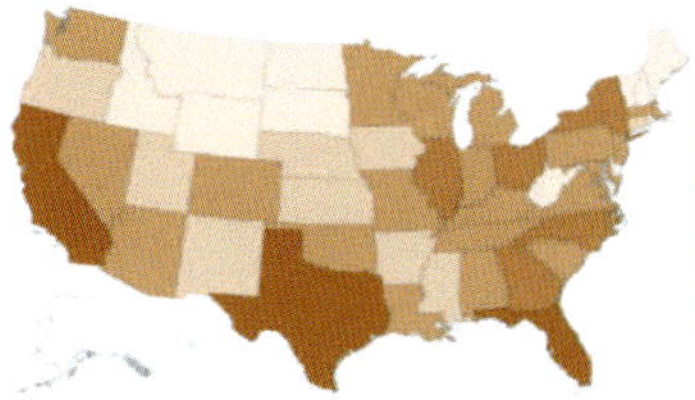

Fig. 9.2. Total estimated technical potential for rooftop PV in the U.S. (Lopez *et al.*, 2012).

Table 9.2. Estimated technical potential of Illinois' total PV solar energy (Lopez *et al.*, 2012).

PV Sector	Land (km^2 (sq.mile))	(GW)	(GWh)
Urban utility-scale	1,325 (511)	64	103, 552
Rural utility-scale	103,524 (39,971)	4,969	8, 090, 985
Rooftop	NA	26	30, 086
Total annual energy output			**8,224,623**

the 5th place in total estimated technical potential for rooftop photovoltaic solar power, estimated at 26 GW power and 30,086 GWh energy output per year (Lopez *et al.*, 2012). A summary of the estimated technical solar power potential of the State of Illinois is listed in Table 9.2.

In 2007, a research study conducted by the DOE reported Illinois' solar electric footprint as 173 m^2 (1,862 ft^2) per person, representing ~1.5% of the total per capita land available in the state. The overall average solar electric footprint for the U.S. during the years evaluated was about 181 m^2 (1,948 ft^2) per person (Denholm and Margolis, 2007). The Illinois figure is promisingly lower than the U.S. national average solar electric footprint. This value, however assumes a long-term availability of energy storage (including seasonal) facilities, and availability of a mixture of tracking and flat-plate PV systems (Denholm and Margolis, 2007). The average solar radiation (energy

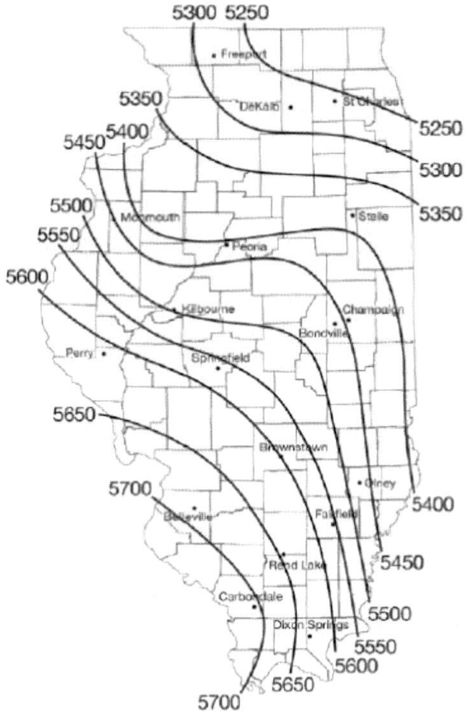

Fig. 9.3. Distribution of annual mean totals of solar radiation (MJ/m^2/year) in Illinois from 1990–2010 (Wendland and Scott, 2011).

per unit area, per unit time) is a function of local climate and module orientation.

Figure 9.3 shows annual mean totals of solar radiation in Illinois from 1990–2010. The data ranges from about 5,250 MJ/m^2/year in the Northeast to about 5,700 MJ/m^2/year in the Southwest, which translates to 4–4.3 kWh/m^2/day. Denholm and Margolis (2007) conducted a study to quantify state-by-state solar footprint in the U.S. The study found that solar module efficiency was strongly dependent on the PV array configuration and in general, 2-axis module had the highest insolation. The study also found the average insolation for flat orientation in Illinois was 3.9 kWh/m^2/day, as compared with New York and New Jersey, both at 3.8 kWh/m^2/day.

Fueled by state and federal incentives, Illinois is also experiencing growth in home and business solar power installations. In 2010, the West Pullman neighborhood in Chicago had the largest urban solar installation (Exelon City Solar) in the U.S. The 10 MW plant is located on a 41-acre (16.6 ha) land in Chicago's West Pullman neighborhood. The solar plant is equipped with SunPower® solar tracking systems and provides 14,000 MWh of electricity per year for 1,500 homes (Exelon, 2014).

In 2014, the number of solar energy companies in the value chain in Illinois exceeded 184, providing a variety of solar products and services ranging from solar system installations to the manufacturing of components used in PV panels (SEIA, 2014). These included Schneider Electric in Palatine, Illinois, a global specialist in energy management with claimed operations in over 100 countries and with more than 110,000 employees; and Chicago-based Earth Wind and Solar Energy that has been installing turnkey energy systems for several years. The latter designs both solar photothermal (PT) and solar PV systems for residential and commercial applications (Craig *et al.*, 2011). Illinois also hosts several CleanTech start-up companies including NextGen Solar and Invenergy, both in Chicago. NextGen Solar is a company that has a Cooperative Research Agreement with the Argonne National Laboratory, with the vision of developing a three-dimensional solar panel technology and maximizing the amount of solar energy capture in a limited footprint (Craig *et al.*, 2011). Invenergy is also a Chicago-based wind and solar energy company claimed to have used a breakthrough technology (new thin-film solar panels) and inverters manufactured by General Electric Co. in a newly completed solar farm (see Sec. 9.4.3.2). The solar farm is adjacent to the company's existing 210-megawatt wind farm at Grand Ridge in LaSalle County, Illinois. Invenergy claims to be using the solar panels to generate 20 MW of energy from the solar farm and sell both wind and solar energy to the Commonwealth Edison Electric (ComEd) Company (Guy, 2011; Business Wire, 2012; Wernau, 2012).

In August 2010, Illinois governor announced a stimulus grant award of over $4 million in value at a ribbon-cutting ceremony of Wanxiang America Corporation's solar panel manufacturing plant in Rockford, Illinois. The grant was intended for the construction of a 62 MW Rockford Solar Project (see Sec. 9.4.3.3), which is to become the largest PV solar development in the Midwest and one of the largest such endeavors in the U.S. (PRN, 2010). The project is sponsored by Rockford Solar Partners, LLC, a joint venture between Wanxiang America and New Generation Power. Wanxiang is the U.S. subsidiary of one of the largest non-state owned companies in China with a claimed $8 billion in revenues worldwide, while New Generation Power is a Chicago-based renewable energy developer. The 3.06 MW Phase-1 of the 62 MW three-phase Rockford PV Solar Project has been operational since 2012. The solar farm is set to be on 70 acres land and is the largest airport-based solar PV electricity generating facility in the U.S. The Rockford Solar Project is claimed to produce sufficient electricity for over 10,000 homes, while reducing carbon dioxide emissions at the rate of 113,000 tons per year (Stuart, 2010).

Illinois grid connected PV installations from 2006 through 2013 are shown in Fig. 9.4.

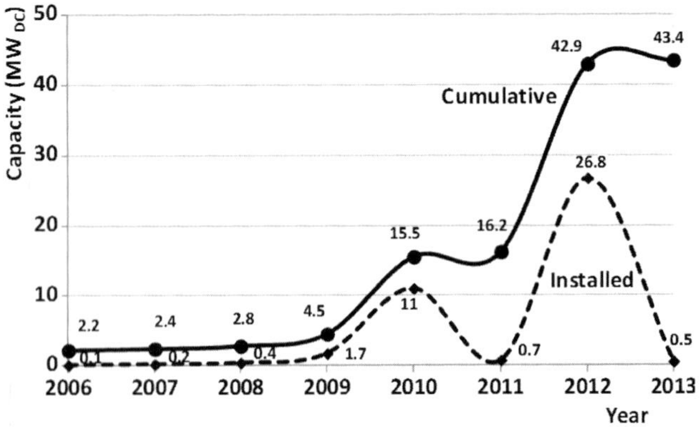

Fig. 9.4. Grid connected PV installations in illinois (Data from Sherwood, 2007–2014).

Table 9.3. Illinois PV solar capacity by sector installed in 2012 and 2013 (in MW_{DC}) (Sherwood 2013, 2014).

	Capacity (MW_{DC})	
Sector	**2012**	**2013**
Residential	1.0	0.1
Non-Residential	2.7	0.4
Utility companies	23.1	(*)
Total	**26.8**	**0.5**

(*) = less than 100 kW_{DC} or data not available.

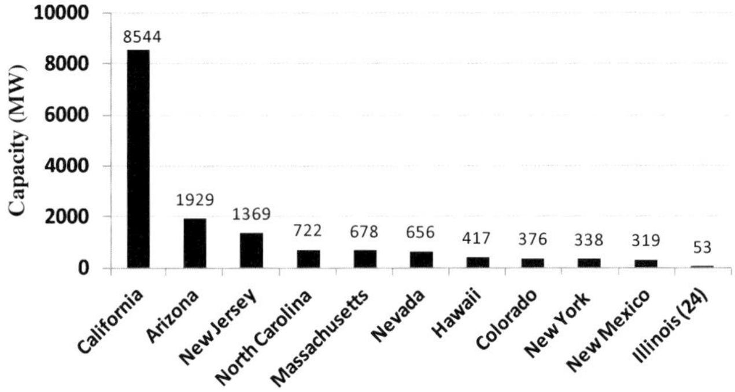

Fig. 9.5. Cumulative solar electricity installed capacity of the top 10 U.S. states and Illinois (rank in parenthesis) as of 2014 (Source: SEIA, 2014).

The rapid growth of solar power installed capacity starting from 2009 was largely due to the federal subsidy from the American Recovery Funding in 2009. The funding expired in 2011, bringing solar installation capacity in the state to almost a halt. After 2011, solar power installation in the state has mostly been originating from the utility sector. Table 9.3 shows Illinois' PV solar power installed capacity by sector in 2012 and 2013.

Figure 9.5 shows cumulative installed solar PV capacity in 2014 of the top 10 U.S. States compared with Illinois, which is ranked 24th.

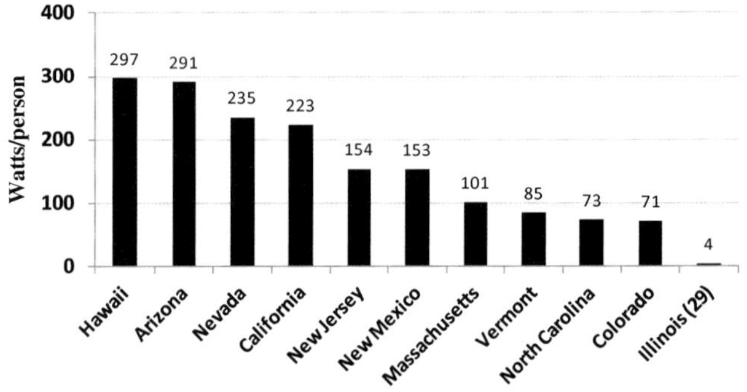

Fig. 9.6. Cumulative solar electricity capacity per capita of the top 10 U.S. states and Illinois (rank in parenthesis) as of 2014 (Sources: SEIA, 2014; US-CB, 2013).

Among all the states, Illinois ranked 24th with 53 MW_{DC} of installed solar PV capacity. California led the states with 8544 MW_{DC} of PV capacity installations (SEIA, 2014). However, on per capita basis, Illinois ranked 29th among all the states with 4 W_{DC}/person (see Fig. 9.6).

Typically, the U.S Southwestern states like New Mexico and Oklahoma have relatively high solar radiation as compared with those of the Northern and Northeastern states, such as Illinois, New Jersey and Pennsylvania. State of Alaska has the lowest solar radiation compared with the other states.

In 2009, the Illinois Power Agency (IPA), which oversees the state's energy procurement and distribution among its other duties, adopted the renewable energy portfolio standards (RPS). The mandates require 1.5% (6% of all renewables) of the energy supplied by the utility companies to be from solar PV by the year 2025 (see Fig. 9.7). According to IPA, the eligible resources must be located in Illinois. Otherwise, if there are not enough cost-effective in-state resources, then the resources can be procured from adjoining states, or in case these are also not cost-effective, the resources can be procured from other regions of the country (PJM and EIS, 2013).

Table 9.4 shows the solar energy percentages required by the electric utilities in Illinois for each successive year in order to meet the RPS mandate of 1.5% by 2025.

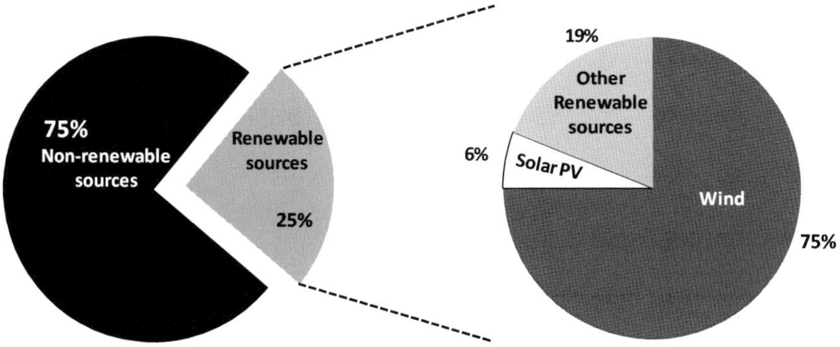

Fig. 9.7. Illinois renewable portfolio standard for electricity generation by 2025 (Loomis and Pagan, 2011).

Solar energy projects in Illinois are funded by incentives and rebates operated by the state. These include Production Incentive, Property Tax Incentive, State Rebate Program, Property Assessed Clean Energy (PACE) template financing and Net Metering. The Production Incentive is a Renewable Energy Credit (REC) aggregation program offered by the State of Illinois through the Illinois Solar Energy Association (ISEA) and is open to residential home owners.

For 2015, the incentives for solar and wind energy projects set by the Illinois Department of Commerce are \$1.50/watt or 25% of project costs for residential applications, \$1.25/watt or 25% of project costs for commercial applications and \$2.50/watt or 40% of project costs for not-for-profits and public applications. The maximum rebate for all solar and wind energy projects will be \$10,000 for homeowners, \$20,000 for businesses and \$30,000 for public sector and non-profit entities. Under the same program, solar thermal projects will be capped at 30% of project costs for residential and commercial applications and 40% for public sector and non-profit entities (IDCEO, 2014).

The property tax incentive is administered by the Illinois Department of Commerce and Economic Opportunity (DCEO). Qualifying applicants must use appliances, such as passive solar space heaters,

Table 9.4. Solar energy percentages requirement for electric utilities in Illinois (PJM and EIS, 2013).

Year	**2012**	**2013**	**2014**	**2015**	**2016**	**2017**	**2018**	**2019**	**2020**	**2021**	**2022**	**2023**	**2024**	**2025**
RPS (%)	0.0035	0.12	0.27	0.60	0.69	0.78	0.87	0.96	1.05	1.14	1.23	1.32	1.41	1.50

RPS = Renewable energy Portfolio Standards.

solar water heaters, active solar space heaters and PV systems. The property owners are required to register with the chief county's assessment office to have their solar energy equipment valued at no more than a conventional energy system (CEA, 2014).

The State Rebate Program is a solar and wind energy rebate program offered by the state of Illinois. In fiscal year 2013, qualifying applicants installing solar space heater, solar water heaters, PV and solar pool heaters received up to 25% of system cost for residential and business applications and 40% for public sector and non-profit entities (IDCEO, 2014; US-DOE, 2013a). The PV systems were required to be not less than 1 kW in size, while the solar thermal systems must be designed to generate not less than 0.61 kW (50,000 Btu per day) (US-DOE, 2013a).

The Property Assessed Clean Energy (PACE) template financing allows homeowners and businesses to save on upfront costs through city loans, which is payable through a voluntary individual increase in property taxes (US-DOE, 2012). In 2009, Illinois government enacted a bill (SB 583) allowing municipalities to develop and operate renewable and energy efficiency incentive programs based on PACE template. However, as of July 2013 and despite the program being popular in several states, there are no municipalities in Illinois offering PACE financing. The reason is that the PACE program is facing federal regulatory challenges nationally.

As mentioned in Chapter 8 (Wind Energy), in August 2007, Illinois enacted a legislation requiring investor-owned utilities to start offering a net metering program by April 1, 2008. The program requires utility companies to offer customer's credit on electric bills for electricity generated using renewable energy systems with nameplate operation of up to 2,000 kW or less (Citizens Utility Board, 2013; IAG, 2010). However, not more than 1% of a utility's peak demand from the previous year may be net-metered. The term 'net-metering' is a policy of giving retail credit to small electricity consumers who own renewable energy systems (solar, wind, etc.) and have electric energy outflow.

According to the Illinois state law, homeowners or small business owners who have a renewable energy system can sell their excess electricity to the utility companies. The net metering system obligates utility companies to offer customers credits on their electric bills for electricity generated by renewable energy systems. Under the Illinois law, renewable energy generators of 40 kW or less can sell excess electricity at the same price that they pay to the utility companies.

In October 2011, a new legislation became effective in Illinois that could affect the solar energy market. The legislation (Public Act 097-0616) requires that the IPA procure at least the following percentages of power from distributed renewable energy generation devices: 0.5% by June 1, 2013, 0.75% by June 1, 2014 and 1.0% by June 1, 2015 (IDCEO, 2012). Additionally, the legislation requires that half of the renewable energy resources procured from distributed renewable energy generation come from devices of 25 kWor less of nameplate capacity (Craig *et al.*, 2011; IDCEO 2012). This legislation paves the way for roof-top solar on homes or businesses to participate in the state's RPS.

In the ensuing sections, we introduce the concepts of solar power generation systems appropriate for Illinois.

9.2. Concepts of Solar Power Generation

Solar radiation can be converted into useful energy using solar collectors made up of PV cells or alternatively, it can be converted into usable thermal energy and subsequently into electricity through a conventional power plant technology (Kalogirou, 2013). PV solar cells typically convert 16–20% of radiation energy into electricity, while solar thermal system can convert close to 90% of sunlight into heat. In what follows, solar power generating systems will briefly be discussed.

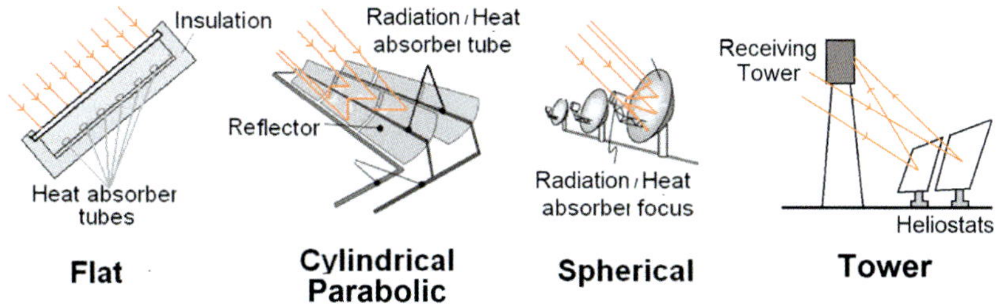

Fig. 9.8. Solar PT collectors.

Table 9.5. Performance characteristics of PT solar collectors.

	Flat-plate	Cylindrical/Parabolic	Spherical (Trough)
Overall Efficiency (%)	40–50$^{(1)}$	30–45$^{(1)}$	56–66$^{(3,4)}$
Temperature (°C (°F))	82 (<180) $^{(1)}$	77–121 (∼170–250)$^{(1)}$	350–377 (662–710)$^{(3,4)}$

(1) Hudon *et al*. (2012); (2) Stine and Geyer (2001); (3) Barton and Winger, (2009); (4) EuroTrough (2007).

9.2.1. *Solar PT*

Solar radiation can be directly converted into thermal energy using solar collectors. Solar collectors are classified as low (flat-plate), medium (cylindrical, parabolic) and high temperature (dish-spherical and tower) collectors (see Fig. 9.8).

Although operating principles of different types of solar PT collectors are the same, their performance varies according to the type. Table 9.5 shows the performance characteristics of flat-plate, cylindrical/parabolic and spherical solar collectors.

The tower collector efficiency is expected to be the same as the flat-plate collector. However, since the radiation of many flat-plate mirrors are focused on a particular point on the tower, it is possible to raise the temperature of the radiation focusing point to very high temperatures, which will depend on the number of heliostats and design of the tower. Contrary to PV technologies that often degrade under concentrated light, solar PT depends upon concentrated light and requires a clear sky to reach suitable temperatures.

The heat collected by the flat plate collectors are generally used in small scale systems such as small building or household use. The rate of heat loss to the environment for cylindrical/parabolic and spherical collectors is much lower than flat plate collectors. However, these collectors have on average a much smaller accumulated solar energy compared to that obtained from flat plate collector working under similar conditions. The main reason for this is the poor solar radiation transmissivity of the cylindrical glass envelopes of evacuated tube collectors (Pluta, 2011).

The concentrating cylindrical/parabolic, spherical and tower collectors produce heat at much higher temperatures, as shown in Table 9.5. Large amount of sunlight is concentrated onto a small area by application of optics, such as lenses (Fig. 9.8). They must be equipped with sophisticated tracking system in order to keep the light focused on the cells. The advantages of concentrating systems are higher efficiency, low system cost and low capital investment compared with flat panel PVs. The drawbacks are additional cost associated with tracking system and low reliability since even small amount of cloud can severely lower their efficiency. Parabolic trough, dish engine and power tower system are the three types of concentrating collectors that can be employed for power generation.

The most common collector system for PT electric energy generation in the U.S. is parabolic trough collectors, schematic of which is shown in Fig. 9.8. The trough system converts thermal energy of the sun into electricity. Because of their parabolic structure, the trough system can increase intensity of sunlight 30–60 times. The receiver pipe located on the focus line of the trough is filled with a heat transfer fluid, such as synthetic oil. The temperature of the heat transfer fluid can reach as high as 390°C (735°F), which is passed through a heat exchanger to generate steam that can ultimately be used to generate electricity in a conventional steam turbine. Solana is the world's largest parabolic trough power plant located on 780 acres (316 ha) of land in Phoenix, Arizona. The 280 MW power plant generates 944,000 MWh/yr of electricity, which is enough to power 70,000 homes and has been operational since 2013 (NREL, 2014).

Power tower concentrating system, shown in Fig. 9.8, uses large numbers of flat, sun-tracking mirrors, known as heliostats, to focus the sun rays onto a receiver that is located at the top of a tower. Similar to the trough system, the focused light can heat a heat transfer fluid which is used to generate steam and ultimately generate electricity. Because the number of mirrors is large, the focused light has higher intensity than the parabolic trough system. The temperature of heat transfer fluid in power tower concentrating system can reach as high as 565°C (1050°F). The world's largest solar power tower system is Ivanpah, which is located in California's Mojave Desert. The 390 MW power plant can generate electricity to serve more than 140,000 homes in California (NREL, 2014).

The dish/engine system uses mirrored dishes, which are roughly 10 times the size of backyard satellite dish to intensify the sunlight (NREL, 1998). The system includes a receiver and an engine/generator. The receiver can be a series of thin tubes filled with working fluids, such as hydrogen or helium. Concentrated sunlight heats up the gas in the tube, causing it to expand and run a heat engine. Temperature of working fluid in dish/engine system can reach as high as 750°C (1380°F). Stirling engine is the most common type of heat engine used in solar dish/engine system (SunLab, 2001).

There is an alternative collector to the flat-plate collector and that is the evacuated tube collector (Fig. 9.9). Evacuated tube collectors absorb solar radiation energy and convert it into thermal energy. Heat pipes conduct the heat from the inner part of the tube to a manifold of copper header pipes where a heat transfer liquid circulates. Heat transfer fluid is circulated through the header pipes, *via* an intermittent pump to conduct the heat. A heat exchanger is used to transfer the heat from the heat transfer fluid to water. Circular design of evacuated tube collectors allows optimal passage of sunlight throughout the day

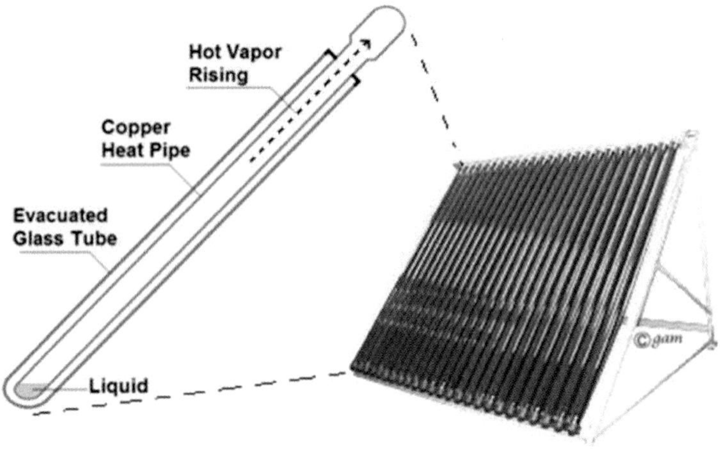

Fig. 9.9. Evacuated tube collector.

and makes evacuated collectors less sensitive to the sun's radiations angle.

9.2.2. *Solar PV cells*

Solar energy may provide a renewable alternative to traditional sources of electricity generation. Solar radiation can be directly converted into DC electricity (and some thermal loss) using PV cells. PV power generation uses solar panels consisting of a number of cells containing one or a few forms of PV material (Bube, 2012). Materials currently used for PV solar panels are monocrystalline silicon, amorphous silicon, cadmium telluride and copper indium gallium selenide/sulfide.

PV solar panels produce DC electric power due to unidirectional flow of electric charge in a conductor. In contrast, AC power is the power produced by alternating current due to the flow of electric charge, which direction is reversing periodically. For electrical power distribution, the utilities prefer AC over DC power. Inverters are used to invert DC power from PV array to AC power through a set of solid state switches (metal–oxide–semiconductor field-effect transistors or

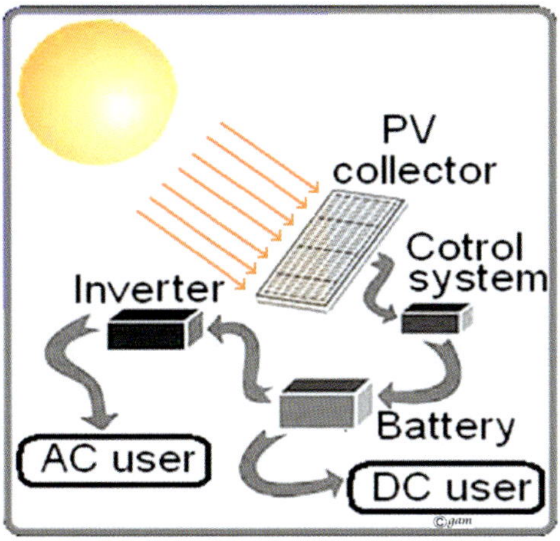

Fig. 9.10. Overall scheme of direct production, storage and consumption of electricity from solar energy PV cells and conversion of DC current to AC current.

insulated-gate bipolar transistors) by flipping the DC power back and forth (Worden and Zuercher-Martinson, 2009). The typical peak efficiencies of inverters vary from about 92–96% (Perez, 2006). This practice began in the late 1880's, when Nikola Tesla, an engineer and advocate of AC power production won the so-called 'The War of Currents' over his adversary Thomas Edison, who was also an engineer, but a proponent of DC power production. The reason for Tesla's victory was that AC power could be stepped up to travel long distances with lower power losses and with minimal resources than DC power (US-DOE, 2013b). In Fig. 9.10, the overall scheme of direct production, storage and consumption of electricity from solar energy PV cells and conversion of DC current to AC current is shown.

In addition to DC/AC conversion, inverters can regulate the PV system and select a voltage that will provide maximum power. Inverters can be used in off-grid or grid-connected PV systems. Off-grid inverters are designed to convert stored battery DC energy to AC energy, typically for home appliances. On the other hand, grid-connected inverters are designed to convert DC power from PV solar panels directly into grid-synchronous AC power. But while most battery-based inverters are designed for only off-grid applications, some are designed for both battery-based off-grid systems and grid-connected solar-electric systems that use battery as backup (Perez, 2006).

A metric that is often used to express development of PV energy generation is the PV energy density. The PV energy density is defined as the energy produced per year per unit of land area (Denholm and Margolis, 2007), expressed as:

$$\text{PV Energy Density} = \frac{\text{Annual PV generation}}{\text{Area of land}} \tag{9.1}$$

The PV energy density not only depends on the number of PV generation modules and availability of solar radiation, but on the deployed PV modules efficiencies.

PV module efficiencies depend on the technology. Advanced silicon modules efficiencies are 14–22%, while thin-film modules efficiencies are presently 7–14% (Sinke, 2014). The PV energy density is in turn used to calculate the solar electric footprint, which is defined as:

$$\text{Solar Electric Footprint} = \frac{\text{Annual Electric Demand}}{\text{PV Energy Density}} \tag{9.2}$$

As mentioned in Sec. 9.1, the estimated average solar incident radiation (insolation) for flat orientation in Illinois is 3.9 kWh/m^2/day, with solar electric footprint of about 1–2% of the total state's area. Also, the highest possible average solar insolation for flat orientation in the U.S. was recorded at 5.6 kWh/m^2/day in Arizona. The electric solar footprint of state of Arizona is 0.1–0.5% of the state's land area (Denholm and Margolis, 2007).

Presently, silicon based PV cells are being replaced with thin-film (CdTe), amorphous silicon and microcrystalline silicon solar cells for better efficiency. In what follows, we present the thin film PV technology.

9.2.2.1. *Thin film PV technology*

Despite the fact that crystalline silicon (c-Si) PV has dominated the solar technology, thin film PV cells have shown tremendous promises. The momentum for thin film PV technology began between 2005 and 2008, when the silicon PV industry faced with scarcity of purified polysilicon, which had been used to manufacture c-Si cells (Shah, 2010). Thin film PV technology requires less active materials, as the thickness of the layers are a few hundreds of nanometers or a fraction of a micrometer, and processing steps are shorter when compared with c-Si technology. Thin film PV cells can be grouped into different categories based on the active absorber materials such as: i) copper-indium-gallium-selenide (CIGS), ii) cadmium telluride (CdTe) or iii) amorphous silicon (a-Si)/microcrystalline silicon (μc-Si) (Shah, 2013). The advantage of thin film technology is fewer requirements of materials and low-cost manufacturing techniques. However, manufacturing scale-up challenges remain an obstacle for development of thin film PV cells. Between 2007 and 2009, share of thin film in the PV market increased from 6.5% to18% , but fell to 9–10% by 2013 (Baker *et al.*, 2013; Burger *et al.*, 2014). Among the three categories of thin film technology mentioned above, CdTe has the most widespread application as 54% of the thin film PV market is captured by this technology. Some of the characteristics, advantages and disadvantages of thin film types and c-Si panels are shown in Table 9.6.

Continuous progress is being made in thin film area as one of the U.S.-based companies, First Solar, revised its target efficiency for CdTe technology, raising it to 25% for research cells and 19% for commercial modules by 2017 (IEA, 2014). Research into materials to replace silicon solar cells is gaining ground. Efficiency of solar cells made from a metal halide perovskite; a compound typically consisting of a precise mixture of lead, iodine and a simple organic component, has jumped from 3 to 20% within five years of research (Grätzel, 2014). In a study conducted by Dong *et al.* (2015), synthesis

Table 9.6. Comparison of c-Si panels with primary types of thin film panels.

		Thin film types		
	c-Si	**a-Si**	**CdTe**	**CIGS**
Best research-cell efficiency (%)	25.6 [(1)]	13.4 [(2)]	19 [(2)]	20.4 [(2)]
Best solar module efficiency (%)	18.5–22.9 [(3)]	8.1–10.9 [(2,3)]	14.4–16.1 [(2,3)]	14.5–15.7 [(2,3)]
2013 Worldwide market share (%)	91 [(3)]	2 [(3)]	5 [(3)]	2 [(3)]
Advantages	High efficiency	Mature technology Excellent for small devices (e.g. pocket calculators)	Low cost manufacturing	High efficiency Glass or flexible substrates
Disadvantages	Energy intensive process, Expensive, Fragile, little degradation	Low efficiency High cost equipment, Degradation over time	Medium efficiency Rigid glass substrates Cadmium is highly toxic, Degradation over time	Costly traditional process, Requires less cadmium than CdTe solar cells, Degradation over time

(1) IEA (2014); (2) Maehlum (2014); (3) Burger *et al.* (2014).

of organolead trihalide perovskites, $CH_3NH_3PbI_3$, showed that balanced electron and hole diffusion lengths of perovskites materials are suitable for highly efficient printable electronics and solar cells.

9.2.2.2. *New generation PV technologies*

The concept of solar electricity generation is based on excitation of electrons in semiconductor materials. Although discussion of physics of solar energy is beyond the scope of this book, it is suffices to say that for conduction to occur, the tightly packed electrons in the valence band must move to a free state (conduction band). The energy difference between valence and conduction bands is called band gap.

Simply stated, energy of light's photon is used to excite an electron from its initial energy level (E_i) to higher energy level (E_f). In order for a semiconductor material to generate electricity, the energy of absorbed photon must be greater than the band gap. But, it should be born in mind that not all the absorbed energy can be converted into electricity and only the free energy (the Helmholtz free energy), which is determined by the second law of thermodynamics, can be extracted (Liao and Hsu, 2012). These intrinsic physical laws set the limit for efficiency of solar energy conversion. Based on a study by Shockley–Queisser (1961), the maximum theoretical efficiency of a solar cell made from a single p-n junction with a band gap of 1.4 eV is around 33.7%.

The new generation PV technologies refer to approaches directed toward overcoming the Shockley-Queisser single band gap limit. Examples of these technologies are multi-junction, quantum dot, nanotubes silicon wires, organic dyes and conductive plastics solar cells. New generation solar cells have not reached commercialization, as most of the work is still done in the laboratory. The solar cells of the near future might be printed from ink, with efficiency twice as high as the current solar cells and producing electricity at the cost well below that of fossil fuels (Hillhouse and Jen, 2014).

9.3. Solar PT Energy Utilization Technologies

Solar PT energy technologies include the use of concentrating solar power and solar thermal collectors for heating and cooling. These technologies are classified as air type and liquid type, and can be used for producing hot water or for space heating, cooling and air conditioning at homes. In general, solar space heating and cooling systems are often designed to provide 40–80% of the annual home heating and cooling requirements. But, they are most economical when designed to meet about 50% of the heating and cooling requirements of a home. In fact, commercial applications of solar water heater can help reducing energy consumption for heating water by 60%, while energy saving can be as high as 70% for home applications (Kalogirou, 2013; EESI, 2014; SEDP-EIS, 2014). The design of solar heating system usually consists of a collector, a heat transfer component, a heat storage device, a heat distribution system, a backup furnace and control mechanisms. Each collector of a solar thermal system can provide a daily supply of 150 liters (~40 gallons) of hot water. Typically, an American family of size four requires 300–450 liters (80–120 gallons) system using 2–3 solar collectors, which requires 4.5–6.7 m^2 (~48–72 square feet) of roof space (City of Chicago, 2005). It is estimated that designing and installation of such a solar hot water system which requires two flat plate solar collectors would cost from $2,500–4,000. The payback time of such a system is 5–10 years. On the other hand, a two-panel system using evacuated tube technology suitable for Illinois cold weather would costs from $5,000–6,000 (Fitzgerald, 2011). For businesses with high-volume hot water usage, such as car washes, restaurants and commercial laundries, the payback period can be expected in as little as 18 months (Fitzgerald, 2011).

Solar Thermal systems are environmentally beneficial compared to electric and gas-fired alternatives (see Table 9.7) (City of Chicago, 2005).

Solar thermal systems have the highest operating performance when installed with an unobstructed southern roof exposure.

Table 9.7. Estimated emissions reductions benefits for a 1kW solar thermal system (City of Chicago, 2005).

	Replacing Electric Heating	**Replacing Gas-fired Heating**
Carbon dioxide (CO_2) (lb/ year)	4,528	1,756
Sulfur dioxide (SO_2) (lb/year)	20	11

However, with an extra collector, east- or west-facing roofs can also support solar thermal heating systems. Early integration of thermal systems into the design phase of new construction will reduce costs and maximize system performance. Solar thermal systems can reduce the cost of building operations by significantly reducing the demand for natural gas resources and traditional electricity. The operating lifetime of a solar thermal system is 30 years and the collectors are warranted for an average of 10 years, which is far better than most home or building appliances.

9.3.1. *Air-based solar heating systems*

The components of air-based solar heating systems include the collectors, fans, ductwork, controls and normally a thermal storage system. Solar air-heating systems are more versatile compared with convective systems because the heat can be directed to rooms farther away from the collector (Kornher and Zaugg, 1984). In general, air-based solar systems are designed for heating homes or buildings without the need for heat exchangers or thermal energy storage devices. However, larger air-based solar systems are often equipped with thermal energy storage, such as rock-filled bins as large as 8 cm (~3 inch). An air-to-water heat exchanger can also be used to store excess thermal energy instead of the rock-filled bin, especially when the system is not used for space heating (Argiriou *et al.*, 1997). This method of energy storage allows the system to be used for producing hot water while avoiding sanitation problems such as bacteria and mold growth on the rocks. In general, these systems are more costly to construct than passive systems, which solely depend on air flow through convection (as heat transfer medium) to operate (Kornher and Zaugg, 1984). Furthermore, with forced-air systems, design problems such as maintaining delicate natural convection airflow are avoided.

9.3.2. *Liquid-based solar heating systems*

Liquid-based solar heating systems are most appropriate for central heating. The components of liquid-based solar heating systems include collectors, pumps, pipes, heat exchangers, storage tanks and controls. Liquid solar heating systems operate in the same way as the air-based solar systems, except that the heat transfer medium is liquid instead of air. Using a circulating pump, a heat transfer liquid such as water or antifreeze (i.e., propylene glycol) moves through the collector to absorb heat and release it into a storage tank. Pressurized or unpressurized storage tanks typically require 1–2 gallons of water per square foot (~41–82 liters per square meter) of collector area (Phillips, 1981; US-DOE, 2012a). Installation of liquid-based solar systems in homes and buildings equipped with forced air heating systems requires a heat exchanger for the transfer of heat from the solar-heated liquid to air (Argiriou *et al.*, 1997). Liquid-based solar heating systems can be classified as *direct* and *indirect* systems, as described below.

9.3.2.1. *Direct or open loop solar liquid-based heating systems*

The open loop solar heating systems circulate potable water through the solar collectors to a storage tank and back to the collectors. They are referred to as direct circulating or open loop systems due to the absence of heat exchangers that normally conduct the heat from the heat transfer fluid (see Fig. 9.11).

These systems are commonly used in areas with temperatures unlikely to be at freezing levels. The main advantage of these systems is the lower cost relative to indirect systems (Argiriou *et al.*, 1997; NCSC, 2000). On the negative side, these systems have little or no overheat protection, unless they are equipped with a heat pump, which normally increases the cost of design. They also tend to provide only limited or no freeze protection, unless the collectors are freeze-tolerant. Furthermore, when hard water is run through the system, an ion-exchange softener is required to avoid accumulation of scales in the collectors. Considering the harsh winters in Illinois and other Midwestern states, this system may be suitable only for Southern and Southwestern states.

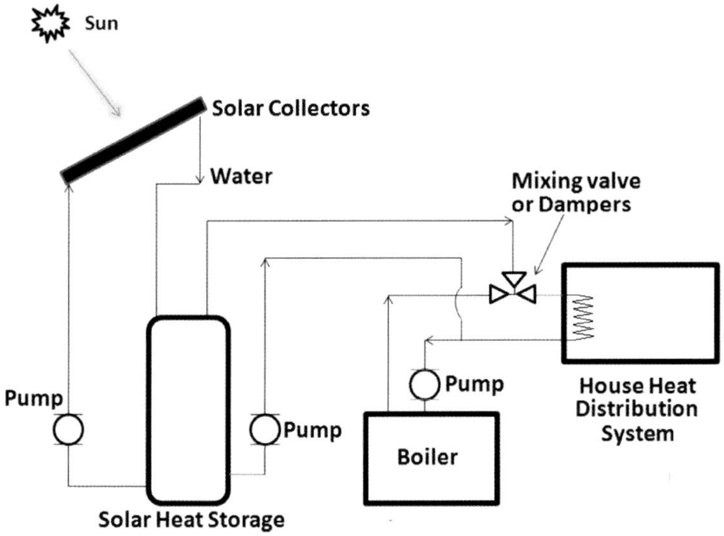

Fig. 9.11. Basic schematic of direct solar space heating system using water (NCSC, 2000).

9.3.2.2. *Indirect or closed loop solar heating systems*

The indirect solar systems circulate a non-freezing heat transfer fluid through solar collectors and a heat exchanger. They are referred to as indirect circulating systems because there is no direct contact between the collectors and potable water (see Fig. 9.12).

The indirect solar systems are commonly used in areas with freezing temperatures. The main advantage of these systems is that potable water does not come in direct contact with solar collector and therefore, the problem of water freezing in the circulating pipes does not exist. Furthermore, these systems are more robust against overheating.

9.3.3. *Examples of solar heating system installations*

Examples of solar heating systems in Illinois include the World's Largest Laundromat and Dry Cleaners in Berwyn, the Governors

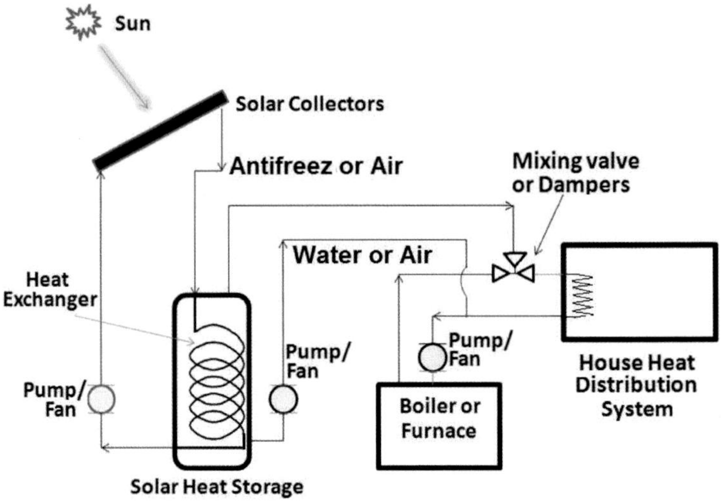

Fig. 9.12. Basic schematic of indirect solar space heating system using antifreeze or air (NCSC, 2000).

State University solar heating system at University Park in Will County and the solar hot water systems on the Oak Park's Leadership in Energy and Environmental Design Silver Public Works building. A brief description of each of these installations is given below.

(a)

In April 14, 2003 an Illinois Laundromat, the 'World's Largest Laundromat and Dry Cleaners' located in Berwyn inaugurated one of the most powerful solar energy systems in the state (see Fig. 9.13).

This solar hot water system was producing more energy than any other solar system in the state by the time of its inauguration. The daily energy savings was about 1.4 million Btu (~410 kWh) of energy, which is equivalent to 1350–1400 cubic feet of natural gas consumption. The solar hot water system has 24 solar collector panels, each measuring 1.22 meters by 3.05 meters (4ft by 10ft) installed on the roof of the Laundromat (PRN, 2003). Currently, the solar system is the primary supplier of hot water for the laundry and

Fig. 9.13. Solar thermal system installed at the 'World's Largest Laundromat and Dry Cleaners' (Renewable Energy World, 2003; ISEA, 2005).

also acts as a pre-heater to the existing water heater. The Laundromat, which occupies a 1.21×10^3 meter square area (13,000 square feet), is a self-service laundry with 145 washing machines. The solar system is capable of heating 4500 liters (1200 gal) of cold water a day to a temperature of 48.9°C (120°F) or more, saving thousands of dollars a year in energy costs. The system is automated, maintenance-free, and has a design life of more than 30 years. The project cost $150,000 with $69,130 matching grant from the Renewable Energy Resources fund administered by the Illinois Department of Commerce and Economic Opportunity (DCEO). The payback time was initially approximated at 11 years (PRN, 2003).

(b)

In 2006, the Governors State University in University Park, Illinois, completed installation of the largest solar heating plant in the State (see Fig. 9.14).

Mounted on top of the university gymnasium, the collection of 64 solar panels covers 2,560 square feet (about 238 m^2) of heating surface and provides heating for the university's 168,000 gallons (~636,000 liters) swimming pool. Additionally, the solar system provides hot water for the locker rooms, including the washrooms throughout a major part of the university. The water is heated by

Fig. 9.14. In 2006, Governors State University installed the largest solar heating plant in the State of Illinois. Situated on top of the physical plant, 64 solar panels encompass about 40 square feet of heating surface. Once the water is heated by the solar panels, it travels to one of two thermal storage tanks. From here it travels to a heat exchanger where temperatures are brought down to accommodate the washrooms, showers and swimming pool (TCE, 2006).

the solar panels and is stored in one of the two available thermal storage tanks. From the storage tank, water passes through a heat exchanger where the temperature is lowered down to appropriate levels for swimming pool, washrooms and showers usages. The flow rate of water circulation in each of the solar panels is approximately 1 gallon (3.79 liters) per minute, with water temperature raised to about 82°C (~180°F). Each panel can provide approximately 65 gallons (~246 liters) of water per day, with overall design capacity of approximately 4,168 gallons (~16,000 liters) of water. The $300,000 project was partially funded via a grant from the Illinois Clean Energy Community Foundation and the Illinois DCEO. The payback time of the project was estimated to be 12 years at the time of construction (TCE, 2006).

(c)

In 2009, the Village of Oak Park, Illinois, installed a 10 panel solar hot water system on its Leadership in Energy and Environmental Design (LEED) Silver Public Works building (see Fig. 9.15).

Fig. 9.15. Village of Oak Park Public Works Building striving for LEED Silver (Holabird, 2008).

To achieve a LEED rating at the Silver level, the building needed to include a restored brownfield site, recycled and low volatile organic compound (VOC) construction materials, a reflective roof with green roof portions and solar domestic water heater. Furthermore, it required to maximize reuse and recovery of materials by diverting more than 87% of construction waste from landfills and achieve energy optimization at a rate of 32.7% better than ASHRAE Standard 90.1-1999 (Mesik and Howery, 2009).

The 260 square feet ($\sim$24.15 m^2) solar thermal system collectors transfer enough thermal energy to heat roughly 300 gallons (1,100 liters) of water per day. In addition, the installation includes two 120-gallon (454 liters) commercial storage tanks for holding the preheated solar water for immediate or future use (ISEA, 2013). This system uses a drain back tank, which allows all the water from solar collectors and related piping connections to drain into the reservoir tank, in order to protect the system from both freezing and overheating. The glycol heat transfer fluid is stored in an indoor vessel until it is needed (ISEA, 2013). This design increases the efficiency of the system and is ideal for solar thermal systems especially in climates that experience very hot or cold temperatures.

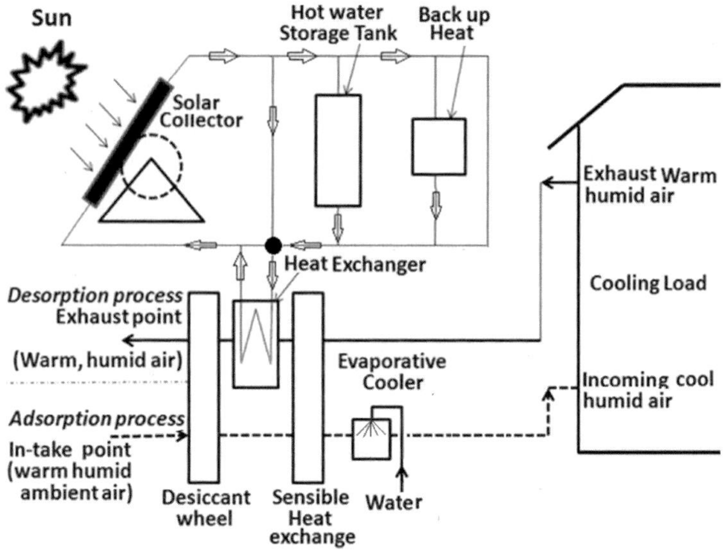

Fig. 9.16. Schematic drawing of an open cycle desiccant cooling system (Modified from RAEE, 2014).

9.3.4. *Potential applications of solar PT cooling*

The potential applications of solar energy for cooling include desiccant systems and absorption chiller systems.

9.3.4.1. *Cooling through evaporation (Desiccant cooling)*

Solar energy can be used for regeneration of sorbent in dehumidifiers of desiccant cooling systems (see Fig. 9.16). These applications are briefly discussed below.

Desiccant cooling systems combine adsorptive or absorption dehumidification of air using a desiccant (hygroscopic) material, as well as humidification of air through evaporation of water to create a cooling process. Most frequently, removal of water from the air is achieved either through absorption in a liquid desiccant or adsorption on a solid desiccant. Open-cycle systems operate close to atmospheric pressure and have mass transfer occurring between the system and the environment, while closed cycle systems usually operate either below or above atmospheric pressure (Collier *et al.*, 1982).

In desiccant cooling systems, heating and cooling is achieved through a physical process, whereby water vapor is collected and released (i.e., adsorbed and desorbed) to the outside environment by the desiccant material. The cooling effect is essentially caused by evaporation of water, while the heating effect is caused by condensation of water. At the in-take point (see Fig. 9.16), warm and humid air from outside enters the slow rotating wheel loaded with desiccant material. The warm and dry air becomes dehumidified due to adsorption of the water molecules (from the air) on the surface of the desiccant material. The warm dry air enters the sensible heat exchanger and loses some of its heat. Then, the (still warm) dry air enters the evaporative cooler, where its temperature is lowered to its wet bulb temperature prior to being delivered to the load (Mei *et al.*, 1992). The wet bulb temperature (commonly known as the wind-chill) is the temperature at which there is no further evaporation of the water, and consequently, there is no further decrease in the air temperature. On the load side, the humidified air close to the saturation point cools the room. This allows full utilization of the cooling potential of the humid air and facilitates air conditioning (heat removal from the room). The humid (exhaust) air from the load flows through an evaporator in order to cool the incoming warm air through a sensible heat exchanger. To allow continuous operation of the dehumidification process, the desiccant material is regenerated using a low temperature (50–75°C) heat source (RAEE, 2014). To achieve this, the exhaust air flows through a heat exchanger heated by a heat source (powered by solar energy) and its temperature is raised to regenerate the desiccant dehumidifier (Mei *et al.*, 1992). The water used in the evaporative cooling system could lower the incoming air temperature to as low as 10°C (50°F) (Mei *et al.*, 1992).

Applications of desiccant cooling systems include large areas, such as hospitals and buildings for housing humidity-sensitive processes (such as photography, microelectronics, printing and archiving) (Gommed and Grossman, 2007). Desiccant cooling can also be used more efficiently and economically in regions of the world where the air is hot and quite dry. Illinois' weather humidity in the summer varies between ~55–100% and as a result, desiccant cooling systems may not be quite effective. Presently, cooling in almost all homes and businesses in Illinois is achieved through compression-expansion (Kelvin) cooling cycle, which is highly energy intensive. However, desiccant cooling is used for special buildings or for demonstration purposes in Illinois. For example in 2009, a desiccant cooling system was installed to maintain the level of humidity at the Norton Building in Springfield, Illinois. The building which was constructed in 1938, houses the library used to preserve Illinois State Archives, a library which contains among other significant documents, a collection of Abraham Lincoln's handwritten papers. However, the library had been experiencing humidity control problem for years, a problem that could have destroyed the building's integrity as a literary preservation place. Currently, the building uses a cooling system customized by Munters Corporation which is based on compression-expansion (Kelvin) cooling cycle system (Munters Corp., 2009).

9.3.4.2. *Solar absorption cooling*

In rural Illinois communities, solar energy can be used for cooling and refrigeration. Applications like this can reduce electricity power grid congesting during the summer peak periods. In this section, we present the principle of a continuous solar absorption cooling system using phase change energy-storage material (PCM) such as alkanes and paraffin wax (see Fig. 9.17) (Agyarko and Mansoori, 2010; Himran *et al.*, 1994; Mansoori and Patel, 1979; Gomez and Mansoori, 1983; Mansoori, 2013).

Solar radiation heat is absorbed by a flat-plate solar collector and the heat is stored in the form of latent heat in the phase change energy storage medium (Himran *et al.*, 1994) to heat the water. The water

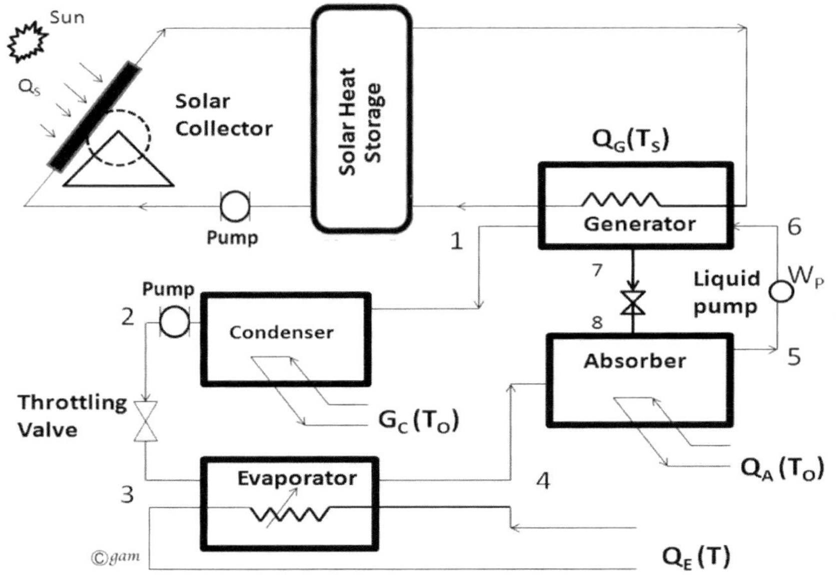

Fig. 9.17. Continuous solar absorption cooling system using thermal energy storage (modified from Mansoori and Patel, 1979).

is transferred to the generator side to vaporize the refrigerant from the absorbent-refrigerant mixture. The vaporized refrigerant enters the condenser (high pressure side), where it is liquefied through heat removal using air or cooling water at room temperature. The condensed refrigerant is vaporized in the evaporator (low pressure side) through an expansion valve to produce cooling. The vaporized refrigerant enters the absorber, dissolving in the weak refrigerant-solvent solution to form a strong refrigerant mixture. Since the mixing of the refrigerant and the absorbent in the absorber is an exothermic process, heat is removed using air or cooling water at room temperature. The refrigerant-rich mixture formed in the absorber is returned to the generator through the liquid pump to repeat the process (Mansoori and Patel, 1979; Gomez and Mansoori, 1983).

The major thermodynamic parameters affecting the performance of solar absorption cooling system are (i) solar collector efficiency, (ii) coefficient of performance $(COP)_{cycle}$ of the cycle, (iii) ratio of mass flow rate of refrigerant throttled per unit mass of solution pumped from the low pressure to high pressure side through the liquid pump, $(\frac{\dot{m}t}{\dot{m}p})$ and (iv) pumping work of the strong solution from low pressure in the absorber to high pressure side in the generator per ton of refrigeration, as shown in Equation 9.3 (Mansoori and Patel, 1979; Gomez and Mansoori,1983; Agyarko and Mansoori, 2011).

$$\frac{|Wp|}{|Q_E|} = \frac{\dot{M}p}{\dot{M}t}\frac{1}{\rho_A} \cdot \frac{(P_G - P_A)}{(h_4 - h_3)} \tag{9.3}$$

In this equation, ρ_A is density of the strong solution pumped from the absorber to the generator, P_A is absorber pressure, P_G is generator pressure and h_3 and h_4 are inlet and outlet enthalpies of the refrigerant passing through evaporator. The ratio $\frac{|\dot{W}p|}{|\dot{Q}_E|}$ is a measure of pumping energy consumed per unit of cooling energy effect. The ratio $\frac{\dot{M}p}{\dot{M}t}$ is a measure of pumping required per unit of refrigerant circulated through the cooling cycle. For optimum design of absorption cooling cycles, smaller values of these ratios are required. In a comparative study of working fluid combinations, the COP, the upper bound of $\frac{\dot{M}t}{\dot{M}p}$ and $\frac{|\dot{W}p|}{|\dot{Q}_E|}$ are used as the criteria (Gomez and Mansoori, 1983).

9.3.4.3. *Other potential applications of solar PT energy*

The other potential applications of solar energy can be found in agricultural and industrial sectors of the economy, since heat represents about 33% of the 'useful' energy requirement in those sectors. These include processes such as drying of agricultural products using unglazed solar collectors, preheating of boiler feed water and sterilization of beverages (Philibert, 2006). Table 9.8 shows the various industrial processes and temperature ranges where solar energy could be used.

Table 9.8. Processes and temperature levels of solar energy applications.

Industry	Process	Temperature [°C]	Temperature [°F]
Food & Beverages	Drying(1)	30–90	86–194
	Heat treatment(1)	40–60	104–140
	Washing(1)	40–80	104–176
	Pasteurizing(1,3)	60–110	140–230
	Boiling/baking(1,2)	90–110	194–230
	Dairy Sterilization(1,3)	100–150	212–302
Chemical	Boiling(1)	95–105	203–221
	Distilling(1)	110–300	230–572
	Boiler feed-water Preheating(3)	60–90	140–194
	Heating of production halls(1)	30–80	86–176
	Preparation of pulp paper(3)	120–170	248–338

(1) Philibert (2006); (2) Garud (2008); (3) Gosselar and Johnson (2011).

9.4. Solar PV Energy Utilization

9.4.1. *Solar PV configurations and types*

In general, components of a solar PV electric system important to zoning policy are the solar collector units typically configured in standard panel assemblies, solar collector mounting structures (includes bracketing hardware, trays, ballasting, etc.), system wiring (includes all cables, conduit, etc.) and other electrical equipment (junction boxes, inverters, disconnect switches, meters, etc.). Figure 9.18 shows a single line diagram of a solar PV system (DHED, 2013).

The four basic configuration types of solar PV systems are the roof-mounted, non-roof-mounted, freestanding and building integrated photovoltaic (BIPV) systems. Each solar PV system configuration type has unique design considerations that will impact the

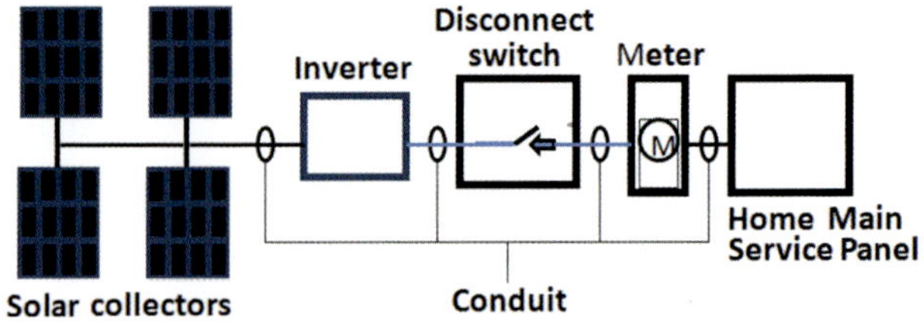

Fig. 9.18. Single line diagram of solar PV system (DHED, 2013).

Fig. 9.19. Solar PV configurations and types (from left to Right): roof-mounted system; non-roof-mounted systems; freestanding system; and BIPV system (Seely, 2013; PESI, 2014; *cityofchicago.org*, 2013).

zoning review (DHED, 2013). Below, we provide a brief description of each of these configurations, as shown in Fig. 9.19.

(1) Roof-mounted systems: this configuration refers to solar PV systems with all the solar collector units mounted on the roof of a structure. In this case, the roof-mounted systems can either be resting on top of the structure through ballasting method (non-penetrating/ballasted) or physically connected to the roof of a structure (penetrating) (DHED, 2013).
(2) Non roof-mounted systems: this type of configuration refers to solar PV systems with all the solar collector units attached to the exterior of a structure at a location apart from the roof. An example is a solar PV system attached to features such as awnings or the exterior walls of a structure.
(3) Freestanding systems: for this type of configuration the solar collector units for the PV systems are mounted independent of

the principal structure or any other accessory structure. Typical examples of freestanding solar PV systems are ground-mounted and pole-mounted systems.

(4) BIPV system: this configuration refers to solar PV systems with the solar collector units integrated into the principal structure, instead of being considered as an accessory to the structure. Typically, the BIPV system replaces or substitutes an architectural or structural element of the structure. A typical example of BIPV electric systems are those of the four Exelon Pavilions in Chicago, which will be mentioned in Sec. 9.4.2.1, that have solar collector units built into building materials such as windows. Depending on the size, application of solar PV technology can be a standalone or a utility scale.

9.4.2. *Standalone solar PV systems*

The size of a standalone PV solar power generation systems typically ranges from 50–1000 W. Location of solar cells may be on the ground or at rooftops (Reyes and Rosen, 2007). In general, the cost for most residential solar PV systems is between $8,000–24,000, with payback time in 10–15 years (Fitzgerald, 2011). The system size may be the most important factor in determining the costs against benefits. For example, the estimated installation cost of a small single PV panel system with built-in inverters that generates about 75 W per hour is about $900 ($12 per watt). On the other hand, installation cost of a 2 kW system to reduce the electricity requirement in a very energy-efficient home is about $16,000–20,000 ($8–10 per watt). But a 5 kW system that will completely offset the energy needs of many conventional homes may cost from $30,000–40,000 ($6–8 per watt) to install. However, these prices are just estimates and actual costs will depend on the system's configuration, equipment options and other factors (NREL, 2013). Standalone PV power generation systems with typical size ranging from 500–5,000 W can also be grid-connected (Reyes and Rosen, 2007). In this case, any electricity

Table 9.9. Estimated installed cost for standalone solar PV systems in the year 2010.

Capacity[1] (kW)	($/Watt)[1]	Total Cost[1] ($ x 1000)	# Payback year[2]
0.075	12	0.9	10–15
2	8–10	16–20	10–15
5	6–8	30–40	10–15

(1) NREL (2003); (2) Fitzgerald (2011).

produced in excess of the household requirement can be fed to the grid through net metering programs. The above-mentioned estimates for buying solar electric systems in the Midwestern states in U.S. are shown in Table 9.9.

There exist numerous ongoing and completed standalone solar PV energy projects in Illinois. A few examples of these projects are described below.

9.4.2.1. *Exelon Pavilions PV project*

In 2005, the four Exelon Pavilions in Chicago, Illinois, were inaugurated as an integral part of the city's efforts to become 'the most' environmentally friendly city in North America (ISEA, 2013). The pavilions, which cost $7 million, were completed in 2004 and are equipped with solar PV technology for converting solar energy into electricity (see Fig. 9.20).

Exelon Corporation and its subsidiary, ComEd, provided the funding of $5.5 million for the Pavilions (Brownsey, 2005). In essence, the Exelon Pavilions are four buildings that are capable of generating electricity from solar energy. Out of the four Exelon Pavilions, two are referred to as the North Exelon Pavilions (i.e., Northeast Exelon Pavilion and Northwest Exelon Pavilion) and are located on the Northern edge of the park along Randolph Street, flanking the Harris Theater. The other two pavilions are referred to as the South Exelon Pavilions (i.e., Southeast Exelon Pavilion and

Fig. 9.20. Exelon Pavilion in Chicago (*cityofchicago.org*, 2013).

Southwest Exelon Pavilion) and are located on the southern edge of the park along Monroe Street, flanking the Lurie Garden. Each of the black-glazed minimalist towers on Randolph Street is covered by 460 solar PV modules on their outside surfaces, with a combined annual electrical energy generating capacity of 16,000 kilowatt hours (Brownsey, 2005). The two south Exelon Pavilions are capable of generating 3,840 kilowatt hours of electricity annually (CPL, 2013; City of Chicago, 2014). Thus, the four pavilions are capable of generating 19,840 kilowatt-hours of electricity each year, with annual value of $2,353 at 2010 average Illinois electricity price, enough to power 14 star-rated energy-efficient houses in Chicago (Brownsey, 2005; City of Chicago, 2014). It is claimed that from April 2005 to March 2008, the power generated by the North Exelon Pavilions offset approximately 130 lb of nitrogen oxide, 64 lbs of sulfur dioxide and 30,000 lbs of carbon dioxide or the equivalent of 624 trees planted (Jalayerian and Eich, 2008). Besides this funding, Exelon and ComEd have been supportive of the city's environmental initiatives through the installation of solar power. By the end of 2004, ComEd had spent $12 million in support of the installation of more than

50 solar systems throughout the City of Chicago that involved eight museums and eight Chicago Public Schools (Brownsey, 2005).

9.4.2.2. *DePaul University PV project*

In 2009, DePaul University completed installation of more than 30 solar powered lights at its Lincoln Park Campus in Chicago, to efficiently reduce electricity usage and illuminate the campus using the sun as the main source of power. The lifespan of the solar module is 25 years, with each light powered by an 85 W solar PV module. It is claimed that these installations could bring emissions reduction of 596 kg (1,314 pounds) of carbon dioxide, 7.6 kilograms (16.8 lbs) of sulfur dioxide and 2.55 kilograms (5.6 lbs) of nitrous oxide each year (Rex Electric, 2009).

9.4.2.3. *Village of Oak Park PV project*

In 2012, the Village of Oak Park in Illinois completed installation of a 99.45 kW solar system at the top of a five-story parking garage, located at the corner of North Boulevard and Oak Park Avenue at the village (ISEA, 2013). The solar electric system was claimed to be one of the largest municipal arrays in Illinois, and the largest garage-mounted system in the state (see Fig. 9.21).

The overall cost of the project was estimated at $765,600 (Lothson, 2012). The system uses three hundred ninety 255-Watt solar PV modules and has an innovative mounting system comprised of 50 tons of galvanized steel and extruded aluminum rails, allowing the array to seamlessly integrate with the garage structure (see Fig. 9.21). The power from the array is expected to produce over 106 megawatt hours of energy per year, providing approximately 20% of the facility's electricity (ISEA, 2013). Additionally, for each year of operation the solar system will help protect the local environment by preventing the release of 76 tons of CO_2, while eliminating an estimated 440 pounds ($\sim$200 kg) of nitric oxide and 1,100 pounds ($\sim$500 kg) of acid rain forming sulfates. The payback period of the system is estimated at 10–12 years (Lothson, 2012).

Fig. 9.21. Solar panels atop the Avenue parking garage are paying dividends in terms of energy conservation (Lothson, 2012).

Fig. 9.22. Solar Sheds Light at Chicago Shedd Aquarium (Zipp, 2013; Anchor, 2013).

9.4.2.4. *Shedd Aquarium PV project*

In October 2013, the Illinois Governor unveiled the first phase of a $1.1 million energy-efficiency project at the John G. Shedd Aquarium in Chicago (see Fig. 9.22).

Since its inauguration in 1930 until 2012, the Shedd Aquarium facility has accommodated more than 32,500 aquatic animals, representing some 1,500 species of fishes, reptiles, invertebrates, amphibians, birds and aquatic mammals from waters around the world (IGNN, 2013). The installation of energy-efficiency solar electric system at the aquarium is an integral part of Shedd's Master Energy Roadmap. This roadmap is an ambitious energy initiative aimed at reducing 50% of the aquarium's energy consumption by 2020, and consequently, transforming it into the first clean energy-powered aquarium in the U.S. (IGNN, 2013). Besides costs reduction and enhancement of the building's energy reserve and various life support systems, the solar energy system supports Shedd's Great Lakes initiative by protecting and preserving the lakes through reduction of its reliance on traditional energy sources (IGNN, 2013). Launched in January 2013, the Shedd's Master Energy Roadmap designed an innovative 'smart building' prototype leading to this project. The project involves installation of 913 PV solar panels on top of the aquarium's world-class marine mammal pavilion. The 265 kW solar energy project is funded through public and private partnership, and is claimed to be the biggest solar installation in any cultural institution in the state of Illinois. The state of Illinois through its DCEO is investing $205,000 in the project, while Schneider Electric, a global specialist in energy management mentioned in Sec. 9.1, has provided its turnkey solution and support to help build the PV system. The other partner of the project is Illinois Clean Energy Community Foundation (IGNN, 2013).

9.4.2.5. *SSA office building PV project*

In 2005, the US-DOE and the U.S. Social Security Administration (SSA) dedicated a new solar array atop the SSA's office building located at 600 West Madison Street in Chicago downtown (see Fig. 9.23).

The SSA contracted with ComEd to complete the 82 kW solar PV modules covering the 7,920 square feet (736 m^2) of roof space on the SSA's building. The modules were claimed to be the largest

Fig. 9.23. Solar array dedicated at Chicago SSA (ISEA, 2005).

Federal solar energy installation and one of the largest solar electric systems in the Midwest when it was completed. According to the U.S. Environmental Protection Agency (EPA) calculations, the solar panels atop the SSA's building displace more than 4 million pounds (1,800 metric tons) of greenhouse gas (GHG) (such as CO_2) over its 30-year design life (i.e., equivalent to CO_2 absorbed by 20 acres of trees). Also, it is claimed that the system will save the equivalent of almost 6,000 barrels of oil (ISEA, 2005).

9.4.2.6. *Village of Downers Grove PV project*

In addition to commercial and residential applications, one potential application of standalone solar PV technology in Illinois is street lighting. Street lighting using solar PV technology is ideal for squares and cross roads in areas that are not connected to the power grid. A standalone PV solar system for street lighting consists of an advanced light source, a rechargeable battery, and PV modules for charging, including pulse-width modulation electronics (for the operation of the lamp and safe charging and discharging). Also, for equipment reliability, solar powered light-emitting diode traffic signal system can be connected to the grid to ensure interruption-free operation of the street lighting system. For this type of system, the connectivity to the grid will enable the signal to draw power from the grid and charge the battery when sunlight is unavailable (ABTI, 2012). In 2010, Downers Grove Village in Illinois completed installation of a hybrid solar and wind powered street lighting system. The 25 hybrid street lights was claimed to be the first of its kind in a residential subdivision in the U.S. (Kozlowski, 2010). It is also claimed that the system can save the village over 500,000 kW of electricity, which can result in carbon emissions reduction of almost 350 tons over a period of 30 years. The cost of the project was estimated at $282,500, with 45% partially funded with a 2009 Community Development Block Grant from the DuPage County Neighborhood Investment Program (Kozlowski, 2010).

9.4.3. *Utility-scale PV projects*

In general, utility-scale PV solar plants range from 10 MW to more than 200 MW in capacity. A 50 MW plant supplies enough energy to power 20,000 homes with clean, renewable energy. As of January 2013, a total of 90 MW of utility-scale solar energy projects were in Illinois. Out of this number, 30 MW was operating, while 60 MW was under development (SEIA, 2013). Among these projects are the 10 MW solar PV power plant located in the West Pullman Industrial Redevelopment Area in Chicago, the 20 MW Grand Ridge solar plant completed in 2012 in LaSalle County, Illinois and a 62 MW Rockford Solar Project under development in Rockford, Illinois (SEIA, 2013). These three projects are described briefly here:

9.4.3.1. *West Pullman PV plant*

As mentioned in Sec. 9.1, in 2010, the West Pullman solar energy plant operated by Exelon with generating capacity of 10 MW was claimed to be the nation's largest urban solar plant (see Fig. 9.24).

Fig. 9.24. Exelon City Solar is a 10-megawatt solar installation located on a 41-acre brownfield in Chicago's West Pullman neighborhood. It captures the sun's rays and converts them into 14,000 megawatt-hours of electricity per year. The facility was completed in 2010 (Exelon, 2014).

The Exelon City Solar is located on a 41-acre brownfield in Chicago's West Pullman neighborhood that was vacant for over 30 years. The plant consists of 32,292 solar PV panels which converts solar radiation into clean electricity. The annual electricity generation of the plant is more than 14,000 MWh, which is enough to meet the annual energy requirements of up to 1,500 homes. The power plant is equipped with SunPower® solar tracking device that can tilt the panels toward the sun as it moves across the sky, thus improving energy generation by as much as 25%. It is claimed that the Exelon City Solar plant displaces more than 1.36×10^4 metric tons (30 million pounds) of GHG emissions annually, the equivalent of removing over 2,500 cars from the road. The estimated cost of the Exelon City Solar project was approximately $60 million, creating over 200 construction jobs, with half of the work hours filled by Chicago residents (Exelon, 2014).

Fig. 9.25. Invenergy's Grand Ridge solar array is the perfect complement to the wind farms (Glantz Design, 2012).

9.4.3.2. *Grand Ridge solar park PV plant*

The Grand Ridge solar energy project in LaSalle County, Illinois, mentioned in the introduction (Sec. 9.1) was completed in 2012 (see Fig. 9.25).

The 20 megawatt project located on a 160-acre solar park has twenty 1 MW solar inverters and over 155,000 PV modules manufactured by General Electric. The solar plant is operated by ComEd, based on a 20-year power purchase agreement. It is claimed that the solar farm can produce 200 MWh per acre, estimated to power about 2,900 average homes annually, compared with wind turbines which produce 13,500 megawatt hours per acre (Borgia, 2011; Barichello, 2011). ComEd will use the solar energy-renewable certificates to meet Illinois state policy which requires that 0.5% of its renewable portfolio be provided by solar energy (Borgia, 2011).

9.4.3.3. *Rockford PV solar plant*

The 62 MW Rockford Solar Project is currently being constructed in Rockford, Illinois (see Fig. 9.26).

Fig. 9.26. Rockford solar project (Glantz Design, 2012; New Generation Power, 2014).

The Illinois DCEO, the administrator of the Illinois Solar Energy Program (SEP) chose Rockford Solar Partners (RSP), LLC as the recipient of a $4 million grant to construct the proposed 62 MW PV solar power generation facility in Rockford, Illinois (US-DOE, 2011). The proposed Rockford Solar Plant project is a joint venture between Wanxiang Investment Corporation of Rockford and New Generation Power, a Chicago-based renewable energy developer.

The EPA earlier approved a 20 MW project to be located on a property within the boundaries of the Rockford International Airport, which has over 70 acres of land available. However, RSP, LLC has committed to scaling up the project to 62 MW. The proposed plant will be operated by Ameren Illinois, based on a 20-year power purchase agreement with RSP, LLC (Stuart, 2010). The PV panels needed for the system will be supplied by a U.S. subsidiary of one of the largest non-state owned companies in China (Wanxiang America). The panels are manufactured at the company's Rockford-based solar panel assembly plant. The facility is claimed to generate enough electricity to power over 10,000 homes and reduce carbon dioxide emissions at the rate of 113,000 metric tons per year (Stuart, 2010). The phase I portion of the three-phase Master Development of this proposed 62 MW project was completed in 2012. The 3.06 MW solar

Table 9.10. Illinois utility-scale solar energy projects capacity and completion status as of November 18, 2014 (SEIA, 2013).

Project	Developer	Electricity Purchaser	Capacity (MW)	Status
W. Pullman Industrial Redevelopment Area	SunPower	Exelon Generation LLC	~10	Operating
Grand Ridge Solar Plant	Invenergy	ComEd	~20	Operating
Rockford Solar Project	Wanxiang America/ New Generation Power	Ameren Illinois	~60	Under Development (Phase 1: 3.06 operating)
Total capacity			**~90**	

PV plant located adjacent to the Chicago-Rockford International Airport (see Fig. 9.26, left) became operational since October 2012. The Second Phase of the 62 MW facility was scheduled for construction during the third quarter of 2015 (New Generation Power, 2014).

In Table 9.10, we present Illinois utility-scale solar energy project capacity and completion status.

9.5. Passive Solar Energy Technology Utilization

Passive solar energy utilization technologies are associated with building designs that incorporate the collection and transformation of solar energy for space heating and cooling, day lighting and natural ventilation, while keeping use of other energies to the minimum. Typically, passive solar systems are cheaper and easier to maintain than active solar systems. Additionally, passive solar systems are more reliable and can last longer than active solar systems. It is mentioned that passive solar systems are usually not as efficient as active systems. The average efficiency of a passive solar system (such as the Trombé wall described in Sec. 9.5.1.2) is defined as the heat delivered to the building, divided by the total solar radiation incident on the exterior of the wall (Torcellini and Pless, 2004). Below, we describe the different passive solar technologies suitable for Illinois.

9.5.1. *Types of passive solar energy technologies*

The complete design of passive solar systems includes a sun-facing glazed window, a double-façade wall construction, an absorber or a hard darkened surface element (to absorb the heat), thermal mass materials (that retain or store the heat produced by sunlight), a distribution system (through which solar heat circulates from the collection and storage points to different areas of the house) and a control roof overhangs (used to shade the aperture area during summer months) (US-DOE, 2000a). Proper use of trees and shrubs around buildings that grow leaves and produce shades in summer and allow the sunshine to enter the building during winter is also a passive solar energy utilization method. In general, passive solar systems distribute heat in living spaces by means of conduction, free convection and radiation. By conduction, heat is propagated through a medium (air, liquid, wall, or floor), as molecules of the medium close to the heat source gain enough thermal energy and begin to vibrate. In the convection process, thermal energy circulates through liquids and gases as the relatively warmer and lighter part of the fluid is displaced by cooler and denser fluid. In radiation heat generation, energy strikes a medium and becomes absorbed, reflected, or transmitted, depending on the properties of the object (AzSC, 2012). There are two types of radiations that are important in the design of passive solar systems. These are solar visible light radiation and infrared radiation. The use of glass in the design of passive solar systems is to allow the transmission of short wavelengths of visible light from solar radiation which is absorbed by the building. However, the longer wavelengths of the infrared radiation emitted by the walls and the floor of the building are unable

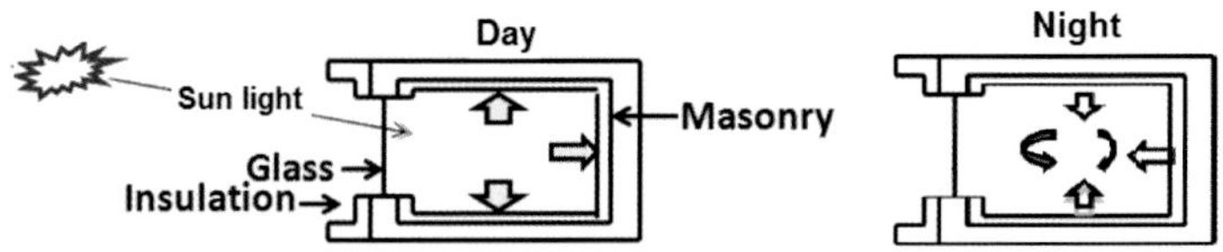

Fig. 9.27. Direct gain design — A direct gain design collects and stores solar radiation heat during the day on the inner walls. At night stored heat on the walls and floor is transferred into the living air spaces through convection (AzSC, 2012).

to pass through the glass, since glass is virtually impervious to high wavelength radiation and therefore, it is re-radiated to warm the interior of the building (AGC, 2013). Passive solar systems can be categorized as direct gain, indirect gain and isolated gain as described below.

9.5.1.1. *Direct gain design of passive solar systems*

The more common design of passive solar energy systems is the direct gain method shown in Fig. 9.27.

In this design, solar irradiation enters the south facing glass aperture and is converted into thermal energy inside the living space of the building. The thermal energy is stored by the walls and the floor of the building through direct interception of the solar radiation and/or through absorption of reflected or reradiated energy (Jaman *et al.*, 2010). At night, when outside temperature is dropping, the heat stored in the thermal mass is retransmitted into the room by convection process. This thermal energy stored in the daytime can be used to maintain a comfortable temperature inside the building during winter cold nights, when temperatures are low. For improved efficiency, the direct gain design must have about 50–66% of the total interior surface area of the building (including walls, floors and ceilings) constructed with thermal storage materials (AzSC, 2012). The most commonly used thermal storage materials are masonry (brick, concrete, adobe, etc.,) and water. In practice, water kept within plastic or metal containers and positioned in the direct path of the solar irradiation has a much better heating efficiency and heat distribution during the convection process, compared with masonry walls.

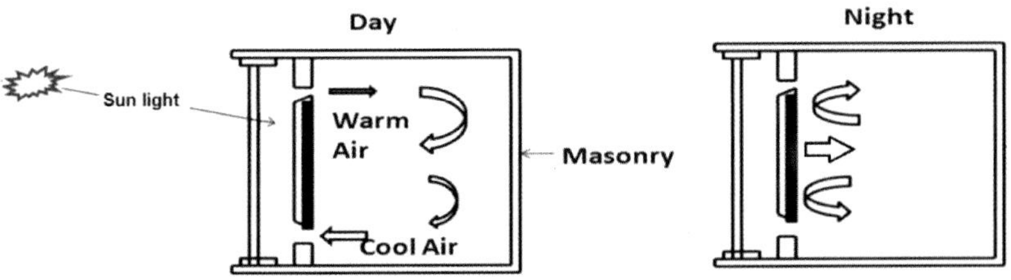

Fig. 9.28. Indirect gain Trombé wall stores heat during the day. Excess heat is vented to the interior space. At night Trombé wall vents are closed and the storage wall transfers heat into the interior space (AzSC, 2012).

9.5.1.2. *Indirect gain design of passive solar systems*

The indirect gain design collects, stores and distributes solar radiation using a thermal storage material such as Trombé wall (US-DOE, 2000a). A trombé wall is a masonry thermal energy storage and delivery system named after Felix Trombé (a French inventor) in the late 1950s (Torcellini and Pless, 2004). The design combines the elements of heat collection and storage with convection process. The thermal storage material is positioned just behind a south-facing glazing window, between the living space of the building and the solar irradiation to avoid direct heating of the room (see Fig. 9.28).

Solar radiation entering through the glass is adsorbed at the surface of the storage wall, where it is stored or possibly convected *via* the material mass to the inside space. But since solar radiation is electromagnetic wave, it gets adsorbed quite fast by a concrete material. This can lead to accumulation of heat and subsequent overheating of the Trombé wall at a temperature above 38°C. This situation will require a venting system to distribute the heat to the living spaces of the building through natural convection at nighttime (AzSC, 2012). Alternatively, a water wall can be placed between the solar radiation and the space within the interior of the building. Thus, as the water is heating up, heat is distributed within the water by convection. Then at night time, the heat in the water wall is transferred more efficiently to the air space.

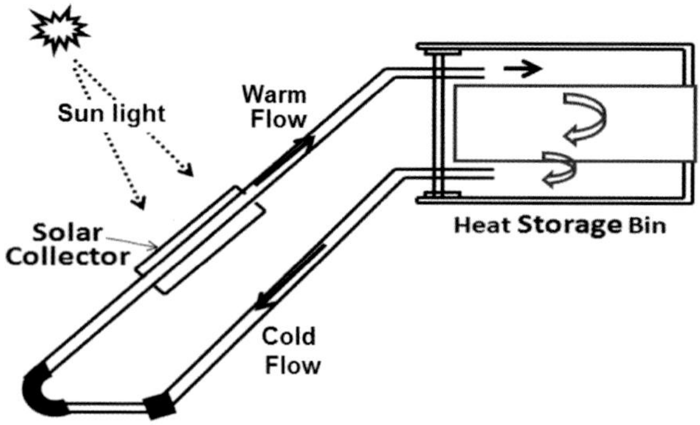

Fig. 9.29. Water or air convection loop (AzSC, 2012).

9.5.1.3. *Isolated gain design of passive solar systems*

The isolated gain systems collects solar radiation from an area that can be selectively closed off or opened to the rest of the house as shown in Fig. 9.29 (US-DOE, 2000a).

This design uses a fluid (air or liquid) to collect heat from a flat plate solar collector attached to a convection loop. The heat is transferred by natural convection *via* ducts or pipes to a storage bin (for air) or a tank (for liquid). Inside the storage area, the incoming heated fluid displaces the cooler fluid, forcing the latter to circulate through the convection loop to be heated to complete the heating cycle. When air is used as the heat transfer medium, the storage bin is normally filled with rocks to absorb the heat from the air (AzSC, 2012). During cold nights, heat radiation from the storage area is used to heat living spaces in the building through natural convection. To make sure the heat transfer in this process occurs by natural convection, proper positioning of the components of the system is very important.

9.5.1.4. *Passive solar cooling*

Passive solar cooling can easily be incorporated into the construction of a new building than modifying an existing building. Passive solar cooling is associated with building designs that can reduce the rate of heat transfer to the building, while improving the removal of unwanted heat in the summer. There are several design techniques that can be used to achieve passive solar cooling in buildings. One method is proper insulation of walls, ceilings and foundations to keep unwanted heat away from the building during the summer (US-DOE, 2001). Another method is the provision of adequate attic ventilation to reduce the interior heat gain in the sunny summer days. Additionally, window orientations can be used to allow just enough useful solar light during summer for the building to stay lighted but not heated. Furthermore, shading or planting of deciduous trees at the north and south of the building can provide shading to windows to cool homes during summer. Planting of deciduous shade tree is especially practical in Illinois because they can absorb a great deal of moisture during the humid days in the summer. However, in winter when the trees are without leaves, sunlight is allowed into the building for heating (Adkins, 1981). Figure 9.30 shows the shading technique for passive solar cooling.

Passive solar cooling systems described above can be economically attractive to Illinois, a state characterized by humid and hot summers (Adkins, 1981). In what follows, we report on the construction of four passive solar houses built in Illinois.

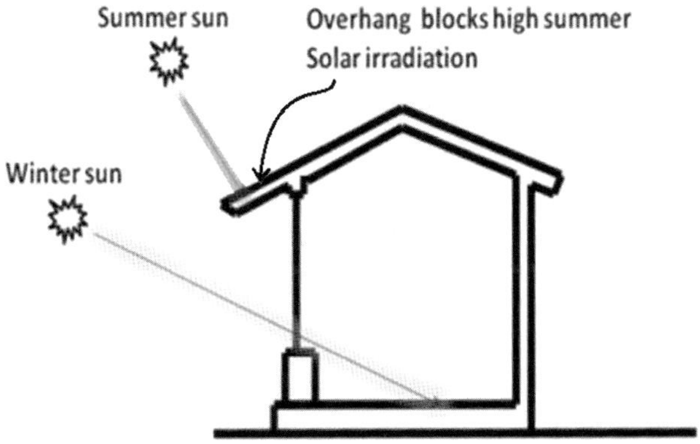

Fig. 9.30. Shading technique for passive solar coding (US-DOE, 2000b).

9.5.2. *Construction of passive solar houses*

A passive house is a set of performance standards representing 70–90% reduction of the energy consumption of a standard residential house. Between the years 1930 and 1940, George F. Keck, an Architect from Illinois and a pioneering designer of passive solar houses, designed an all-glass house known as the 'House of Tomorrow' for the 1933 Century of Progress Exposition in Chicago. Following the success of his passive solar design, Keck designed a passive solar home for real estate developer Howard Sloan in Glenview, Illinois in 1940 (see Fig. 9.31).

The Chicago Tribune referred to the Sloan House as 'solar house', which was claimed to be the first use of the term in contemporary society (Collins and Nash, 2002). Also, during the world's energy crisis in the 1970's, the US-DOE was actively involved in some of the original research works on passive houses. But in later years, as

Fig. 9.31. The 'House of Tomorrow' at the Century of Progress International Exposition, 1933–1934 (Carli Digital Collections, University of Illinois at Chicago, 2014).

the oil price in U.S. was declining compared to those in Europe, the passive house frenzy in the U.S. subsided. The later passive house designs developed in Germany in the late 1980s, originated from the super-insulation of buildings during the 1970s, many of which were built in North America (Kernagis, 2008).

In 1996, Wolfgang Feist, a German physicist, established the 'Passivhaus Institute' in Darmstadt in Germany, which led to the creation of European Passivhaus Standards which is also used in the U.S. (Collins and Nash, 2002). The passive house designs place emphasis on the economic benefits of improving the building envelope (i.e., an area that separates conditioned space from unconditioned space of the building) to avoid the use of furnace, as well as the importance of eliminating all thermal bridging in construction. Additionally, the passive house designs lay emphasis on compact designs that minimizes exterior surface area and extremely air tight construction, with small heating systems integrated into the ventilation air distribution system. Equally important in the design of passive houses, is the designation of specific performance targets, along with software calculation of the levels needed to reach those targets (Klingenberg, 2004).

The 'Passivhaus Institut Standard' currently adopted for Central Europe stipulates that an energy-efficient house can be certified as a passive house if it meets the following three targets:

(i) The design of the house must have an annual load demand not exceeding 15 kWh/m^2 (4755 Btu/ft^2) per year in cooling or heating energy, or a peak demand of 10 watts per square meter (0.93 watts/ft^2);
(ii) The total primary energy consumption for heating, hot water, and electricity must not exceed 120 kWh/m^2 (38,000 Btu/ft^2) per year; and
(iii) The building must not leak air in quantities more than 60% of house volume per hour at a pressure of 50 pascals (7×10^{-3} psi), as tested by a blower door (Bell, 2011). Such low pressure

air leakage requirement ensures that the building's location and its surrounding wind conditions will have a minimal impact on the heating/cooling requirement of the building (Krylov, 2010).

The passive house concept, which is of European design strategy, allows very low energy consumption in buildings, through optimization of the building design using better insulation to reduce the initial and operational costs of the building. The ultimate goal is to obtain sufficient reduction of cooling/heating loads, while maintaining comfortable temperatures in the house through passive means, such as solar gain, waste heat from occupants and appliances, and a small back-up cooling/heating system. With proper design, it has been noted that even without a back-up system, temperatures in a passive house will neither drop below 10°C in winter nor rise above 27°C in summer.

Below, we describe some of the model passive solar houses in Illinois.

9.5.2.1. *Lo-Cal House*

In 1974, the Lo-Cal House (Low Calorie House) in Illinois was designed by engineers and architects from the University of Illinois at Urbana-Champaign. Several houses based on this design were built in Urbana and Champaign, Illinois (Henry, 2012). It has been claimed that a young architect known as Mike McCulley, who worked with the University of Illinois Small Homes Council (currently known as the Building Research Council) at that time, built four Lo-Cal houses in Urbana and Champaign. These houses were subsequently monitored and evaluated by the council for performance, which drew attention from several newspapers across the U.S., including the Louisville Courier-Journal in 1982 (Klingenberg, 2013). The Lo-Cal house relied essentially on 'super insulation' and was notably one of the more successful passive house designs in the U.S. The Lo-Cal house design consists of a ceiling of at least 30.48 cm (12 inches) thickness, wall insulation of 21.59 cm (8.5 inches) thickness;

and a 15.24 cm (6 inches) thickness of insulation below the floor and along the border of the crawl space (Adkins, 1981). Included in the Lo-Cal House design are features such as double-glazed windows facing south, triple glazed windows facing north, a 76.2 cm (30 inches) roof overhang allowing winter sun (but excluding most of the summer solar irradiation), a window area less than or equal to 10% of the house floor space; and an east-west stretching rectangular shape designed to leave more space for south wall windows. In the 1980s, it was estimated that the Lo-Cal House design could reduce fuel bills by as much as two-thirds, without the need for solar heat-collection equipment. In the 1980s, the estimated cost of the super insulation and solar features of the Lo-Cal House was $3,000 (added to the regular cost of a new home). However, zoning code variances are sometimes required for this type of design, which could result in payments of hundreds or even thousands of dollars in additional expenses (Adkins, 1981). The code affects homes prices by setting the rules and procedures for the use of land, as well as the mass and the nature of buildings regarding the height, density, parking standards, etc. within the city.

9.5.2.2. *Smith's Passivhaus*

From 2002 to 2003, the Smith's Passivhaus in Urbana, Illinois was constructed in North America (see Fig. 9.32).

The Smith's *Passivhaus* is a single-family residence with a finished floor area of 111.5 m^2 (1200 ft^2). The roof insulation is made of blown-in fiber glass measuring 40.6 cm (16 inches) in thickness. The wall insulation is made of a 30.5 cm thick (12 inches thick) blown-in fiber glass with a 10.2 cm (4 inches) thick exterior polystyrene. The large amount of unshaded south glazing of the Smith Passivhaus was shaded by growing deciduous plants to prevent overheating of the house in summer (Parker, 2009). The annual total electricity consumption of the Smith Passivhaus was 39 kWh/m^2 (12,363 Btu/ft^2). It was claimed that from February 2005 to February 2006 (i.e., within

Fig. 9.32. The Smith House, in Urbana, Illinois, 'has a simple, compact shape that conserves energy,' says its architect, Katrin Klingenberg. Photo: Passive House Institute U.S. (Sauter, 2009).

a 12 month period), the annual electricity consumption for heating the Smith PassivHaus was 10.7 kWh/m^2 (~3400 Btu/ft^2), a value which was below the 15 kWh/m^2 (4755 Btu/ft^2) standard set by the 'Passivhaus Institut'. But at $1,015 per meter square ($94.3/ft^2), the price of the Smith's house was more than the highest average price for new construction houses in the region (Wostzak, 2009). It was estimated that the incremental cost of the house was approximately $18,000 or $162 per meter square ($15.05/ft^2).

Based on the highly predictable nature of the performance of PV systems, it can be demonstrated that with a 3.8 kW PV system around an area in Chicago, Illinois, the Smith Passivhaus would meet the Net Zero Energy Home requirement (i.e., a home with no energy bill) (Klingenberg, 2004). In general, the initial construction cost of a Passivhaus can be 10–20% higher than conventional well-built custom construction. But, the absence of a furnace and ductwork means an immediate cost savings. The payback time comes within the first decade from an annual heating bill which represents 10–15% of that of a conventional house (Bell, 2011).

Fig. 9.33. (Left) Model of south wall of Carbondale passive house during summer solstice and (Right) actual house (Mathias and Mathias, 2010).

9.5.2.3. *Carbondale passive house*

In 2009, a passive house was constructed in a mixed humid climate region in Carbondale, Illinois to demonstrate the cost effectiveness of a passive house (see Fig. 9.33).

The building construction was based on the best current construction practices using commonly available materials. The house was made up of insulated concrete forms for foundation walls, a 15 cm (~6 inches) thick insulated walls, insulated rim joist, a ground-source heat pump, Energy-Star windows, compact fluorescent bulbs, clothes washer and a refrigerator (Mathias and Mathias, 2010). To obtain good passive solar characteristics, the house was properly oriented to have many south-facing windows with proper overhangs. Thus 23% of the heating requirement of the house was obtained from the sun's energy. The electricity consumed by the house (i.e., for heating, air conditioning, hot water, lights and appliances) was compared to that of a similar house of the same size, which construction was based on the standards of International Energy Code Council 2004 Residential Energy Code. It was observed that the passive house used 7,809 kWh, which was 50% less energy than the house built using the standard code, representing an annual cost savings of $826 (Mathias and Mathias, 2010). The cost of the energy-efficient improvements in the passive house was $7,990, while the yearly on-site electricity usage by the house was 19.8kWh/m^2 (~6280 Btu/ft^2). The payback

Table 9.11. Comparison of the Smith's *Passivhaus* (see Sec. 9.5.2.2) in Urbana, Illinois and the Passive House in Carbondale, Illinois with performance standards.

	Performance Standards[1]	**Smith's Passivhaus**[2] **in 2003**	**Carbondale Passive House**[3] **in 2009**
Heating ($kWh/m^2/year$)	≤15	10.7[2]	19.8
Primary energy usage ($kWh/m^2/year$)	≤120	39[2]	59.3
Air tightness (@ 50 pascal)	≤0.6 ACH	0.6 ACH	NA
Total cost (US$)	NA	112,800[4]	$213,143
Floor area (m^2)	NA	111.5[4]	200.7
Cost per unit floor area ($/m^2)	NA	1012[4]	1062

(1) Bell (2011); (2) Parker (2009); (3) Mathias and Mathias (2010); (4) Wostzak (2009).
NA = not available; ACH = air changes per hour.

time of the passive house was 9.7 years. However, when interest is included without accounting for inflation, it results in a payback time of 10.5 years. The total cost of the passive house including all material, land, on-site general contractor and subcontractor costs resulted in a final price of $1062/m^2 ($99/ft^2) of finished floor area. A similar house in the same city would cost approximately $1345–1507 per square meter ($125/ft^2–140/ft^2) (Mathias and Mathias, 2010). In Table 9.11, we compare the Smith's Passivhaus (see Sec. 9.5.2.2.) and the Passive House in Carbondale with the passive house performance standards.

9.6. Problems Associated with Solar Energy Utilization

The major problems facing the solar energy industry, in general, include the relatively high cost of the solar systems, low capacity factor of solar power plants due to their variable power output, unavailability of capital for financing the technology, zoning rules, as well as the public lack of knowledge about the benefits of the technology and its application.

Despite the constant performance improvement and declining cost of solar PT and PV technologies during the past decade, the cost of solar power in most part of U.S., and specially in Illinois, is still not competitive enough, compared with conventional electricity generation using fossil fuels and nuclear power (IEA, 2010). For example, the U.S. total levelized cost of energy (LCOE) including subsidy in 2012 dollars of new generation resources for 2019 is projected at $118.6 per MWh for solar PV power and $223.6 per MWh for solar PT systems (US-EIA, 2014a).

The reason for the relatively high cost of solar PV power is the variable operation of solar PV power plants. For example, the 'capacity factor' of solar PV power plants is just 25%, compared to 85% for conventional coal-fired power plant or 90% for advanced nuclear power plant (US-EIA, 2014a). The capacity factor of a power plant is defined as the ratio of its actual output over a period of time to its potential output if it were to operate at full capacity indefinitely. The variable operating capacity of solar PV power plants results in low capacity factor. This low capacity factor translates into high electricity generation cost, requiring appropriate grid management and energy storage or backup power for off-grid installations (Wilson, 2013).

Another problem that can stall expansion of the solar energy industry is the restrictive policies for public use of land, as well as the opposition from some neighbors and environmentalists as will be discussed in Secs. 9.7.2 and 9.7.3. In early 2013, the City of Chicago created an expedited permitting process and introduced a new set of standards designed to facilitate and promote installation of solar PV panels in Chicago. The details of the new process and standards can be found in the City of Chicago Department of Building's New Photovoltaic Panel Permitting Guidelines. Additionally, the process offers training sessions to help solar PV design professionals to become familiar with specific requirements for zoning, as well as the electrical and structural aspects of PV installations (DHED, 2013).

Also, there is growing pessimism in the state that the mandatory benchmarks in the RPS will not be met in 2025, due the relatively small amount of renewable power generation in the state. This situation could require that the renewable power currently being purchased by utilities for Illinois customers through short-term contracts for renewable energy credits be purchased from other states in the future (Lydersen, 2013). To avoid this situation, Illinois energy experts have been asking for reforms in the state's RPS. However, with more renewable power on the market, some utility companies can anticipate keen competition and depressed prices (Lydersen, 2013).

As mentioned in Sec. 9.1, lack of funding can negatively impact solar power installation in the state. Between 2009 and 2011, Illinois received over $100 million from the US-DOE's State Energy program through the 2009 federal stimulus package. The funding was used to finance over 100 energy projects in the state, with $67 million allocated to solar, wind and other green projects, mostly to cover much of the projects' initial equipment costs (Rogal, 2011). The expiration of the federal funding in 2011 nearly brought solar power installations in the state to a halt.

Moreover, lack of public awareness about the various incentives for solar energy utilization in the state has resulted in fewer people using solar power. This situation can be reversed by using effective public education; however, the associated cost can be quite high.

9.7. Sustainability of Solar Energy

As mentioned in Chapter 1, economics, environmental issues and social acceptance are the three interconnected pillars of sustainability in any renewable energy, including solar energy, which must be taken into account for successful deployment of that technology. Therefore, to evaluate the overall burden of solar energy utilization systems, the economic analysis, impacts of solar PT and PV manufacturing emissions to the environment as well as and solar energy system water usage and land footprint must be evaluated.

9.7.1. *Economics of solar power conversion*

One of the major advantages of solar energy is that the energy source (sunlight) is free. Although this feature is not specific to solar energy, it lowers associated variable costs close to zero. The good feature of PT solar energy utilizations is production of heating and cooling without the release of any pollutants and with limited use of electricity. But the unique feature of PV solar energy utilizations is their ability to directly convert the energy of photons into electricity without release of any pollutants and without a need for any intermediate conversion process and long-distance transportation for residential PV systems between the source and the end use.

Some of the drawbacks of solar power generation are variability in power generation and intermittency. Variability of solar energy is due to its diurnal and seasonal cycles, while intermittency is due to unpredictable weather conditions. The cost associated with both solar PT and PV can be divided into two parts: module part and balance of system (BOS). BOS typically includes the cost of inverters, mounting hardware, labor, permit fees, etc. While the cost of modules is decreasing, BOS costs do not show the same trend. For residential and commercial solar systems, the BOS for PV systems can be up to 66% of the system cost (Baker *et al.*, 2013).

A study by the US-EIA (2013) evaluated the capital cost of various power generating technologies including solar power technology. It was shown that the capital cost of solar power technology was higher than natural gas-powered technologies, but lower than advanced nuclear and advanced pulverized coal equipped with carbon capture and sequestration (CCS), as reported in Table 9.12.

To measure the overall competiveness of different power generating technologies, an analysis of LCOE was performed by the EIA

Table 9.12. Comparison of capital cost for various power generating technologies (US-EIA, 2013).

Power Generating Technology	Nominal Capacity (MW)	Overnight Capital Cost ($/Kw)	Fixed O&M ($/kW/yr)	Variable O&M (Excluding Fuel) ($/MWh)
IGCC	600	4400	62.25	7.22
IGCC with CCS	520	6599	72.83	8.45
NG-ACC	400	1023	15.37	3.27
NG-ACC with CCS	340	2095	31.79	6.78
Solar PV w/tracking	150	3873	24.69	0
Solar PV	20	4183	27.75	0
Solar PT	100	5067	67.26	0

IGCC: Integrated Coal-Gasification Combined Cycle.
CCS: Carbon Capture and Sequestration.
NG-ACC: Natural Gas-Advanced Combined Cycle.

Table 9.13. Comparison of estimated LCOE* for power generating technologies (US-EIA, 2014a).

Power Generating Technology	Capacity Factor (%)	Levelized Capital Cost ($/MWh)	Fixed O&M ($/MWh)	Variable O&M (including fuel) ($/MWh)	Transmission Investment ($/MWh)	Total system LCOE ($/MWh)
IGCC	85	76.1	6.9	31.7	1.2	115.9
IGCC with CCS	85	97.8	9.8	38.6	1.2	147.4
NG-ACC	87	15.7	2	45.5	1.2	64.4
NG-ACC with CCS	87	30.3	4.2	55.6	1.2	91.3
Solar PV	25	114.5	11.4	0	4.1	130
Solar PT	20	195	42.1	0	6	243.1

(*)U.S. average levelized costs based on the value of USD in 2012 for plants entering service in 2019.
IGCC: Integrated Coal-Gasification Combined Cycle.
CCS: Carbon Capture and Sequestration.
NG-ACC: Natural Gas-Advanced Combined Cycle.

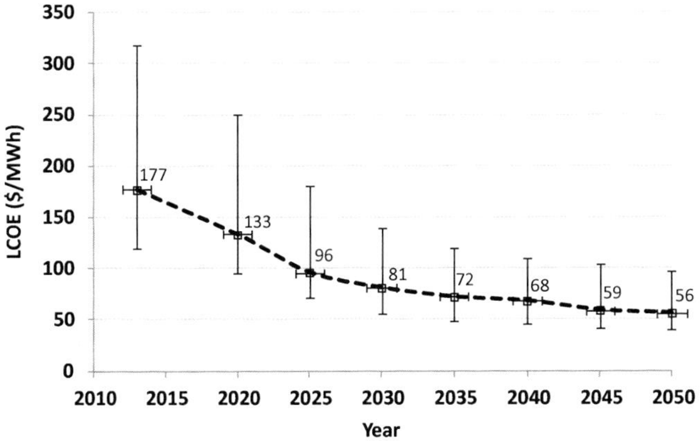

Fig. 9.34. Estimated LCOE for future new-built utility-scale PV plants (IEA, 2014).

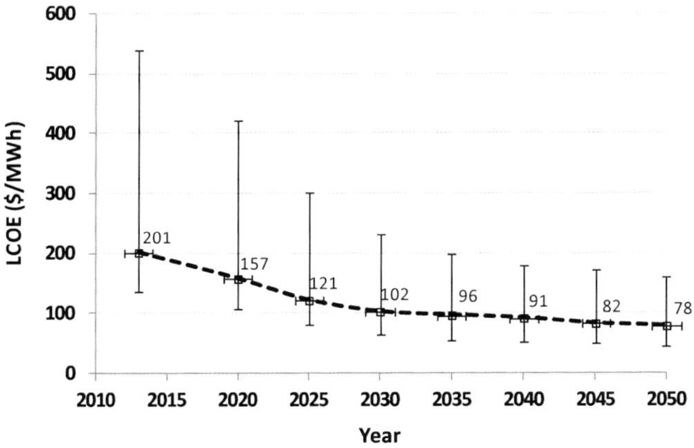

Fig. 9.35. Estimated LCOE for future new-built rooftop PV systems (IEA, 2014).

(2013). Based on this analysis, shown in Table 9.13, solar PV technology is only competitive with integrated coal gasification with CCS.

As of 2014, the cost per unit amount of electric generation, in $/MWh, of solar module is more than other power generating technologies. As the solar PT system seems much closer to its maturity, it is expected that the cost of new PV systems for both utility-scale and rooftop to decrease in the future (see Figs. 9.34 and 9.35).

The cost reduction of solar PV cells can be achieved by reducing the cost of raw materials (through more optimized processing) and increasing cell efficiency through advances in the technology. For example, SunPower technology claims to have reached an efficiency of 21% for its c-Si modules and is targeting 23% efficiency by 2015. Similarly, in April 2014, Panasonic announced a record efficiency of 25.6% for its heterojunction with intrinsic thin-layer cell (IEA, 2014). It is expected that by 2020, the average cost of newly built utility-scale and rooftop solar modules will drop 25% and 22%, respectively.

The cost of solar power generating systems varies with the type according to a life cycle analysis by Dale (2013), which indicated that among solar power technologies, CdTe thin film PV system has the lowest and PT power tower concentrating system has the highest cost.

Cost breakdown of 10 kW and 50 kW solar PV power systems was done in a study conducted by Friedman *et al.* (2014) and is shown in Table 9.14.

The 'soft costs' appearing in Table 9.14 include sales and marketing direct costs, cost of system-design, application for permitting with utility provider, inspection fees and interconnection process fees. Based on Friedman *et al.* (2014) analysis, soft costs are the

Table 9.14. Comparison of residential (10 kW) and small commercial (50 kW) PV system costs in ($/W) in U.S. Numbers in parenthesis are % of total cost (Friedman *et al.*, 2014).

		10 kW	50 kW
Hardware Cost	Inverter	0.38 (8)	0.22 (5)
	Module	0.76 (15)	0.8 (18)
	Other hardware	0.49 (10)	0.5 (11)
Soft Cost	Mech. install. labor	0.25 (5)	0.24 (5)
	Elect. Install. labor	0.23 (5)	0.2 (4)
	Other soft costs	2.82 (57)	2.55 (57)
TOTAL COST		**4.93 (100)**	**4.51 (100)**
Time Requirement (hours/kW)	for Mechanical installation	6.22	3.34
	for Electrical installation	3.76	5.93

major cost associated with both residential and commercial solar system, accounting for more than 65% of the total cost. As the size of solar PV system increases, the cost is expected to decrease. For utility-scale solar PV system the cost is expected to be 26% less than that of commercial-scale, reaching \$3.3/W (IEA, 2014). The most recent study by Hoen *et al.* (2015) analyzed the cost of PV system for 3951 homes across the U.S. and found that the average net cost (after incentives) for a 3.6 kW system was approximately \$4.14/W, while the gross cost (before incentives) was \$6.9/W. The study also found that the average gross cost of a PV system was higher for new homes (\$7.34/W) than for existing homes (\$6.65/W).

Availability of raw materials is an important factor affecting the cost of a solar panel. Securing access to minerals and precious rare earth metals used in solar cells is of vital importance. With emergence of new technologies, demand for some of the rare earth metals is expected to multiply. As an example, demand for gallium, which is used in thin film solar cells, is expected to increase six-fold by 2030. Among the transition metals used in solar cell manufacturing, only the demand for silver is expected to reduce by 2030, as shown in Table 9.15.

Total energy consumption in 2014 in the U.S. was estimated to be approximately 100 quadrillion Btu (quads) and 40% of the total primary energy was consumed in residential and commercial buildings (US-EIA, 2014b). The share of energy consumption for space and water heating in residential and commercial buildings are 63% and 32%, respectively (Hudon *et al.*, 2012). Out of the 110 million households in the U.S., 39% use electricity and 54% use natural gas for water heating (Cassard *et al.*, 2011). These statistics indicate the great potential market for alternative energy sources that can replace the current sources for electric or gas heaters in residential or commercial buildings. However, the current high cost of solar PT prevents successful market penetration of such systems. Based on the analysis done by Hudon *et al.* (2012), a factor of three to five reduction in the cost of solar PT water heating systems is required

Table 9.15. Estimated global demand for rare earth metals and minerals used in solar PV cells technologies in 2030 as a multiple of production in 2006 (Sathaye *et al.*, 2011).

Element	Application	Multiple
Gallium	TFS	6
Indium	Transparent electrode, TFS	3.3
Germanium	TFS	2.4
Platinum	DSSC	1.6
Silver	DSSC, Electric contacts	0.8
Selenium	TFS	NA.
Tellurium	TFS	NA.
Ruthenium	DSSC	NA.

DSSC = Dye-sensitized solar cell.
NA = Data not available.
TFS = Thin film semiconductor.

for this technology to be economically viable. A LCOE of solar PT water and air heater systems done by MCG consulting (MCG, 2013) for Minnesota Department of Commerce indicted that the average LCOE for residential solar water heating system is twice as electric and almost six times as natural gas water heaters. A summary of the results of MCG study is shown in Table 9.16.

Similar to PV technology, increase in efficiency of solar PT systems can reduce the cost and economically, make it a more viable option. Efforts are underway toward increasing the efficiency of solar PT using nanotechnology. Mahian *et al.* (2013) reviewed the effect of nanofluids on efficiency improvement of solar collectors and economic and environmental considerations of using such systems. Nanofluids are nano-size (1–100 nm) solid particles suspended in fluids in order to improve their thermal conductivity. It was found that addition of up to 2% volume fraction nanoparticles to the heat transfer fluid remarkably increases the efficiency of solar collector, but further addition of nanoparticles does not show added improvement of efficiency.

Table 9.16. Characteristics and cost estimates of solar thermal technology (MCG, 2013).

	Residential	Commercial	Residential heating
	Solar hot water	Solar hot water	Solar air
Gallons/day (Liters/day)	63 (238)	1,432 (5,420)	NA
Collectors area in ft^2 (m^2)	64 (6)	767 (71)	65 (6)
Storage in gal (liter)	81 (306)	969 (3,668)	NA
Heat delivered in 10^6Btu (MWh)	10 (3)	151 (44.2)	7(2.1)
Average cost in $/ft^2$ ($/m^2$)	182 (1,960)	132 (1,420)	97 (1,048)
Ave. LCOE in $/10^6$Btu ($/MWh)	71 (244)	47 (160)	53 (181)

9.7.2. *Land and water use in solar power technology*

One of the concerns of application of utility-scale solar power is its effect on land degradation and habitat loss. Land requirement for development of large-scale solar power generation is location-dependent, as topography and sunlight intensity varies from one site to another (Rule, 2014). It is estimated that utility-scale PV systems require 3.5–10 acres per MW (1.4–4 ha/MW) of land, while concentrating solar PT power (CSP-PT) systems land requirement ranges from 4–16.5 acres per MW (1.6–6.7 ha/MW) (Union of Concerns Scientists, 2014). Ong *et al.*, (2013) studied more than 100 small (>1 MW, <20 MW), large (>20 MW) and CSP plants to evaluate the land requirement for solar technologies. Their study indicated that not only the type of technology, but the choice of tracking system influences the land requirement, as addition of tracking system lessens the area of land requirement. A summary of the results obtained by Ong *et al.*, study is shown in Table 9.17.

Water use is an important issue and as Macknick *et al.* (2011) indicated, highly depends on the technology and choice of the cooling

Table 9.17. Land use requirement for solar power technologies (Ong *et al.*, 2013).

	Direct Area		Total Area	
	Weighted Average Land Use			
Solar Power Technology	**acres/ GWh/yr**	**GWh/yr/km^2**	**acres/ GWh/yr**	**GWh/yr/km^2**
Small PV (1–20 MW)	3.1	81	4.1	61
Large PV (>20 MW)	3.1	80	3.4	72
CSP-PT	2.7	92	3.5	71

Table 9.18. Comparison of water consumption for solar, wind and natural gas technologies in gal/MWh (liters/MWh) [Macknick *et al.*, 2011].

Power Source	Technology	Cooling	Median use	Maximum use
PV	Utility Scale PV	NA	26 (98)	33 (125)
CSP-PT	Trough	Tower	865 (3,274)	1057 (4,000)
		Dry	78 (295)	79 (300)
	Power Tower	Tower	786 (2,975)	860 (3,255)
		Dry	26 (98)	26 (98)
Wind	Wind turbine	NA	0	1 (3.8)
Natural gas	Combined cycle	Tower	198 (749)	300 (1,136)
		Dry	2 (7.6)	4 (15)
Nuclear	Fission	Tower	673 (2547)	845 (3198)
Coal	IGCC w/ CCS	Tower	540 (2,040)	558 (2,112)

system. For a PV system, operational use of water is at minimum since water is occasionally used for panel washing. For CSP-PT system, water is used for steam cycle processes and cleaning of mirrors. Depending on the cooling type, operational water use for CSP can vary more than one order of magnitude, as shown in Table 9.18.

Life cycle analysis of water use for several technologies was conducted in a study by Harto *et al.* (2010). In their analysis, Harto and

Table 9.19. Comparison of life cycle water use of solar power Generation technologies (Harto *et al.*, 2010).

	Average water use	
Process	**(gal/MWh)**	**(liter/MWh)**
Solar PV		
Quartz mining	0.5	2
Cell manufacturing	10	38
Other manufacturing	95	360
O&M	13	49
Transmission losses	10.5	40
TOTAL	**129**	**489**
Concentrated solar PT		
Cooling water	845	3198
Plant construction	50	189
O&M	11.5	44
Transmission losses	90	341
TOTAL	**996.5**	**3772**
Coal + carbon sequestration		
Coal mining	42.5	161
Coal washing	12.5	47
Coal transportation	35	132
Plant operation/cooling	850	3217
Plant construction	19	72
Transmission losses	90	341
TOTAL	**1049**	**3970**

co-workers considered water use for mining, manufacturing, operation and transmission losses of silicon-based solar PV cells and compared that with water use in CSP-PT and a coal power plant. A summary of their results is shown in Table 9.19.

As shown in Table 9.19, for solar PV system 73% of water is used for manufacturing of other components used in the system, while for CSP-PT system majority of water is used for cooling purpose. To lower water footprint, especially in arid areas, it would be beneficial to use other cooling methods, such as dry cooling.

9.7.3. *Environmental issues of solar cell manufacturing*

One of the great attributes of renewable energies, such as solar and wind energies is the fact that they do not emit toxic or GHG during their operation. While this is true, the same cannot be said about manufacturing of the components used in these systems.

Currently, most of the solar PV cells are silicon-based. The starting material of silicon is quartz, which has to be extracted from mines. Further steps convert quartz to metallurgical-grade silicon and then into a purer form called polysilicon. Manufacturing of high purity polysilicon requires use of hydrochloric acid in the process. Combining hydrochloric acid and metallurgical-grade silicon produces trichlorosilanes. Reaction of trichlorosilanes and hydrogen produces polysilicon and toxic liquid compound silicon tetrachloride. The amount of by-product silicon tetrachloride produced in the reaction is three to four times as the main product (Mulvaney, 2014). Similarly, production of CIG thin film solar PV cells requires use of metallic cadmium, which is highly carcinogenic and a weakly genotoxin (Bertin and Averbeck, 2006).

The above-mentioned examples necessitate a careful LCA for evaluation of environmental impacts of solar power generating systems, as large amount of GHG are emitted during solar PV cell manufacturing, PV module assembly, installation and system disposal, or recycling.

Kim *et al.* (2012) conducted a review of LCA studies of GHG emissions by several thin film PV technologies. Their analysis showed that after harmonization of major parameters of PV life cycle GHG emissions, emission rates of a-Si, CdTe and CIGS were 20, 14 and 26 gCO_2-eq/kWh, respectively, for groundmount system under solar radiation of 2400 kWh/m^2/yr.

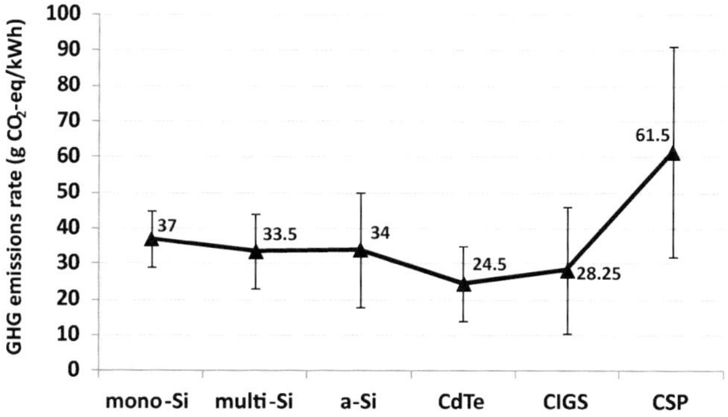

Fig. 9.36. Average GHG emission rates by various PV technologies and CSP-PT system (Peng *et al.*, 2013; Union of Concerns Scientists, 2014).

A similar review of LCA studies of GHG emissions by several PV technologies, including monocrystalline silicon (mono-Si), multicrystalline silicon (multi-Si), a-Si, CdTe and CIGS, was done by Peng *et al.* (2013). Their review indicated variations in GHG emissions not only by different technologies, but within each technology as well and in good agreement with the data reported by Kim *et al.* (2012). A summary of GHG emissions by different solar power technologies is shown in Fig. 9.36.

The rate of GHG emissions by a CSP system is the highest when compared with other solar technologies, but even at these levels, the rate of GHG emissions by solar power technologies is much less than GHG emissions by natural gas (272–907 gCO_2eq/kWh) and coal (635–1633 gCO_2eq/kWh) power technologies (Union of Concerns Scientists, 2014).

According to US-DOE (2013c), the state of Illinois electricity consumption grew at an annual average rate of 1.2% between 1980 and 2010. Assuming continuation of the same growth rate, it is estimated that Illinois electricity consumption will reach 163,000 GWh by 2025. RPS mandate requires Illinois to provide 1.5% of its electricity from solar and therefore, the state needs to generate 2,445 GWh of electricity by solar energy. Assuming average capacity factor for solar energy at 18%, Illinois needs 1,550 MW capacity of solar energy to meet its RPS mandate by 2025.

As of 2014, a total of 53 MW of solar energy has been installed in Illinois, enough to power 7,500 homes. Based on the above estimate and current installation, there is a huge gap of solar electricity power generation that needs to be filled by 2025. Such scenario illustrates the potential opportunities that exist in solar technology development in Illinois.

As mentioned in Sec. 9.1, the average insolation for Illinois is between 4–4.3 kWh/m^2/day. Denholm and Margolis (2007) reported the average solar intensity for New York and New Jersey at 4–4.1 kWh/m^2/day. According to NREL-GIS (2014), insolation in New York is less than New Jersey and Illinois. However, both states of New York and New Jersey have surpassed Illinois in their total installed solar energy. Trailing behind these two states is not related to limitations of solar intensity, and with public awareness and encouragement, Illinois can reach its RPS mandate by 2025.

In this section, we examined sustainability of solar energy by reviewing economic and environmental issues related to this renewable energy source. Although the economics of solar energy may not be viable without proper financial incentives, with advancement of technology the cost of solar power is expected to become competitive with fossil energy technologies in a near future.

9.8. Concluding Remarks

An important merit of solar energy utilization is its potential to mitigate GHG emissions. For example, as compared with coal, electricity generated by a PV system can eliminate about 1,000 tons of CO_2 per GWh, 10 tons of SO_2 per GWh, 4 tons of NO_x per GWh and 0.7 tons of particulate matter per GWh of electricity produced (Fthenakis and Moskowitz, 2000). Reduction of tremendous amount of GHG emissions due to fossil fuels combustion is one of the major factors in the growth of PT and PV industries. According to an estimate by the IEA (2014), global PV installed by the end of 2019 will be generating about 2% of global electricity or 530–580 terawatt-hours (TWh). There are issues with utilization of solar energy since its availability is not continuous and the cost is not competitive with traditional fossil energy. Despite the fact that solar energy generation is intermittent, this feature is not a serious disadvantage since solar energy utilization for electricity generation, heating and cooling is most productive during high-demand day-time hours (Baker *et al.*, 2013). The cost of solar technology continues to decrease as advances are made in processing and manufacturing of the necessary components.

Applications of nanotechnology or advanced organic molecules has opened new avenue in solar energy conversion. For example, recent work by Zhao *et al.* (2014) resulted in the development of a transparent solar cell that can be placed on windows. Although efficiency of this developed cell is low, researchers are optimizing the cell to raise the efficiency to 5%.

State of Illinois has the potential to become a major player in renewable energy development. Despite advancement in wind energy installations, the state lags behind other major U.S. cities in solar power installations. A recent report by Schneider and Sargent (2014) emphasized that Illinois has sufficient insolation to be utilized for solar power generation.

Solar energy is a renewable form of energy that can complement biofuel, wind and geothermal energy utilizations to help attain energy independence, cleaner environment and an overall sustainability for Illinois and any other community. Solar energy is an abundant heat and radiant energy resource on earth, which is inexhaustible and environmentally friendly. It is estimated that about 9.02×10^{23} calories (3.77×10^{24}J) of energy from the sun strike the earth surface every year, compared to 9.8×10^{19} calories (4.1×10^{20}J) of all the energy consumed on earth each year (US-DOE, 2005). As we mentioned in Sec. 9.1, Illinois total estimated technical potential for urban utility-scale solar PV power, for 1,325 km^2 land with 64GW power capacity, is 8.9×10^{16} calories per year (103,500 GWh per year). Additionally, the total estimated technical potential for rural utility-scale solar PV energy in the state for 103,524 km^2 land with 4,969 GW power capacity, is 6.96×10^{18} calories per year (8,094,480 GWh per year) (Lopez *et al.*, 2012). But, despite the abundance of this energy resource in the state, Illinois solar power cumulative installed capacity in 2014 was 53 MW. This installation was almost five times lower than New York and twenty times less than New Jersey (SEIA, 2014).

However, Illinois solar energy sector is making progress due to partnerships forged between private and government sectors. In 2005, the four Exelon Pavilions in Chicago, Illinois, was inaugurated as an integral part of the city's efforts to become a model of environmentally friendly city in North America (ISEA, 2013). The pavilions are capable of generating 19,840 kilowatt-hours of electricity each year, enough to power 14 star-rated energy-efficient houses in Chicago (Brownsey, 2005; City of Chicago, 2008). Also, as mentioned in Sec. 9.1, in 2010 the West Pullman neighborhood in Chicago had the largest urban solar installation in the U.S. The 10 MW solar power plant is located on a brownfield site and can power 1,500 homes. By the end of 2012, Illinois was leading the Midwest in solar energy resources development with a total of 90 MW of utility-scale solar energy projects. Out of these projects, 30 MW is currently operational, while 60 MW is under development (SEIA, 2013). According to the Solar Energy Industries Association, in 2014, the number of solar energy companies in the value chain in Illinois exceeded 184. These companies employed 2,100 people and provided a variety of solar products and services ranging from solar system installations to the manufacturing of components used in PV panels (SEIA, 2014).

Also, according to a 2013 research study, a total of 2,292 MW of solar PV installations will be needed in Illinois to achieve the goal of the solar carve-out rate of 6% of the RPS by the year 2025. Solar carve-out is a portion of the renewable energy needs in the state that is expected to be met by 2025 using solar power. It is anticipated that these installations will result in annual environmental benefits (in the form of GHG reductions) of 28,146 tons of nitrous oxide (N_2O), 33,062 tons of methane (CH_4) and 1,713,517 tons of carbon dioxide (CO_2) (Jo *et al.*, 2013).

Solar PT and passive solar utilization technologies can provide an alternative cost-effective solution to volatile natural gas prices, including traditional electricity usage. As mentioned in Sec. 9.3, early integration of solar thermal systems into the design phase of new construction of buildings will reduce costs and maximize system performance (City of Chicago, 2005).

Apart from electrical power generation, the other potential applications of solar energy in Illinois include desiccant cooling systems, absorption cooling chiller systems, passive solar heating and drying of agricultural and industrial products.

Furthermore, the International Energy Agency (IEA) has projected a 70% drop in the prices for turn-key PV solar systems from the current price of $4,000–6,000 per kW to $1,200–1,800 per kW by the year 2030, with a major price reduction of over 50% expected to be achieved by 2020 (IEA, 2010).

In Table 9.20, we present a summary of Illinois solar energy technical potential, generation and milestone.

Economically, prices for both PT and PV solar collector panels have been declining and the trend is expected to continue because of economies of scale and improved technology. The estimated U.S. average levelized cost of electricity (2012 dollars) for power plants entering service is 2019 is $130 per MWh for PV solar and $243 per MWh for PT solar (US-EIA, 2014a). Environmentally, solar energy utilization does not emit pollutants during operation, and the largest of all carbon-neutral energy sources is supplied by solar radiation (US-DOE, 2005).

We are so much used to seeing gas stations and parking lots. It is not unusual to join the two together and call them SCSP 'solar charging sheltered parking'!

Table 9.20. Summary of Illinois solar energy technical potential, current installations and milestone.

Total PV Solar Energy Estimated Technical Potential[1]	**(GWh/yr)**
Urban utility-scale	103,552
Rural utility-scale	8,090,985
Rooftop	30,086

Milestone for solar PV installations and current installations		
Anticipated year to reach solar carve-out of RPS [2]		2025
Solar carve-out rate of RPS [2] (% of all renewable)		6%
Total capacity of utility-scale solar projects in Illinois in 2012[4]		90.0 MW
Solar energy investments for home, business and utility applications in 2013[7]		\$6,000,000
Total solar PV installations needed to achieve the goal of solar carve-out rate of 6% RPS by 2025 [5]		2,292 MW
Trend in grid connected cumulative PV solar installed capacity (MWdc)]	in 2012[3]	42.9
	in 2013[6]	43.4
	in 2014[7]	53.0

Annual environmental benefits from GHG reductions using current carve-out of 6% RPS by 2025[5]	**(Tons/year)**
Nitrous Oxide (N_2O) Reduction	28,146
Methane (CH_4) Reduction	33,062
Carbon Dioxide (CO_2) Reduction	1,713,517
Fresh Water Saving (liters)	4.78×10^9

(1) Lopez *et al.* (2012); (2) Loomis and Pagan (2011); (3) Sherwood (2013); (4) SEIA (2013); (5) Jo *et al.* (2013); (6) Sherwood (2014); (7) SEIA, (2014). RPS = Renewable energy portfolio standard; GHG = Greenhouse gas.

Fig. 9.37. Solar charging sheltered parking (Source: http://inhabitat.com/envision-plans-to-install-2300-rotating-solar-tree-car-shelters-in-south-carolina/. Accessed on 2 May 2014).

Bibliography

[ABTI] Am. Building Technol. Inc. (2012). Energy saving lighting series, Renewable Energy Technical Brochure, pp. 1–8.. ABTI, Rockford, IL.

Adkins, G. (1981). Solar energy, *Illinois Issues*, Univ. of Illinois at Springfield, **7**(4): 12.

[AGC] AGC Glass Company, Inc. (2013). Your Glass Pocket — AGC. Alpharetta, GA. *us.agc.com*.

Agyarko, L. and Mansoori, G. A. (2010). Cooling and air conditioning through solar energy. In *Proceedings of Alternative Energy Symp.* Chicago, 7 October, IL.

Agyarko, L. and Mansoori, G. A. (2011). Solar absorption cooling with alkanes as phase change energy storage medium. In *Proceedings of AIChE 2011 Spring Meet. & 7th Global Congress on Process Safety*. Chicago, 14–17 March, IL.

Agyarko, L. B. and Mansoori, G. A. (2013). Non-renewable Energy Options in Illinois, *Int. J. Oil, Gas and Coal Technol.*, **6**(3): 288–347.

Anchor (2013). Project: John G. Shedd Aquarium, Anchor Products, LLC., Colleyville TX.

Argiriou, A., Klitsikas, N., Balaras, C. A. and Asimakopoulos, D. N. (1997). Active Solar Space Heating of Residential Buildings in Northern Hellas — a Case Study, *Energy and Buildings*, **26**(2): 215–221.

AzSC (2012). Passive Solar Heating & Cooling Manual: Part 2 of 4, Technol. & Sci. Database, Solar Architecture. Arizona Solar Center, Tucson, AZ.

Baker, E., Fowlie, M., Lemoine, D. and Reynolds, S. S. (2013). The Economics of Solar Electricity, *Ann. Review of Resource Economics*, **5**: 387–426.

Barichello, D. (2011). Streator solar farm construction begins. News section, *The Times*, Ottawa, IL.

Barton, S. A. and Winger, I. L. (2009). Parabolic-dish solar concentrators of film on foam, Cornell Univ. Lib., Ithaca, NY.

Bell, T. E. (2011). Aggressive engineering for passive houses . *The Bent of Tau Beta Pi*, Knoxville, TN. *tbp.org/pubs*.

Bertin, G. and Averbeck, D. (2006). Cadmium: Cellular Effects, Modifications of Biomolecules, Modulation of DNA Repair and Genotoxic Consequences (A Review), *Biochimie*, **88**(11): 1549–1559

Borgia, K. (2011). Invenergy's Grand Ridge solar plant begins construction. News section, *Wind On The Wires*, St. Paul, MN.

Brownsey, A. (2005). The Millennium Park Welcome Center opens in the Northwest Exelon Pavilion, Exelon Corp., Chicago, IL.

Bube, R. (2012). *Fundamentals of Solar Cells: Photovoltaic Solar Energy Conversion*. Academic Press.

Burger, B., Kiefer, K., Kost, C., Nold, S., Philipps, S., Preu, R., Schindler, R., Schlegl, T., Stryi-Hipp, G., Willeke, G., Wirth, H., Brucker, I., Haberle, A., Schacht, V. and Warmuth, W. (2014). Fraunhofer Institute for Solar Energy Systems ISE-Photovoltaics Report, Fraunhofer-Gesellschaft , Munich, Germany.

Business Wire (2012). GE Solar Tech. to Bring Power of the Sun to Invenergy's Grand Ridge Solar Project. News section, *Business Wire*, San Francisco, CA.

Butti, K. and Perlin, J. (1980). *A Golden Thread: 2500 Years of Solar Architecture and Technology*. Cheshire Books, UK.

Cassard, H., Denholm, P. and Ong, S. (2011). Break-even cost for residential solar water heating in the United States: Key drivers and sensitivities, Report No. NREL/TP-6A20-48986, NREL, U.S. DOE, Golden, CO.

[CEA] Clean EnergyAuthority (2014). Illinois solar rebates and incentives. CEA, Evergreen, CO.

City of Chicago (2005). Solar thermal collector systems in Free eBooks Download—Farmer: "Solar collector". *farmer3.org/solar-collector.*

City of Chicago (2014). Millennium park — art & architecture: Exelon pavilions, Chicago, IL. *cityofchicago.org.*

Collier, R. K., Barlow, R. S. and Arnold, F. H. (1982). An Overview of Open-cycle Desiccant-cooling Systems and Materials, *J. Solar Energy Eng.*, **104**(1): 28–34.

Collins, J. and Nash, A. (2002). Preserving yesterday's view of tomorrow the Chicago world's fair houses. *CRM Magazine*, **25**(5): 27–37.

[CPL] Chicago Public Library (2013). Millennium Park, Inc., Exelon pavilions series. CPL Special Collections and Preservation Div., Chicago, IL.

Craig, A., Learner, H. P. and Dobies, J. (2011). The clean energy supply chain in Illinois: Wind, solar and geothermal, Environ Law and Policy Center, Chicago, IL.

[CUB] Citizens Utility Board (2013). Net Metering, C.U.B., Chicago, IL. *citizensutilityboard.org*.

Currentresults.com (2014). Days of sunshine per year in Illinois", Current Results, BC, Canada.

Dale, M. (2013). A Comparative Analysis of Energy Costs of Photovoltaic, Solar thermal, and Wind Electricity Generation Technologies, *Appl. Sci.*, **3**: 325–337.

Denholm, P. and Margolis, R. (2007). The regional per-capita solar electric footprint for the United States, Report No. NREL/TP-670-42463, NREL, U.S. DOE, Golden, CO.

[DHED] Dept of Housing and Economic Development (2013). City of Chicago solar zoning policy. DHED, City of Chicago, IL. *cityofchicago.org*.

Dong, Q., Fang, Y., Shao, Y., Mulligan, P., Qiu, J., Cao, L. and Huang, J. (2015). Electron-hole Diffusion Lengths > 175 mm in Solution-grown $CH_3NH_3PbI_3$ Single Crystals, *Science*, **347**(6225): 967–969.

[EESI] Environmental and Energy Study Institute (2014). Solar energy. EESI,Washington, DC. *eesi.org/topics/solar.*

Einstein, A. (1905). Concerning an Heuristic Point of View Toward the Emission and Transformation of Light, *Ann. Phys.* **17**: 132–148.

EuroTrough. (2007). Concentrated solar power from research to implementation, ISBN 978-92-79-05355-9, Euro. Commission, Luxembourg.

Exelon (2014). Exelon city solar, Power plants. Exelon Corp., Chicago, IL.

Fitzgerald, K. (2011). Renew your home with solar and wind—now is the time to invest in renewal. *Illinois Times*, Springfield, IL.

Friedman, B., Margolis, R. and Seel, J. (2014). Comparing photovoltaic (PV) costs and deployment drivers in the Japanese and U.S. residential and commercial markets, Report No. NREL/TP-6A20-60360, NREL, U.S. DOE, Golden, CO.

Fthenakis, V. M. and Moskowitz, P. D. (2000). Photovoltaics: Environmental, Health and Safety Issues and Perspectives, *Prog. Photovolt. Res. Appl.*, **8**(27): 27–38.

Garud, S. (2008). Solar thermal energy technologies for industrial applications — India's experience: 2008. In *Proceedings of Global Renewable Energy Forum*. 18–21 May, Brazil.

Glantz Design (2012). Illinois solar is hot, Ecowren, Winnetka, IL.

Gomez, A. L. and Mansoori, G. A. (1983). Thermodynamic Equation of State Approach for the Choice of Working Fluids of Absorption Cooling Cycles, *Solar Energy*, **31**(6): 557–566.

Gommed, K. and Grossman, G. (2007). Experimental Investigation of a Liquid Desiccant System for Solar Cooling and Dehumidification, *Solar Energy*, **81**(1): 131–138.

Gosselar, J. and Johnson, M. (2011). Issue Briefs: Solar thermal energy for industrial uses, Env, Energy Study Inst., Washington, DC.

Grätzel, M. (2014). The Light and Shade of Perovskite Solar Cells, *Nature Materials*, **13**: 838–842.

Guy, S. (2011). Invenergy plans solar farm in Illinois. News section, *Chicago Sun-Times*, Chicago, IL.

Harto, C., Meyers, R. and Williams, E. (2010). Life Cycle Water Use of Low-carbon Transport Fuels, *Energy Policy*, **38**: 4933–4944.

Henry, M. (2012). The common origins of superinsulation, Passivhaus, and Net Zero homes. News section, *The Sustainable Home. thesustainable-home.net.*

Hillhouse, H. and Jen, A. K. Y. (2014). The next revolution in solar energy: High efficiency printable tandem solar cells, Clean Energy Inst., Univ. Washington, Seattle, WA.

Himran, S., Suwono, A. and Mansoori, G. A. (1994). Characterization of Alkanes and Paraffin Waxes for Application as Phase Change Energy Storage Medium, *Energy Sources*, **16**(1): 117–128.

Hoen, B., Adomatis, S., Jackson, T., Graff-Zivin, J., Thayer, M., Klise, G. T. and Wiser, R. (2015). Selling into the sun: Price premium analysis of a multi-state dataset of solar homes, LBL, Berkeley, CA.

Holabird (2008). Oak Park Public Works Building striving for LEED Silver, Design Cost Data, Holabird, Chicago, IL.

Hudon, K., Merrigan, T., Burch, J. and Maguire, J. (2012). Low-cost solar water heating research and development roadmap, Report #NREL/TP-5500-54793, NREL, U.S. DOE, Golden, CO.

[IAG] Illinois Attorney General (2010). Net metering and interconnection. Office of Ill. Att. Gen., Chicago, IL.

[IDCEO] Illinois Dept of Commerce and Economic Opportunity (2012). Renewable energy resources program report: January through December 2011. IDCEO, Springfield, IL.

[IDCEO] Illinois Dept of Commerce and Economic Opportunity (2014). Solar and wind energy rebate program. IDCEO, Springfield, IL.

[IEA] Intl Energy Agency (2010). Technology Roadmap: Solar photovoltaic energy. IEA, Paris, France.

[IEA] Intl Energy Agency (2014). Technology Roadmap: Solar photovoltaic energy. IEA, Paris, France.

[IGNN] Illinois Gov. News Network (2013). Governor Pat Quinn, Shedd announce largest solar panel installation at cultural institution in Illinois. IGNN, Springfield, IL.

[ISEA] Illinois Solar Energy Assoc. (2005). Solar array dedicated at Chicago Social Security Administration,ISEA, *Heliographs*, **24**(4): 5–12.

[ISEA] Illinois Solar Energy Assoc. (2013). SPI13 Chicago solar tour — Village of Oak Park parking facility. ISEA, Wheaton, IL.

Jalayerian, M. and Eich, S. (2008). More than skins deep: A case study North Exelon pavilion. *High Performing Buildings Magazine*.

Jaman, M. T., Rahman, M. M., Rahman, A. M. and Khulna, B. (2010). Design of a direct gain passive solar heating system. In *Proceedings of 2010 Intl Conf. on Industrial Eng and Operations Management*. Dhaka, 9–10 January, Bangladesh.

Jo, J., Loomis, D. and Aldeman, M. (2013). Technical potential for solar photovoltaics in Illinois, Center for Renewable Energy, Ill. State Univ., Normal, IL.

Kalogirou, S. A. (2013). *Solar Energy Engineering: Processes and Systems Hardcover*, 2nd Edition. Academic Press.

Kernagis, M. (2008). Passive house takes root in the United States, Home Energy, Berkeley, CA.

Kim, H. C., Fthenakis, V., Choi, J-K. and Turney, D. E. (2012). Life Cycle Greenhouse Gas Emissions of Thin-film Photovoltaic Electricity Generation, *J Ind Ecol.*, **16**(S1): S110–S121.

Klingenberg, K. (2004). Energy Design Update: An Illinois Passivhaus, *Energy Design Update*, **24**(5): 1–5

Klingenberg, K. (2013). Passive house history (PHistory) Part 1-North American roots, The Klingenblog.

Kornher, S. and Zaugg, A. (1984). *The Complete Handbook of Solar Air Heating Systems*. Rodale Press, Emmaus, PA.

Kozlowski, D. (2010). Downers Grove, IL installs solar and wind powered street light system, Intl City/County Management Assoc. (IMCA), Washington, DC.

Krylov, I. (2010). Cost optimization of passive house external walls, B.Sc. Thesis, Saimaa Univ. of Applied Sci. Technol., Lappeenranta, Finland.

Lewis, N. S. and Nocera, D. G. (2006). Powering The Planet: Chemical Challenges in Solar Energy Utilization, *PNAS*, **103**(43): 15729–15735.

Liao, B. and Hsu, W-C. (2012). An investigation of Shockley-Queisser limit of single p-n junction solar cells, Bolin Liao's Notes, 2.997 Project Report, MIT.

Loomis, D. and Pagan, A. (2011). The Illinois RPS: Context, structure and history of the policy, Center for Renewable Energy, Ill. State Univ., Normal, IL.

Lopez, A., Roberts, B., Heimiller, D., Blair, N. and Porro, G. (2012). U.S. renewable energy Tech. potentials: A GIS-based analysis, Report No. NREL/TP-6A20-51946, NREL, U.S. DOE, Golden, CO.

Lothson, A. (2012). Solar panels on parking garage will offer Oak Park long-term savings, Oak Park, IL. *oakpark.com*.

Lydersen, K. (2013). Fix for Illinois renewable energy law faces. *Midwest Energy News*, St. Paul, MN.

Macknick, J., Newmark, R., Heath, G. and Hallett, K. C. (2011). A review of operational water consumption and withdrawal factors for electricity generating technologies, Report No. NREL/TP-6A20-50900, NREL, U.S. DOE, Golden, CO.

Maehlum, A. M. (2014). Best thin film solar panels — amorphous, cadmium telluride or CIGS?Energy Informative. *energyinformative.org*.

Mahian, O., Kianifar, A., Kalogirou, S. A., Pop, I. and Wongwises, S. (2013). A Review of the Applications of Nanofluids in Solar Energy, *Int J Heat Mass Tran.*, **57**(2): 582–594.

Mansoori, G. A. (1976). Prospects for Solar Energy Utilization — Photothermal Methods, *IJST*, **5**(2): 55–73.

Mansoori, G. A. and Patel, V. (1979). Thermodynamic Basis for the Choice of Working Fluids for Solar Absorption Cooling Cycles, *Solar Energy J.*, **22**(6): 483–491.

Mansoori, G. A. (2013). Entropy Production Gives Rise to Upper- and Lower-bounds to Efficiency and COP of Cycles, *J. Thermodynamics & Catalysis*, **4**(2): 1–5.

Mathias, J. A and Mathias, D. M. (2010). Energy Efficient, Cost Effective, Passive Solar, *ASHRAE Transactions*, **115**(1): 419–426

[MCG] Meister Consultants Group, Inc. (2013). The value of solar heating and cooling (SHC) in Minnesota, Contract No. 70685. MCG, Inc., Boston, MA.

Mei, V. C., Chen, F. C., Lavan, Z., Collier, R. K. and Meckler, G. (1992). An assessment of desiccant cooling and dehumidification Technology, Doc.No. ORNL/CON-309, ORNL, Oak Ridge, TN.

Mesik, J. and Howery, D. (2009). Energy Conservation at the Oak Park Public Works Center, *ASHARE Transactions*, **115**(1): 415.

Mulvaney, D. (2014). Solar energy isn't always as green as you think, IEEE Spectrum, New York, NY.

Munters Corp. (2009). Humidity problem solved at Illinois State archives. Munters Corp., Kista, Sweden.

[NCSC] North Carolina Solar Center (2000). Space heating with active solar energy systems, Solar Center Info. NCSC, Raleigh, NC.

New Generation Power (2014). Rockford solar. NGP Co., Chicago, IL.

[NREL] Natl Renewable Energy Lab. (1998). Solar trough systems, Fact sheet, Doc. No. DOE/GO-10097-395. NREL, U.S. DOE, Golden, CO.

[NREL] Natl Renewable Energy Lab. (2003). Consumer's guide: Get your power from the sun, Report No. DOE/GO-102003-1844. NREL, U.S. DOE, Golden, CO.

[NREL] Natl Renewable Energy Lab. (2014). Concentrating solar power projects. NREL, U.S. DOE, Golden, CO.

[NREL-GIS] Natl Renewable Energy Lab., Geographic Info. System (2014). Dynamic maps, GIS data and analysis tools. NREL, Golden, CO.

Ong, S., Campbell, C., Denholm, P., Margolis, R. and Heath, G. (2013). Land-use requirements for solar power plants in the United States, Report No. NREL/TP-6A20-56290, NREL, U.S. DOE, Golden, CO.

Parker, D. S. (2009). Very low energy homes in the United States: Perspectives on Performance From Measure Data, *Energy Buildings*, **41**(5): 512–520.

Peng, J., Lu, L. and Yang, H. (2013). Review on Life Cycle Assessment of Energy Payback and Greenhouse Gas Emission of Solar Photovoltaic Systems, *Renewable and Sustainable Energy Reviews*, **19**: 255–274.

Perez, R. (2006). Off-grid Inverter Efficiency, *Home Power Magazine*, **113**: 36–37.

PESI (2014). Recently Installed Residential Solar Systems: Residential Installations. Pure Energy Solar Intl, Inc., Gainesville, FL.

Philibert, C. (2006). Barriers to Technology diffusion: The case of solar thermal Technologies, Doc. No. COM/ENV/EPOC/IEA/SLT (2006) 9, OECD, IEA, Paris, France.

Phillips, W. F. (1981). Integrated Performance of Liquid-based Solar Heating Systems, *Solar Energy*, **26**(4): 287–295.

[PJM and EIS] PJM & Environ Info. Services (2013). Comparison of Renewable Portfolio Standards (RPS) Programs in PJM states. PJM & Environ Info. Services.

Pluta, Z. (2011). Evacuated Tubular or Classical Flat Plate Solar Collectors? *J Power Technol.*, **91**(3): 158–164.

PRN (2003). Most powerful solar energy system in Illinois is dedicated: Solar service and state officials celebrate earth day. PR Newswire. *prnewswire.com.*

PRN (2010). Governor Patrick Quinn signs law to ramp up solar energy in Illinois. PR Newswire. *prnewswire.com.*

RAEE (2014). Solar cooling: solid desiccant cooling with rotating wheels. Regional Agent of Energy and the Environment. *raee.org.*

Renewable Energy World (2003). Laundromat incorporates solar thermal. *Renewable Energy World*, Nashua, NH.

Rex Electric (2009). Rex Electric & Technologies dedicated to going green. Rex Electric Co., Chicago, IL.

Reyes, J. and Rosen, M. (2007). Active solar energy technologies-existing and new development, New Jersey Dept of Env. Protection, Trenton, NJ.

Rogal, B. (2011). High-profile Illinois solar project languishes due to lack of funding. *Midwest Energy News*, St. Paul, MN.

Rule, T. A. (2014). *Solar, Wind and Land-Conflicts in Renewable Energy Development*. Routledge, Taylor & Francis Group, New York, NY.

Sathaye, J., Lucon, O., Rahman, A., Christensen, J., Denton, F., Fujino, J., Heath, G., Kadner, S., Mirza, M., Rudnick, H., Schlaepfer, A. and Shmakin, A. (2011). Renewable energy in the context of sustainable development. In *IPCC Special Report on Renewable Energy Sources and Climate Change Mitigation*, Edenhofer, O., Pichs-Madruga, R., Sokona, Y., Seyboth, K., Matschoss, P., Kadner, S., Zwickel, T., Eickemeier, P., Hansen, G., Schlomer, S., von Stechow, C. (eds.). Cambridge Univ. Press, Cambridge, UK.

Sauter, U. (2009). Passive house: Super energy efficient housing. *Ode Magazine*, San Francisco, CA.

Schneider, J. and Sargent, R. (2014). Lighting the way, The top ten states that helped drive America's solar energy boom in 2013, Env. America Research & Policy Center, Boston, MA.

[SEDP-EIS] Solar Energy Development Programmatic EIS (2014). Concentrating solar power (CSP) Technologies. SEDP-EIS, Washington, DC.

Seely, O. (2013). Some observations on photovoltaic cell panels, Cal. State U., Dominguez Hills, CA.

[SEIA] Solar Energy Indus. Assoc. (2013). Major solar project list, Research and Resources. SEIA, Washington, DC. *seia.org*.

[SEIA] Solar Energy Indus. Assoc. (2014). Illinois solar, State solar policy. SEIA, Washington, DC. *seia.org*.

Shah, A. (ed.) (2010). *Thin-film Silicon Solar Cells*. CRC Press.

Sherwood, L. (2007). U.S solar market trends 2006, Interstate Renewable Energy Council, Inc, Latham, NY.

Sherwood, L. (2008). U.S. solar market trends 2007, Interstate Renewable Energy Council, Inc, Latham, NY.

Sherwood, L. (2009). U.S. solar market trends 2008, Interstate Renewable Energy Council, Inc, Latham, NY.

Sherwood, L. (2010). U.S. solar market trends 2009, Interstate Renewable Energy Council, Inc, Latham, NY.

Sherwood, L. (2011). U.S. solar market trends 2010, Interstate Renewable Energy Council Inc., Latham, NY.

Sherwood, L. (2012). U.S. solar market trends 2011, Interstate Renewable Energy Council, Inc., Latham, NY.

Sherwood, L. (2013). U.S. solar market trends 2012, Interstate Renewable Energy Council, Inc., Latham, NY.

Sherwood, L. (2014). U.S. solar market trends 2013, Interstate Renewable Energy Council, Inc., Latham, NY.

Shockley, W. and Queisser, H. J. (1961). Detailed Balance Limit of Efficiency of p-n Junction Solar Cells, *J. of Applied Physics*, **32**: 510–519.

Sinke, W. (2014). Cost-effective high-efficiency 1-sun PV modules. In *Proceedings of IEA PV Roadmap Workshop*. Paris, 2 April, France.

Stine, W. B. and Geyer, M. (2001). Solar Collectors, Chapter 5. In *Power From The Sun. powerfromthesun.net*.

Stuart, B. (2010,). U.S.: Over $4 million grant awarded for 62 MW project. News section, *PV Magazine*, Photovoltaic Markets & Tech.

[SunLab] (2001). Solar dish-engine systems, Doc. No. SAND2001-2534P. SunLab, NREL/Sandia National Lab., U.S. Department of Energy, Washington, DC.

[TCE] (2006). Governors State University harnesses the power of the sun. News section, The Chief Engineer, Crestwood, IL. *chiefengineer.org*.

Tchernev, D. I. (2001). Natural Zeolites in Solar Energy Heating, Cooling, and Energy Storage, *Reviews in Mineralogy and Geochemistry*, **45**(1): 589–617.

Tweed, K. (2014). Can distributed storage help expand Midwest solar markets? *Midwest Energy News*, St. Paul, MN.

Torcellini, P. and Pless, S. (2004). Trombe walls in low-energy buildings: Practical experiences, Doc. No. NREL/CP-550-36277, National Renewable Energy Lab., Golden, CO.

[UIC] Univ. of Illinois Chicago (2014). Century of Progress World's Fair, 1933–1934, CARLI Digital Coll., UIC, Chicago, IL.

Union of Concerns Scientists (2013). Environmental impacts of solar power, Cambridge, MA. ucsusa.org.

[US-CB] U.S. Census Bureau (2013). Population Estimates. U.S. Dep't Commerce, Washington, DC. *census.gov.*

[US-DOE] U.S. Dept of Energy (2000a). Passive solar design: Increase energy efficiency and comfort in homes by incorporating, Factsheet, Doc. No. DOE/GO10099-790. U.S. DOE, Washington, DC.

[US-DOE] U.S. Dept of Energy (2001). Passive solar design for the home, Doc. No. DOE/GO-102001-1105 FS12, NREL. U.S. DOE, Washington, DC.

[US-DOE] U.S. Dept of Energy (2005). Basic research needs for solar energy utilization, Report on the basic energy science workshop on solar energy utilization, Office of Sci. U.S. Dept of Energy, Washington, DC.

[US-DOE] U.S. Dept of Energy (2009). The history of solar energy. U.S. Dept of Energy, Washington, DC.

[US-DOE] U.S Dept of Energy (2011). Final environmental assessment for Rockford Solar energy project: Chicago-Rockford International Airport, Doc. No. DOE/EA-1823. U.S. Dept of Energy, Washington, DC.

[US-DOE] US Dept of Energy (2012). Illinois incentives/policies for renewables & Efficiency: Local option — contractual assessments for renewable energy and/or energy efficiency. U.S. DOE, Washington, DC.

[US-DOE] U.S. Dept of Energy (2012a). Active solar heating. U.S. DOE, Washington, DC. *energy.gov/energysaver*.

[US-DOE] U.S. Dept of Energy (2013a). Solar and Wind Energy Rebate Program. U.S. DOE, Washington, DC.

[US-DOE] U.S. Dept of Energy (2013b). The war of the currents: AC vs. DC Power. U.S. DOE, Washington, DC.

US-DOE], U.S. Dept of Energy (2013c). Clean Energy in My State: Illinois Electricity Generation. U.S. DOE, Washington, DC.

[US-EIA] U.S. Energy Info. Adm. (2013). Updated capital cost estimates for utility scale electricity generating plants. U.S. EIA, Washington, DC.

[US-EIA] U.S. Energy Info. Adm. (2014a). Annual Energy Outlook 2014: Levelized Cost and Levelized Avoided Cost of New Generation Resources in the Annual Energy Outlook 2014. U.S. EIA, Washington, DC.

[US-EIA] U.S. Energy Info. Adm. (2014b). Annual energy review (Data). U.S. EIA, Washington, DC.

U.S. Gov. (1979). Public papers of the presidents of the United States, Jimmy Carter, Book 1: January 1 to June 22, 1979. Presidential Docs., General Services Adm., Natl Archives, Office of the Fed. Register, GPO Stock # 069-000-00115-7, pp.1092–1094.

Wendland, W. M. and Scott, R. W. (2011). Measured solar radiation over Illinois, Ill. State Water Survey, Miscellaneous Pub.No. 198, UIUC, Champaign, IL.

Wernau, J. (2012). Money meant for wind, solar power loses steam as customers shift from utility giants. News section, *Chicago Tribune*, Chicago, IL.

Wilson, R. (2013). Low capacity factors: Challenges for a low carbon energy transition. Editorial, *The Energy Collective. theenergycollective.com*

Worden, J. and Zuercher-Martinson, M. (2009). How inverters work: What goes on inside the magic box. Solar Pro Magazine, **2**(3): 1–8.

Wostzak, R. (2009). The first U.S. passive house shows that energy efficiency can be affordable. News section, *Green Bldg Advisor. greenbuildingadvisor.com.*

Zhao, Y., Meek, G. A., Levine, B. G. and Lunt, R. R. (2014). Light Harvesting: Near-infrared Harvesting Transparent Luminescent Solar Concentrators, *Adv. Optical Materials*, **2**(7): 599.

Zipp, K . (2013). Solar shedds light on Chicago aquarium. News section, *Solar Power World. solarpowerworldonline.com.*

10

Geothermal Energy

"Another way of getting motive power without consuming any material would be to utilize the heat contained in the earth, the water, or the air for driving an engine. The difficulties of sinking shafts and placing boilers at depths of, say, twelve thousand feet, corresponding to an increase in temperature of about 120°C, are not insuperable, and we could certainly avail ourselves in this way of the internal heat of the globe. [The] internal heat of the earth is great and, in comparison with the demands which man can make upon it, is practically inexhaustible; since the heat contents of the earth are sextrillions of tons." **Nikola Tesla**, The Problem of Increasing Human Energy, 1900.

10.1. Introduction

Geothermal energy is the heat from below the earth's surface that can be used as an alternative to other forms of energy. Geothermal energy has a high potential to help energy self-sufficiency. Geothermal energy is a sustainable form of energy, and the second most abundant energy source on earth after solar energy that can be used for cooling and heating (Lee *et al.*, 2007; Dickson and Fanellim, 2013). The benefits of geothermal energy over fossil energy include energy reliability and security, as well as economic development and improved air quality (Glassley, 2015). Geothermal energy is usually in the form of steam, hot water and hot rocks concentrated in

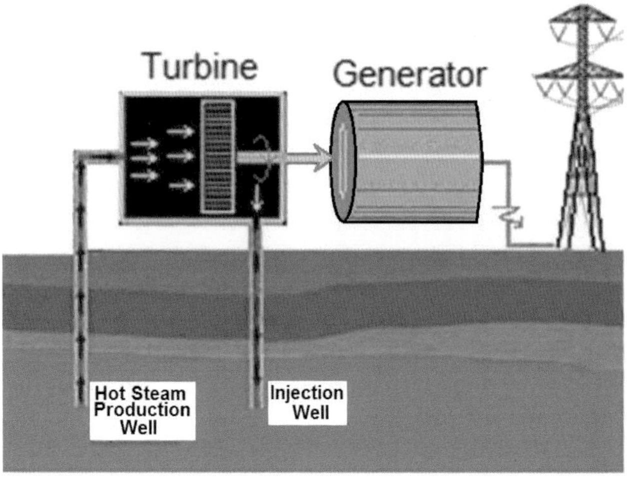

Fig. 10.1. General concept of electric power generation using high temperature geothermal energy resource.

underground reservoirs. The three applicable technology categories are geothermal heat pumps, direct-use and electric power plants (GHC, 2012).

The application of geothermal energy resources can be categorized as low temperature (less than 90°C or 194°F), moderate temperature (90–150°C or 194–302°F) and high temperature (greater than 150°C or 302°F) (GHC, 2012). In general, the high temperature resources are used primarily for electric power generation (see Fig. 10.1).

The high temperature geothermal resources are very common in tectonically active regions, such as areas near geological plate boundaries, rift zones, mantle plumes or hot spots. An example of such regions is the State of California, where geological activities have resulted in the formation of geysers or hot springs that may be useful for electrical power production. Applications of low to moderate geothermal resources may be direct or indirect. The direct applications of geothermal resources include bathing and balneology, residential and district heating, industrial processes, as well as shallow geothermal applications such as heat pump systems. The indirect usage involves conversion of geothermal resources into electrical power using geothermal power plants. Geothermal energy has been used in ancient times by the Romans, as well as in modern times by the Icelanders, Japanese, Koreans, Persians, Turks and others for bathing and space heating. Geothermal energy utilization could reduce the dependence of any community on fossil and nuclear energy appreciably, while improving the air quality and allaying the concerns about nuclear waste problems.

The first application of geothermal energy for electricity production occurred in 1904 at the Larderello field in Tuscany, Italy. In a span of 10 years, the Larderello field project subsequently developed to a capacity of 250 kW. In 1919, the Beppu site in Japan was used for experimental geothermal work, leading to a pilot plant that produced 1 kW of electricity in 1924. But prior to this, the Japanese had been using geoheat to warm their greenhouses. Since the 1930's, hot geothermal water has been used to provide municipal heating in Iceland and remains the major source of heating in this country to date. In the U.S. it was not until the early 1920s that commercial applications of geothermal steam were investigated (Lee *et al.*, 2007). However, the keen competition from hydroelectric power technology at that time made it impossible to encourage further development of geothermal power generation technology in this country. Currently, one of the larger geothermal power plants in the world is located at the Geysers in California (Lee *et al.*, 2007).

The high capital cost of geothermal power plants makes them less attractive for power generation in Illinois. For example, the 2012 capital cost of a binary cycle geothermal power plant was $4362/kW, compared with $917/kW for Natural Gas Combined Cycle (NGCC) power plant (US-EIA, 2013). Department of Energy in the U.S., through its GeoPowering the West (GPW) program has been working to further geothermal energy efforts in this country. Consequently, the Illinois geothermal industry has seen some success, especially in southern and central Illinois. At the same time, the industry is

Table 10.1. ARRA funded geothermal projects in Illinois (Jennejohn, 2011).

Recipient	Focus	DOE $ Amount	$ Cost Share
Ill. Dep. of Military Affairs	GHP	1,200,000	400,000
Indie Energy Systems Co.	GHP	2,459,971	2,515,250
Total		**3,659,971**	**2,915,250**

GHP = geothermal heat pump.

beginning to realize some progress in the Northern part of the state. By the end of 2011, the Environmental Law and Protection Center (ELPC), an Illinois group responsible for protecting the Midwest Environment and Natural Heritage, could identify 89 businesses in the Illinois geothermal energy supply chain. These included installers, drillers and several large-scale commercial geothermal businesses (Graig *et al.*, 2011; Battocletti and Glassley, 2012).

In 2009, the American Recovery and Reinvestment Act (ARRA) provided the U.S. geothermal industry with $360.8 million of federal funding. Out of this amount, the Illinois Department of Military Affairs received $1.2 million (see Table 10.1) and Indie Energy Systems Company, based in Illinois, received $2.46 million. However, the bulk of the money went to states such as California, where $32.3 million was allocated to 12 research, development and demonstration projects. Moreover, $21.9 million was allocated to the Arizona Geological Survey for the development of the National Geothermal Data System. Oregon and Texas states received $36.5 million and $25.5 million, respectively, for 16 research, development and demonstration projects in these states (Jennejohn, 2011; PRN, 2009).

Geothermal systems in Illinois are installed in a variety of buildings ranging from K-12 schools to colleges and universities to community health centers (ICE, 2013). In Fig. 10.2, we report the geothermal energy consumption estimates in Illinois from 1990–2012. The sharp drop from 2010–2011 was due to expiration of the federal tax credit for renewable energy production, which affected all states in the U.S., including Illinois, as discussed in Chapter 7.

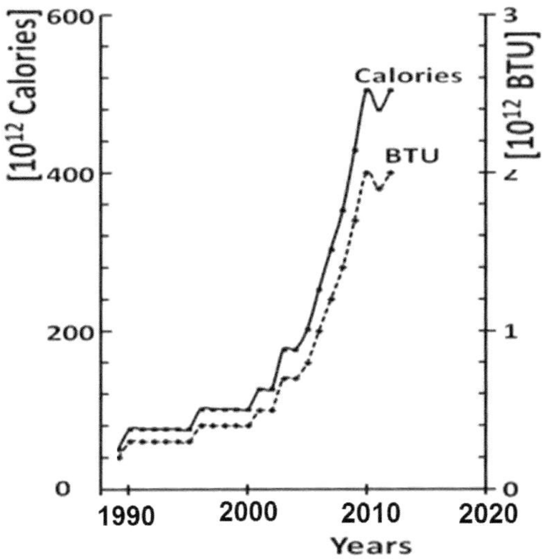

Fig. 10.2. Geothermal energy consumption in Illinois, from 1990–2012 ([Data from US-EIA, 2014a]).

To increase the growth of geothermal industry across the states, the U.S. Energy Improvement and Extension Act of 2008 (H.R. 1424) was passed by the U.S. Congress. The bill provides a tax credit of 30% (caped at $2000 prior to 2009) to homeowners who choose to install geothermal heat pump (ground source heat pump) equipment on their property (Congress U.S., 2008). However, the property excludes geothermal systems used purposely for swimming pools and/or hot tubs. Additionally, the geothermal system must be for a residential property within the U.S., but not necessarily a primary residence. To qualify for the tax credit, the geothermal system must meet or exceed the current Energy Star (energy savings) requirements, and must be operating after 31 December 2007. The 30% tax credit is applicable towards purchasing of heat pumps, ground loop, and associated equipment. Included in the tax credit are labor costs for excavation and installation of loopfield, installation of the heat pump and any other plumbing, wiring and/or ducting required for operation of the system. However, effective January 1, 2009, the ARRA eliminated the $2000 Cap to the 30% tax credit, but with everything in the

Bill, H.R 1424 remaining intact. The implication of this change is that homeowners can now enjoy the benefit of taking full advantage of the 30% tax credit without a cap. This legislation is valid for installations which have been in operation from January 1, 2009 through December 31, 2016. Also, as part of the ARRA, the Commercial Building Tax Deduction was extended till 2013. This extension was to allow building owners to deduct up to $20 per square meter ($1.80 per square foot) for improvement in overall performance of heating and cooling systems using geothermal energy (GAOI, 2012).

10.2. Geothermal Energy Resources

A naturally occurring geothermal system also known as a hydrothermal system is defined by three key elements: heat, fluid and permeability at depth.

Deep geothermal systems or enhanced geothermal systems (EGS) are defined as man-made reservoirs created to extract heat from economically unproductive geothermal resources. The EGS system extracts heat from the earth by creating a subsurface fracture system in which water can be added through injection wells. The water is heated by contact with the hot rock and then returns to the surface through production wells (Green and Nix, 2006).

According to a 2006 assessment (US-DOE, 2006), Illinois geothermal resource base is 186, 123 $\times$ 10^{18} Joules (5.17 $\times$ 10^{10} GWh) (see Table 10.2).

Also, a DOE report published in July 2012, estimated Illinois total technical potential for enhanced geothermal systems (see Sec. 10.3.3) to be 2.43 $\times$ 10^{18} J (675,540 GWh), ranking the state 18th place among the 50 U.S. states. Texas leads all the states with 10.91 $\times$ 10^{18} J (>3,000,000 GWh) of estimated total technical potential for enhanced geothermal systems (Lopez *et al.*, 2012). Additionally, Illinois estimated average subterranean temperature at the depth of 6 kilometers ranges from 100–150°C, with thermal heat flow rate of 144–216 Joules/hour (40–60 milliwatts) per square meter. Nevada

Table 10.2. Illinois geothermal resource base (US-DOE, 2006).

Depth km (mile)	Temperature °C (°F)	Energy (GWh)
6.5 (4)	150 (302)	5.57×10^8
7.5 (4.7)	150 (302)	1.68×10^{10}
8.5 (5.3)	150 (302)	1.70×10^{10}
	200 (392)	1.06×10^8
9.5 (5.9)	150 (302)	1.64×10^{10}
	200 (392)	8.57×10^8
Total geothermal energy resources		5.17×10^{10}

Fig. 10.3. Total estimated technical potential for enhanced geothermal systems, as described in Sec. 10.3.3, in the U.S. provided by NREL (Lopez *et al.*, 2012).

and Idaho are among the states with the highest average subterranean temperatures of over 200°C, and thermal heat flows greater than 360 Joules/hour (100 milliwatts) per square meter (Green and Nix, 2006).

Figure 10.3 shows the State of Illinois standing as compared with other U.S. states in geothermal resource base suitable for electricity production, using existing and emerging technologies.

In general, average temperature of the earth at depths ranging from 6 meters (~20 ft) to 46 meters (~150 ft) remains fairly stable for most part of the year, under normal geothermal gradient of ~30°C per kilometer (1.64°F per 100 ft). The fairly stable temperature at this depth in Illinois and elsewhere is due to the complex interaction between heat fluxes from the sun and the earth's upper mantle (Chiasson, 1992). This temperature gradient (~30°C per kilometer) is about 1–2°C higher than the local annual mean temperature of

Table 10.3. Estimated maximum and minimum mean monthly soil temperatures for the city of Urbana, Illinois (Hendrie and Wendland, 1981).

Depth		**Minimum Temperature**		**Maximum Temperature**	
(ft)	**(m)**	**(°F)**	**(°C)**	**(°F)**	**(°C)**
10	3.05	47	8.3	61	16.1
15	4.6	50	10.0	58	14.4
20	6.1	52	11.1	54	12.2

Illinois or other parts of the world. Table 10.3 shows the estimated maximum and minimum mean monthly soil temperatures for the City of Urbana, Illinois (Hendrie and Wendland, 1981).

A DOE sponsored study has led to the most current geothermal database of the Illinois Basin, as shown in Fig. 10.4 (Proffitt *et al.*, 2012).

This new geothermal database indicates that the mean geothermal gradient is 1.64°F/100 ft, which is higher than the often assumed gradient of 1.0°F/100 ft. Higher geothermal gradients can be expected for thick sequences of sedimentary rocks that have lower thermal conductivity than underlying crystalline rocks.

Geothermal resources are usually classified as convective hydrothermal resources, other hydrothermal resources and hot rock resources (see Table 10.4).

The convective hydrothermal resources include sedimentary basins and regional aquifers (hot fluid in sedimentary rocks).

The other hydrothermal resources include regional aquifers (hot fluid in sedimentary rocks), geopressured geothermal systems (or hot fluid under geological pressure that is greater than hydrostatic), radiogenic (heat generated by radioactive decay), as well as sedimentary basins. The hot rock resources include solidified magma (hot dry rock) and partly molten rock (magma).

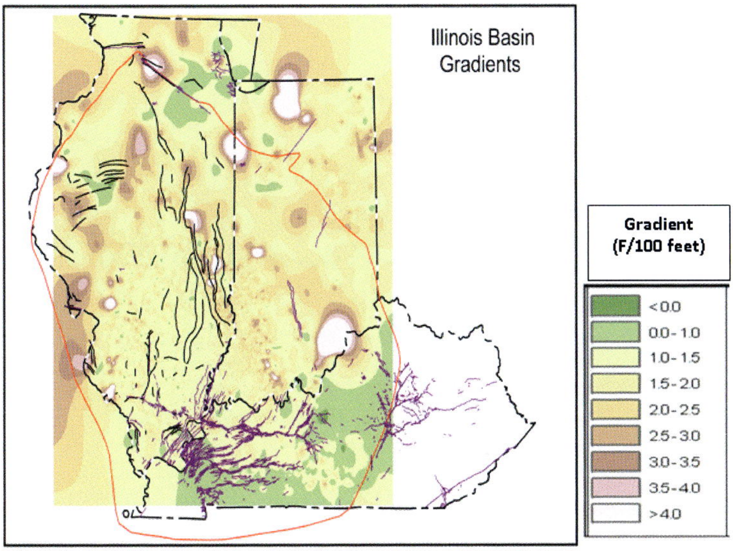

Fig. 10.4. Illinois Basin geothermal gradients and significant faults (purple) and folds (black) (Proffitt, *et al.*, 2012).

Table 10.4. Geothermal resource types and their approximate temperatures (Lund, 2007).

	Temperature	
Type of Geothermal Resource	(°C)	(°F)
Convective hydrothermal resource:		
Vapor dominated systems	240	464
Hot-water-dominated systems	20–350+	68–662
Other hydrothermal resources:		
Sedimentary basins/regional aquifers(1)	20–150	68–302
Geopressured(2)	90–200	194–392
Radiogenic(3)	30–150	86–302
Hot rock resources:		
Solidified (hot dry rock)	90–650	194–1202
Part still molten (magma)	>600	>1112

(1) Hot fluid in sedimentary rocks.
(2) Hot fluid under pressure that is greater than hydrostatic pressure.
(3) Heat generated by radioactive decay.

In general, resources above 150°C may be used for electric power generation, although power has also been generated recently at Chena Hot Springs Resort in Alaska using geothermal resource at 74°C (Lund, 2006). Resources below 150°C are usually used in direct-use projects for heating. Ambient temperatures in the 5–30°C range may be used with geothermal (ground-source) heat pumps to provide both heating and cooling (Duffield and Sass, 2003; US-DOE, 1997).

Geothermal resources in Illinois that can be used for electric power generation are hydrothermal (hot fluid in sedimentary rocks) and hot dry rocks intrinsic thermal energies (Clark and Veil, 2009; Hinze *et al.*, 1986), as described in more details below.

10.2.1. *Hot dry rock (HDR) intrinsic energy resources*

This type of geothermal energy resource occurs at depths ranging from 8–16 kilometers (5–10 miles) beneath the surface of the earth, and where geothermal gradient (the vertical profile of changing temperature) are significantly above average (>50°C/km), but it can also occur at shallow depths (US-DOE, 1997; EPRI, 2010).

In the past, access to this energy resource was practically impossible because of lack of adequate technology. However, recent technological innovations such as enhanced geothermal systems (see Sec. 10.3.3), have touched off a national interest in geothermal development, especially in certain areas across the U.S. with no history of tectonic activity or volcanism. An enhanced geothermal system is an artificial reservoir created in areas where there is hot dry rock, but insufficient or little natural permeability or fluid saturation for thermal energy extraction (US-DOE, 2012b).

The core of the earth has a high temperature which can reach more than 4,000°C (>7,200°F) and continuous flow of this thermal energy to the surface of the earth can be harnessed as a sustainable source (Green and Nix, 2006). The U.S. has abundant resources of geothermal energy and a high potential of this resource for energy

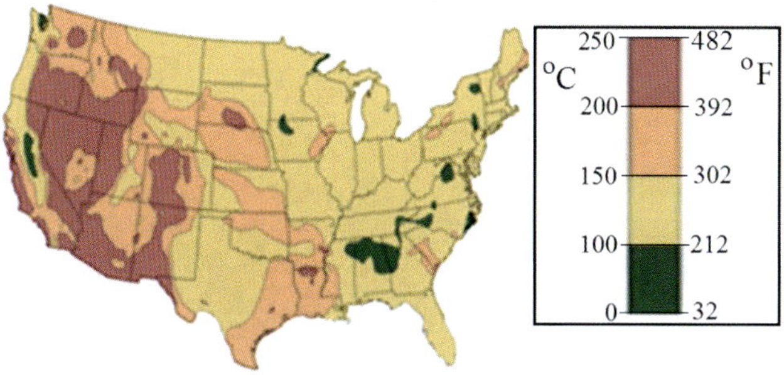

Fig. 10.5. Based on temperature at 6 km depth, the U.S. geothermal resource base suitable for electricity production using existing and emerging technologies appears to span almost the entire country. (Credit: DOE Geothermal Technologies Program) (EPRI, 2010; ITO, 2015).

production exist in most part of the country. The average annual weather temperature across the U.S. varies from the low of 39°F (~4°C) in upper Midwest to 72°F (~22°C) in Southeastern part of the country. The subterranean temperatures at the depth of 6 km (3.7 miles) for most part of the country varies from 100°C (212 °F) to more than 200°C (392°F) (see Fig. 10.5). This massive source of energy can be utilized as a practical substitute for fossil energy use, which is producing massive amount of greenhouse gases as discussed in Chapter 4 of this book and by Agyarko and Mansoori (2013). Among the areas of particular geothermal development interest in Illinois is the radioactive basement granites beneath sediments in Northwestern Illinois (see Fig. 10.5).

Such areas have high temperature geothermal resources, but lack underground water reservoir for heat extraction (SMU, 2011). Hot dry rock as an energy source in the U.S. mid-continent, including Illinois, is particularly appealing because of the high population density and the high demand for industrial and domestic energy applications in the region (Hinze *et al.*, 1986). In the 1980's, the Illinois deep drill project sponsored by Commonwealth Edison, led to measurement of the thermal conductivity and radioactivity of core samples of the Illinois hot dry rock deposits in Stephenson County, Northwestern Illinois. The project was part of an effort by the consolidated Edison Company in Illinois to use the unusually high temperatures in the region for temporary storage of water to maximize energy efficiency of nuclear power plants. A total of three wells (UPH-1, UPH-2, UPH-3) were drilled. The measured average heat flow of the granitic rock from a well (UPH-3) at a depth of 1.6 kilometers was 75 milliwatts/m^2, at a temperature gradient of 23°C per kilometer, with heat generation rate of 16 microwatts/m^3 (Hinze *et al.*, 1986). Since granite contains on the average 3–4 ppm uranium, 10–15 ppm thorium and 3–4% potassium, with an equivalent heat generation from radiogenic sources of 2–4 microwatts/m^3, the measured average heat flow of 75 milliwatts/m^2, represented a higher than normal heat flow measured in most part of the eastern U.S., which is estimated at 57 ± 17 milliwatts/m^2 (Hinze *et al.*, 1986). However, the heat flow measurements of the remaining wells were considerably less than what were predicted for the eastern part of the U.S. That discrepancy was attributed to the limited vertical extent of the granite in such areas, as confirmed by interpretation of gravity data (Hinze *et al.*, 1986).

10.2.2. *Hydrothermal energy resources*

These are geothermal environments where available fluid freely circulates (see Fig. 10.6). These subterranean geothermal reservoirs of steam or hot water are formed by water seeping into the earth and getting heated up by fractured or porous hot rock. There is upward transfer of heat energy in these reservoirs caused by vertical circulation of fluids driven by differences in fluid density which is a function of temperature difference. Depending on whether the fluid is steam or liquid water, hydrothermal systems can be classified as vapor or hot water dominated, respectively (see Fig. 10.6).

Also, several high-temperature geothermal resources occur when magma (molten rock) emerges at the upper crust of the earth to heat the surrounding rock. When the surrounding rock is porous enough to allow passage of water, the resulting heated water or steam is

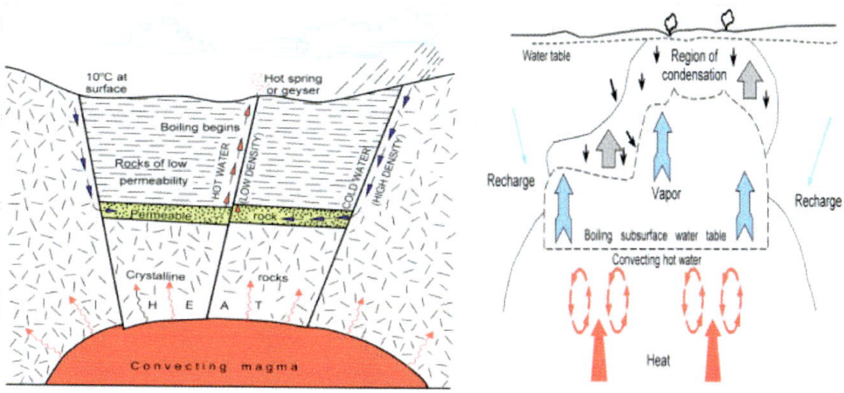

Fig. 10.6. Hot water dominated geothermal system (left); vapor dominated geothermal system (right) (White *et al.*, 1971).

referred to as a hydrothermal resource. Presently, such resources are used for commercial production of geothermal power elsewhere. The energy supply is continuous because there is continuous recharge of the energy, as heat flows into the reservoir from greater depths (Green and Nix, 2006). According to a DOE report by the National Energy Research Laboratory, Illinois' total estimated technical potential for hydrothermal power is less than 1 GW (Lopez *et al.*, 2012). Nationwide, the conventional hydrothermal resources potential is estimated at 30 GW. These reservoirs can be tapped by drilling wells to deliver hot water to the surface for generation of electricity or for direct use. The process technologies for tapping hydrothermal resources are commercially proven (Duffield and Sass, 2003).

Low-temperature geothermal waters at depths less than one kilometer appear to be widely available throughout much of the U.S. and have the potential for significant utilization in space heating and agriculture on a local basis (Sammel, 1978). Low-temperature and co-produced resources from sedimentary basins (see Fig. 10.7), also referred to as non-conventional hydrothermal resources, represent a small but growing sector of hydrothermal development in geothermal resources (US-DOE, 2010a; SMU, 2011).

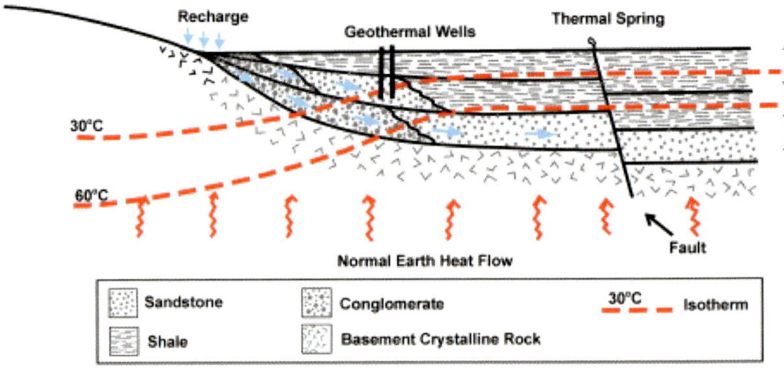

Fig. 10.7. Sedimentary basin geothermal resource (Anderson and Lund, 1979).

Co-produced fluids geothermal energy resources are derived from oil and gas wells at temperatures over 150°C (302°F) (SMU, 2011). It is estimated that 10 barrels of hot water are produced for every barrel of oil produced from an oil well in the U.S. (Clark and Veil, 2009). This amount results in an average of 20–25 billion barrels of hot water produced every year from oil and gas wells within the U.S. (Clark and Veil, 2009; US-DOE, 2010a). For years, co-produced water has been a problem in the oil and gas industry, because it requires expensive budget for its handling and disposal. However, due to technological innovations, this water is currently being considered as geothermal resource that can be used to produce electricity for field operations or be sold to the power grid. These emerging technologies will allow expansion of geothermal energy application using co-produced fluids at temperatures as low as 73°C (163°F). It is estimated that a water flow rate of 1,000 gpm is needed to produce 1.5 MW of electricity from a reservoir temperature of 149°C (300°F) (McKenna *et al.*, 2005). In 2004, the water production rate from the production wells of Illinois' existing hydrocarbon fields was ~200,000 gallons (4.76 × 10^3 barrels) per minute, enough to generate 199–954 MW of power, as shown in Table 10.5 and compared with those of selected states with hydrocarbon fields.

Table 10.5. Illinois equivalent geothermal power from co-produced hot water from the production wells of Illinois existing hydrocarbon fields compared with those of selected other states.

State	2004 Total Processed Water[1,2] (10^3 bbl)	Processed Water Cut (%)	Rate of Water Production[1] (10^3 gpm)	(10^3 bpm)	Power at 100°C (MW)[1]	Power at 140°C (MW)[1]	Power at 150°C (MW)[1]	Power at 180°C (MW)[1]
Illinois	2,197,080	~98.0[3]	199	4.70	200	506	688	954
Mississippi	592,518	96.7[2]	54	1.3	54	136	185	257
Oklahoma	12,423,264	99.5[2]	1124	26.8	1,129	2,860	3,888	5,393
Wyoming	3,809,087	~97.6[4]	344	8.2	346	877	1,192	1,654

(1) US-DOE (2006); (2) McKenna *et al.*, (2005); (3) TOD (2011); (4) Wo (2011).

In 2004, Illinois annual total processed water from its oil and gas fields was estimated at ~2.2 billion barrels, compared to that of Wyoming which was ~3.8 billion barrels or to that of Oklahoma, which was the highest producer of processed co-produced water in the nation estimated at ~12.4 billion barrels. The benefits of co-produced fluid geothermal resources include near-term energy savings, reduction of greenhouse gas emissions, extension of the economic life of oil and gas fields and profitable utilization of abandoned oil and gas field infrastructure (Clark and Veil, 2009). Table 10.5 shows Illinois co-produced fluids and water cut, as well as the equivalent geothermal power compared with other states in the U.S. For example, the water cut of the Lawrence field wells in Illinois basin is estimated at 98%, meaning that for every 100 barrels of fluid pumped out of a well, just 2 barrels is oil, which must be separated from the water (TOD, 2011).

Figure 10.8 shows the geothermal resource of the U.S., including the locations of identified sites for both hydrothermal and enhanced geothermal resources in Illinois. According to this map, Illinois' favorability of deep enhanced geothermal system is between most and least favorable.

In the following sections, we introduce the concepts of electrical power generation using geothermal energy, a technology not

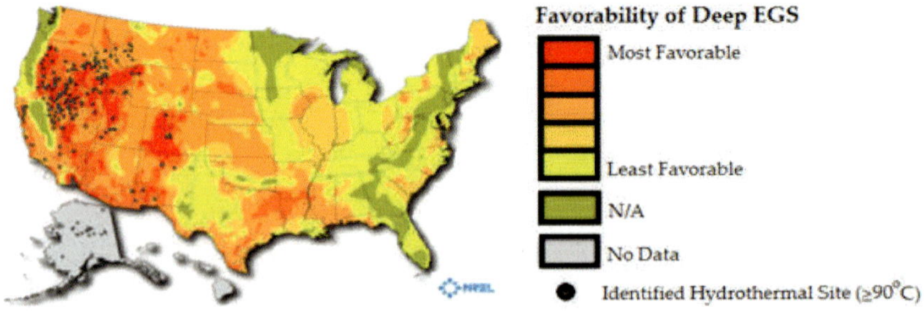

Fig. 10.8. Geothermal resource of the U.S. Locations of identified hydrothermal sites, the co-located near-hydrothermal field EGS resource and the relative favorability of the deep EGS resource. Undiscovered hydrothermal resources and other geothermal resources, such as co-produced fluids, are not represented (Augustine, 2011).

currently deployed in Illinois, but which could be deployed in the future in the Illinois Basin, as well as the Northwestern part of Illinois. These include the flash steam power plant, the binary cycle power plant and enhanced geothermal power generation systems (EGS). We also discuss the geothermal energy technologies using geothermal heat pumps that are currently used in the state. These are the horizontal closed loop geothermal heat pump (GHP) systems, vertical closed-loop GHP systems and open-loop GHP systems, including pond/lake loop GHP systems. Also, in Sec. 10.5 we present the geothermal energy projects in the State of Illinois, with specific references to the Grant's Hall vertical closed-loop GHP system at the University of Illinois at Chicago, the Walgreens geothermal energy cooling and heating system at the Village of Oak Park (a suburb of Chicago) in Illinois, the pond/lake closed loop GHP system serving the Advocate Sherman Hospital in Elgin, in Kane County, Illinois, and the geothermal retrofit of Illinois National Guard State Headquarters building project at Camp Lincoln, Illinois. Furthermore, the problems associated with geothermal energy resource development are discussed in Sec. 10.6.

10.3. Power Generation Using Geothermal Energy Resource and its Advantages

Power generation using geothermal energy involves the use of commercially available geothermal power plants such as flash steam power plant, binary cycle power plant and enhanced geothermal systems. As base load power generation sources, geothermal power plants are more reliable because they may be dispatched to meet fluctuating loads to avoid many of the grid integration problems faced by large-scale solar and wind energy projects. Also, compared with coal-fired power plants, geothermal power plants emit fewer or no pollutants, such as carbon dioxide, nitrogen oxides, or sulfur dioxide. It is estimated that emissions offset by geothermal power plants are the equivalent of 22 million metric tons of carbon dioxide, 200,000 metric tons of nitrogen oxides and 110,000 tons of particulate matter from conventional coal-fired plants (Green and Nix, 2006). Geothermal power plants are designed for a life span of 20 to 30 years, but with proper resource management strategies, the life span can exceed the design period (EPRI, 2010). In addition, compared to fossil-fueled power generation plants, geothermal power plants are much less complex, and operate at relatively lower pressure and temperature. Besides, geothermal power plants have less auxiliary equipment, and in many instances they have capacity factors of 85% or higher (EPRI, 2010). The capacity factor compares how much electricity a power plant actually produces with the maximum it could produce at continuous full power operation during the same period. Below, we describe each of these geothermal power generation processes.

10.3.1. *Flash steam power plant*

The flash steam power plant is the most common type of geothermal power plant. The flash steam power plant (see Fig. 10.9) extracts energy from high enthalpy geothermal resources at temperatures greater than 182°C (~360°F) to generate electricity (IEA, 2010). As indicated in Table 10.2, Illinois geothermal energy resources at the depth of 8.5–9.0 kilometers (5.3–5.6 miles) is at temperature of ~200°C (~390°F). However, harnessing this energy resource for electricity generation in the future would require a more advanced drilling and a flash steam power plant. In essence, the flash steam power plant separates the steam from the produced hot water in a cyclone separator under atmospheric conditions (flashing). The steam is directed to turbines and the resulting condensate is routed for re-injection or further flashing at lower-pressure. Once the energy is extracted, the fraction of the fluid exiting the separators, including the steam condensate (with the exception of condensate evaporated in the wet cooling system), is usually re-injected to earth. In the U.S., the largest flash steam power plant is the Navy 1 geothermal power plant near Coso Hot Springs, in California (IEA, 2010). The capacity

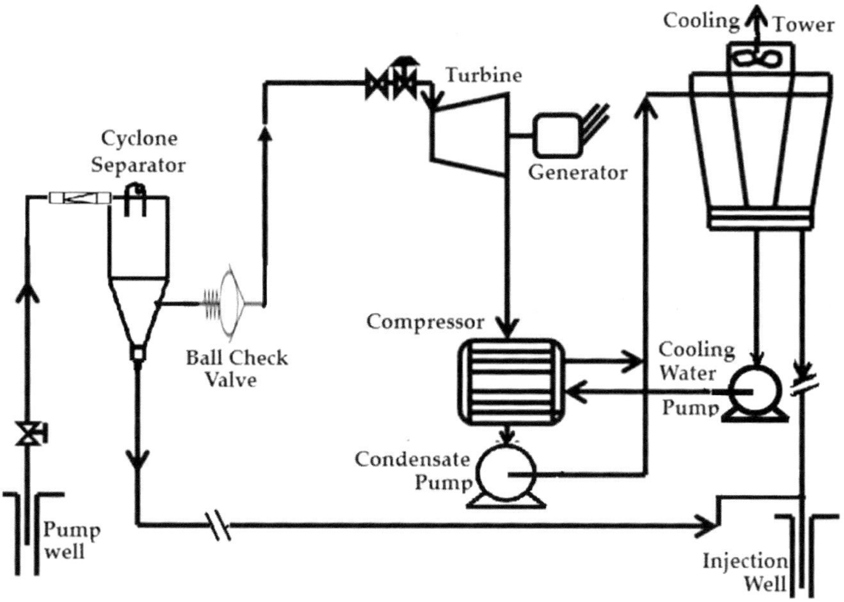

Fig. 10.9. A single-flash power plant (US-DOE, 2006).

of a typical flash steam power plant ranges from 2–45 MW (IEA-ETSAP, 2010). According to a 2013 study by EIA, the capital cost of a 50 MW geothermal dual-flash steam power plant development is $6,243/kW (US-EIA, 2013).

10.3.2. *Binary cycle power plant*

Binary cycle power plants (see Fig. 10.10) are used to generate electricity with low or medium enthalpy geothermal fluid at temperatures usually ranging from 107–182°C (225–360°F).

Illinois geothermal resource indicates a temperature of 150°C at the depth of 6.5–9.0 kilometers below the surface of earth, which is in the range of temperature requirement for this technology. Additionally, the presence of low-temperature co-produced fluid geothermal resource in Illinois makes the binary cycle power plant a suitable choice for power generation in the state in the future. Binary cycle

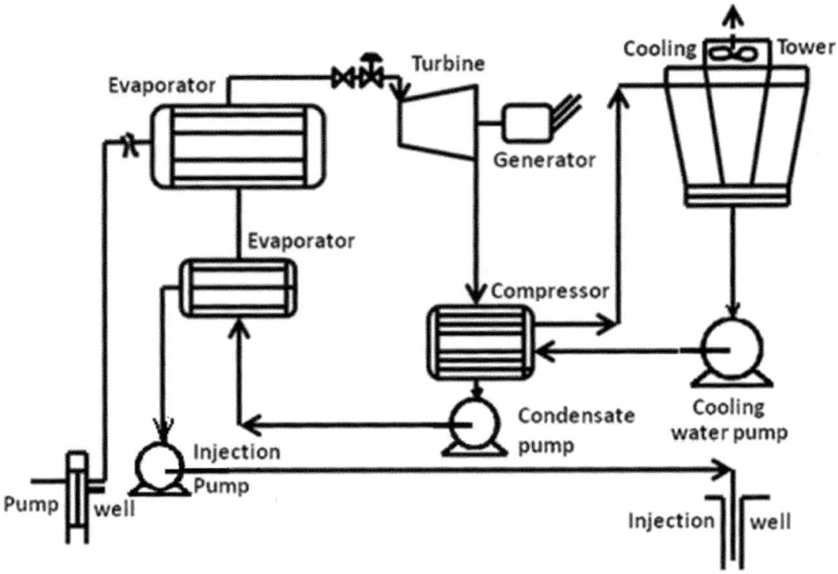

Fig. 10.10. A binary cycle power plant (US-DOE, 2006).

power plants use Organic Rankine Cycle (ORC) to circulate geothermal water in heat exchangers to heat an organic process fluid or secondary fluid (1,1,1,2-Tetrafluoroethane or R134a). An organic process fluid with low boiling point is used because of the relatively low temperature of the geothermal fluid. The inlet temperature required to obtain electrical power in binary plants is a function of the well-bore flow rate, as well as the type of fluid in the binary generator heat exchanger. Such a relationship has been demonstrated by using isopentane in the secondary loop as process fluid, which accounts for in-plant losses, but not the energy costs of pumping fluid from the subsurface (McKenna *et al.*, 2005). Isopentane is a very volatile liquid at room temperature and pressure, with normal boiling point at 27.8°C (82°F), which is just a few degrees above room temperature. This process fluid, which has relatively lower boiling point than water, expands into gaseous vapor, which is used to turn a turbine to drive a generator to produce electricity. Once the energy is extracted, the geothermal fluid is re-injected into the reservoir. The power generation capacity of binary plants is typically 25 MW and with thermal efficiency of 8–15% (IEA-ETSAP, 2010; DiPippo, 1999). In 2006, Chena Power, LLC in Alaska developed a mobile geothermal power plant partially funded by DOE (Clark and Veil, 2009). This power plant was used to demonstrate that low-temperature co-produced fluid resource across the U.S. can be accessed and used for power generation. The 200 kW binary cycle power plant in Chena Hot Springs, in Alaska was used to generate electricity at the world's lowest fluid temperature of 74°C (165°F) (GHC, 2012; Aneke *et al.*, 2011). Subsequently, the Dixie Valley Bottoming Binary Power Plant in Terra-Gen in Nevada, funded by the American Recovery and Reinvestment Act of 2009 was also used to demonstrate the technical and economic feasibility of electricity generation from non-conventional geothermal energy resources at 106°C (223°F). In essence, Dixie Valley Bottoming Binary Power Plant employed the first commercial use of a supercritical cycle (using new biphasic metal organic working fluids) in a geothermal power plant, at an inlet temperature less than 149°C (300°F), and has been producing 6 MW gross electricity since September 2012 (McDonald, 2010; OpenEI, 2011b). Binary cycle power plants are currently used for energy production in Alaska, Oregon, Idaho and Utah (SMU, 2011).

10.3.3. *Enhanced geothermal system (EGS) for power generation*

An EGS is a man-made reservoir with available thermal energy but insufficient or little natural permeability or fluid saturation. However, it is possible to achieve commercial-scale power production through hydraulic fracturing of the system (see Fig. 10.11) (US-DOE, 2012b; Jennejohn, 2011).

As mentioned in Sec. 10.2, the estimated total technical potential for EGS in Illinois is 2.43×10^{18} J (675,540 GWh) (US-DOE, 2006). The EGS technology can be used to create engineered geothermal reservoirs by injecting water into boreholes reaching depths of

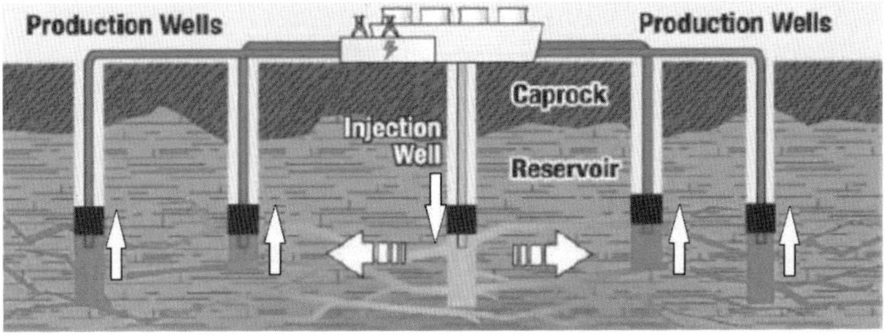

Fig. 10.11. General schematic of an EGS for power generation (EERE, 2004).

~1,500–3,000 meters (5,000–10,000 ft). At such depths, the water temperature is raised to over 150°C (302°F) by the extremely hot underground layers of dry rock. The injected hot water is pumped to the surface of earth through secondary boreholes and is used to generate electricity (Cross *et al.*, 2011). According to a 2006 research study (US-DOE, 2006), the EGS technology could provide 100,000 MW of electricity in the U.S. by the year 2050. This energy would be enough to provide 10% of the nation's energy needs (US-DOE, 2006). Currently, this emerging technology is not deployed in Illinois even though it seems suitable for the state. However, it is vigorously being pursued in places such as France, Australia, and in some western states in the U.S., such as California (GEA, 2012). In the U.S., the production cost for EGS based on current power plant technology ranges from $100/MWh (for 300°C resource at 4 km depth) to $190/MWh (for 150°C resource at 5 km) (IEA, 2011).

10.4. Geothermal Energy Technologies Currently Used in Illinois

Geothermal energy technologies currently used in Illinois are GHP systems (see Fig. 10.12).

GHPs take advantage of the constant temperature of the earth as the exchange medium, as opposed to the outside air temperature, and use this heat source/sink for heating/cooling. Despite the fact that seasonal temperatures vary widely, a few feet below the

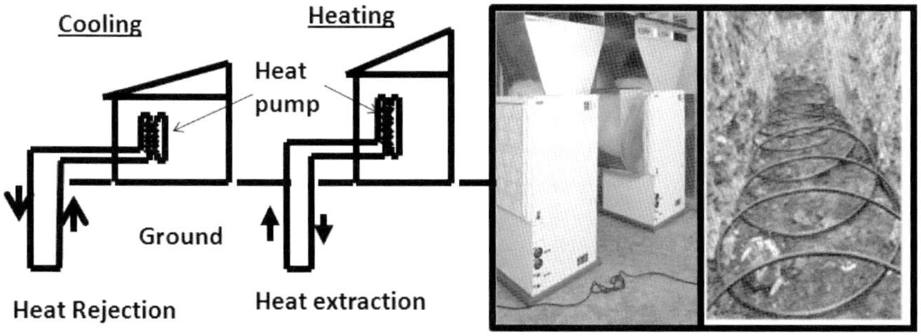

Fig. 10.12. Schematic diagram for operating principles of geothermal heat pump for cooling in summer and heating in winter (US-DOE, 2012a) and an example of the heat exchanger installed underground (US-DOE, 2011).

earth's surface the ground remains at a relatively constant temperature between 50°and 60°F (10°–16°C). The advantage of GHP system is that the heat source is at higher temperature than the outside temperature during winter and therefore, less energy is required to heat a building. In contrast and during summer, the heat sink is at lower temperature than outside and consequently, less energy is required to cool a building.

The basic components required for a GHP system are the ground heat exchanger, the heat pump unit and a thermal distribution system. The heat exchanger is a system of connected pipes called a loop that are buried in the ground near a building. A heat exchanger fluid (usually water or a mixture of water and antifreeze) circulates through the pipes to absorb or release heat to the ground.

The different types of geothermal heat pump systems are the horizontal closed-loop systems, vertical closed-loop systems, pond/lake systems and opened-loop systems. For closed loop systems, the same heat transfer fluid circulates through the system during each cycle of the process, while for open loop systems the fluid is discharged after each process cycle. The choice of any of these systems depends on criteria such as climate, soil conditions, available land and local installation costs at the site (US-DOE, 2012a). A brief description of each geothermal system is given below.

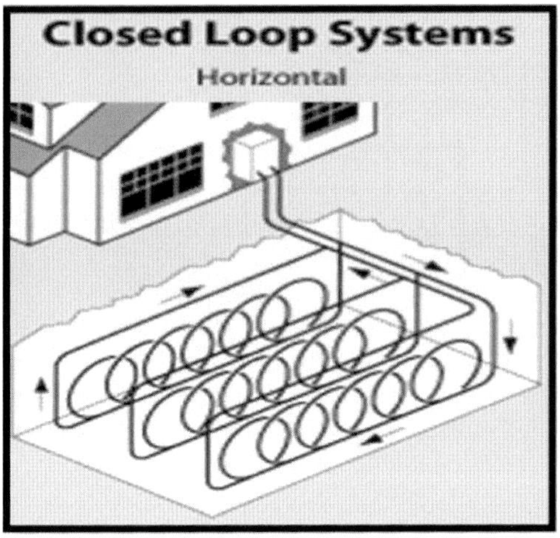

Fig. 10.13. Horizontal closed loop GHP system (US-DOE, 2012a).

10.4.1. *Horizontal closed loop GHP system*

The horizontal closed loop system (see Fig. 10.13) is the most commonly used because it is efficient, reliable and economical for small residential buildings. However, since the installation requires more land space, it is usually used for new building constructions where extra land is available. For example, the system may use horizontal polyethylene pipes, which are laid in trenches of over 1.2 meters (over 4 ft) of depth. More frequently, two pipes are used, where one is placed at 1.8 meters (about 6 ft) deep and the other at 1.2 meters (about 4 ft) deep. Alternatively, two pipes can be placed parallel to each other in a 1.5 meters (5 ft) deep trench measuring 0.62 meter (about 2 ft) wide. A maximum of six pipes, usually in parallel connections are laid in each trench, with minimum separations of 0.3 meter (about one foot) between pipes and 3.1–4.6 meters (about 10–15 ft) between trenches. The length of the loop piping can vary, depending on thermal conductivity and temperature of the ground, soil moisture and system design. Another type of horizontal close loop system is the Slinky™. This method of loop piping allows the arrangement of more pipes within a short trench. The Slinky™ reduces the installation

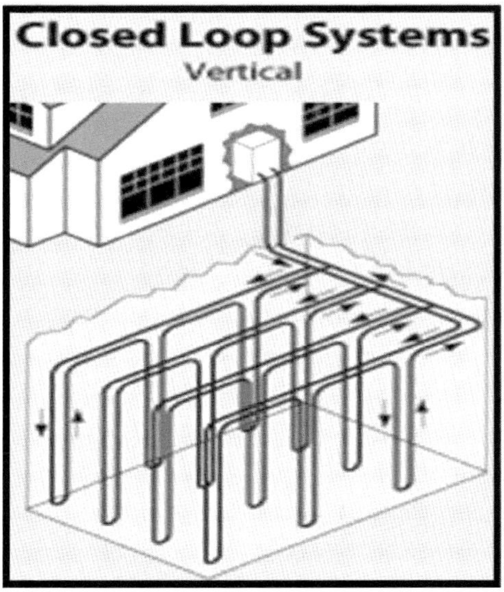

Fig. 10.14. Vertical closed-loop GHP system (US-DOE, 2012a).

cost for horizontal close loop systems, especially in areas that would otherwise be difficult to install conventional horizontal close loop systems (US-DOE, 2012a).

10.4.2. *Vertical closed-loop GHP systems*

The vertical closed-loop system (see Fig. 10.14) is recommended for large commercial buildings and schools, where the land area requirement for horizontal loops would be expensive.

Also, vertical loops are used in places where the soil is too shallow to be trenched. Besides, the vertical arrangement of the pipes in a limited area minimizes the effect on the existing landscape. The vertical closed-loop system usually requires vertical boreholes of approximately 10 cm in diameter (~4 inches) drilled about 6 meters (20 ft) apart, and reaching a depth of 30–122 meters (100–400 ft) or more (US-DOE, 2012a). But, vertical closed-loop systems reaching a depth of 213 meters (~700 ft), which is more than double the industry standard of ~61–76 meters (200–250 ft), is not uncommon. Such depths allow much higher energy volume within a smaller footprint to be achieved, but at a relatively high drilling cost.

For example, a recent 9.6 acres (~420,000-square-foot) office project in Glenview, Illinois, reaching a depth of ~213.6 meters (700 ft) cost nearly $2 million more than a conventional heating and air-conditioning system (Lepeska, 2011). The boreholes were laid with two straight polyethylene pipes with U-bend tube connection at the bottom to form a loop. The vertical looped pipes were hooked with horizontal pipes placed in underground trenches and were connected to the heat pump located in the building. There are several vertical closed-loop GHP systems currently in operation in Illinois. These include the Grant's Hall GHP system at the University of Illinois at Chicago, as mentioned in Sec. 10.2.2.

10.4.3. *Open-loop GHP systems*

The open-loop GHP system (Fig. 10.15) uses wells or surface body water as the heat exchange fluid that circulates directly through the GHP system. The open loop GHP systems were the most commonly used geothermal heat pumps on the market from 1946 until around

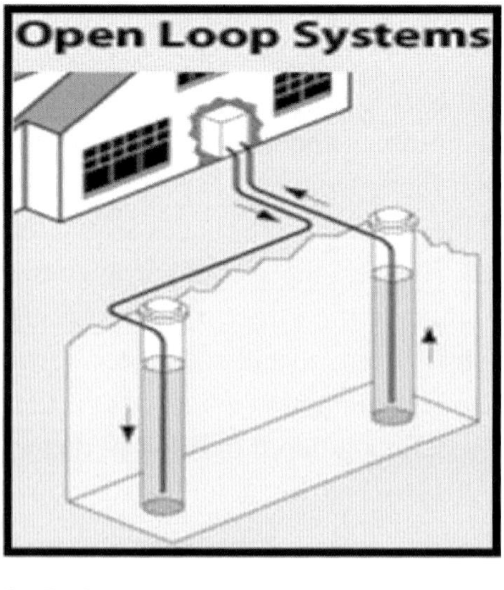

Fig. 10.15. Open loop GHP system (US-DOE, 2012a).

1980, when vertical and horizontal closed-loop systems began dominating the market (Bloomquist, 1999). Whenever wells are used, a large number of open loop systems employ one or more wells. Water is withdrawn from the well or other surface water body and once circulated through the system, the water returns to the ground through surface discharge or it can be returned to the well or water body. The open loop option is used in areas where there is adequate supply of relatively clean water and where all local codes or statutory regulations regarding groundwater discharge can be met (US-DOE, 2012a). The wells of open-loop GHP systems are usually drilled much deeper than the boreholes of closed loop systems, ranging from 305 to 610 meters in depth (1,000–2,000 ft). In places with limited surface space, such as cities and urban campuses, standing column wells are preferred (Cross *et al.*, 2011). The disadvantages with the open-loop GHP system include insufficient flow in the well, as well as plugging of the loop caused by scaling (due to calcium carbonate formation), which can create pressure build-up in the injection well. This scaling problem can be minimized using chemical treatment, but in severe conditions, some reworking of the well on frequent basis may be needed. Also, excessive drop in the aquifer level due to severe drought or well overuse can result in insufficient flow in the well. Insufficient well flow can also be attributed to sedimentation at the bottom of the well, as well as a colony of iron bacteria. These problems can be minimized by back pumping of the well and frequent maintenance, such as well chlorination in every 3–6 months. Other problems with this system are excessive injection pressure, which may be due to poor well completion or inadequate injection horizon or pump failures, due to problems with bearings and seals (Bloomquist, 1999).

10.4.4. *Pond and lake loop GHP systems*

The pond and lake loop system (Fig. 10.16) can be installed at homes or buildings located close to a body of surface water, such as a pond

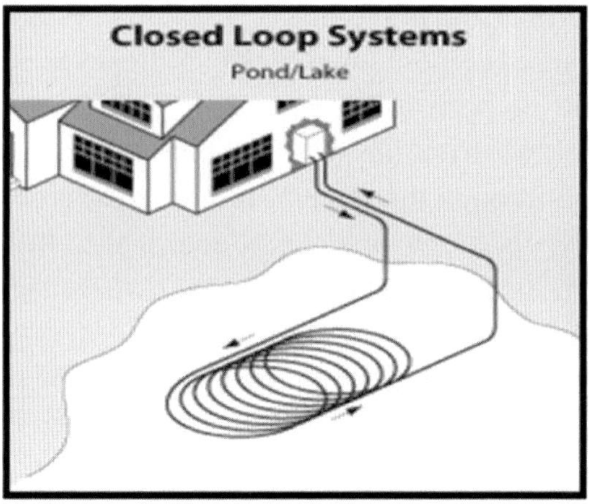

Fig. 10.16. Pond/lake closed loop GHP system (US-DOE, 2012a).

or lake (Geothermal-House, 2010). There are several natural and artificial lakes in Illinois that can be used to provide heating and cooling in buildings and homes. It is estimated that these sources of water can provide 66–75% geothermal energy for human comfort during summer and winter (Geothermal-House, 2010). The basic operating principle of the pond/lake loop GHP system is that it uses the constant temperature under the bottom of lakes/ponds for cooling and heating. The heat capacity or thermal conductivity of water is higher than those of many fluids, but not unduly high, when compared to metals or semi-metals. During summer, the temperature at the bottom of most lakes and ponds remains constant, despite the high ambient temperatures. Similarly, the thermal conductivities of ice and snow are very low, which explains why the temperature at the bottom of lakes/ponds remains constant, irrespective of the amount of snow or ice that accumulates on the surface of the pond during the cold weather. This constant temperature can be harnessed for home heating during winter, when the ambient temperature is far lower than the bottom of the lake, or for cooling in the summer when the temperature at the bottom of the lake is lower than the ambient temperature. A polyethylene pipe supply line runs underground from the building, where the heat pump is located, to the pond or lake. The pipe is coiled into circles, not less than 2.4 meters (8 ft) below the surface of the water to avoid freezing. To obtain enough heat extraction, the pond loops are used strictly under situations where there is no tendency for the water to drop below 1.8–2.4 meters (~6–8 ft) at its lowest level (Geothermal house, 2010). It is recommended that the coils be placed in a water source that meets minimum volume, depth and quality criteria (US-DOE, 2012a). The pond and lake system is the most economical configuration of the geothermal heat pump systems and can be a closed or open loop system.

The selection of the required ground loop and the size of the geothermal heat pump will depend on cooling and heating needs of a building, which will in turn determine the performance and efficiency of the system (Energy Homes, 2008). Also, depending on the type of loop system (i.e., vertical or horizontal), the cost of geothermal heating systems can vary. It is estimated that on the average, installation cost of a typical home occupying 2,500 square feet of area (with a heating and cooling load of 17.58 kWh or 60,000 Btu) can range from \$20,000 to 25,000 (Energy Homes, 2008). Expressed in \$US/m^2, this average cost for home cooling and heating using GHP installation represents about \$86–\$108/m^2. In Table 10.6, we report the average capital costs of using GHP system for the various ground loop types that are already discussed in this section. Among these loop systems, the vertical closed-loop ground loop systems are considered

Table 10.6. Average GHP heating-ventilation and air conditioning (HVAC) system capital costs by ground loop type (Bloomquist, 2001).

Ground Loop Type	**\$/m^2 (# of data points)**	**\$/m^2/kW (# of data points)**
Vertical Closed	117.90 (50)	1,106.9 (39)
Horizontal Closed	55.10 (8)	717.7 (6)
Vertical Open/Groundwater	55.00 (7)	853.4 (5)
Hybrid (Vertical closed + cooling tower)	112.70 (1)	1,472.0 (1)

Table 10.7. Installed-cost estimates for ground loops (Goetzler *et al.*, 2009).

	Ground-Loop Installed Cost			
	(1995 dollars)		(2008 dollars)	
Ground Loops Type	**\$/ton**	**Relative Cost**	**\$/ton**	**Relative Cost**
Vertical	2,999	1.11	4400	1.32
Slinky	2,875	1.06	4200	1.26
Horizontal	2,712	1.00	4000	1.20
Open	—	—	3300	1.00

Relative cost = Cost compared with that of open loop system.
1 ton = 3000 kcal/hr or 12000 Btu/hr (cooling capacity of 1 ton of ice)

to be the most expensive, due to the associated high cost of deeper drilling. However, it has been claimed that the drilling cost can be reduced through the use of enhanced thermal conductivity grouts that will reduce the total depth of vertical drilling (Bloomquist, 2001).

In Table 10.7, we report a comparison of the installed-cost estimates for ground loops system in \$/ton of refrigeration (Goetzler *et al.*, 2009).

Based on the average data from Tables 10.6 (\$/m^2) and 10.7 (\$/ton in 2008 dollars), it can be estimated that about one ton of refrigeration will be required to cool a 46.4 m^2 (500 ft^2) area, using a GHP system.

The dimensionless coefficient of performance (COP) of a geothermal heat pump is a measure of the efficiency of the system in the heating mode, expressed as the ratio of total heating capacity in Btu per Btu of energy input to the system (Puttagunta and Shapiro, 2012).

$$\text{COP} = \frac{\text{Useful Heating Energy}}{\text{Net Energy Input}} \tag{10.1}$$

The Energy Efficiency Ratio (EER) is a measure of the efficiency of geothermal heat pump system in the cooling mode, representing the ratio of total cooling capacity per electrical energy input to the system (Puttagunta and Shapiro, 2012).

$$\text{EER} = \frac{\text{Useful Cooling Energy}}{\text{Net Energy Input}} \tag{10.2}$$

Currently, residential geothermal heat pumps on the market have standard COPs ranging from 2.4 to 5.0 and EERs ranging from 10.6 to 30 (Energy Star, 2008). During cold winters, geothermal heat pump systems can have coefficient of performance (COP) in the range of 3 to 6, compared to that of air-source heat pumps which can range from 1.75 to 2.5 on cool days (Puttagunta and Shapiro, 2012; Rafferty, 1997). However, the actual COP of a geothermal system, including the energy needed to pump the fluid through the underground tubes can be less than 2.5. For a geothermal heat pump to be rated as energy star (based on version 3.1 tier 3 rating, effective January 1, 2012) the COP and EER must have values as indicated in Table 10.8.

The Energy Star (2012) GHPs provide the maximum energy savings compared to all other designs of GHPs in the market. The installation cost for GHP systems is higher than conventional cooling and heating systems, but the high cost is normally offset through energy savings in 5–10 years. The loop fields of GHP systems are estimated to last for approximately 50 years, while the indoor components such as heat pumps can last for about 25 years, which is comparable to the life-span of any well-built heating and cooling system (US-DOE, 2011). The advantages of GHPs include less electricity consumption (i.e., about 25–50% less than conventional heating or cooling systems), reduction of greenhouse gas (GHG) emissions by as much as 44% compared to air-source heat pumps and up to 72% compared to electric resistance heating with standard air conditioning equipment. The other benefit of GHP systems is the relative indoor humidity control which the system can maintain at about 50% (Cross *et al.*, 2011). Additionally, GHPs have less maintenance costs, since the high-density polyethylene pipe typically used in loop fields is made to be flexible to resist breaking.

Table 10.8. Energy efficiency requirements based on Energy Star program (Energy Star, 2012).

Product Type	EER	COP
Closed Loop Water-to-Air	17.1	3.6
Open Loop Water-to-Air	21.1	4.1
Closed Loop Water-to-Water	16.1	3.1
Open Loop Water-to-Water	20.1	3.5

Water-to-Air: A geothermal heat pump model providing space conditioning primarily by the use of an indoor air heat exchange coil. These models may also provide domestic water heating and hydronic space heating by using desuperheater and/or demand water heating functions.

Water-to-Water: A geothermal heat pump model that provides space conditioning and/or domestic water heating by the use of indoor refrigerant-to-water heat exchanger(s). These models may provide domestic water heating by using desuperheater and/or demand water heating functions. The partial hot water generating component of the system is referred to as the "desuperheater".

Desuperheater: A pump and a heat exchanger installed in a way to remove waste heat (superheat) from the refrigeration cycle and transport it to the water heater before the refrigerant gas reaches the condenser (Gocze, 2013).

10.5. Geothermal Energy Projects in Illinois

In this section, we report four well-known geothermal energy projects in Illinois.

10.5.1. *Project 1*

In 2007, The University of Illinois at Chicago (UIC) renovated some of its buildings, which included installation of GHP systems. These renovations became necessary in order to make the buildings reflect the 21st century building innovations, which includes energy sustainability (Cross *et al.*, 2011). As a public institution of higher learning

Fig. 10.17. Pictures of the Grant Hall (left), Stevenson Hall (middle) and Lincoln Hall (right) at UIC (UIC 2008; Chicago Architecture Today, 2006).

with over 25,000 students, the UIC campus consists of 15 different colleges, encompassing over 100 buildings spanning over 240 acres of land, with most of the buildings constructed in the 1960's. The renovations began with the Grant Hall, a 15,000 feet square office and classroom building located at the Northeast side of the campus (see Fig. 10.17).

The cost of the Grant Hall geothermal project was about \$190,000, mainly funded through a \$153,813 grant received from the Illinois Clean Energy Community Foundation (ICECF). The ICECF is a foundation that awards grants to non-profit organizations, schools, municipalities and other government agencies for programs that improve energy efficiency or develop renewable energy resources to improve environmental quality in the state of Illinois (Cross *et al.*, 2011). The rest of the money was from UIC funds and the Port Family (a philanthropist). The geothermal field at UIC Grant Hall consists of 14 wells drilled in a vertical configuration that reaches a depth of 152 meters (~500 feet). The wells are spaced 6.1 meters (~20 feet) apart, each with a diameter of 6.4 centimeters (2.5 inches). The geothermal field occupies a small parcel of land and the wells are hardly visible because they are completely covered. In essence, the UIC Grant Hall geothermal heat pump has made it easier to maintain an annual constant temperature of 22.2°C (72°F) throughout the building. Besides, the building can now realize about 17–18% of energy savings with the geothermal cooling and heating systems, compared to the previous method of cooling and heating (Cross *et al.*, 2011). In the past years, the method of cooling and heating the Grant Hall relied on hot and chilled water from a centralized heating and cooling plant on the UIC campus. The success of the UIC's Grant Hall geothermal energy system has led to expansion of the GHP system, with additional 50-borehole loop field drilled at a depth of 154.2 meters (500 ft) in the same area. This expansion became necessary, in order to provide similar cooling and heating to the newly renovated 2,322 square meters (25,000 square ft) Lincoln Hall at UIC. The approximate installation cost of the additional GHP system was $650,000. An additional loop field was also designed to accommodate the nearby Douglas Hall, a 2,322 square meters (25,000 square ft) business school which was renovated in 2010 (Cross *et al.*, 2011). The university plans to upgrade some other UIC campus buildings with this technology, which would include Stevenson, Jefferson and Henry Halls.

10.5.2. *Project 2*

In 2009, an ordinance was passed in the Village of Oak Park, a suburb of Chicago in Illinois, requiring any retailer seeking to build a commercial property within the boundaries of the village to explore geothermal energy as an option. Following the passage of the ordinance, a Walgreens store located in the village was claimed to become the first drug store retailer in the nation to use geothermal energy for heating and cooling (Fig. 10.18).

The Walgreens heating and cooling system makes use of the geothermal energy of the earth, through a network of four closed-loop boreholes sunk to a depth of 198 meters (650 ft), and connected to a system of heat exchangers within the building (Buczynski, 2010). The water-based heat transfer fluid exchanges energy with the earth, providing a constant temperature of 55°F inside the building. The system was installed by Indie Energy Systems Company, LLC and uses EnergyLoop technology also designed by the Company for monitoring and optimization of the heat exchangers of the system in real-time to provide maximum energy efficiency. The geothermal system at the

Fig. 10.18. A rendering of Oak Park Walgreens, Illinois (GreenBiz, 2010).

Fig. 10.19. Geothermal Lake at Advocate Sherman Hospital, Elgin, Illinois (Sherman Health, 2014).

Walgreens store is expected to reduce energy use of the 12,825 square feet (∼1,200 m^2) building by 46%, equivalent to removal of nine cars from the road or planting 43 acres of trees (Frederick, 2010).

10.5.3. *Project 3*

In 2006, Advocate Sherman hospital located in Elgin in Kane County, Illinois was claimed to be the first hospital to build a lake loop GHP system in the U.S. (see Fig. 10.19).

The Illinois Clean Energy Foundation provided a $400,000 grant to reduce the higher installation cost of the system. The Foundation's $8.5 million support for geothermal installations systems in the state has led to over 150 installations in 57 Illinois counties. With 2,400 tons of cooling and heating capacity, the system was claimed to be the largest geothermal system in Illinois. Additionally, it was claimed to be one of the only two medical centers in the U.S. using lake-based geothermal system for heating and cooling. The Sherman Hospital's building is provided with cooling and heating from a 15-acre geothermal lake, which brings an estimated cost savings of $1 million in electricity and natural gas per year. The energy from the lake is exploited by a heat pump system connected to over 298 kilometers (185 miles) of pipes under water. Although the lake loop GHP system at Sherman Hospital cost 13% more to install compared to a HVAC system, the payback time for the investment is estimated to be in just few years (ICE, 2013).

10.5.4. *Project 4*

Another utilization of low-temperature geothermal energy is conversion of mine shafts into geothermal boilers to provide heating and hot water for buildings and homes (Rodríguez and Diaz, 2009). As mentioned in the Introduction of this chapter (Sec. 10.1), the Illinois Department of Military Affairs was awarded $1.2 million from the ARRA of 2009 (US-DOE, 2010b). With this award money, the Department is demonstrating that building retrofit costs for low level geothermal energy can significantly be lowered by using abandoned subsurface coal mines in Illinois, a resource which is in abundance in the state. This project originated as part of the Energy Policy Act (EPAct) of 2005, President Bush's 2007 Executive Order 13423, and Illinois Governor Quinn's Executive Order Number 11 (2009) including the Governor's own green energy initiative. Under the EPAct of 2005, the Illinois National Guard is required to significantly reduce its energy usage (OpenEI, 2011a; IGNN, 2013a). The project involved

Fig. 10.20. Left: Mine shaft shows the location w.r.t. where the geothermal energy could be taped and be delivered to local towns (Plataforma SINC, 2009); Right: Geothermal retrofit of Illinois National Guard State Headquarters building using low-temperature heat from abandoned coal mine (right)(US-DOE, 2010b).

retrofitting of aging boiler/chiller equipment with a geothermal system at the 74,000 ft^2 ($\sim$6,900 m^2) Illinois National Guard State Headquarters building located at Camp Lincoln, Illinois. Monitoring of the project was scheduled in August 2014 (see Fig. 10.20).

The geothermal system would utilize ground water trapped inside the Panther Creek #2 Coal Mine located at about 200 ft ($\sim$61 meter) directly underneath the building. The Panther Creek #2 Coal Mine was considered to be a good candidate for the project because of its stability over the last 82 years without any incidence of mine collapse. Also, important to the selection of this mine is the depth of the formation and its location within a major floodplain (water source), including its location which is directly underneath the target building (and campus). Between 1927 and 1948, about 1.76 million tons of coal was reportedly removed from the Panther Creek #2 Coal Mine (see Fig. 10.21) which surface area spans over three square miles.

The anticipated cost savings from the geothermal system is almost $40,000 per year or more than 275,600 kWh in energy savings over the existing HVAC system, which is currently confined to the Headquarters building. Furthermore, the amount of energy potential extractable from the mine will make it necessary to expand the project to other facilities on Camp Lincoln in the future. These include a

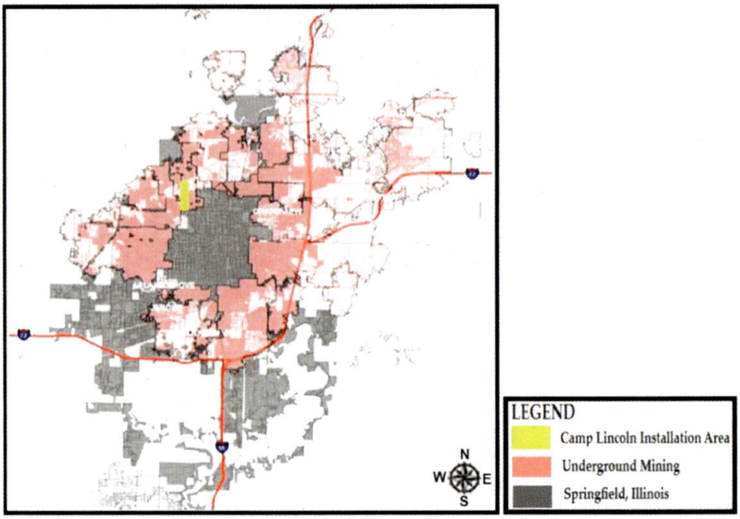

Fig. 10.21. Locations of Panther Creek #2 coal mine and the Illinois National Guard Building at Camp Lincoln, Illinois (Modified from US-DOE, 2010b).

replacement armory currently on the facility's master plan and a new "Combined Service Maintenance Support" structure scheduled for construction in 2016. It is anticipated that both projects will have a much greater energy savings advantage over conventional HVAC system (OpenEI, 2011a).

10.6. Problems Associated with Geothermal Energy Resource Development

The major problems associated with geothermal energy resource development include risk, high initial capital costs, management problems, and their likelihood of triggering earthquakes in areas with seismic gaps (i.e., a segment of active fault zones that has triggered significant earthquakes in the past years, but now remains dormant relative to other segments along the same fault zone).

To manage risks, it is crucial to obtain accurate identification, characterization and confirmation of the resource (Grant, 2013). Additionally, land prospecting and exploration may be hampered by permitting and lease restrictions, especially on the U.S. federal lands where geothermal resource developers require at least 2,000 acres of lease rights prior to resource exploration (EPRI, 2010). Moreover, low success rate of well drilling, significant depth requirements and insufficient productivity of the reservoir remain the major problems in geothermal energy resource development.

The high initial capital cost of geothermal energy development is a hindrance to any expansion of this technology. For example, well drilling alone represents a greater proportion of the capital cost of geothermal projects, often accounting for as much as one-third to one-half of the total cost (US-DOE, 2006). For this reason, geothermal energy projects are generally developed in incremental stages, despite the fact that large scale power plants are cheaper due to economies of scale. The project begins with construction of smaller generating capacity with provision for future expansion, instead of a single large-scale facility that with up-front cost can be very prohibitive. This approach is claimed to reduce the capital investment, allow faster cash flow and provide valuable cost-performance data for the development of additional capacity in future (EPRI, 2010).

Geothermal energy development can present unique challenges due to the fact that underground infrastructure is not well understood (EPRI, 2010). Normally, the reservoirs of geothermal power plants are managed to ensure maintenance of adequate water flow rates and sustained production. However, severe down-hole conditions and deteriorating hydrothermal fluid constituents can cause corrosion to power plant components, wells, pumps and piping (EPRI, 2010). Such conditions can contribute to high operating and maintenance costs for geothermal power plants. Also, depending on the location, geothermal power plants may compete for water that must be injected in to wells to sustain reservoir productivity or create artificial reservoirs.

There are also concerns about induced seismic activity mostly associated with the EGS technology, as described in Sec. 10.3.3. For example, land subsistence or sinkholes may occur due to extraction of hydrothermal underground fluids, as has occurred in some parts of the U.S. due to water withdrawal for municipal, agricultural and industrial usage (EPRI, 2010). According to a 2012 report released by the National Research Council (NRC), in recent decades, at least 13 U.S. states and 20 additional countries have experienced earthquakes linked with subsurface energy development (Weinhold, 2012). These states include Alabama, Arkansas, California, Colorado, Illinois, Louisiana, Mississippi, Nebraska, Nevada, New Mexico, Ohio, Oklahoma and Texas (Weinhold, 2012). It should be mentioned that most part of Southern Illinois is located on top of one of the two major fault zones in U.S. These are the New Madrid Seismic Zone (see Fig. 10.22) and the Wabash Valley Seismic Zone. In 1811 and 1812, the most powerful earthquakes in the U.S. history occurred along the New Madrid Fault zone, which includes a portion of Southern Illinois (Rafferty, 2014). Therefore, with Southern Illinois already inundated with oil and gas companies ready to begin a large-scale hydraulic fracking in the region, it is feared that these activities may trigger a major earthquake in the region. A 2008 research study by the Mid-America Earthquake Center at the University of Illinois projected that in case there is a repeat of the 1811–1812 earthquakes in Southern Illinois of the same magnitude, "there would be 3,500 fatalities, 2.6 million people without electricity and $300 billion in direct economic losses. Bridges, docks, highways and water infrastructure would be in shambles" (IGNN, 2013b). EGS projects involve high-pressure hydraulic fracturing that can cause seismic events similar to induced seismicity caused by shale gas production (US-EIA, 2014c).

An example of possible geothermal utilization causing earthquake is the 2006's 3.4 magnitude earthquake which occurred below the city of Basel in Switzerland, during hydraulic stimulation in an injection well of an EGS demonstration project. Following the

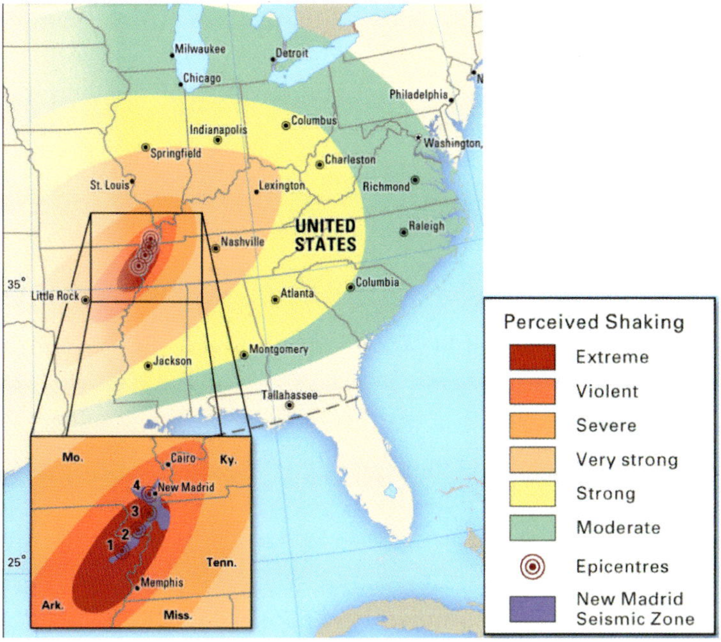

Fig. 10.22. Map of the New Madrid earthquakes of 1811–1812 (Encycl. Brit., 2014).

seismic activity, the project was halted, however what triggered the earthquake has not been conclusively attributed to the hydraulic stimulation that occurred slightly deeper than 5 km, located on a major fault line (EPRI, 2010). While such events are rare (US-DOE, 2014c), their occurrence can be catastrophic. Therefore, out of abundance of caution, on February 7, 2013, the Illinois governor called on every resident in the state, schools and businesses to participate in what was referred to as the "largest earthquake drill in the history of the Midwest". This third annual "Great Central U.S. Shake Out" was a nine-state emergency preparedness drill that was designed to encourage residents to think about what to do before, during and after an earthquake (IGNN, 2013b). Similarly, to allay the fears of the public about earthquake events, and without hampering the efforts in the development of subsurface energy projects, the US-DOE has developed a protocol to deal with seismicity issues (Majer *et al.*, 2012; US-EIA 2014c). As mentioned earlier in Sec. 10.3.3, EGS plants are currently being developed in several countries. The first commercial-scale plant in the U.S. (the Desert Peak East pilot project in Nevada) began operating in 2013 (US-EIA 2014c).

10.7. Sustainability of Geothermal Energy

The origin of geothermal energy is the tremendous amount of heat stored within the earth: 99% of the earth's volume has temperatures above 1000°C (>1,800°F), with only 0.1% at temperatures below 100°C (<212°F). The total heat content of the earth is estimated to be about 10^{13} EJ (one exajoule = 10^{18} Joules) or 94.782×10^{12} Quad, or Quadrillion (one Quad = 10^{15} Btu) (Rybach, 2007). Considering the current world energy consumption of ~560 Quad per year, it would take billions of years to completely exhaust the energy content of the earth.

GHP systems are generally risk free and sustainable and will help reduce the consumption of fossil fuels. However, one of the major risks associated with development of large-scale and deep geothermal energy systems is the high up-front cost of exploration and drilling (DiPippo, 2012). Even if drilling is successful, there is no guarantee that heat-carrying fluid has adequate temperature or has a sustainable flow rate. Moreover, drilling may cause cross-contamination of aquifers. Development of economically viable large-scale and deep geothermal system requires (i) sufficient amount of heat at accessible depth; (ii) existence of proper soil with enough porosity to allow transport of heat-carrying fluid and (iii) presence of impermeable levels to prevent flow of geothermal fluids to leak to the surface of the earth (Bruni, 2014). These characteristics of geothermal source are crucial for a successful development of large-scale and deep geothermal energy systems. The science and technology of underground exploration and drilling, developed by the petroleum and natural gas production industry (see for example Theodoropoulos, 2011), can be utilized for geothermal energy application.

It should be pointed out that factors such as exploration costs, sustainability of the reservoir, type of cooling system and market factors (generated-electricity sell price and raw material cost) can have significant effect on the economics of large-scale and deep geothermal power generation systems (Hillesheim and Mosey, 2013).

Other than economic aspects of large-scale and deep geothermal energy, environmental issues related to this power source have to be considered as well. The main environmental effects of such geothermal development are related to physical effects of fluid withdrawal, heat effects and discharge of acid gases and salts. These issues will be briefly discussed in the following sections.

10.7.1. *Water use and quality*

One of the main concerns of development of large and deep geothermal power generation systems is its water withdrawal, which may cause subsidence and lowering of groundwater level. In addition, reinjection of fluid into the well may induce seismicity (Ármannsson and Kristmannsdóttir, 1992). Subsidence takes place when mass of fluid withdrawn exceeds the natural inflow. This in effect causes compactness of the loose formations (i.e., clays) on top of the withdrawal site and causes the land to sink. The largest subsidence on record is at Wairākei, New Zealand, where the center of the subsidence is sinking at a rate of almost half a meter every year (Stewart, 2012). Lowering of groundwater level may cause mixing of fluids between aquifers and hence, cause deterioration of drinking water.

Water usage in geothermal energy depends on the type of technology used for energy conversion. Most geothermal power generation facilities use closed-loop water systems, in which extracted water is pumped back into the geothermal reservoir after it has been used for power production. However, over time, efficiencies of geothermal power generation facilities may decline and may require fresh water

Table 10.9. Comparison of average water consumption estimates in (liters/kWh) for power generation technologies over lifetime energy output (Clark *et al.*, 2012).

		Fuel Production	**Plant Construction**	**Plant Operations**	**Total Life Cycle**
G.T. PP*	EGS	—	0.0380	1.1–2.72	1.13–2.76
	Binary cycle	—	0.0038	0.3–1.02	0.30–1.02
	Flash	—	0.0038	0.02–0.038	0.038
Fossil-Fuels PP	Coal-fired	0.98	—	0.015–4.50	0.98–5.50
	Coal-CCS	0.38–0.64	0.49–0.95	1.9–4.5	2.16–5.80
	NG	1.1	—	0.34–2.60	1.44–3.70
	NG-CC	0.83	—	0.076–1.900	0.9–2.70

(*) In life-cycle analysis, plant lifetime of 30 years was assumed.
Coal-CCS: Coal-fired power plant with carbon capture and sequestration.
NG-CC: Natural Gas combined Cycle; PP = power plant.

to manage dissolved solids, reduce scaling and provide makeup water (Clark *et al.*, 2011).

Clark *et al.* (2012) conducted a Life Cycle Analysis (LCA) to examine water usage of geothermal power plants. In their analysis, they compared three different geothermal plants with fossil fuel technologies, as summarized in Table 10.9.

The results of Table 10.9 indicate that among geothermal technologies, water consumption per kilowatt hour of EGS during construction is an order of magnitude greater than Binary and Flash technologies. The same trend, but with less intensity, is observed in water consumption of EGS technology during operation. The higher water requirements for the EGS technology are not only influenced by depth of well, but caused by a combination of factors, including lower flow rates per well, which necessitates higher number of wells to be drilled (Clark *et al.*, 2010). Figure 10.23 shows water consumption of geothermal systems as a function of well's depth.

Make-up water is often used to compensate for evaporative losses or added to the re-injected steam condensate. Make-up water does not need to be fresh water, but low-quality water requires more cycling and larger volumes (BLM, 2011).

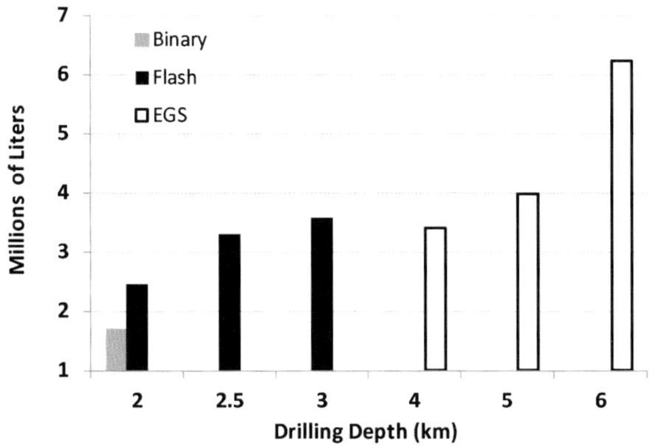

Fig. 10.23. Water consumption in drilling and construction of a binary, flash and EGS geothermal well (Clark *et al.*, 2011).

Changes in temperature affect the chemical equilibrium of minerals in the ground and influence mixing process. For example, carbonate precipitation, salt solubility, dissolution of silicate minerals, oxygen saturation and gas solubility can be affected by temperature variations (Hähnlein *et al.*, 2013) and may change water quality.

Overall, geothermal power plants water consumption per kilowatt hour of lifetime energy output is less than fossil fuel power generating technologies. Among the geothermal technologies, flash power plants use the least amount of water since they use the geothermal fluid (geofluid) for cooling. However, the long-term sustainability of such an approach is unknown with average geofluid losses estimated at 2.7 gal/kWh (10.2 liter/kWh) (Clark, 2010). The US-EPA calculates a range of about 0.75–1.15 m^3/MWh (0.2–0.3 gal/kWh) of total water volume consumed for electricity generation from geothermal resources (Bayer *et al.*, 2013), which is in line with the numbers reported in Table 10.9.

In order to reduce the impact of geothermal energy generation on water resources, strategies can be implemented for water conservation. Among these strategies, reuse of water during construction, optimization of operational water consumption, use of non-freshwater makeup and use of binary system are recommended. Binary systems, in particular, can enhance sustainability of the reservoir by preventing evaporation of geofluid cooling losses. By implementing these strategies, improvements in efficiency of geothermal power plants can be made across the plant life cycle to further reduce the impacts of geothermal power on freshwater resources.

Cooling system plays an important role in development of efficient and sustainable geothermal power generation, as the choice of a proper cooling system can have substantial environmental impacts on local water resources (Macknick *et al.*, 2012).

The major factor in sustainability of geothermal power generation is preventing depletion of water that transfers the thermal energy of the earth to the surface. Use of field-proven air-cooled Organic Rankine Cycle-based geothermal power plant can help achieving this goal and extend the lifespan of the wells and reducing emissions (Hammons, 2011).

10.7.2. *Emissions related to geothermal energy*

The geothermal power generating technologies presented in Sec. 10.3 may be categorized into two classes based on their GHG emissions. (i) Binary plants, in which the geofluid is not exposed to the atmosphere and hence, have negligible or no operational GHG emissions. (ii) Geothermal flash plants, in which the geofluid is exposed to the atmosphere and releases GHGs (Sullivan and Wang, 2013).

Overall, compared to fossil-fuel-burning power generating technologies, geothermal power plants are more environmentally benign, but nonetheless, fluids pumped from geothermal reservoirs often contain high levels of chemicals such as sulfur and its compounds, salt and minerals and a fraction of dissolved non-condensable gases (NCG). These NCG at normal temperature often include nitrogen, carbon dioxide, methane, argon, oxygen and hydrogen sulfide. For efficient operation of a geothermal system, these NCGs have to be removed and, with the exception of hydrogen sulfide, can be released to the atmosphere (DiPippo, 2012).

Gas emissions from geothermal power plants vary in compositions and concentrations, but methane (CH_4), hydrogen sulfide (H_2S) and carbon dioxide (CO_2) are often discharged from geothermal plants. The toxic gases like H_2S are heavier than air and would collect in pits, depressions or confined spaces (Stewart, 2012).

Geothermal plants do not emit sulfur dioxide (SO_2) directly, but when H_2S is released as a gas into the atmosphere, it disperses in the air and oxidizes to sulfur dioxide and sulfuric acid. Some of these sulfur compounds have been identified as components of acid rain. However, no scientific evidence exists to directly link H_2S emissions to acid rains (Hunt, 2000).

Emissions of greenhouse gases from different power generating technologies are shown in Table 10.10. Emission values for geothermal energy vary from technology to technology, but instead of specifying each technology in Table 10.10, they are shown as a range. The reported range of SO_2 and CO_2 emissions from coal power plants is wide and it is believed this wide range depends on the age of the plant and type of coal burned.

The operating geothermal power plants can be classified into dry-steam, flash-steam, or binary power cycles. Among these, dry- and flash-steam plants are likely to emit measurable amounts of GHGs. The non-condensable gases emitted from geothermal power plants include carbon dioxide, ammonia, nitrogen, methane, hydrogen sulfide and hydrogen, but the amount of these gases can vary greatly from one resource to another.

It is reported that non-condensable gases typically make up less than 5% (by weight) of total gases, while volume of CO_2 is more than 97% of non-condensable gases (Holm *et al.*, 2012). However, despite the low amount of non-condensable gases in geothermal fluids, their measurement is imperative during preliminary stage of well testing. As discussed earlier, characteristics of each geothermal site is unique and careful analysis of reservoir's temperature, pressure, fluid composition, amount and type of non-condensable gases is required since these characteristics impact the design and type of a power plant.

Table 10.10. Emissions levels by energy source all in gram per kWh of energy produced.

	Geothermal	CNG-CC	Adv.NGCC +CCS	Hydro*	Adv. Nuclear	IGCC	IGCC+CCS
CH_4	0.75[(1)]	0.008[(5)]	N/A	N/A	N/A	0.114 [(5)]	N/A
H_2S	0.03–6.4[(1,2)]	N/A	N/A	N/A	N/A	N/A	N/A
$PM_{2.5}^{‡}$	N/A	0.05[(5)]	N/A	N/A	N/A	0.267 [(5)]	N/A
$PM_{10}^{‡‡}$	N/A	0.054[(5)]	N/A	N/A	N/A	0.325 [(5)]	N/A
SO_2	0.0001–0.3[(3,4)]	0.0015[(4)]	0.0015[(4)]	0[(4)]	0[(4)]	0.038–8.5[(4,5)]	0.0385[(4)]
NO_x	0[(4)]	0.0116[(4)]	0.0116[(4)]	0[(4)]	0[(4)]	0.0116[(4)]	0.0116[(4)]
CO_2	27–185[(1),(5)]	180.2–597[(1,4)]	18.5[(1,4)]	0[(4)]	0[(4)]	317–994[(4,5)]	31.7[(4)]

(1) Bloomfield *et al*. (2003); (2) Hunt (2000); (3) Kagel *et al*. (2007); (4) US-EIA (2013); (5) Matek (2013).
CNG-CC: Conventional Natural Gas Combined Cycle.
Adv. NGCC+CCS: Natural Gas Advanced Combined Cycle with Carbon Capture and Sequestration.
IGCC: Integrated Coal-Gasification Combined Cycle
IGCC+CCS: Integrated Coal-Gasification Combined Cycle with Carbon Capture and Sequestration.
(*) Hydroelectric is assumed to have seasonal storage so that it can be dispatched within a season, but overall operation is limited by resources available by site and season.
(‡)Particulate matter, with size less than 2.5 micrometers in diameter.
(‡‡)Particulate matter, with size less than 10 micrometers in diameter.

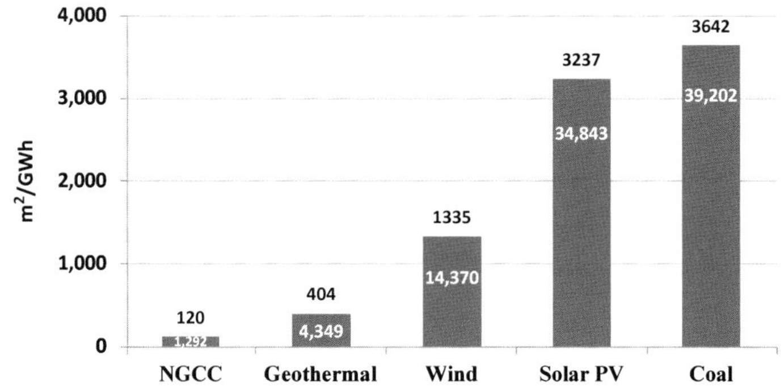

Fig. 10.24. Land footprint for different power generating technologies (numbers inside the bar are ft^2/GWh) (GeoBasics, 2015; Florin and Fennell, 2010).

10.7.3. *Land footprint of geothermal energy*

Land footprint of geothermal power plants is small; that is geothermal power facilities require less land per unit of power generation when compared with other technologies, such as nuclear or coal power plants of similar capacities (see Fig. 10.24).

According to Fig. 10.24, estimate of land requirement for geothermal power plants is 404 m^2/GWh, as opposed to 3,642 m^2/GWh for coal-burning power plants. However, land requirement for natural gas combined cycle power plants is about 70% less than geothermal power plants.

U.S. bureau of land management has published the results of their comprehensive study (BLM, 2008) of land disturbances by different phases of development of a typical geothermal power plant. The study considered a 30 MW and 50 MW plant, but the type of geothermal plant was not specified. Table 10.11 shows the results of the study done by BLM.

As seen from Table 10.11, most of the land disturbances are attributed to installation of transmission lines, since 45–65% of the total land dedicated to development of geothermal technology is used for transmission lines.

One of the problems of development of geothermal power generation is changes to the original landscape. The visual intrusion from

Table 10.11. Typical land disturbances during development of a 30–50 MW geothermal power generation (BLM, 2008).

	Disturbance Estimate	
Development Phase	**(acres)**	**(1000 m^2)**
Exploration	**2–7**	**8.1–28.3**
Geologic mapping	~0	~0
Geophysical surveys	30 ft^2	2.8 m^2
Gravity & magnetic surveys	~0	~0
Seismic surveys	~0	~0
Resistivity surveys	~0	~0
Shallow temperature measurements	~0	~0
Road/access construction	1–6	4–24
Temperature gradient wells	1	4.0
Drilling Operations and Utilization	**51–350***	**206–1416**
Drilling & well field development	5–50	20–202
Road improvement/construction	4–32	16–130
Powerplant construction	15–25	61–101
Installing wellfield equipment & pipelines	5–20	20–81
Installing transmission lines	24–240	97–971
Well workovers, repairs & maintenance	~0	~0
TOTAL	**53–367**	**214–1485**

~0 = Negligible

(*)For unexplained reasons the total for "Drilling Operation and Utilization" as reported in the reference in not the summation of individual numbers reported.

installations of geothermal power plants may irreversibly alter a great scenic quality. However, small geothermal power plants with a size of 100 MW can be designed and built to have minimal disturbances even in recreational areas. Flash steam geothermal plants have lower land footprints per MW as compared with small-size binary geothermal plants, because cooling tower installation in the latter increases the land footprint.

10.7.4. *Costs associated with geothermal energy*

The average geothermal gradient is about 2.5–3°C/100 meter depth (1.37–1.65°F/100 ft depth), but a reservoir temperature less than 110°C is considered too low for cost-effective electricity production (Augustine *et al.*, 2012). Shallow wells at depth of less than 400 m (1300 ft) is an attractive source for heating and cooling of buildings (Hähnlein *et al.*, 2013), but for electricity production higher temperature at higher depth needs to be reached.

The cost of a geothermal power plant depends on many factors including the type of plant, temperature of the well and the type of cooling system. While it is true that geothermal plants require significant amount of capital start-up, the overall cost of geothermal plants are very competitive with other power generating technologies.

Bruni (2014) reports that surface activities for geothermal development in the U.S. is typically between 1 and 1.5 million dollars (US$), while 1.5–3 million dollars should be allocated for each slim hole and 2.5–3.5 million dollars for each kilometer depth (0.62 mile) of commercial well dug. Despite these apparent high costs, the levelized cost of a geothermal power plant is indeed less than other power generation facilities, as shown in Table 10.12.

As expected, majority of the cost associated with development of geothermal power generation is for construction. The second highest cost of the process is for exploration and production drilling, which together comprise 42.1% of the total cost, as shown in Table 10.13. Because of requirement of larger diameter, presence of hard rocks and operation at higher temperature, the cost of geothermal wells is 2–5 times greater than oil and gas wells of comparable depth and increases non-linearly with depth (Tester *et al.*, 2011).

In addition to environmental benefits, there are direct economic benefits associated with development of geothermal power generation. Close to 50% of geothermal plants are on public lands, generating revenue for state government. In addition, geothermal industry

Table 10.12. Comparison of estimated average levelized cost of electricity (LCOE) of geothermal energy (GE) with other power generating technologies (based on the values of $US in 2012) for plants entering service in 2019 (US-EIA, 2014b).

		GE	CNG-CC	Adv. NG-CC +CCS	Hydro*	Adv. Nuclear	IGCC	IGCC+CCS
	Capacity factor (%)	92	87	87	53	90	85	85
($/MWh)	Levelized capital cost	34.2	14.3	30.3	72	71.4	76.1	97.8
	Fixed O&M	12.2	1.7	4.2	4.1	11.8	6.9	9.8
	Variable O&M ‡	0	49.1	55.6	6.4	11.8	31.7	38.6
	Transmission invest.	1.4	1.2	1.2	2	1.1	1.2	1.2
	Total system LCOE	47.9	66.3	91.3	84.5	96.1	115.9	147.4

CNG-CC: Conventional Natural Gas Combined Cycle.
Adv. NGCC w/CCS: Natural Gas Advanced Combined Cycle with Carbon Capture and Sequestration.
IGCC: Integrated Coal-Gasification Combined Cycle.
IGCC w/CCS: Integrated Coal-Gasification Combined Cycle with Carbon Capture and Sequestration.
(*)Hydroelectric is assumed to have seasonal storage so that it can be dispatched within a season, but overall operation is limited by resources available by site and season.
(‡)The variable cost includes the fuel cost.

Table 10.13. Estimated breakdown costs for a typical 50 MW geothermal power plant (Cross and Freeman, 2009).

Developmental Stage	Installed Cost ($/kW)	% of total cost
Exploration	14	0.4
Permitting	50	1.4
Transmission	100	2.7
Exploratory Drilling	169	4.6
Steam Gathering	250	6.8
Production Drilling	1,367	37.5
Plant & Construction	1,700	46.6
TOTAL	**3,650**	**100**

can generate 1.7 permanent jobs per MW of power capacity for plant operation, maintenance, research and consulting. This is in addition to 3.1 temporary jobs per MW installed for manufacturing (Matek and Gawell, 2014).

Despite the fact that most of geothermal power plants are in the Western part of the U.S., Illinois has resources available for development of such technology. Illinois geothermal resource base is estimated at 5.17×10^{10} GWh, as shown in Table 10.2 (US-DOE, 2006). Geothermal energy resources available in Illinois are in the form of hydrothermal energy and HDR energy (Lopez *et al.*, 2012). Based on an assessment by Vaught (1980), the subsurface geothermal gradient in Illinois varies from 1.2 to 2°F/100 ft (2.2–3.6°C/100 meters), as shown in Fig. 10.25. However, assessment of data by Vaught indicated that shallow wells (less than 3000 ft, or 914 meters) seem to have higher gradient values than the deep wells. Vaught attributed this variation to movement of groundwater, errors in bottomhole temperature measurements or variation in the heat flow and thermal conductivity of the rocks. Vaught suggested that bottomhole temperatures may not be reliable indicators of temperature at depth when data are taken from shallow oil and gas wells. A more reliable estimate of the average gradient for Illinois seems to be 1.0–1.2°F/100 ft (1.8–2.2°C/100 m),

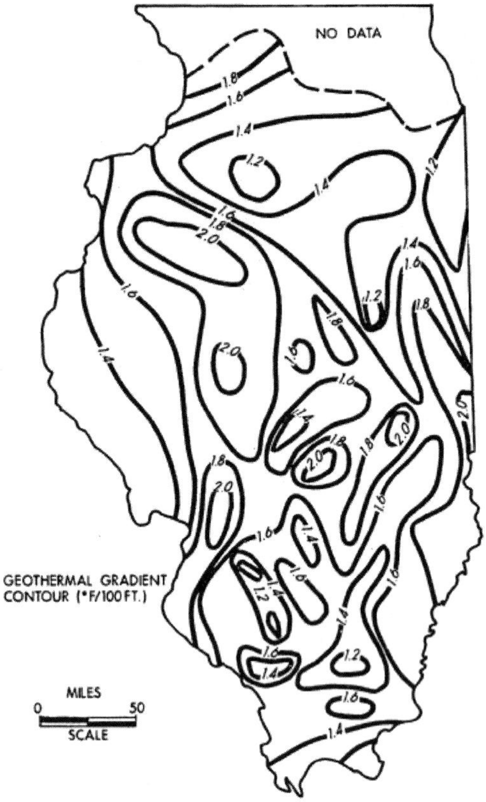

Fig. 10.25. Geothermal gradients map of subsurface temperature in Illinois (Vaught, 1980).

which seems to be consistent for deep wells throughout the state.

High-temperature, convective hydrothermal systems are usually associated with tectonic activity. Although the Eastern U.S. is generally regarded as tectonically stable, some seismicity remains. Based on the data gathered by Vaught, most seismic activity in Illinois between 1800–1972 occurred in the Southern part of the state. This suggests that Central part of the state might be safer for development of deep and large geothermal energy projects.

Although physical limits for development of geothermal energy are not fully established, provisional limits for depth and temperature

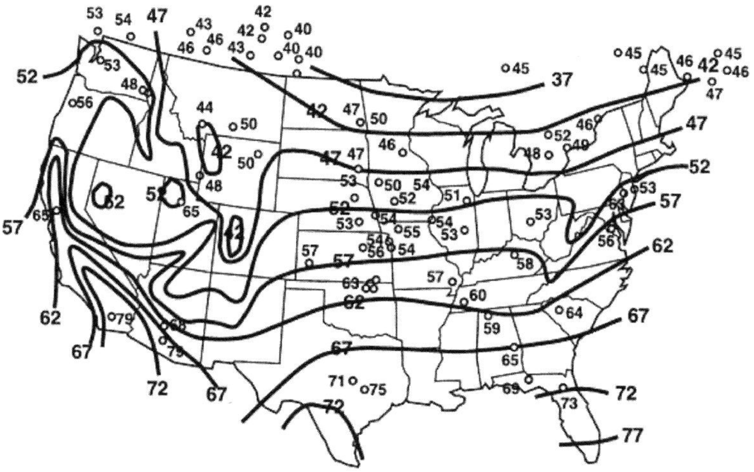

Fig. 10.26. Mean annual and well-water temperature (°F) contours in the U.S. (Earth River Geothermal, 2015).

can be suggested. Figure 10.26 shows the mean annual and shallow groundwater temperatures in the U.S. and indicates that shallow groundwater temperature spans from 37°F (2.8°C) in upper Midwest to 77°F (25°C) in Southeastern part of the country. Illinois's mean annual temperature varies from 51 to 53°F (10.5–11.7°C) and its shallow groundwater temperature varies from 52 to 57°F (11.1–13.9°C) (US-EPA, 2014). Average temperature in Illinois is below 50°F (10°C) half of the year and geothermal energy can be utilized as a reliable source of energy for heating.

As of 2014, large-scale geothermal power generation in Illinois does not exist. This may be due to high up-front cost of geothermal energy. With advancement of drilling technology and lowering the cost of drilling, it would be economical to develop geothermal energy for power generation in Illinois. Meanwhile, shallow-depth geothermal energy for heating and cooling of buildings has been developed successfully in the state, as mentioned in Sec. 10.5.

In the above sections, we have presented the economics and environmental issues related to geothermal energy utilization, which are the two out of the three pillars of sustainability. The third pillar of sustainability, i.e. social acceptance, must also be taken into account for successful development of geothermal energy utilization. There have been some detailed studies made in the social acceptability of geothermal energy in several countries including Australia, Italy, Indonesia, etc. (Carr-Cornish and Romanach, 2012; Dowd *et al.*, 2010; Pellizzone *et al.*, 2013). The overall conclusion is (as with any new project related to technology and land management) that geothermal acceptability by citizens is quite related to success of the project implementation. Also, public education about the potential advantages of geothermal energy will contribute to its social acceptance. Mass media, like newspapers, radio and television can also have a strong influence on shaping the minds of the public towards utilization of geothermal energy.

10.8. Concluding Remarks

Diversity and abundance of the U.S. renewable energy sources, such as geothermal, biomass, solar, hydropower and wind, has put the U.S. in a unique position to utilize these renewable energy sources and reduce its dependence on fossil energy.

Geothermal power generation is growing internationally at the sustained rate of 4–5% and close to 700 geothermal projects are under development in 76 countries around the world. The claimed world's largest geothermal power plant began commercial operation in Olkaria, Kenya in 2015. This 280 MW power plant is expected to provide approximately 20% of Kenya's total power capacity (PEI, 2015). In contrast to the global market, the U.S. market has not experienced the same growth in the geothermal energy production (GEA, 2014). As of April 2014, the U.S. had a total of 3,442 MW installed and 978 MW of planned capacity of geothermal energy.

Major and deep geothermal energy resources are concentrated in the Western part of the U.S. and are relatively limited to hydrothermal energy. With advancement in research and development in geothermal energy technology or co-production from oil and gas wells, the potential of this renewable energy can be expanded (Mai *et al.*, 2012). For example, research has been done to develop a process by which the electricity generated from geothermal power plant is exploited to produce hydrogen through water electrolysis (Rahmouni *et al.*, 2014).

Illinois has a great potential to become one of the centers of renewable energy in the U.S. From 2006 to 2013, Illinois, California, Texas, Iowa, Arizona and Oregon attracted over half of all U.S. renewable energy investment (Bloomberg, 2015).

Geothermal energy is a renewable form of energy resource that can contribute to energy and environmental sustainability and help a state to attain energy independence. It is the second most abundant energy source on earth after solar energy, and can be used for cooling and heating (Lee *et al.*, 2007).

An examination of the full life cycle of geothermal energy developments indicates that the overall environmental impacts of geothermal energy are relatively lower than coal-fired power plants in terms of CO_2 and NO_x emissions (US-DOE, 2006). For example, geothermal binary cycle power plants emit 0.2 pounds of SO_2 per MMBtu, 120 pounds of CO_2 per MMBtu and zero emissions of nitrogen oxides. On the other hand, advanced pulverized coal-fired power plants emit 0.1 pounds of SO_2 per MMBtu, 206 pounds of CO_2 per MMBtu and 0.06 pounds of NO_x per MMBtu (US-EIA, 2013). Moreover, geothermal power plants provide dispatchable base-load capacity without the need for storage or backup-power equipment.

Geothermal energy resources available in Illinois and most other parts of the U.S. and the rest of the world are in the form of earth's energy, hydrothermal energy and HDR energy. As mentioned in Sec. 10.2, the geothermal resource base in Illinois is estimated at 5.17×10^{10} GWh (186, 123 $\times 10^{18}$ Joules). Also, the estimated total technical potential for enhanced geothermal system in Illinois is 676,056 GWh, ranking the state 18th place among the 50 states in the U.S. (Lopez *et al.*, 2012).

In recent years, the geothermal industry has seen some successes, especially in southern and central Illinois. At the same time, the industry is beginning to realize some success in the northern part of the state. By the end of 2011, the ELPC, an Illinois group responsible for protecting the Midwest's Environment and Natural Heritage, had identified 89 businesses in the Illinois geothermal energy supply chain (Graig *et al.*, 2011).

Currently, there is no geothermal power plant in Illinois. But, the co-produced hot water from the oilfields in southern Illinois, which is estimated at $\sim 2.2 \times 10^9$ barrels (in 2004) as mentioned in Sec. 10.2.2, can be used for electricity production (Clark and Veil, 2009). The geothermal energy technologies currently in Illinois are GHP systems. These include horizontal closed-loop systems, vertical closed-loop systems, pond/lake systems and opened-loop systems. Capital cost and emissions of some of the power generating plants are shown in Table 10.14.

According to Table 10.12, the fixed O&M cost of geothermal power plants are higher than all other power plants. Economically, geothermal power plants are predicted to be cheaper with technology improvement. It is predicted that for geothermal power plants entering service in 2016, the levelized cost of energy will be $91.0–115.7 MWh (based on 2009 U.S. dollar), which could make geothermal energy more competitive with coal energy in the future (US-EIA, 2014b). The decreasing trend of geothermal capital cost will continue and it is estimated that U.S. average levelized cost (in 2012 dollars) for geothermal power plants entering service in 2019 will be $47.9 per MWh, compared with $95.6 per MWh for conventional coal-fired power plants, as shown in Table 10.15.

But currently, the high capital cost of geothermal power plants makes them less attractive for power generation. It is worth mentioning that small-scale geothermal heating and cooling systems for individual buildings, whether residential or commercial, will be quite

Table 10.14. Comparison of characteristics of geothermal power plants with other power plants technologies.

	($lb/10^6 Btu$)					
Technology	**SO_2**	**NO_X**	**CO_2**	**Nominal Capacity (MW)**	**Capital Cost ($/Kw)**	**Fixed O&M Cost ($/kW-yr)**
Single Unit Advanced Pulverized Coal[(1)]	0.1	0.06	206	650	3,246	37.80
NG Conventional CC[(1)]	0.001	0.0075	117	620	917	13.17
Advanced Nuclear[(1)]	0	0	0	2,234	5,530	93.28
Onshore Wind[(1)]	0	0	0	100	2,213	39.55
Photovoltaic[(1)]	0	0	0	20	4,183	27.75
Geothermal — Binary cycle[(1)]	0.2	0	120	50	4,362	100
Geothermal — Dual Flash cycle [(1)]	0.2	0	120	50	6,243	132
*Enhanced Geothermal Systems[(2)]	NA	NA	NA	NA	9,625	0
*Hydrothermal Power Plant[(2)]	NA	NA	NA	NA	5,940	0

(1) US-EIA (2013); (2) Black and Veatch (2012).
(*) 2020 Cost and performance projection for an enhanced geothermal systems and hydrothermal power plants.
Dual Flash = A steam power cycle which includes a second low-pressure flash to the single-flash cycle.
NA = Not Available.
Fixed O&M Cost = expenses incurred at a power plant that do not vary significantly with generation (i.e., staffing and monthly fees, plant support equipment, routine preventive and predictive maintenance etc.)

Table 10.15. Estimated levelized cost for plants entering service in 2019 (US-EIA 2014b).

Plant Technology	U.S. Average LCOE (2012$/MWh)
Conventional Coal	95.6
IGCC with CCS (++)	147.4
AG NGCC	64.4
AG NGCC with CCS (++)	91.3
Advanced Nuclear	96.1
Biomass Combined Cycle	102.6
Geothermal	47.9
Onshore-Wind	80.3
Offshore-Wind	204.1
Solar Photovoltaic	130.0
Solar Thermal (+)	243.1

AG NGCC = Advanced Generation Natural Gas-fired Combined Cycle.
IGCC with CCS = Integrated gasification and combined cycle with Carbon capture and sequestration.
(+) Costs are expressed in terms of net AC power available to the grid for the installed capacity.
(++) Assuming 100% capture.

beneficial, especially if it is planned to be installed upon construction of the building. In Table 10.15, we present the estimated total levelized cost (excluding subsidy) of new generation resources in the year 2019.

Table 10.16 shows the capital costs of using the GHP system for different types of buildings. The differences in the capital costs can be attributed to the type of building, the ground loop type, the ground or water temperature, including the rock or soil type that can determine the level of difficulty in drilling (Bloomquist, 2001). For unknown reasons, the data points (shown in parenthesis) in Table 10.16 do not add up to 72 and 55, as reported in the literature source.

The average ground water temperature in Illinois is about 54°F (Lund, 1988). The average capital cost of GHP systems installation in

Table 10.16. GHP HVAC system capital costs by the type of building (Bloomquist, 2001).

Building Type	$/m^2 (# of data)	$/m^2/KW (# of data)
Retail	35.90 (05)	1,097.4 (03)
Medical Centers	84.10 (02)	811.1 (02)
Office Buildings	85.00 (13)	1,005.1 (11)
Apartment/Multi-Residential	100.00 (02)	1,059.1 (02)
Schools	115.40 (32)	1,020.6 (24)
Retirement	126.20 (03)	1,119.1 (02)
Prisons (correctional facility)	134.90 (03)	1,320.6 (02)
Gas Station/Convenience Store	232.40 (01)	1,952.3 (01)
All Case Studies & References	**100.10 (72)**	**1,026.6 (55)**

buildings is about $100+/m^2, representing about two times the cost of a conventional heating, cooling and hot water system installations (Bloomquist, 2001; Energy Homes, 2008). However, as mentioned in Sec. 10.4.4, geothermal heating/cooling systems consumes about 25–50% less electricity than conventional heating or cooling systems (Cross *et al.*, 2011). The high capital cost of GHP system installation for correctional facilities is due to security concerns considered in the system design, while the high capital cost for gas station/convenience stores is due to the small floor area compared to the load (Bloomquist, 2001).

In Table 10.17, we report the rated efficiencies and installed-cost estimates for a range of small-commercial space-conditioning technologies as of 2007 (Goetzler *et al.*, 2009).

The discussion presented throughout this chapter highlighted many facets of geothermal power technology. While geothermal energy can be considered as a reliable, sustainable and secure source for power generation, many challenging issues ought to be tackled to ensure successful and long-lasting operation of this technology. It is an indisputable fact that geothermal energy is environmentally cleaner than fossil fuels; however the high initial capital cost of this

Table 10.17. Competing commercial space-conditioning technologies (Goetzler *et al.*, 2009).

Technology	Typical Rated Efficiencies		Typical Installed Cost[(a)] $/kBtu]	Typical Installed Cost [$/kWh]
	Cooling	Heating		
Gas-Fired Furnace	NA	80%	8.1–8.8	27.6–30
Oil-Fired Furnace	NA	81%	8.1	27.6
Roof-Top Air Conditioner	10.1 EER	NA	65.6–85.0	224–290
Roof-Top Heat Pump	10.3 EER	NA	73.0–97.0	249–331

(a) Based primarily on retrofit installations, as this is the general case. Figures are mid-range values from EIA 2007.
Heat-pump costs are per nominal ton of cooling capacity. NA = not applicable
EER: Energy Efficiency Ratio, EER=E_c/P_a, where E_c: net cooling capacity(Btu/hour), P_a: applied electrical power (Watts).

technology at the present time might not justify its application in some areas. However, it is estimated that with advancement of manufacturing and drilling technologies, the issue of cost will be less pronounced in the near future.

Bibliography

Agyarko, L. B. and Mansoori, G. A. (2013). Non-renewable Energy Options in Illinois, *Int. J. Oil, Gas and Coal Tech.*, **6**(3): 288–347.

Anderson, D. N. and Lund, J. W. (eds.) (1979). Direct utilization of geothermal energy: A Tech. handbook, Doc. No. DOE/ET/27256-T9, Geothermal Resources Council, Davis, CA.

Aneke, M., Agnew, B and Underwood, C. (2011). Performance Analysis of the Chena Binary Geothermal Power Plant, *Applied Thermal Eng'g*, **31**(10): 1825–1832.

Ármannsson, H. and Kristmannsdóttir, H. (1992). Geothermal Environmental Impact, *Geothermics*, **21**(5/6): 869–880.

Augustine, C. (2011). Updated U.S. geothermal supply characterization and representation for market penetration model input, Tech. Report No. NREL/TP-6A20-47459, NREL, U.S. DOE, Washington, DC.

Augustine, C., Bain, R., Chapman, J., Denholm, P., Drury, E., Hall, D. G., Lantz, E., Margolis, R., Thresher, R., Sandor, D., Bishop, N.,A., Brown, S. R., Cada, G. F., Felker, F., Fernandez, S. J., Goodrich, A. C., Hagerman, G., Heath, G., O'Neil, S., Paquette, J., Tegen, S. and Young, K. (2012). Renewable electricity generation and storage technologies, Vol. 2 of Renewable Elect. Futures Study, NREL/TP-6A20-52409-2, Nat'l Renewable Energy Lab, U.S. DOE, Golden, CO.

Battocletti, E. C. and Glassley, W. E. (2012). Measuring the costs and benefits of nationwide geothermal heat ump deployment, Geothermal Resources Council Transactions, Vol. 36, Geothermal: Reliable, Renewable, Global, GRC 2012 Annual Meeting.

Bayer, P., Rybach, L., Blum, P. and Brauchler, R. (2013). Review on Life Cycle Environmental Effects of Geothermal Power Generation, *Renewable and Sustainable Energy Reviews*, **26**: 446–463.

Black and Veatch (2012). Cost Rep.: Cost and performance data for power generation technologies, Nat'l Renewable Energy Lab., U.S. DOE, Washington, DC.

[BLM] Bureau of Land Management (2008). Final programmatic environmental impact statement for geothermal leasing in the Western United States. News release, U.S. Dept of Interior, Washington, DC.

Bloomberg (2015). Sustainable energy in America — Fact Book. Bloomberg New Energy Finance, The Business Council for Sustainable Energy.

Bloomfield, K., Moore, J. N. and Neilson, R. M. (2003). Geothermal energy reduces greenhouse gases, GRC Bull., Geothermal Research Council, Davis, CA. *geothermal.org*.

Bloomquist, R. G. (1999). Geothermal Heat Pumps: Four Plus Decades of Experience, *GHC Bull.*, **20**(4): 13–18.

Bloomquist, R. G. (2001). The economics of geothermal heat pump systems for commercial and institutional buildings. In *Proceedings of Int.'l Course on Geothermal Heat Pumps*, Bad Urach, Germany.

Bruni, S. (2014). Geothermal energy: A sustainable source of power, Energy Innovation Center, Inter-American Dev. Bank, Washington, DC. *publications.iadb.org*.

Buczynski, B. (2010). Illinois Walgreens converts to geothermal energy, Care2.Com, Inc. Redwood City, CA.*Care2.com*.

Carr-Cornish, S. and Romanach, L. (2014). Differences in Public Perceptions of Geothermal Energy Tech. in Australia, *Energies*, **7**(3): 1555–1575.

Chiasson, A. D. (1992). Advances in modeling of ground-source heat pump systems, Master's Thesis, Univ. Windsor, Ontario, Canada.

Chicago Architecture Today (2006). Firm Foundation. Chicago, IL. *chicagoarchitecturetoday.com/firmfoundation*.

Clark, C. E. and Veil, J. A. (2009). Produced Water Volumes and Management Practices in the United States, Doc. No. ANL/EVS/R-09/1, NETL, Office of Fossil Energy, U.S. DOE, Washington, DC.

Clark, C. E, Sullivan, J., Harto, C., Han, J. and Wang, M. (2010). Water use in the development and operation of geothermal power plants, ANL/EVS/R-10/5, Env.'l Sci. Div., Argonne Nat'l Lab., Argonne, IL.

Clark, C. E., Harto, C. B. and Troppe, W. A. (2011). Water Resource assessment of geothermal resources and water use in geopressured geothermal systems, Env.'l Sci. Div., Argonne Nat'l Lab., Argonne, IL.

Clark, C. E, Sullivan, J., Harto, C., Han, J. and Wang. M. (2012). Life cycle environmental impacts of geothermal systems. In Proceedings of 37th Workshop on Geothermal Reservoir Eng'g. Stanford Univ., 30 January–1 February, Stanford, CA.

Congress, U.S. (2008). Emergency Economic Stabilization Act, Vol. 3. In *HR 1424, 110th Congress of the United States of America*, Second Session.

Cross, J. and Freeman, J. (2009). 2008 Geothermal technologies market report, Efficiency and Renewable Energy, U.S. DOE, Washington, DC.

Cross, S., Eagan, D. J., Tolmé, P., Keniry, J., Kelly, J. and Koester, R. J. (2011). Going underground on campus: tapping the earth for clean, efficient heating and cooling: a guide to geothermal energy and underground buildings on campus, Nat'l Wildlife Federation, Reston, VA. *nwf.org*.

Dowd, A. Boughen, N., Ashworth, P. and Carr-Cornish, S. (2010). Geothermal Tech. in Australia: Investigating Social Acceptance, *Energy Policy*, **39**(10): 6301–6307.

Dickson, M. H. and Fanellim M. (2013). *Geothermal Energy: Utilization and Technology*. Routledge.

DiPippo, R. (1999). Small Geothermal power plants: Design, performance and economics. *GHC Bull.*, Geo-Heat Center, Oregon Inst. Tech., Klamath Falls, OR.

DiPippo, R. (2012). Geothermal power plants: principles, applications, case studies, and environmental impact, 2nd Edn. Butterworth-Heinemann, Elsevier.

Duffield, W. A. and Sass, J. H. (2003). Geothermal energy: Clean power from the earth's heat, Circular 1249, U.S. Geol. Survey, Washington, DC. *usgs.gov.*

[ERG] Earth River Geo. (2015). Geothermal HVAC systems — An in depth overview. Earth River Geothermal, Inc., Annapolis, MD. *earthriver-geo.com.*

[EERE] Energy Efficiency and Renewable Energy (2004). Geothermal technologies program: Enhanced geothermal systems. Doc. No. DOE/GO-102004-1958, NREL, U.S. DOE, Washington, DC.

Encycl. Brit. (2014). New Madrid earthquakes of 1811–12, Encyclopedia Britannica, Inc., Chicago, IL. *britannica.com.*

Energy Homes (2008). What is a Geothermal Heat Pump? Energy Environmental Corp., Centennial, CO. *energyhomes.org.*

Energy Star (2008). Energy Star Building Manual: 9, Heating and Cooling. U.S. DOE. *energystar.gov.*

Energy Star (2012). Energy Star Program Requirements for Geothermal Heat Pumps — Partner Commitments, U.S. EPA, Washington, DC. *energystar.gov.*

[EPRI] (2010). Geothermal power: Issues, technologies, and opportunities for research development, demonstration, and deployment. EPRI Tech. Innovation White Paper (Product ID: 1020783), Electric Power Research Inst., Orland Park, IL.

Frederick, J. (2010). Walgreens taps geothermal energy to power new store in Chicago suburb. *Drug Store News. drugstorenews.com.*

[GAOI] (2012). Geothermal Tax Credits. Geothermal Alliance of Illinois, Springfield, IL. *gaoi.org.*

[GEA] (2015). Geothermal Basics. Geothermal Energy Assoc., Washington, DC. *geo-energy.org.*

[GEA] (2014). 2014 annual U.S. & global geothermal power production ReportGeothermal Energy Assoc., Washington, DC. *geo-energy.org.*

[GeoBasics] (2015). Geothermal basics. Office of Energy Efficiency and Renewable Energy, U.S. DOE, Washington, DC. *Energy.gov*/eere.

Geothermal-House (2010). Geothermal pond loops. Geothermal House, Burr Ridge, IL. *geothermal-house.com.*

[GHC] (2012). Geothermal information and technology transfer. Geo-heat Center, Oregon Inst. of Tech., Klamath Falls, OR. *geoheat.oit.edu.*

Glassley, W. E. (2015). *Geothermal Energy: Renewable Energy and the Environment*, 2nd Edn. CRC Press.

Gocze, T. (2013). Heat-Pump Water Heaters. *Home Power Magazine.*

Goetzler, W., Zogg, R., Lisle, H. and Burgos, J. (2009). Ground-source heat pumps: Overview of market status, barriers to adoption, and options for overcoming barriers, Geothermal Tech.'s Program, Energy Efficiency and Renewable Energy, U.S. DOE, Washington D.C.

Graig, A., Learner, H., Gray, P. and Dobies, J. (2011). The clean energy supply chain in Illinois: Wind, solar and geothermal, Env.'l Law and Policy Center, Chicago, IL. *elpc.org.*

Grant, M. (2013). *Geothermal rRservoir Engineering*. Academic Press.

Green, B. D. and Nix, R. G. (2006). Geothermal-the energy under our feet geothermal resource estimates for the United States, Tech. Report No. NREL/TP-840-40665, Nat'l Renewable Energy Lab., U.S. DOE, Golden, CO.

GreenBiz (2010). Walgreens taps geothermal energy for Illinois drugstore, GreenBiz Group, Inc., Oakland, CA. *greenbiz.com.*

Hähnlein, S., Bayer, P., Ferguson, G. and Blum, P. (2013). Sustainability and Policy for the Thermal Use of Shallow Geothermal Energy, *Energy Policy*, **59**: 914–925.

Hammons, T. J. (ed.) (2011). Geothermal power generation: Global perspectives, tech., direct uses, plants, drilling and sustainability worldwide, Chapter 5, Elect. Infrastructures in the Global Marketplace, pp. 195–236. *intechopen.com.*

Hendrie, K. and Wendland, M. W. (1981). Illinois solar and wind climate (miscellaneous publication), *Ill. Dept. of Energy and Natural Resources*, State Water Survey Div., Climatology Section.

Hillesheim, M. and Mosey, G. (2013). Feasibility study of economics and performance of geothermal power generation at the Lakeview Uranium

Mill site in Lakeview, Oregon, Tech. Report No. NREL/TP-6A10-60251, Nat'l Renewable Energy Lab., U.S. DOE, Golden, CO.

Hinze, W. J., Braile, L. W., von Frese, R. R. B., Lidiakt, E. G., Denison, R. E., Morgan, P., Keller, G. R., Roy, R. F., Swanberg, C. A., Aiken, C. L. V. and Gosnold, W. D. (1986). Exploration for hot dry rock geothermal resources in the Midcontinent USA, Hot dry rock conceptual models for exploration, HDR test site investigations, and the Illinois Deep Drill Hole Project, Vol. 2, Doc. No. LA-10659-HDR), Los Alamos Nat'l Lab., Los Alamos, NM.

Holm, A., Jennejohn, D. and Blodgett, L. (2012). Geothermal energy and greenhouse gas emissions, Geothermal Energy Assoc., Washington, DC. *geo-energy.org.*

Hunt, T. M. (2000). Five lectures on environmental effects of geothermal utilization, U.N. Univ., Geothermal Training Prog., Reykjavík, Iceland.

[ICE] (2013). Projects highlights: Energy efficiency-geothermal HVAC installation at New Sherman Hospital. , Illinois Clean Energy (Community Foundation), Chicago, IL. *illinoiscleanenergy.org.*

[IEA] Int'l Energy Agency. (2010). Renewable energy essentials: Geothermal. OECD/IEA, Paris, France.

[IEA] Int.'l Energy Agency (2011). Technology roadmap-geothermal heat and power. OECD/IEA, IEA, Paris, France.

[IEA-ETSAP] Int.'l Energy Agency and Energy Tech. Systems Analysis Prog. (2010). Geothermal heat and power, IEA ETSAP — Tech. Brief E07. IEA Secretariat, ETSAP, Paris, France.

[IGNN] Illinois Gov.s News Network (2013a,). Governor Quinn announces more than $800,000 for Camp Lincoln Geothermal System Project will create jobs and save energy. Illinois Government News Network, Office of the Governor, Springfield, IL. *illinois.gov.*

[IGNN] (2013b). Governor Quinn urges Illinoisans to join biggest earthquake drill in Midwest history. *Illinois Government News Network*, Office of the Governor, Springfield, IL. *illinois.gov.*

[ITO] Intellectual Takeout (2015). Estimated subterranean temperatures at a depth of 6 kilometers. ITO, Bloomington, MN. intellectualtakeout.org.

Jennejohn, D. (2011). Annual geothermal power production and development Report, Geothermal Energy Assoc., Washington, DC. *geo-energy.org.*

Kagel, A., Bates. D. and Gawell, K. (2007). A guide to geothermal energy and the environment. Geothermal Energy Assoc., Washington, DC. *geo-energy.org*.

Lee, S., Speight, J. G. and Loyalka, S. K. (eds.) (2007). Handbook of Alternative Fuel Technologies. CRC Press.

Lepeska, D. (2011). Deep Impact: Indie Energy pioneers geothermal alternatives in dense urban environments. *Metropolis Magazine*, New York, NY.

Lopez, A., Roberts, B., Heimiller, D., Blair, N. and Porro, G. (2012). US renewable energy Tech. potentials: A GIS-based analysis, Tech. Report No. NREL/TP-6A20-51946, Nat'l Renewable Energy Lab., U.S. DOE, Golden, CO.

Lund, J. W. (1988). Geothermal heat pump utilization in the United States. *Geo-Heat Center Quarterly Bull.*, **11**(1): 1–4, Oregon Inst. of Tech., Klamath Falls, OR. *geoheat.oit.edu/bullet.*

Lund, J. W. (2006). Chena hot springs. *Geo-Heat Center Quarterly Bull.*, **27**(3): 2–4, Oregon Inst. of Tech., Klamath Falls, OR.

Lund, J. W. (2007). "Characteristics, development and utilization of geothermal resources. *Geo-Heat Center Quarterly Bull.*, **28**(2):1–9, Oregon Inst. of Tech., Klamath Falls, OR.

Macknick, J., Newmark, R., Heath, G. and Hallett, K. C. (2012). Operational Water Consumption and Withdrawal Factors for Electricity Generating Technologies: A Review of Existing Literature, *Environ. Res. Lett.* **7**(4).

Mai, T., Wiser, R., Sandor, D., Brinkman, G., Heath, G., Denholm, P., Hostick, D. J., Darghouth, N., Schlosser, A. and Strzepek, K. (2012). Exploration of High-Penetration Renewable Electricity Futures, Vol. 1. Renewable Elect. Futures Study, Doc. No NREL/TP-6A20-52409-1, Nat'l Renewable Energy Lab., U.S. DOE, Golden, CO.

Majer, E., Nelson, J., Robertson-Tait, A., Savy, J and Wong, I. (2012). Protocol for addressing induced seismicity associated with enhanced geothermal systems, Doc. No. DOE/EE-0662, Geothermal Tech.'s Program, Energy Efficiency and Renewable Energy, U.S. DOE, Washington, DC.

Matek, B. (2013). Promoting geothermal energy: air emissions comparison and externality analysis, Geothermal Energy Assoc., Washington, DC. *geo-energy.org*.

Matek, B. and Gawell, K. (2014). The economic costs and benefits of geothermal power, Geothermal Energy Assoc., Washington, DC. *geo-energy.org*.

McDonald, D. (2010). Dixie Valley bottoming inary cycle,Geothermal Tech.'s Program, Energy Efficiency & Renewable Energy, U.S. DOE, Washington, DC.

McKenna, J. R., Blackwell, D. D., Moyes, C. P. and Patterson, P. D. (2005). Geothermal Electric Power Supply Possible from Gulf Coast, Midcontinent Oil Field Waters, *Oil & gas J.*, **103**(33): 34–40.

[OpenEI] (2011a). Retrofit the 74,000 sq.ft Illinois Nat'l Guard State Headquarter's building's aging boiler/chiller equipment with a geothermal system Geothermal Project. Open Energy Info. *en.openei.org*.

[OpenEI] (2011b). Dixie valley bottoming binary project: Geothermal project. Open Energy Info. *en.openei.org*.

[PEI] Power Engineering International (2015). World's largest geothermal plant opens in Kenya as global development steams ahead. PEI, Tulsa, OK. *powerengineeringint.com*.

Pellizzone, A., Allansdottir, A., De Franco, R. *et al.* (2013). Assessment of social acceptance of geothermal energy exploitation in southern Italy. In *Proceedings of European Geothermal Congress*.

Plataforma SINC (2009). Mines Could Provide Geothermal Energy. *Sci. Daily*, Rockville, MD. *sciencedaily.com*.

[PRN] (2009). Indie Energy's breakthrough geothermal Tech. for buildings earns U.S. Dep't of Energy Funding. *PR Newswire*. *prnewswire.com*.

Proffitt, T. A., Ellett, K. M., Zuppann, C. W., Naylor, S. C., Barrett, M. E., Huff, B. G., Korose, C. P., Lecouris, A. B. and Noakes, T. C. (2012). Development and application of a new geothermal database for the Illinois Basin, AAPG Search and Discovery Article No.90154©2012, AAPG Eastern Section Meeting, Cleveland, Ohio, 22–26 September 2012.

Puttagunta, S. and Shapiro, C. (2012). An in-depth look at ground source heat pumps and other electric loads in two GreenMax homes, U.S. Dep't of Energy, Office of Scientific and Tech. Info., Oak Ridge, TN.

Rafferty, K. (1997). An information survival kit for the prospective residential geothermal heat pump owner. *Geo-Heat Center Quarterly Bull.*, **18**(2): 1–11, Oregon Inst. of Tech., Klamath Falls, OR. *geoheat.oit.edu/bullet.*

Rafferty, J. P. (2014). New Madrid earthquakes of 1811–12. Encyclopedia Britannica, Chicago, IL. *britannica.com.*

Rahmouni, S., Settou, N., Chennouf, N., Negrou, B. and Houari, M. (2014). A Technical, Economic and Environmental Analysis of Combining Geothermal Energy with Carbon Sequestration for Hydrogen Production, *Energy Procedia*, **50**: 263–269.

Rodríguez, R. and Diaz, M. B. (2009). Analysis of the Utilization of Mine Galleries as Geothermal Heat Exchangers by Means a Semi-empirical Prediction Method, *Renewable Energy*, **34**(7): 1716–1725.

Rybach, L. (2007). Geothermal sustainability. Geo-Heat Center Bull., **28**(3), Oregon Inst. of Tech., Klamath Falls, OR. *geoheat.oit.edu.*

Sammel, E. A. (1979). Assessment of geothermal resources of the United States — 1978, Muffler, L. J. P (ed.) — *pubs.usgs.gov.*

Sherman Health (2014). Advocate Sherman Hospital: Geothermal lake. Advocate Sherman Hospital, Elgin, IL. *advocatehealth.com.*

Skone, T. J., Littlefield, J., Marriott, J., Cooney, G., Jamieson, M., Hakian, J. and Schivley, G. (2014). "Life cycle analysis of natural gas extraction and power generation" Doc. # DOE/NETL-2014/1646, NETL, Office of Fossil Energy, U.S. DOE, Washington, DC.

[SMU] Southern Methodist Univ. (2011). First Google.Org-funded geothermal mapping Rep. confirms vast coast-to-coast clean energy source. Roy M. Huffington Dep't of Earth Sci.'s, SMU, Dallas, TX.

Stewart, C. (2012). Geothermal energy — Effects on the environment, Te Ara — the Encyclopedia of New Zealand. *TeAra.govt.nz.*

Sullivan, J. L. and Wang, M. Q. (2013). Life Cycle Greenhouse Gas Emissions from Geothermal Electricity Production, *J. Renewable and Sustainable Energy*, **5**(6).

Tester, J., Stutz, G., Augustine, C., Mooreand, M. and Livesay, B. (2011). Oil and gas well drilling, ChemE 6666, Shale Gas Module, Coop. Extension, Cornell Univ, Ithaca, NY. *cce.cornell.edu.*

Theodoropoulos, T. E. (2011). *Oil — Gas Exploration & Drilling*, 1st Edn. BookBaby.

[TOD] (2011). Tech Talk — chemical floods to enhance oil recovery, The Oil Drum. *theoildrum.com*.

[UIC] (2008). Historical Netsch campus at UIC. Univ. of Illinois at Chicago, Chicago, IL. *uic.edu*.

[US-DOE] U.S. Dep't of Energy (1997). Geothermal energy: Power from the depths, Doc. No. DOE/GO-10097-518-FS 188, Energy Efficiency and Renewable. U.S. DOE, Washington, DC.

[US-DOE] U.S. Dep't of Energy (2006). The future of geothermal energy: Impact of enhanced geothermal systems (EGS) on the United States in the 21st century, U.S. DOE, Washington, DC.

[US-DOE] U.S. Dep't of Energy (2010a). Geothermal energy production with co-produced, geopressured resources, Doc. No. DOE/GO-102010-3004, Geothermal Tech. Programs. U.S. DOE, Washington, DC.

[US-DOE] U.S. Dep't of Energy (2010b). Geothermal retrofit of Illinois Nat'l guard state headquarters building, Geothermal Tech.'s Program, U.S. DOE, Washington, DC.

[US-DOE] U.S. Dep't of Energy (2011). Guide to geothermal heat pumps. Doc. No. DOE/EE-0385, U.S. DOE, Washington, DC.

[US-DOE] U.S. Dep't of Energy (2012a). Geothermal heat pumps, Geothermal Tech.'s Office. U.S. DOE, Washington DC.

[US-DOE] U.S. Dep't of Energy (2012b). What is an enhanced geothermal system (EGS)? Doc. No. DOE/EE-0785, Geothermal Tech.'s Office. U.S. DOE, Washington, DC.

[US-EIA] U.S. Energy Info. Admin. (2013). Updated capital cost estimates for electricity generation plants. U.S. EIA, Washington, D.C.

[US-EIA] U.S. Energy Info. Admin. (2014a). State energy consumption estimates, 1960 through 2012, Doc. No. DOE/EIA-0214(2012). U.S. EIA., Washington, DC.

[US-EIA] U.S. Energy Info. Admin. (2014b). Annual Energy Outlook 2014: Levelized cost and levelized avoided cost of new generation resources in the Annual Energy Outlook 2014. U.S. EIA.,Washington, DC.

[US-EIA] U.S. Energy Info. Admin. (2014c). Geothermal resources used to produce renewable electricity in western states. U.S. EIA, Washington, DC.

[US-EPA] U.S. Env'l Protection Agency (2014). Average temperature of shallow groundwater, Ecosystems Research. U.S. EPA, Athens, GA.

Vaught, T. L. (1980). An assessment of the geothermal resources of Illinois based on existing geologic data, Doc. No. DOE/NV/10072-2. U.S. DOE, Washington, DC.

Weinhold, B. (2012). Energy development linked with earthquakes Env'l Health Perspectives, **120**(10): A388. *ehp.niehs.nih.gov*.

White, D. E., Muffler, L. J. P. and Truesdell, A. H. (1971). Vapor — dominated hydrothermal systems compared with hot-water systems, *Economic Geology*, **66**: 75–97.

Wo, S. (2011). Reservoir modeling & simulation at EORI: 2011. In *Proceedings of 3rd Wyoming IOR/EOR Conf.*, pp. 1–2. Jackson, 12–13 September, WY.

11

Energy Storage

"Concerns about the availability of oil and other peaking fuels in this period was so great that an international conference (including the U.S. National Academy of Sciences) on the subject in 1979 described energy storage as "a vital element in mankind's quest for survival and progress" (Silverman, 1980).

"It continues to go down [solar and wind energy cost] *year after year after year, and the more of it we use, the cheaper it gets. . . . So the handwriting is on the wall, and with the advances in cost-effective energy storage, the intermittency problem is going to be solved."* Al Gore, the 2007 Nobel peace prize laureate for his work on climate change activism, 45th Vice President of the U.S. (1993–2001). [National Journal (2014, Sep. 18) "In the interview given in "24 Hours of Reality" broadcast on climate change in Duggal Greenhouse in Brooklyn's Navy Yard". *nationaljournal.com,* New York].

11.1. Introduction

Continuous increase in the level, and concern over greenhouse gas (GHG) emissions are the main impetus behind efforts to find better ways to utilize energy. We are looking for new and renewable energy sources to alleviate problems associated with GHG emissions. Electricity must be generated at the same rate as it is being used, so that demand and supply are always kept in balance. This is a critical issue with respect to intermittent sources like wind and solar energy. For such sporadic energy sources, electricity load management is a critical factor and there are methods to match energy sources with electricity demand. One of the methods is energy storage, where energy can be absorbed and stored for a period of time before it is released for future use.

Storage of natural energy resources and energy forms are a part of any energy production and utilization activity. In utilizing any energy resource like coal, biofuels, natural gas, nuclear, petroleum, solar and wind, and any energy form like electrical, mechanical or thermal, etc. whether in large- or in small-scale, some kind of a storage must be considered. We have already presented several cases of such energy storage systems and their implications in the previous chapters. In the present chapter, we introduce the concept and technology for energy storage systems, which will help in reducing our dependence on fossil and nuclear energy. Accordingly, such energy storage may be considered as an energy resource which is sustainable, environmentally friendly and economically viable for small- and large-scale application.

In general, energy storage systems fall into four main categories; these include **mechanical** (such as compressed air, flywheel or pumped hydroelectric); **electrochemical** (such as batteries or capacitors); **chemical** (such as renewable energy conversion to substitute natural gas and hydrogen) and **thermal** (such as phase change materials or aquifer thermal energy storage systems) processes (US-DOE, 2013a). The ones that are suitable for grid-scale energy storage include batteries, compressed air, flywheels, pump hydroelectric and electrochemical capacitors.

Applications of grid-scale energy storage systems include energy management, frequency regulation, backup power, load leveling, voltage support and grid stabilization.

Since the major challenge facing renewable power generation is integration of variable and uncertain renewable power to the grid, there is a need for efficient energy storage systems to facilitate their penetration to the power transmission lines.

According to the U.S. Department of Energy (DOE), about 70% of power transmission lines and power transformers in the U.S. are 25 years or older, while 60% of all circuit breakers are more than 30 years old (US-DOE 2013a). Therefore, development of efficient energy storage systems can reduce the need for major new transmission grid and construction upgrades, including increase in performance of the existing transmission and distribution infrastructure. Indeed, the lifespan and capabilities of the transmission grid can be extended by pre-installing energy storage devices on the load side of transmission points, to make the grid more reliable, responsive and secure (US-DOE, 2013a).

In August 2013, a report prepared by the U.S. President's Council of Economic Advisers, in conjunction with the US-DOE's Office of Electricity Delivery and Energy Reliability and with assistance from the White House Office of Science and Technology, stressed the significance of energy storage in playing a role in emergency preparedness and increasing the overall grid resilience (US-DOE, 2013a). According to the report, the four challenges related to the widespread deployment of energy storage are cost competitive energy storage technologies (including manufacturing and grid integration), validated reliability and safety, equitable regulatory environment and industry acceptance (US-DOE, 2013a).

In what follows, we present the concepts and technologies for the above-mentioned grid-scale energy storage systems.

11.2. Large-scale Energy Storage (Grid Energy Storage) Systems

11.2.1. *Pumped hydro storage (PHS)*

PHS storage is a large, mature and commercially available utility-scale technology. PHS is currently used in many locations around the

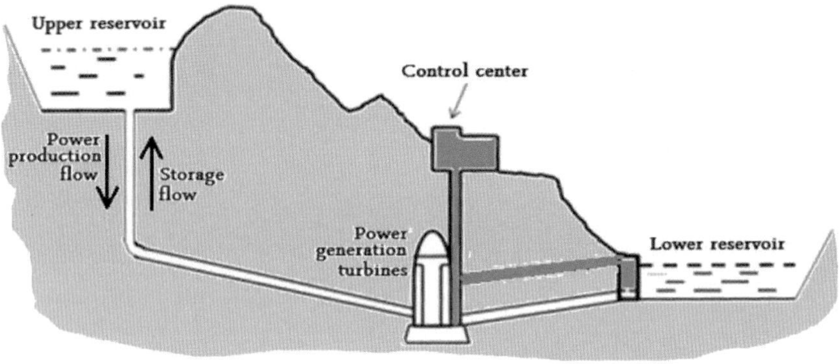

Fig. 11.1. An example of pumped storage hydroelectric technology.

world with the total of 26.6 GW installed facilities across the U.S. (U.S. DOE, 2014e) and is one of the economically viable options for large-scale electricity storage, accounting for about 99% of all large-scale electricity storage systems (see Fig. 11.1).

Applications of pumped hydro storage systems to the grid include energy management, backup and seasonal reserves services, including regulation service available through variable speed pumps (US-DOE, 2013a). This technology produces electricity to meet high peak demands, using low price off-peak electricity from the grid by pumping water between reservoirs at different elevations. This well-proven technology is found in areas where the local topography allows its application. The plant can also store electricity generated by photovoltaic (PV) solar and wind power systems during off-peak periods, in which case, the excess electricity from the renewable power plants can be used to operate a pump (Peltier, 2011). The claimed conversion efficiency of this technology ranges from 75 to 80%. The 2010 capital cost for a 500 MW pumped-storage hydropower plant was estimated at \$2,004 /kW + 50% (Black and Veatch, 2012).

In 2014, the Federal Energy Regulatory Commission (FERC) issued a preliminary permit for the development of a proposed Elmhurst Quarry Pumped Storage Project (EQPS) in the City of Elmhurst in DuPage County, Illinois. The initial design capacity of the EQPS ranges from 50–250 MW. However, the final project capacity

Fig. 11.2. The proposed hydro development (in Elmhurst, Illinois) will utilize a quarry that's already in place as part of the pumped storage system. The powerhouse will be excavated below the reservoir. The quarry's location outside the City of Chicago gives the project a developmental appeal. Its success could potentially spur additional urban pumped-storage development (Charlton and Haag, 2012).

will be determined based on long- or short-term energy storage requirements, the selected method of project pump and the generation cycle operation (Charlton and Haag, 2012). The conceptual project, located at just 20 miles from downtown Chicago (see Fig. 11.2), will test the feasibility of locating such facilities in urban areas.

The site was selected for the project because it has several important pumped-storage development characteristics. These include proximity to a major electric load center and its suitability for pumped storage development. Also, existence of upper and lower reservoirs at the site could decrease cost and construction time. In addition, the site has ideal transmission access, with only 3.2 kilometers (2 miles) link to a substation owned by Commonwealth Edison (Charlton and Haag, 2012). The normal minimum operating level of the lower reservoir will be at 152.4 meters (500 ft), while the normal maximum operating level of the upper reservoir will be at a height of 203.3 meters (667 ft). The water surface area will vary from 33.9 acres (13.7 ha) at minimum to 55.1 acres (22.3 ha) at maximum. But at normal maximum operating level, the total reservoir storage capacity will be 8,145 acre-feet, while the total active capacity will be 7,465 acre-feet (an acre-foot is a unit of volume frequently used in the U.S. in reference to large-scale water resources, such as reservoirs and aqueducts and is equivalent to 1.2335×10^6 liters). The average net head of the system will be 132.6 meters (435 ft), with total annual energy production of 708.5 GWh (Charlton and Haag, 2012).

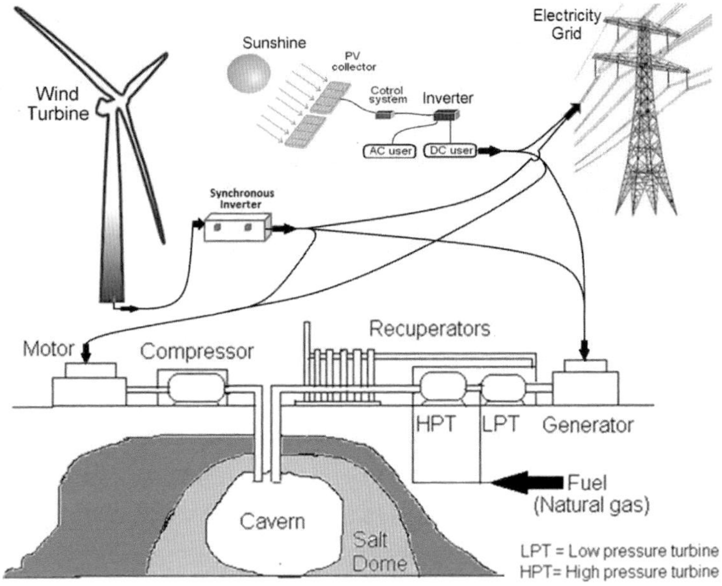

Fig. 11.3. An example of compressed air energy storage system (modified from US-DOE/EPRI 2013).

11.2.2. *Compressed air energy storage (CAES)*

CAES systems can be used to store electricity generated from wind and PV solar power plants (see Fig. 11.3).

Applications of CAES as a grid storage system include energy management, backup and seasonal reserves, and renewable power integration (US-DOE, 2013a). The technology is based on conventional gas turbine technology using elastic potential energy of compressed air (Denholm *et al.*, 2010). During off-peak periods of electricity generation by utility power plants, the system stores energy through isothermal compression of air in an air-tight underground storage cavern. The stored energy is extracted by drawing the compressed air from the storage vessel. The air is heated, followed by isothermal expansion through a high-pressure turbine that uses part of the energy in the compressed air. After mixing the air with fuel, the fuel/air mixture is combusted. The exhaust gas is then expanded through low-pressure gas turbines or hydroelectric motors connected to an electrical generator to produce electricity (Denholm *et al.*, 2010). The compressed air can be stored in a facility that is usually located below ground level such as porous rock formations, depleted gas/oil fields and salt or rock cavern formations. Alternatively, the air can be stored in aboveground or near-surface pressurized air vessels/pipelines, including those used to transport high-pressure natural gas (Al-Khoury and Bundschuh, 2014). The claimed efficiency of the CAES is estimated at 45–70%. CAES plants with aboveground air storage reservoir would typically be smaller than the ones with underground storage and have capacities in the order of 3–50 MW and discharge times of 2–6 hours (Akhil *et al.*, 2013). The aboveground CAES systems are easier to site, but are more expensive to build.

In 1983, the DOE and Electric Power Research Institute jointly sponsored a program to study an aquifer site development by Public Service of Indiana. In this program, literature review of regional geology was used to identify sites for compressed air energy storage in a study area comprising parts of Illinois, Indiana, Kentucky and Tennessee (Allen *et al.*, 1983). A total of 14 potential sites were identified, with 10 in Illinois and 4 in Indiana (see map in Fig. 11.4).

But despite the existence of these potential sites in Illinois, there is no compressed air energy storage in the state, while currently two of such facilities exist in the world (Haugen, 2011). These are the 320-megawatt Huntorf facility operating in Germany since the late 1970s and the 110-megawatt McIntosh power plant in Alabama brought online in 1991 by an Alabama electric co-operative (Haugen, 2011; IEA, 2014).

11.2.3. *Flywheel energy storage (FES)*

A flywheel is a simple form of mechanical energy storage device (Fig. 11.5).

In general, flywheels are described as low-speed or high-speed. The low-speed flywheels have rotating speed of up to 10,000

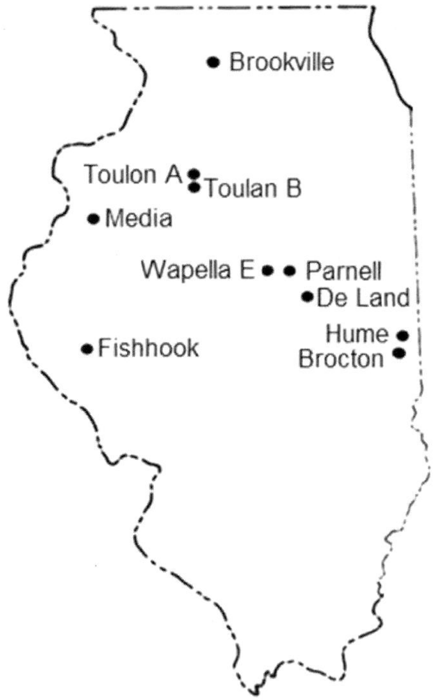

Fig. 11.4. Potential site locations for CAES in Illinois (Allen *et al.*, 1983).

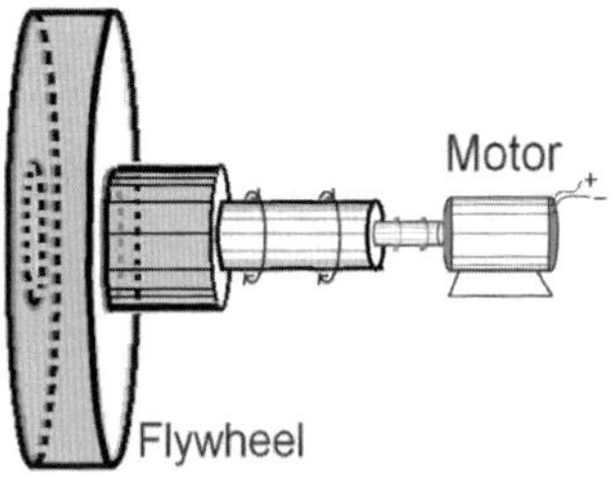

Fig. 11.5. Illustration of the FES system.

revolutions per minute (rpm) (Genta, 1985). On the other hand, the high-speed flywheels operate at speed measured more than 100,000 rpm. The applications of flywheels for large scale energy storage systems include load leveling, frequency regulation, peak shaving and off peak storage and transient stability. In essence, flywheels convert electrical energy into kinetic energy, which is stored in a rotating wheel or a rotating cylinder. For grid frequency regulation or load leveling, the rotating wheel provides mechanical power to drive a generator that generates the required electricity.

Flywheels can serve as replacement for batteries for uninterruptible power supply (UPS), due to their long service life (US-DOE, 2003). Caterpillar Company in Peoria, Illinois is one of the manufacturers of flywheels used for UPS. The company's product "Cat Flywheel UPS" is claimed to provide their highest (>97%) efficiency of continuous power solution (Caterpillar, 2014).

In 2011, the DOE National Energy Technology Laboratory (NETL) completed the final environmental assessment of two proposed Beacon Power Corporation flywheel frequency regulation plants. One of these plants is to be located in Chicago Heights in Illinois and the other at Hazle Township in Pennsylvania. Chicago Heights is a major industrial area in Illinois and was selected for the project after DOE had determined the impact of the project on air quality, socioeconomics, environmental justice and occupational health and safety in the area (US-DOE, 2013b). The proposed projects are part of DOE's nationwide effort to demonstrate its smart grid program, which aims to verify the technological and business viability of new technologies. Also, the smart grid program will show how fully integrated smart grid systems can be readily adapted to extend its application in the U.S. (US-DOE, 2013b). In essence, the goals for installation of the proposed flywheel frequency regulation plants are: achieving a better balance between network load and generated power, obtaining more efficient maintenance of Pennsylvania, Jersey, Maryland (PJM) grid frequency performance to grid reliability and increasing the use of intermittent renewable wind and solar power. PJM stands for the Pennsylvania, Jersey, Maryland Power Pool. It is the electricity control area (the electric grid) for New Jersey and all or parts of Delaware, Illinois, Indiana, Kentucky, Maryland, Michigan, New Jersey, North Carolina, Ohio, Pennsylvania, Tennessee, Virginia, West Virginia and the District of Columbia.

Additionally, the projects aim to reduce carbon dioxide and other air emissions, lower the cost of frequency regulation to ratepayers and increase the regional peak power generation capacity, while reducing national dependence on fossil fuels. As part of a cost-sharing agreement, Beacon Power Corporation will receive $24 million financial assistance grant from DOE through the American Recovery and Reinvention Act (ARRA) of 2009. The award is to facilitate installation and operation of the proposed Illinois project, which is estimated at $48.1 million (US-DOE, 2013b). The proposed 20 MW plant for Illinois will be located on a 25-acre parcel of land located in Chicago Heights in Cook County. The land was previously occupied by a 60 MW oil-fired power generator owned by Midwest Generation, an independent power producer. Beacon Power Corporation will use the plant in cooperation with PJM Interconnection (the operator of the regional electrical grid) and Exelon Corporation (the transmission system owner) (US-DOE, 2013b). During electricity off-peak demand periods, the plant would convert the excess electricity from the grid and store it as kinetic energy in the flywheels. During peak electricity demand periods, the stored kinetic energy will be reconverted into electricity and returned to the grid.

11.2.4. *Battery energy storage systems (BESS)*

BESS have the potential to provide a long term energy storage solution to wind and solar energy industries. Applications of batteries as energy storage systems to the power grid include load leveling and regulation, grid stabilization, power quality, congestion relief, renewable power integration, frequency regulation, ramping, peak shaving and time shifting (US-DOE, 2013a). Additionally, a possible breakthrough technology for battery energy storage systems could help the State of Illinois transition its automotive fleet to plug-in hybrid and electric vehicles, while increasing the use of renewable energy to reduce fossil fuels consumption. Therefore, in 2012 the US-DOE established a Joint Center for Energy Storage Research

(JCESR) at the Argonne National Laboratory, located at 40 km (25 miles) Southwest of Chicago in Illinois. The JCESR (pronounced "J-Caesar") is one of the five Energy Innovation Hubs in the nation established by DOE since 2010. The projects aim to combine basic and applied research with engineering to speed up scientific discoveries that could provide solutions to critical energy issues. In essence, these Energy Innovation Hubs are integrated research centers modeled after the strong scientific management characteristics of the Manhattan Project and the AT&T Bell Laboratories that led to the creation of atomic bomb and transistors in the 20th century (US-DOE, 2012a). The JCESR draws on the skills of researchers from the five DOE national laboratories, which include Argonne, Lawrence Berkeley, Pacific Northwest, Sandia and SLAC National Accelerator. Also included, are four private companies that have been working independently to advance battery technology, as well as students and researchers at University of Chicago, University of Illinois at Chicago (UIC), Northwestern University, University of Illinois at Urbana-Champaign and the University of Michigan. The JCESR is funded with $120 million from the DOE and a $35 million commitment from the State of Illinois. The center aims to develop lighter and cheaper batteries that are capable of storing more power and charging faster. These future batteries would be powering almost every device, including smartphones and electric vehicles (Wernau, 2012).

In 2013, researchers at the Illinois Institute of Technology (IIT) and Argonne National Laboratory received a $3.4 million federal grant to develop a prototype of a modified "flow" battery using liquid nanoelectrofuel electrodes instead of solid electrodes employed in a typical solid state battery. In essence, success of this technology could allow electric cars to travel 500–1,000 miles on a single electric charge, instead of the current 40–100 miles range (Guy, 2013). Additionally, the technology would allow fast refueling of motorists by providing an alternative to charge the battery in a conventional plug-in setup or quickly switch from the discharged battery fluid to a charged one. Other advantages of the rechargeable liquid nanoelectrofuels include zero emissions, easy temperature control, and non-flammability. Currently, most new plug-in hybrids and all-electric vehicles owners would prefer lithium-ion batteries to the nickel-metal hydride batteries used in many hybrid electric vehicles.

In 2010, the DOE NETL completed the final environmental assessment for construction of a manufacturing plant to produce lithium hexafluorophosphate ($LiPF_6$), a critical battery material. The proposed project is a cooperative agreement between DOE and Honeywell International Inc. The plant would be located on a 10-acre parcel of land within the boundaries of a 1,010-acre tract of land owned by Honeywell, in Massac County, near the City of Metropolis, Illinois (US-DOE, 2010). Besides boosting Illinois economy, the project aims to support development of the Electric Drive Vehicle (EDV) Industry in the U.S, while substantially reducing the nation's dependence on petroleum. The project would accelerate development and production of various EDV systems by building or increasing the U.S. production capacity for advanced automotive batteries, recycling facilities and components (US-DOE, 2010). The project would help introducing various vehicle technologies to the market and reduce the cost of battery packs, batteries and electrical propulsion system for EDVs by using high-volume production (US-DOE, 2010). The proposed project would produce a maximum of 1,500 metric tons of $LiPF_6$ per year for making high-quality Li-ion batteries. Additionally, the project would create about 34 permanent jobs in Illinois. Also under the terms of the cooperative agreement, DOE would provide about 50% of the cost of the project (US-DOE, 2010).

Also in 2013, a research group at the University of Illinois at Urbana-Champaign claimed to have developed the most powerful microbatteries ever documented. The high performance of the batteries is due to their internal three-dimensional microstructure. The nanoporous microbattery electrodes is made of an electrolytically active layer coated on an electrically conductive bicontinuous (from one layer to another layer) nickel scaffold. The nickel scaffold acts as the current collector attached to an outside circuit. A nickel-tin alloy is used as the anode and lithiated manganese oxide as the cathode. It is claimed that the battery can be used to power sensors or power electronic devices to produce radio signals that can broadcast 30 times farther than the present radio signals. Also, these rechargeable batteries are claimed to be capable of charging 1,000 times faster than competing technologies, such as conventional lithium ion batteries. Therefore, there is an anticipation that availability of such power sources could boost technology development for consumer electronics, medical devices, lasers, sensors and other applications (Gowda, *et al.*, 2012; Pikul *et al.*, 2013).

Table 11.1 shows the technology types, applications and challenges facing the development of batteries for advanced energy storage systems.

Currently, lithium-ion batteries are used in most portable consumer electronics, such as cell phones and laptops. Additionally, most plug-in hybrid electric vehicles and all-electric vehicles use lithium-ion batteries, with exact chemistry often different from that of consumer electronics batteries (US-DOE, 2013c). Lithium-ion batteries have high energy per unit mass relative to other electrical energy storage systems. Besides, they have high power-to-weight ratio, high energy efficiency, good high-temperature performance and low self-discharge. Furthermore, most components of lithium-ion batteries can be recycled. However, further research and development of these batteries are needed to reduce the cost and extend their useful life cycle (US-DOE, 2013c). Research on lithium-ion batteries is ongoing and progress is continuously made in this area. Yang *et al.* (2015) have synthesized Li_2MnSiO_4C porous nanoboxes *via* a wet-chemistry associated solid-state reaction method that could more than double the energy capacity of lithium-ion batteries, when compared with conventional phosphate-based cathodes.

Another advanced battery system is Sodium Sulfur (NaS), which are available in multiples of 1-MW/6-MWh units and have capacity in the range of 2 to 10 MW. The largest single installation is the 34-MW Rokkasho wind-stabilization project in Northern Japan that has been

Table 11.1. Applications and challenges of different types of batteries for energy storage (US-DOE, 2013a).

Battery Type	Applications	Challenges
Advanced lead-acid	Grid stabilization Load leveling & regulation	Early electrode corrosion limits useful life. Large footprint. Limited depth-of-discharge(1) Low energy density.
Sodium sulfide (NaS)	Congestion relief Power quality Renewable source integration	High operating temperature (250–300°C). Liquid containment issues corrosion and brittle glass seals
Lithium ions	Frequency regulation Power quality	High production cost and problems linked with scale-up. Extremely sensitive to overheating, overcharging and internal pressure build-up. Intolerance to deep discharge.(2)
Flow	Frequency regulation Peak shaving Power quality Ramping Time shifting	Developing Technology, not mature for commercial scale development Complicated design Lower energy density

(1) Depth-of-discharge is a ratio expressed as the percentage of the ampere-hours discharged from a battery to the available battery capacity under the same specified conditions.
(2) Deep discharge is a qualitative term that indicates the discharge of a large proportion of the available capacity of a battery.

operational since 1 August 2008. (Akhil *et al.*, 2013). Batteries in general cannot operate at high power level for a long time because of chemical kinetics involved in their operation. Heat generation as a result of rapid discharges may lead to early replacement of batteries. In addition, generation of toxic gases as a result of charge/discharge is an environmental concern (Ribeiro *et al.*, 2001).

In December 2012, Invenergy LLC, a leading clean energy generation company in Chicago, Illinois and Xtreme Power, the claimed world leader in real-time power management and energy storage solutions, made an announcement to launch a joint venture. The aim of the joint venture is to deploy a fast-acting energy storage system in Illinois (Xtreme Power, 2012). The 1.5 MW Regulation Power Management™(RPM) system owned by Xtreme Power is a versatile and scalable equipment. The system will supply clean renewable power to the new frequency response market administered by PJM (the regional transmission organization) to ensure grid reliability. The energy storage installation will be located at the site of the Grand Ridge Wind Project in La Salle County owned by Invenergy. The system will ultimately employ long-life lithium-titanate battery technology, and an automatic gain control (AGC) signal to harness full four-quadrant power in less than one second. It is expected that such instantaneous energy delivery would enable Invenergy to balance the supply and demand of energy in the PJM market (Xtreme Power, 2012).

In December 2011, the City of Chicago began installation of its first publicly accessible fast charging stations for electric vehicles. These stations are capable of charging vehicle batteries within 30 minutes. Of the total 73 charging stations scheduled for installation in the city and its suburbs by the end of 2011, the first was installed at the Midway International Airport at 61st Street and Cicero Avenue, and the second was installed at the UIC (Wernau, 2011). However, three years after the ~$2 million initiative purported to begin the electric-vehicle (EV) charging infrastructure development in Chicago, a dispute between 350 Green (the former EV charging

Fig. 11.6. Sign for EVCS.

station vendor) and JNS Power & Control Systems Inc. (the current operator of the EV charging station, based in Arlington Heights) has brought the project to a standstill (Wernau, 2014).

In 2012, the City of Naperville, Illinois launched its first publicly accessible electric vehicle charging station (EVCS) (Fig. 11.6) at a parking lot located at 44 W. Van Buren Avenue in the city.

Also in February 2014, the Naperville City Council approved three additional charging spots. To promote the use of the EVCS, the city made charging at the station free until October 2013, after which a 75-cent hourly fee for using the EVCS became effective and it was increased to $1.50 hourly fee in 2014. The EVCSs are level 2 charging stations employing 220–240 volts of electricity and requiring a shorter charging time than a wall-plug residential unit (Naperville, 2013; Jenco, 2014).

In 2014, the Village of Skokie, Illinois installed its first EV charging station through the North Shore Electricity Aggregation Consortium of which the village is a member. The charging station is located at a temporary parking lot at the northwest corner of Searle Parkway and Skokie Boulevard. To promote the use of EVCSs in the village, charging of electric vehicles at the station is free for visitors. In addition, two parking spaces exempted from a 15-munite parking regulation in the village are available at the station. Moreover, the community consortium is applying to the State of Illinois for EV Charging station rebates. These rebates will return 50% of the charging station and installation costs to the Village of Skokie (Village of Skokie, 2014).

The other Illinois cities that are joining Naperville and the Village of Skokie in acquiring EVCS are St. Charles and Lake Forest. In 2013, aldermen of St. Charles in Illinois gave the city officials permission to purchase and install the first dual vehicle battery charging station of the city. The charging station is planned to be located at the downtown area of the city and would charge electric cars for free. Similarly, officials in the City of Lake Forest, Illinois announced the same measure; however, the city intends to charge a fee for the service offered at the charging station (Baer, 2013).

11.2.5. *Electrochemical capacitors*

One of the significant developments in the field of energy storage and conversion is supercapacitors (Yu *et al.*, 2013). Supercapacitor, formerly known as electric double-layer capacitor or ultracapacitor, is the generic term for a family of electrochemical capacitors. Currently, these energy storage systems are experiencing rapid annual growth because of their potentially broad applications in electric grid, transportation, renewable energy, industrial processes and consumer electronics (Jiang *et al.*, 2013). These energy storage systems have electrodes made from activated carbon. The activated carbon provides a very large surface to volume ratio of electrodes, with an extremely small separation of distance between them. The very large surface to volume ratio makes the electrodes capable of absorbing large amount of ions, leading to high capacitance and high energy density supercapacitors (Gebbia *et al.*, 2013). In 2013, a research group at the University of Illinois at Urbana-Champaign reported that wood-biochar could serve as a potential replacement for activated-carbon electrodes used in supercapacitors (UIUC, 2013). Wood bio-char is prepared through the pyrolysis of wood in absence of oxygen at a temperature of about 400–500°C (752–932°F). The performance results of bio-char materials are claimed to be comparable to advanced carbon materials used for electrodes in supercapacitors, including carbon nanotubes and graphenes (UIUC, 2013).

11.2.6. *Renewable energy conversion to substitute natural gas (SNG)*

Excess electricity generated from PV solar and wind turbine systems can be converted to SNG for storage (Peltier 2011). In this process, the electricity generated from the variable renewable energy sources is used to split water into hydrogen and oxygen through electrolysis reaction according to Eq. (11.1), below (Peltier, 2011).

$$H_2O + \text{Elecricity} \rightarrow H_2 + \left(\frac{1}{2}\right) O_2 \tag{11.1}$$

The resulting hydrogen is reacted with carbon dioxide to produce SNG according to Eq. (11.2), which can be stored, as shown in Fig. 11.7 (Peltier, 2011).

$$CO_2 + 4H_2 \rightarrow CH_4 + 2H_2O \tag{11.2}$$

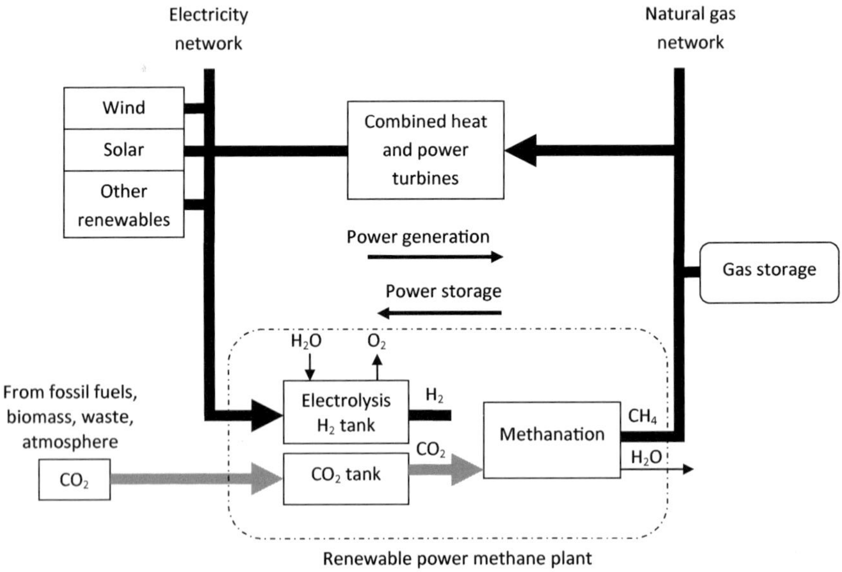

Fig. 11.7. A two-step natural gas process by electrolysis of water using surplus wind or solar energy followed by methanation of the hydrogen with carbon dioxide to produce a synthetic natural gas (modified from Peltier, 2011).

The claimed energy efficiency (i.e., kWh of SNG produced per kWh of electricity for the electrolysis) of this process is greater than 60% (Specht *et al.*, 2009).

11.2.7. *Hydrogen production, storage and applications using renewable energy*

Presently, in the industry hydrogen is produced from steam reforming which uses fossil fuels such as coal or natural gas as feedstock. However, hydrogen can be produced by electrolytic dissociation of water using the electricity generated from renewable energy sources such as wind, solar, geothermal and biomass to avoid burning of fossil fuels, which emit pollutants. In this process, electricity generated from the renewable energy sources is used to split water into hydrogen and oxygen through electrolysis reaction, as shown in Eq. (11.1) (in Sec. 11.2.6). Genovese *et al.*, (2009) reported a lower heating value (LHV) efficiency of greater than 67% for the electrolysis process. This conversion efficiency is the ratio obtained by dividing the amount of chemical energy in the form of hydrogen produced from the electrolysis process (based on the lower heating value of hydrogen), by the amount of electrical energy input to the process, usually expressed as a percentage. The hydrogen produced through the electrolysis process may be stored and subsequently used to power vehicles or generate electricity (see Fig. 11.8).

Currently, fuel cells are being developed to power homes, commercial buildings, passenger vehicles and small devices such as laptop computers (US-DOE, 2014a). A fuel cell is an electrochemical device used for the direct conversion of chemical energy in a fuel to electrical energy. A typical fuel cell consists of an electrolyte medium serving as interface between two electrodes (cathode and anode) containing platinum catalyst that speeds up the electrochemical reactions in the cell. For a reversible fuel cell using hydrogen and oxygen (or air), the following electrochemical reactions occur at the anode and cathode (Ramani *et al.*, 2004):

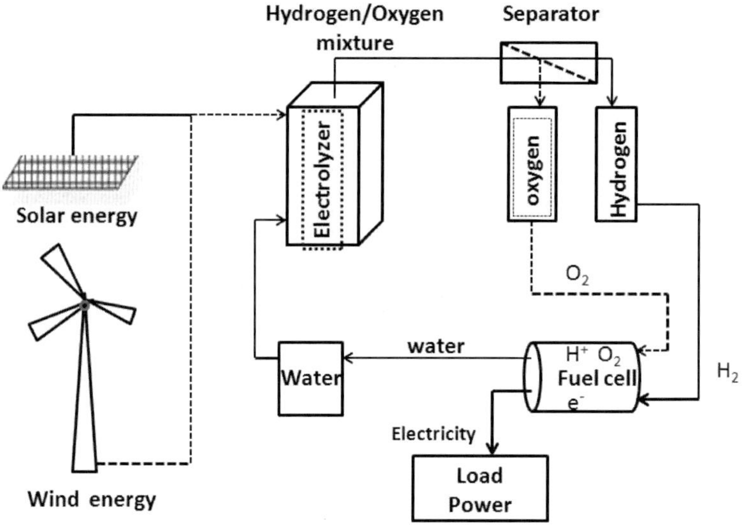

Fig. 11.8. Regenerative renewable electrolyzer/fuel cell system (modified from Wee, 2007).

Anode:

$$2H_2(g) \rightarrow 4H^+(aq) + 4e^-(aq) \tag{11.3}$$

Cathode:

$$O_2(g) + 4H^+(aq) + 4e^-(aq) \rightarrow 2H_2O(l) \tag{11.4}$$

For the above chemical reactions, protons are produced by electrochemical oxidation of hydrogen in Eq. (11.3), which are transported from the anode to the cathode through the ionically conducting (but electronically insulating) electrolyte (Ramani *et al.*, 2004). The electrons produced by the oxidation reaction travel through an external circuit to the cathode to complete the electrochemical reduction of oxygen to produce water, as shown in Eq. (11.4). As a result, the overall cell reaction is given as shown in Eq. (11.5) (Ramani *et al.*, 2004).

$$2H_2 + O_2 \rightarrow 2H_2O + \text{Electricty} \tag{11.5}$$

The current problems with the large (industrial) scale hydrogen production are mainly the leak-proof pipeline transportation and storage. Below, we present a brief description of hydrogen storage methods.

11.2.7.1. *Compressed hydrogen*

This method for hydrogen storage involves compression of the gas to pressures that can range from 5,000–10,000 psi (34.5–68.9 MPa), in order to obtain smaller volume for storage. The design and construction materials for hydrogen storage tanks have been a major challenge. In recent years, carbon fiber-reinforced tanks that are capable of withstanding these high pressures without leak are being considered (US-DOE, 2014c; Ozakia *et al.*, 2014).

11.2.7.2. *Cryogenic liquid hydrogen*

This method involves cooling of hydrogen gas to its liquid state in order to obtain a smaller volume for on-board vehicles application. But since the boiling point of hydrogen at atmospheric pressure is −252.8°C (−423°F), storage of the gas in liquid state at these very low cryogenic temperatures requires a process which consumes appreciable amount of mechanical energy (Syed *et al.*, 1998). Typically, about 30% of the heating value of hydrogen is required for liquefaction of the gas (Ozakia *et al.*, 2014). But, the advantage is that given the same volume of fuel, liquid hydrogen tank can store more hydrogen than compressed hydrogen gas tank. As an example, using a 10,000 psi gas tank, the volumetric capacity of liquid hydrogen is 0.070 kg+/L compared to 0.030 kg+/L for compressed hydrogen gas (From: (http://energy.gov/eere/).

Consequently, this method of hydrogen storage improves the energy density of hydrogen. However, liquid hydrogen tanks have problems such as hydrogen boil-off, weight control and tank cost. Because of these problems, a hybrid tank concept integrating both high-pressure gaseous and cryogenic storage of hydrogen is being studied for improved designs (US-DOE, 2014c).

11.2.7.3. *Material-based hydrogen storage*

Besides compression and liquefaction of gaseous hydrogen for storage, gaseous hydrogen can be stored using material-based hydrogen storage technologies. The three underlying mechanisms used for this storage are absorption process, adsorption process and chemical reactions, as described below.

a. *Hydrogen storage through absorption/desorption mechanism*

In this process, molecular hydrogen dissociates into atomic hydrogen on the surface of a reactive metal and diffuses through the interstitial sites in the lattice of the reactive metal, producing metal hydride (Broom, 2011; Kalantzopoulosa *et al.*, 2014; Maeland and Flangan, 1966). Hydrogen can be released from the reactive metal through the desorption process.

The two possible routes for the synthesis of metal hydrides are the direct dissociative chemisorptions, which is a gas phase reversible reaction shown in Eq. (11.6) and electrochemical splitting of water, which is another reversible reaction shown in Eq. (11.7) below (Lim *et al.*, 2010; Riis *et al.*, 2005).

$$2M + xH_2 \leftrightarrow 2MH_x + \text{heat} \tag{11.6}$$

$$M + xH_2O + xe^- \leftrightarrow MH_x + xOH^- \tag{11.7}$$

In the above chemical equations, M is the metal and x is the valence electron of the metal. The commonly used materials for these metal hydrides are alloys (solid solutions and intermetallic hydrides) of light metals such as lithium, beryllium, sodium, magnesium, boron, palladium and aluminum (Dobbins *et al.*, 2007). It has been claimed that previous research work on hydrogen storage using intermetallic hydrides such as $LaNi_5$ and TiFe demonstrated very good sorption/desorption kinetics. But, it was also discovered that hydrogen storage capacities of these metals were lower than 2% of hydrogen by weight (Soloveichik *et al.*, 2009). The exceptions include magnesium hydride (MgH_2), which contains 7.6 wt% hydrogen (Kalantzopoulosa *et al.*, 2014; Lim *et al.*, 2010; Soloveichik *et al.*, 2009). Hydrogen storage capacity is defined as the mass of hydrogen stored, divided by the mass of the fully hydrided storage material expressed as a percentage. In general, metal hydrides have the potential for reversible on-board hydrogen storage and release at low temperatures and pressures. The advantages of these storage materials are lower volume, lower pressure (with greater energy efficiency) and higher purity of hydrogen output (Riis *et al.*, 2005). But the current problems with these storage materials include the cost of manufacturing, low hydrogen capacity, slow hydrogen uptake and release kinetics, as well as how to manage the heat generated by the chemical reaction during refueling (US-DOE, 2014c).

b. *Hydrogen storage through adsorption mechanism*

In addition to the absorption mechanism, hydrogen storage can be achieved through adsorption as well. Adsorption involves attachment of the hydrogen atoms or molecules to the surface of solid materials, instead of binding to the interstitial sites of the materials, as is the case in absorption. A possible group of solid materials that are being studied for this technology are carbon-based materials such as aerogels, carbon nanotubes, carbon nanofibers (including metal-doped hybrids), as well as organometallic frameworks and conducting polymers (US-DOE, 2014c). Among these materials, the one that may look promising in the future are the high-surface-area hydrogen sorbents using microporous metal-organic frameworks (MP-MOFs). The MP-MOFs consist of metal/oxide groups linked together by organic struts that are synthetic, crystalline and microporous in nature.

Many studies have been conducted to evaluate hydrogen storage using MP-MOFs material through an adsorptive mechanism. For example, Furukawa *et al.*, (2007) reported hydrogen uptake using MOF-177 Zn_4O (1,3,5-benzenetribenzoate)$_2$ samples was 7.5 wt%

at −196°C and 70 MPa (−321°F and 10,152 psi), which corresponds to an absolute adsorbed amount of 11 wt%. The results at room temperature were less encouraging and maximum hydrogen uptake using MIL-101 ($Cr_3F(H_2O)_2O[(O_2C)–C_6H_4-(CO_2)]_3nH_2O$) was 0.51 wt% at 10 MPa (Li and Yang, 2008). While these data may appear encouraging, further research work is still needed to optimize and improve functionality of MP-MOFs. Although highly porous material should be a better storage medium, it is claimed that the highly porous nature of this material may present a significant volumetric capacity problem (Zhao *et al.*, 2008).

Another promising group of materials for hydrogen storage based on adsorption process are the clathrates. The clathrates hydrates are basically hydrogen-bonded H_2O frameworks that bear resemblance of ice in physical appearance (Mahajan *et al.*, 2007). But clathrate hydrates have small non-polar molecules (such as methane and ethane) or polar molecules with large hydrophobic moieties trapped inside the hydrogen bonded water molecules (Hu and Ruckenstein, 2006; Mahajan *et al.*, 2007). Preliminary studies have shown that large amounts of hydrogen molecules can be incorporated into the structure II (sII) of clathrate hydrate (Mahajan, *et al.*, 2007). Materials like this may be particularly appealing for off-board storage of hydrogen that will ultimately circumvent the need for high pressure or liquid hydrogen tanks (US-DOE, 2014c).

c. *Hydrogen storage through chemical reaction*

This method of hydrogen storage is achieved through the synthesis of complex metal borohydrides (hydrogen containing compounds), which can undergo dissociation to release the hydrogen. For example, a metal borohydride, sodium borohydride ($NaBH_4$), can be prepared by reacting methylborate ($B(OCH_3)_3$) with sodium hydride (NaH) in mineral oil, according to the following chemical equation (Soloveichik, 2007).

$$4NaH + B(OCH_3)_3 \rightarrow NaBH_4 + 3NaOCH_3 \qquad (11.8)$$

The reaction in Eq. (11.8) does not need pressurized hydrogen and occurs at the temperature range of 250–280°C (482–536°F). The hydrogen storage capacity (i.e., mass of hydrogen stored divided by the mass of sodium borohydride at fully hydrided value expressed as percentage) of the $NaBH_4$ solution is 7.9 wt%. The hydrogen can be released through hydrolysis of the solution in accordance with the chemical equation shown below (Wu, 2003).

$$NaBH_4 + 2H_2O(\mathit{liquid}) \xrightarrow{cat} NaBO_2(aq) + 4H_2(\mathit{gas}) + {\sim}300\,kJ \tag{11.9}$$

Several methods have been studied in order to regenerate sodium borohydride ($NaBH_4$). But, the most favorable approach is through the reaction of sodium metaborate ($NaBO_2$) with calcium hydride (CaH), according to the chemical equation (Soloveichik, 2007)

$$NaBO_2 + 2CaH_2 \rightarrow NaBH_4 + 2CaO \tag{11.10}$$

A major advantage with the use of $NaBH_4$ for hydrogen storage is the safe and controllable application of the material onboard vehicles (Riis *et al.*, 2005). However, since the hydrolysis reaction, Eq. (11.9), is not reversible, efficient regeneration of the $NaBH_4$ from $NaBO_2$ off-board vehicles is needed in order for this technology to become a viable option for hydrogen storage on-board vehicles (NREL, 2007). Currently, the research focus on this technology is the regeneration energy requirements, including cost and life-cycle impacts. There exist many other similar chemical reactions involving other metal borohydrides, but this reaction system is more commonly studied than other similar reactions for hydrogen storage (US-DOE, 2014c).

11.2.8. *Use of hydrogen as a transportation fuel*

Use of hydrogen as a transportation fuel has many problems, which include the design of high pressure on-board hydrogen storage systems and safety aspects of hydrogen. In addition to these problems, the same volume of hydrogen as combustion fuel contains less energy than gasoline and natural gas (i.e., 8 MJ/liter for liquid hydrogen versus 32 MJ/liter for gasoline) (US-DOE 2014d) and about 10 MJ/liter for compressed natural gas (CNG) (Cohen, 2012). This means that more space is required for hydrogen gas storage on-board of hydrogen vehicles than is required for gasoline and natural gas vehicles, a condition that complicates the design of hydrogen storage systems in automobiles. Due to these problems and the need for much better feasibility of design of more efficient batteries for automobiles, an economic approach for mass-production of hydrogen-powered vehicles in the near future is in doubt.

For hydrogen to become a major transportation fuel, it must be cost-competitive on per mile basis, compared with conventional fuels like gasoline and CNG. This means that hydrogen must be available (untaxed) at a cost ranging from \$2–4 per gallon gasoline equivalent (in \$2007) (Ruth and Joseck, 2011; US-DOE, 2012c). One kilogram of hydrogen is approximately one gallon (3.785 liters) of gasoline equivalent (gge). The gge is the amount of alternative fuel with an approximate energy content of one gallon of liquid gasoline.

According to DOE, during the last decade, the production cost of automotive fuel cell has declined from \$275/kW (2002) to \$55/kW (2013) (US-DOE, 2014b; Spendelow and Marcinkoski, 2013). The ultimate goal is to reach a target of \$30/kW in 2017, based on projections to high-volume manufacturing of 500,000 vehicles per year (US-DOE, 2014b; James and Spisak, 2012; US-DRIVE, 2013). At \$30/kW, an automotive fuel cell system is projected to be more economically favorable compared to internal combustion engine vehicles, which could make fuel cell vehicles more affordable in future (US-DOE, 2014b). Recently, automakers have made some important progress using prototype vehicles that are capable of traveling over 300 miles on a single fill of hydrogen fuel. However, it must be pointed out that this driving range must be obtained consistently, using different models of vehicles and without sacrificing cost, space, or performance (US-DOE, 2014a).

Illinois is one of the several states in the U.S. where practical application of fuel cell technology has been demonstrated. In 2006, the Federal Transit Administration (FTA), through its National Fuel Cell Bus Program (NFCBP), began the demonstration of hydrogen fuel cell buses in the nation. Three demonstration buses were assigned to the Chicago Transit Authority (CTA) (US-DOE, 2008). These projects are aimed to demonstrate the suitability of fuel cell buses in a large transit agency, with the potential for larger fleets of fuel cell bus procurements, including development and demonstration of new technology that will enable operation of fuel cell buses in cold climates (US-DOE, 2008). In Table 11.2, we present the progress report of the national fuel cell program reported by the U.S. FTA in 2014.

There are already facilities in Illinois using fuel cell-powered forklifts. They include Central Grocers in Joliet, Illinois, Golden State Foods in Lemont, Illinois and Testa Produce in Chicago, Illinois (Curtin *et al.*, 2012). Also, the AT&T Company has two sites using fuel cells to back up their telecommunication sites in Illinois (Curtin *et al.*, 2012).

In 2007, the first hydrogen gas station in Illinois was opened by the Gas Technology Institute (GTI) in Des Plaines, Illinois. The hydrogen station is located alongside the GTI's two natural gas filling stations. The 2007 fee at the hydrogen pump was $3.49 per kilogram of gaseous hydrogen. As of October 2014, the GTI's hydrogen pump station remains probably the only one in Illinois open to the public. The reason given is that there are not many hydrogen vehicles on the road, due to high production cost of hydrogen. The station was initially used by few prototype vehicles, including two hydrogen shuttle buses that were licensed to the city of Chicago (Kevin, 2007).

In 2011, the City of Naperville, Illinois, unveiled its new gasifier facility (Green Fuels Depot), located at the City's Springbrook Waste Facility (Curtin *et al.*, 2012). The new facility converts wood chips from the City's yard waste into three fuels (electricity, hydrogen and ethanol) and can process wood chips at the rate of one ton per day to produce enough electricity to power about 12–15 suburban homes. The project uses synthetic gas (syngas) to produce green electricity

Table 11.2. Progress toward achieving technical performance objectives (US-FHWA, 2014).

NFCBP performance objective	Progress through fiscal year 2011
Less than five times the cost of a conventional (commercial diesel) transit bus	Cost reductions from more than \$3.0 million per bus in 2006 to \$2.3 million for last bus ordered. Battery dominant bus with smaller fuel cell power system significantly less than \$2.3 million.
Four to six years or 20,000 to 30,000 hours of durability for the fuel cell power system	10,000+ hours achieved on fuel cell power system, with durability warranties at 10,000–12,000 hours.
Double the fuel economy compared to commercial (diesel transit) bus	Exceed two times conventional (diesel transit) bus fuel economy, but depends on route.
Bus performance equal to or greater than equivalent commercial (diesel transit) bus	Operated up to 19 hours/day, with good availability and miles between road calls[(1)] at 4,000 miles. Better acceleration. Quiet operation. Weight is still high.
Exceed current emissions standards	Zero emissions.
Foster economic competitiveness in FCEB technologies	Multiple manufacturers and platforms demonstrating buses.
Increase public acceptance for FCEB technologies	Continued progress.

(1) A road call (RC) or revenue vehicle system failure is defined as a failure of an in-service bus that causes the bus to be replaced en route or results in a significant schedule delay (Chandler and Walkowicz, 2006).

for charging plug-in hybrid and electric vehicles (PHEVs). The facility also includes a hydrogen separation and storage process used to fill up fuel cell-powered vehicles. The fuels are used to serve the City's vehicle fleet and other municipal uses (Government Fleet, 2011).

Current DOE-sponsored research and development (R&D) projects in Illinois are fuel cell and hydrogen-related R&D. These projects are distributed at several locations in the state, which include Northwestern University, University of Chicago, University of Illinois at Urbana-Champaign, Southern Illinois University at Carbondale, Illinois Institute of Technology (IIT), Argonne National Laboratory, and Illinois-based companies, including, Orion Industries, Gas Technology Institute and RCF Economic and Financial Consulting (Curtin *et al.*, 2012).

11.3. Thermal Energy Storage (TES) Systems

TES systems may be used to store cold or hot thermal energy to reduce fossil and nuclear energy consumption, as well as fossil fuel power plant emissions (Cabeza, 2014). Energy demand for refrigeration and air conditioning in Illinois is mostly covered by electricity that requires burning of coal, which is a major source of pollution in the state as discussed in Chapter 2 and by Agyarko and Mansoori (2013), or nuclear energy, which has its own major problems as discussed in Chapter 6. While TES technology is not common in Illinois, in some temperate regions around the world such as Asia, some parts of the U.S., Canada and Europe, TES systems are frequently used to store snow and ice in winter for refrigeration and/or cooling during the year. Similarly, waste heat or solar thermal energy is stored in appropriate TES systems in summer to warm buildings during the cold season.

We must point out that the technology for thermal energy storage systems in Illinois is not new. For example, Underwriters Laboratories (UL), Inc. which is headquartered in Northbrook, Illinois is already using thermal energy storage to take advantage of low-cost, off-peak electric rates to produce cooling energy. UL Inc. claims that they have been using a hybrid cooling system for energy efficiency and operating flexibility since 2001 to save ~\$350,000 annually on their utility bill (Calmac, 2009).

While UL Inc. system apparently reduces their utility bill, there is no indication that it also reduces their electric energy consumption. We are of the opinion that the company may save much more funds, if they modify their system to a seasonal energy storage system. Such modification will also save the state of Illinois in fossil and nuclear energy resources consumption and will reduce emissions of carbon dioxide and other pollutants. The principal author of this book has developed the design of a seasonal energy storage system, which if implemented, will make the buildings in the East Campus of the University of Illinois at Chicago self-sufficient in air-conditioning, not needing electrical energy for summer season cooling. Also, the University of Illinois at Urbana-Champaign has recently added a thermal energy storage tank to the Campus Chilled Water System. The purpose of the 6.5 million gallon tank is to help meet the ~7,000 tons of additional cooling requirements for building projects on the campus (UIUC, 2011).

There are different types of TES systems. These include the aquifer thermal energy storage (ATES open loop) systems, borehole thermal energy storage (BTES) systems, pit storage systems, rock-cavern storage systems, closed loop pipe systems and open loop gravel water pit storage systems (Roth and Brodrick, 2009). Among these, the most common ones are the ATES or the open-loop systems, the BTES systems, including ice storage systems, which are described below.

11.3.1. *ATES system*

The system consists of a thermal storage medium, a heating or cooling source, a thermal distribution system, heat exchangers and thermal loads (Roth and Brodrick, 2009). The geographical location of Illinois makes this technology very attractive for seasonal energy storage because of its abundance of snow in winter and high temperatures in summer. Several ATES systems exist worldwide and the technology has proven to be viable, cost-competitive and energy-efficient. An example of application of an ATES technology is the large aquifer used for both heating and cooling of a building at Oslo airport in Gardemoen in Norway. The estimated area of the airport building is

180,000 m^2 (~2,000,000 ft^2), with a cooling demand of 5,000 MWh per year (Bellù, 2009). Depending on the form of energy storage, three main types of ATES systems can be identified. These are the chilled water storage systems, heat storage systems and integrated heat and cold storage systems. Below, we describe the different types of the ATES systems which can be useful to Illinois.

11.3.1.1. *Chilled water storage systems*

For these systems, chilled (or cold) water at temperatures ranging from 6–12°C (~43–54°F) is collected, injected and stored in a suitable underground aquifer storage for durations that spans from several hours to several months, depending on the cold source and application, followed by subsequent recovery of the chilled water for cooling purposes (see the "Cooling mode" on Fig. 11.9). Depending on the properties and energy requirements of the aquifer, the system can use one or more wells (Bridger and Allen, 2005). The storage efficiency

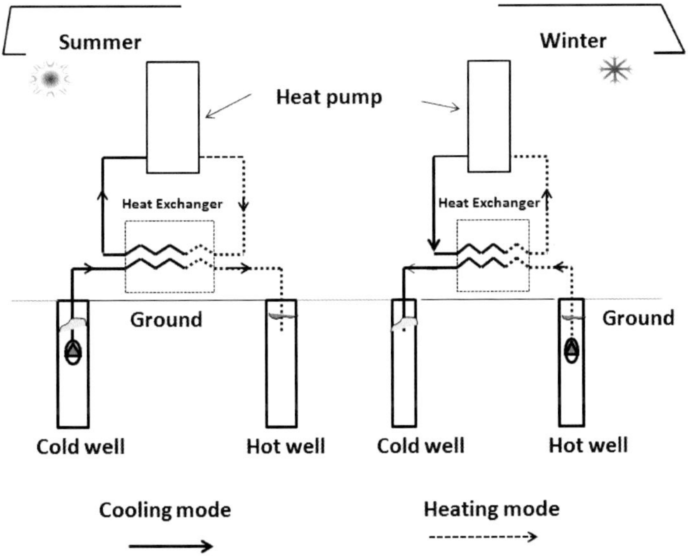

Fig. 11.9. An example of aquifer seasonal thermal energy storage system for cooling and heating (Modified from IEA, 2009).

of the aquifer cold storage systems is about 50–90% (IEA, 2011a). The storage thermal efficiency of a chilled water storage system is defined as the amount of energy taken from an aquifer divided by the amount of energy injected into the aquifer. Applications of aquifer cold storage systems include air conditioning and equipment cooling, as well as industrial process cooling.

11.3.1.2. *Heat storage systems*

For these systems, heated water is injected and stored in suitable underground aquifer storage and subsequently recovered for use (see the "Heating mode" on Fig. 11.9). These heat storage systems are classified as low and medium temperatures (10–40°C) or high temperatures (40–150°C) and high pressures (Bridger and Allen, 2005). The sources of hot water could be waste heat from coal-fired power plants, solar energy, etc. Applications of aquifer heat storage systems include space heating, industrial heating, agricultural purposes such as green houses, roadway de-icing and snow-melting. Similar to chilled water storage system, efficiency of the aquifer heat storage systems is about 50–90% (IEA, 2011a).

11.3.1.3. *Integrated ATES systems*

The integrated ATES systems use the same components as the aquifer heat and cold storage systems, and are for large-scale applications of combined space cooling and heating. For this technology, cold water is produced by the evaporator of a heat pump, which is stored in cold wells for cooling in summer. The returning warm water from the air conditioned rooms is stored in hot wells and is used for space heating in winter. Since the integrated ATES systems are for large-scale applications, they may offer a much better improved level of efficiency than using only the cold or heat storage systems. However, a disadvantage of the integrated heat and cold aquifer systems is development of thermal plumes around the wells. This can result in increased risk of heat transfer between the wells. However, this

Table 11.3. Example of an aquifer hydraulic properties and design assumption of an ATES system in Columbus, Ohio (Worthington, 2014).

Parameter	Value
Range of aquifer saturated thickness (ft/m)	20–65/6–19.8
Aquifer depth below ground (ft/m)	115/35
Below surface groundwater depth in aquifer (ft/m)	40/12
Groundwater flow velocity (ft/day) — (m/day)	∼2–0.6
Groundwater temp. (°F/°C)	55/12.8
Depth of well (ft/m)	110/33.5
Range of max. well yield (gpm/lpm)	180–230/680–870
Range of injection Temp. cold wells, ATES storage temperature during loading operation (°F/°C)	39.2–41/4–5
Range of extraction Temp. cold wells, supply from cold storage during cooling operation (°F/°C)	41–46.4/5–8
ATES system size (tons cooling)	380
Range of ATES system flow rate (gpm/lpm)	360–700/1363–2650
Reduction in electricity use	28–49%
Total cost per ton cooling ($)	5,600–6,300
Construction cost ($1,000)	1,780–2,380
Soft cost ($1,000)	360–450
Total cost ($1,000)	2,140–2,830

problem can be eliminated through adequate spacing of the cold and hot wells (Bridger and Allen, 2005). The storage capacity of an ATES system mainly depends on the permissible range of temperature, thermal conductivity and the flow rate of the underground water (Bellù, 2009).

In Table 11.3, we report an example of an aquifer hydraulic properties and design parameters of an ATES system in Columbus, Ohio.

Typically, ATES systems are more difficult to construct and require more maintenance. But under favorable conditions, they have relatively shorter payback time (2–5 years) compared to BTES systems (NER, 2004). The other major problems with ATES systems are wells clogging, leakages and production of sand from the wells (NER, 2004). Among these problems, the most common one is clogging of

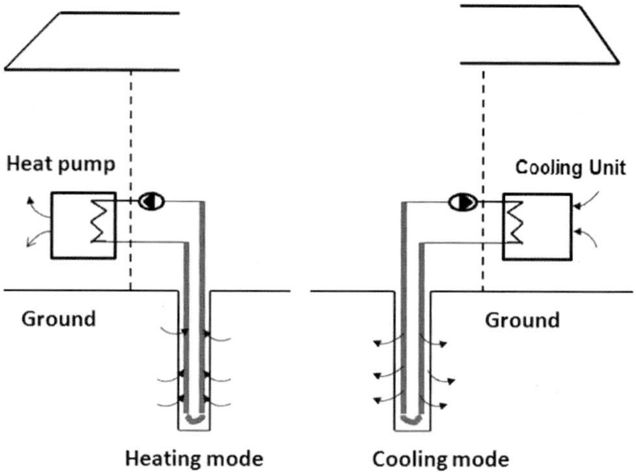

Fig. 11.10. BGEU system (Modified from Nordell and Ahlström, 2007).

wells, usually due to iron precipitation, which can cause restrictions to sections of the wells. This problem can lead to decreased system flow rate resulting in lower capacity of the system.

11.3.2. *Borehole geothermal energy utilization (BGEU) system*

The BGEU systems use vertical heat exchangers inserted underground for transfer of thermal energy to and from the ground (see Fig. 11.10).

BGEU systems usually consist of several multiple bore holes with approximate diameter of 0.1–0.15 meter, reaching the depth of about 50–200 meters (Qing *et al.*, 2006; NER, 2004; Nordell and Ahlström, 2007; Nordell *et al.*, 2007). Different kinds of heat exchangers can be inserted into the boreholes to make them appear like a large heat exchanger between the system and the ground (NER, 2004). The U-pipe borehole heat exchangers (BHEs) represent the most common method to exchange heat with the ground (Acuña and Palm, 2013). It is a single polyethylene U-pipe with standard dimensions of 40 mm of outer diameter and 2.4 mm of wall thickness. However, their thermal performance is poor due to "thermal shunt flow" between the pipes and poor placement of pipes relative to the borehole wall (Acuña and Palm, 2013). The use of more effective heat exchangers, such as double U-tube systems made of polyethylene or polypropylene is not uncommon (Acuña and Palm, 2013). Extraction and delivery of heat and cold to and from underground are achieved through circulation of a heat transfer fluid in a closed-loop through the boreholes. The heat transfer fluid is normally water-based anti-freeze (ethylene glycol or other heavy alcohols), which enables the system to operate below the freezing point of water when needed. The storage thermal efficiency of a BGEU system is defined as the amount of energy taken from the ground, divided by the amount of energy injected into the ground. This efficiency depends on several factors, including ground temperature, operational temperature of the storage system and thermal properties of the ground. The largest BGEU system in the world is located at Fort Polk Army Base in Fort Polk, Louisiana, USA. The system has 8,000 boreholes which supply both cold and warm thermal energy to residents of the Army Base. BGEU systems are also used for de-icing of frozen roads in Japan and Switzerland (Qing *et al.*, 2006).

In Table 11.4, we report the technical specification of a large BTES system located at Lund University, Sweden.

But unlike ATES systems, BGEU systems have no clogging problem. However, since BGEU systems use cooling fluids, in case of a poorly installed, damaged or aged system, the fluid can leak into aquifers. The coolant is often a mixture of an antifreeze agent such as glycol, a biocide and a corrosion inhibitor (Bonte *et al.*, 2011).

11.3.3. *Snow and ice seasonal storage systems*

The concept of natural ice and snow storage in winter season and its subsequent use in the summer season for cooling purposes has long been recognized in ancient cultures and societies in various parts of the world, dating back to the pre-industrial era (Kreider *et al.*,

Table 11.4. Technical specification of a large BTES system in Lund University, Sweden (Harris and Playford, 2012).

Date Inaugurated	**2005**
Construction period (yrs)	1.5
Size (m^2)	60,000
Number of Boreholes	153
Depth (m)	230
Total piping (km)	70
Heating/cooling needs met (%)	60–80
Additional heating	District heating
Total cost (USD)	~4.8 Million
Estimated payback period (yrs)	10

2002). Illinois, having a highly diverse weather is a suitable state where natural snow can be stored in winter and used in summer for air conditioning and refrigeration (see Fig. 11.11). The snow can be stored in underground cavities, such as abandoned coal mines, tunnels, or rock caverns or artificially constructed caverns in rocks or deep pits in soil. Cold water is extracted from the melting snow for cooling. But since the collected snow is contaminated by dirt, filters should be installed between the snow pit and the heat exchanger to prevent the dirt in the cold water plugging the pipes (NER, 2004).

11.3.4. *Thermal storage using ice harvester*

Thermal energy storage using natural sources such as air, ground and water evaporation dates back to ancient history. However, for the first time in 1775, William Cullen, a Scottish physician, chemist and agriculturist made ice through evacuation of air and water vapor in a water tank (Kreider *et al.*, 2002). During the same period in history, it was known that certain chemicals, such as calcium chloride, could lower temperatures when mixed with water and even with snow and

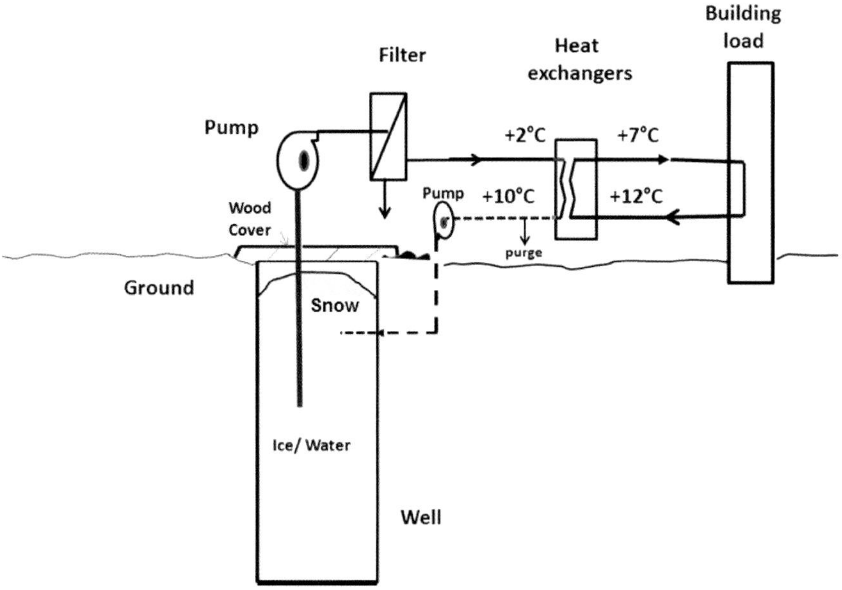

Fig. 11.11. Snow and ice system for cooling (Modified from NER, 2004).

ice (endothermic mixing). Currently, technologies such as ice harvesters and snow guns have made it possible to produce artificial ice and snow that are easily used for cooling (see Fig. 11.12). To produce ice at 0°C, the recirculation pump draws water from the storage tank and transports it to the ice harvester. Inside the ice harvester, the water flows into a drain pan located above a series of refrigerated plates. The water then exits the drain pan and flows downwards by gravity along the sides of the refrigerated plates to form ice. The ice crystals continue to grow until it reaches a thickness of about 4.8–6.4 mm (0.189–0.252 inch) (Knebel, 1986). Depending on the timing of a control system, a hot gas defrost cycle periodically causes blocks of ice to fall from the plates to the water tank under gravity. In cooling mode, the chilled water pump draws iced water from the storage tank and transports it to the building heat exchanger for cooling. These systems can be used to produce solid ice and snow at night during electricity off-peak periods. Late during the night, when most people

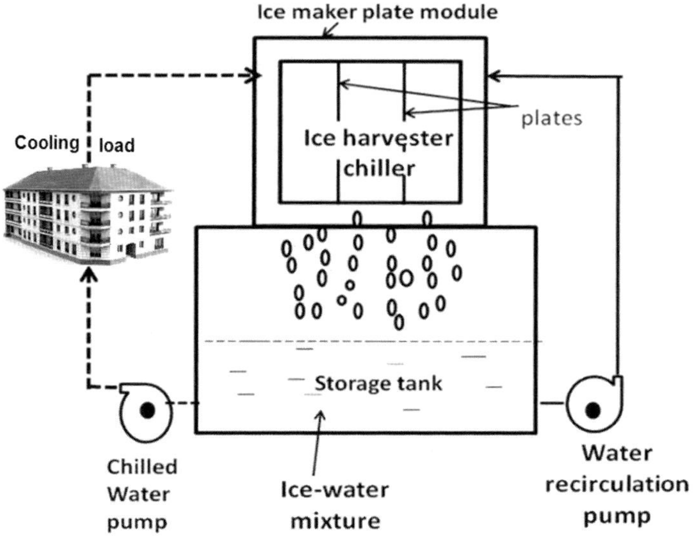

Fig. 11.12. Ice harvester (modified from Knebel 1986).

are sleeping, electricity loads of buildings are at their minimum and as a result, the generating capacities of utility power plants are utilized at their lowest rates because of low energy consumption. In essence, ice harvesters can serve as good replacements for conventional condenser units used for cooling of buildings and consumers could end up paying less money for energy usage (Sohn, 1991).

As mentioned earlier in Sec. 11.3, in 2010, the University of Illinois completed construction of a 6.5 million gallon (24.6×10^6 liters) TES tank at its Urbana campus. The TES tank has a spiral stair casing and measures 36.1 meters (118.5 ft) in diameter by 23.8 meters (78.1 ft) in height (Andrepont and Reifsteck, 2012). The estimated cost of the tank was $3,700,000, and was designed to meet the cooling load requirement of the campus, but without expanding the existing campus chilled water system, which consists of 18 chillers that use steam and electricity (UIUC, 2011). The TES tank is used to store reserve chilled water at night during off-peak hours when electricity costs are low. In night time, the TES tank is charged by operating the chillers. But during afternoon peak cooling periods when electricity costs are higher, the tank is discharged to allow the chillers to be turned off. The TES tank is filled using pumps located at an adjacent pump house. These pumps and the tank are connected to the campus chilled water system (CCWS) through a 30-inch diameter piping. This connection allows cold water to be pumped from the chillers to the bottom part of the TES tank during the charging operation, while warm water is being returned from the top of the tank to the chiller plants (UIUC, 2011). In essence, the tank provides approximately 7,000 tons of additional cooling requirements for the campus buildings. The estimated annual energy cost avoidance from this project is $860,000 (UIUC, 2011).

In 2013, the cities of St. Charles, Naperville and Evanston, in Illinois were selected as demonstration sites for energy storage projects by Verde Energy USA (Verde Energy, 2014), an independent retail power supplier and energy solutions provider. The three cities partnered with Verde in launching the intelligent smart grid-enabled distributed energy storage system (Ice Energy's "Ice Bear"™ technology) in the Illinois market to demonstrate the energy efficiency potential of the thermal energy storage technology. Verde installed energy storage system (Ice Energy's "Ice Bear"™ technology) at one city-owned facility in each of the three cities (Whelpley *et al.*, 2013). Also in conjunction with the three cities, Verde gathered energy efficiency data and monitored the performance of the Ice Bear technology system (Verde Energy, 2014).

In St. Charles, the energy storage system was installed at the Municipal Water Treatment Laboratory building. With the peak load consumption of 4.2 MWh, approximately 2.0 MWh of load was shifted to the night because of installation of the energy storage system (Verde Energy, 2014).

In Evanston, the storage system was installed at Chandler-Newberger Community Center of the city. Approximately, 1.9 MWh of the load was shifted to night because of installation of the energy

Table 11.5. Performance of Ice Bear technology in the cities of St. Charles, Naperville and Evanston (Verde Energy, 2014).

Illinois City	Approximate Electric Loads Shifted to Night (MWh)	Peak Periods Electrical Energy Consumption (MWh)	Reduction in HVAC[(1)] Electricity Usage in 6 hr peak Demand Period (%)
St. Charles	2.0	4.2	95
Naperville	3.5	6.3	95
Evanston	1.9	3.8	95

HVAC = heating, ventilation and air-conditioning.

storage system and avoided peak load of 3.8 MWh (Verde Energy, 2014).

In Naperville, the energy storage system was installed at the City's Water Service Center located at 1200 W. Ogden Ave. Similar to St. Charles and Evanston, the City of Naperville avoided peak load of 6.3 MWh consumption by shifting roughly 3.5 MWh of the load to the night because of installation of the storage energy system (Verde Energy, 2014). It is also claimed that the energy storage system allows the City of Naperville more efficient and effective utilization of air-conditioning, while upgrading its aging infrastructure at this location (Ice Energy, 2013).

In Table 11.5, we summarize the performance of the Ice Bear technology in the three cities (St. Charles, Naperville and Evanston) as claimed by Verde. The percentage reduction in heating, ventilation and air conditioning (HVAC) electricity usage (in a 6-hour peak demand period) for each of the three installations was claimed to be 95%. According to Verde, these data are consistent with the results achieved by similar installations in other regions across the U.S., where large-scale deployments of the technology produced significant peak load demand reductions (Verde Energy, 2014). The estimated cost of labor for the equipment installation was 30% of the

Table 11.6. Ice thermal storage technology projects in Illinois (US-DOE, 2014e). The projected project lifetime for these projects ranges from 20–25 years.

Location	Utility Type	Rated Power (kW)	Status
1200 W. Ogden Ave, Naperville, IL	Public owned	10	Operational
1028 Central Street, Evanston, IL	Public owned	10	Operational
200 Devereaux Way, St. Charles, IL	Utility-owned	10	Operational
Underwriters Laboratories, Northbrook, IL	Investor owned	1250	Operational

project cost. But according to Verde, this amount was underwritten and reinvested in the economy of the local area through hiring of local contractors.

Table 11.6 shows the Illinois' cities using thermal storage technology as reported by DOE's Global Energy Storage Database (US-DOE, 2014e).

11.3.5. *Phase-change energy storage mediums*

Thermal energy from solar panels can be stored using phase change materials (PCM) as energy storage medium (Himran *et al.*, 1994a, 1994b; Mansoori *et al.*, 2003; Agyenim *et al.*, 2010). PCM can provide higher storage capacity and storage efficiencies ranging from 75 to 90% (Hauer, 2012). Latent heat is absorbed by the PCM as it is melting, and heat is released during solidification or cooling. During this phase transformation of the PCM, temperature of the medium remains constant, as latent heat is being stored. In addition to the latent heat, the PCM utilizes some sensible heat below and above its melting point temperature to store energy (Q), which is defined by Eq. 11.11 (Himran *et al.*, 1994a, 1994b; Mansoori *et al.*, 2003; Agyarko and Mansoori, 2011).

$$Q = m\left[\int_{T1}^{Tm} Cps\,dT + L + \int_{Tm}^{T2} C_{pl}\,dT\right] \qquad (11.11)$$

In Eq. (11.11), m is the mass of the PCM, L is the latent heat of melting, T_1, T_2 and T_m are the initial, final and melting absolute temperatures, respectively. Also, C_{ps} and C_{pl} are the specific heats of solid and liquid of the PCM phases, respectively. Therefore, if we assume that the PCM is a mixture of compounds such as paraffin-wax containing several alkanes, and we also assume that the PCM is an ideal mixture, then the total heat storage in the paraffin-wax is defined as Eq. (11.12) (Himran *et al.*, 1994a, 1994b; Mansoori *et al.*, 2003).

$$Q = \sum_i m_i \left\{ \int_{T_1}^{T_{mi}} C_{ps_i}\, dT + L_i + \int_{T_{mi}}^{T_2} C_{pl_i}\, dT \right\} \tag{11.12}$$

In Eq. (11.12), the summation is over the components of the mixture, and every term with subscript i is a property of the ith component (alkanes) of the mixture. The cumulative thermal energy is an important thermal energy storage characteristic of a mixture. The cumulative thermal energy is defined as the thermal energy stored in a mixture, by raising its temperature from a reference temperature (i.e., 298 K) to a higher temperature. The economic viability of thermal energy storage systems is strongly dependent on application and operational needs, as well as the number and frequency of the storage cycles (Himran *et al.*, 1994a, 1994b; Mansoori *et al.*, 2003).

In Table 11.7, we present the comparison of the capital cost and efficiencies of the various thermal energy storage systems.

11.4. Sustainability of Energy Storage Systems

Energy storage can provide substantial benefit to energy users, especially in cases where energy source is subjected to variation or intermittency. It is reported that in the U.S., not more than 20% of the electricity demand can be provided by intermittent renewable without energy storage (Cavallo, 2001). The concept of energy storage and its success are based on the principle that it should be more cost effective to store energy from one time than to produce it later, when

Table 11.7. Capital cost and efficiencies of the various energy storage systems.

Energy Storage Technology	Efficiency (%)	Capital Cost ($/kWh)
Phase change material storage (general) (PCM)	75–90 (2)	13–65 (2)
Advanced lead-acid batteries (50 MW)	85–90 (5)	425–475 (5)
Lithium-ion batteries (1–100 MW)	87–92 (5)	4340–6200 (5)
Flywheels (20 MW)	85–87 (5)	7800–8800 (5)
Compressed air energy storage (CAES) (250 MW capacity above the ground)	45–60 (4)	390–430 (5)
CAES (135 MW capacity underground)	45–60 (4)	60–125 (1)
Pumped storage power plants (Pumped hydroelectric) 0.9–1.4 GW	80–82 (1)	250–270 (1)
Renewable energy conversion to substitute natural gas (SNG)	>60 (3)	NA

(1) EPRI (2010); (2) Hauer (2013); (3) Zuberbühler (2010); (4) IRENA (2012); (5) Carnegie *et al.* (2013)
N/A = not available.

it is needed (Hyman, 2011). In other words, the cost of energy storage must be cheaper when it is stored than the value of energy when it is recovered. The difference must be big enough to cover the cost of investment, maintenance, operation and energy losses.

Energy storage systems are not just storage facilities since their operation requires application of energy. For example, in CAES system the air is compressed and heated (to prevent freezing of water vapor) and then, electricity is generated through expansion of heated air in expanders (Safaei *et al.*, 2013). Such expenditure of energy requires energy storage systems to be economically and environmentally sustainable.

Energy storage systems can be applied on small or large scales in distributed or centralized systems and while some storage technologies are mature or in demonstration phase, most are still in early stages of development, as shown in Table 11.8.

Table 11.8. Status of energy storage technologies (IEA, 2014a).

Storage Technology Type	Capital Requirement × Technology Risk	Maturity Stage	Output
Cold water	low	com	Thermal
Pit	low	com	Thermal
Pumped hydropower storage (PHS)	low	com	Electricity
Thermochemical	med	R&D	Thermal
Synthetic Natural Gas (SNG)	med	R&D	Electricity
Hydrogen	med	R&D	Electricity
Adiabatic Compressed Air Energy	med	R&D	Electricity
CAES	med	demo	Electricity
Residential hot water	med	demo	Thermal
Underground thermal energy storage (UTES)	med	com	Thermal
Superconducting magnetic energy storage (SMES)	high	R&D	Electricity
Supercapacitor	high	R&D	Electricity
Flywheel (high speed)	high	R&D	Electricity
Flow batteries	high	R&D	Electricity
Li-based batteries	high	demo	Electricity
Molten salt	high	demo	Thermal
Flywheel (low speed)	high	demo	Electricity
Ice	high	demo	Thermal
Sodium — sulfur (NaS) batteries	high	demo	Electricity

Com: Commercialization stage; R&D: Research and Development stage; Demo: Demonstration stage.

As discussed in Secs. 11.2 and 11.3, there are varieties of energy storage technologies, but PHS, CAES and BESS are more suitable for bulk energy storage systems. A study conducted by the Electric Power Research Institute (EPRI, 2010), showed that underground compressed air energy storage technology is the most cost effective when compared with other energy storage technologies such as PHS, BESS, or flywheel. Their result also showed that for CAES, location of the storage reservoir is very crucial in cost ($/kWh) of the

Table 11.9. Performance of CAES with different compressor, turbine and reservoir ratings (Das and McCalley, 2012).

Com-pressor Power (MW)	Turbine Power (MW)	Reservoir Volume ($\times 10^3$ m^3)	Charging Time (hrs)	Dis-charging Time (hrs)	Cost (M$)	Payback Period (yrs)	Capacity Factor
50	50	50	11.03	6.72	30.94	12	0.385
50	50	100	22.06	12.71	42.91	13	0.392
50	50	150	33.08	18.75	55	16	0.395

technology and an aboveground system is six times more expensive than the underground system.

Das and McCalley (2012) conducted a simulation study to ascertain the impact of CAES sizing on its overall performance and cost. In their simulation, they considered three ratings of turbine (50 MW, 125 MW and 200 MW), compressor (50 MW, 125 MW, and 220 MW) and reservoir volume (50,000 m^3, 100,000 m^3 and 150,000 m^3) and studied combination of scenarios using the three levels. Considering the payback period as an economic indicator, the best three scenarios are shown in Table 11.9.

To better determine the cost of storage technologies, Akhil *et al.*, (2013) conducted a survey of more than 40 vendors in the U.S. to determine the levelized cost of energy storage technologies, but their analysis did not include hydrogen storage. Based on their research, the cost of the system was influenced by size and amount of time the system is capable of storing energy. In their analyses and for each technology, a variety of scenarios and different sizes and variable amount of energy storage times were considered. In Table 11.10 and for each technology, we only present the ones with the lowest installed cost. Akhil *et al.* defined levelized cost as the total cost of the initial capital investment, interest, replacement costs and fixed and variable operating costs over the lifetime of the facility, divided by the total yearly energy output from the system.

Table 11.10. Installed and levelized cost of some energy storage technologies[‡] (Akhil *et al.*, 2013).

Energy Storage Technology	Capacity (MW)	Efficiency[¶] (%)	Hours of Energy Storage at Rated Capacity	Present Value Installed Cost* ($/kW)	Levelized Cost of Energy* ($/MWh)	Levelized Cost of Capacity* ($/kW-yr)
PHS	280	80–82[(1)]	8	5600	220	420
CAES	408	45–60[(2)]	8	4470	120	347
NaS	50	75–85[(1)]	6	5750	294	642
VRB	1.2	75[(1)]	3.33	6050	560	680
Adv. Lead acid	50	70–90[(2)]	5	3800	220	410
Li-ion	1	85–95[(1)]	0.25	1900	170	215
Flywheel	20	85–90[(1)]	0.25	4250	380	480
Hydrogen FC	NA	40–85[(3)]	N/A	1500–3000[(3)]	170–600[(4)]	NA

(1) Viswanathan *et al.* (2013); (2) IRENA (2012); (3) Meena *et al.* (2014); (4) Steward *et al.* (2009).
(‡) Other than levelized cost of hydrogen, the rest of the numbers shown are approximations based on reading from graphs.
(*) Based on the value of USD in 2012.
(¶) In this context efficiency is defined as the ratio of energy discharged from storage to the energy input into storage.
PHS: based on round-trip efficiency of 81%, 365 cycles per year and plant life of 60 years.
CAES: Below ground, 365 cycles per year and plant life of 30 years.
NaS (Sodium sulfur battery): 365 cycles per year for 15 years.
VRB: 365 cycles peak-shaving annually for 15 years.
Lead acid: 365 cycles annually for 15 years.
Li-ion: 365 cycles annually for 15 years.
Flywheel: based on 5000 cycles per year.

To determine the cost of hydrogen, Steward *et al.* (2009) used HOMER, the National Renewable Energy Laboratory's distributed-generation economic model. In their analysis, a water electrolyzer was used to convert electricity into hydrogen, which was later to be used in a fuel cell to generate electricity. A summary of the levelized cost analysis of some of the energy storage technologies are shown in Table 11.10.

Based on a study published by Black and Veatch (2012), turbine and cavern constitute 70% of the capital cost of CAES, while powerhouse and battery share is 37% and 48% for PHS and BESS technologies, respectively. As technology improves, the cost of components required for energy storage is reduced. The reductions are more pronounced for new technologies such as flywheel and battery, as presented in Table 11.11.

As mentioned above (Sec. 11.4), energy storage technologies require energy expenditure and therefore sustainable application of these technologies has to offer environmental benefits. The life cycle GHG emission rate from storage systems must be advantageous when compared with fossil fuel-derived electricity. To ensure environmentally advantageous nature of energy storage technologies, precise accounting of all energy and emissions related to construction,

Table 11.11. Range of system capital cost for years 2011 and 2020 (Viswanathan *et al.*, 2013).

	2011 Range		2020 Range	
Technology	**($/kWh)**	**($/kW)**	**($/kWh)**	**($/kW)**
PHS	10	1500–2300	10	1640–2440
CAES	3	850–1140	3	500–1140
Flywheel	148	965–1590	81–148	200–820
Li-ion battery	850–1000	NA	290–700	NA
NaS battery	257–491	NA	181–331	NA
Redox flow Battery	173–257	942–1280	88–173	608–942

NA: Not applicable.

operation and decommissioning of an energy storage plant must be done. Denholm and Kulcinski (2004) conducted a life cycle analysis to examine energy requirement and GHG emissions of PHS, CAES and BES energy storage systems. In their analysis and for CAES systems, Denholm and Kulcinski considered a salt dome for air storage and for BES systems, they considered VRB. They found that BESS systems require more energy (roughly 4–8 times) for plant construction compared to equivalent size PHS and CAES systems. The energy media for PHS and CAES are water and air, which are essentially free. In contrast, the electrolyte for BESS requires intensive mining and ore processing. Summary of the findings of Denholm and Kulcinski is shown in Table 11.12.

Studies indicate that over 75% of the U.S. land has geologic conditions that are potentially favorable for underground air storage (Succar and Williams, 2008). Domal salt formations appear to be the most feasible geology for CAES development (NYSEG, 2009), but Illinois does not have any salt cavern formations. Instead, Illinois has a total of 11 sites of depleted natural gas/oil fields and 18 sites of aquifer storages with the total working capacity of 301 billion cubic feet (US-EIA, 2014a), which are suitable for CAES technology. A study by NYSEG (2009) showed that porous rock or aquifers may offer a more economical choice for large scale CAES development, as shown in Table 11.13.

The cost, efficiency and life cycle of energy storage devices vary widely across technologies. With the exception of a few advanced batteries, typical lifetime cycle capacity of batteries are in the range of 5,000 to 10,000, while PHS, CAES, flywheels and capacitors are generally rated at 10,000 to 100,000 cycles.

Despite economic advantages of CAES, their response time, similar to PHS, are on the order of minutes. In contrast, batteries, flywheels and capacitors have much quicker response time on the order of milliseconds (Carnegie *et al.*, 2013).

The economic performance of an energy storage system depends greatly on its specific application and operational needs. For example,

Table 11.12. Characteristics of selected energy storage technologies (Denholm and Kulcinski, 2004).

Technology	**Life Cycle Energy* (GJ_t/MWh Storage Capacity)**	**GHG Emissions (Tones CO_{2e}/MWh Storage Capacity)**	**Estimated Plant Life (Years)**	**Estimated Capacity Factor (%)**	**Construction and O&M Energy Ratio (GWh_{in}/GWh_{out})**	**Life Cycle Efficiency§ (%)**
PHS	373	35.7	60	20	0.01	74
CAES‡	265.7	19.4	40	20	0.07**	65
VRB	2253	161.4	20	20	0.06	72

(*) Energy associated with storage plant construction and decommissioning per unit of storage constructed.
(‡) CAES project using a salt dome for air storage.
(**) Includes delivery of natural gas fuel for CAES.
(§) Lifetime electrical conversion efficiency of storage.

Table 11.13. Estimated capital cost of CAES by different geologies (NYSEG, 2009).

Geology	Capital Cost of Storage
Hard Rock/Excavated and Existing Mines	\$30/kWh
Abandoned Limestone or Coalmines	\$10/kWh
Salt Cavern/Dry Mined	\$10/kWh
Salt Cavern/Solution Mined	\$1/kWh
Porous Rock/Aquifer	\$0.10/kWh

storage of hydrogen in aboveground tanks is \$17–27/kWh, which is potentially cost competitive with Li-ion and advanced lead battery systems, but not competitive with CAES systems.

In general, PHS and CAES are suitable for applications in scales above 100 MW and with long output durations. Flywheels, batteries, capacitors and supercapacitors can be utilized for power quality such as the instantaneous voltage drop and short duration UPS (uninterruptible power supply) due to their fast response time. Fuel cells, similar to batteries and flow batteries, have a fast response time and long discharge time and can be utilized for bridging power (Chen *et al.*, 2009).

11.5. Concluding Remarks

While energy storage technologies do not represent energy sources, they provide valuable added benefits to improve stability, power quality and reliability of supply (Ribeiro *et al.*, 2001). Lack of effective storage has often been cited as a major hurdle to substantial introduction of renewable energy sources into the electricity supply network (Ter-Gazarian, 2011). The supply of energy from primary sources is not constant and rarely matches the pattern of demand from consumers. Electricity is also difficult to store in significant quantities. Therefore, secondary storage of energy is essential to increase generation capacity efficiency and to allow more substantial use of renewable energy sources that only provide energy intermittently (Ter-Gazarian, 2011). In today's world, there is a continuous global need for more energy which, at the same time, has to be cleaner than the energy produced from the traditional generation technologies (Hadjipaschalis *et al.*, 2009). The extensive use of such energy sources in today's electricity networks can indisputably minimize the threat of global warming and climate change. However, the power output of these energy sources is not as reliable and as easy to adjust to changing cycles of demand, compared to the traditional power sources. Therefore, in order for these new energy sources to become completely reliable as primary sources of energy, energy storage becomes a crucial factor (Hadjipaschalis *et al.*, 2009).

The major challenge facing renewable power generation is integration of the variable and intermittent renewable energy to the electrical power grid. According to DOE estimates, about 70% of transmission lines in the U.S. are 25 years or older, while 60% of all circuit breakers are more than 30 years old (US-DOE, 2013a). Therefore, development of efficient energy storage systems can reduce the need for major new transmission grid and construction upgrades and would increase the performance of existing transmission and distribution system (US-DOE, 2013a). Application of energy storage system would result in grid congestion minimization during peak times by reducing the need for more power generation. Additionally, distributed energy storage systems can reduce power line losses during transmission, while extending the life of existing infrastructure. In Figure 11.13, all the possible ways of electrical energy storage technologies and their readiness status are shown.

Energy storage systems will also play a significant role in emergency preparedness by increasing the overall grid resilience (US-DOE, 2013a). For example, energy storage systems can make the grid more efficient, reliable and resilient and prevent power outages. Such efficiency would be critical in times of weather related emergencies (such as tornados), which frequency is expected to increase in the future as a result of global warming.

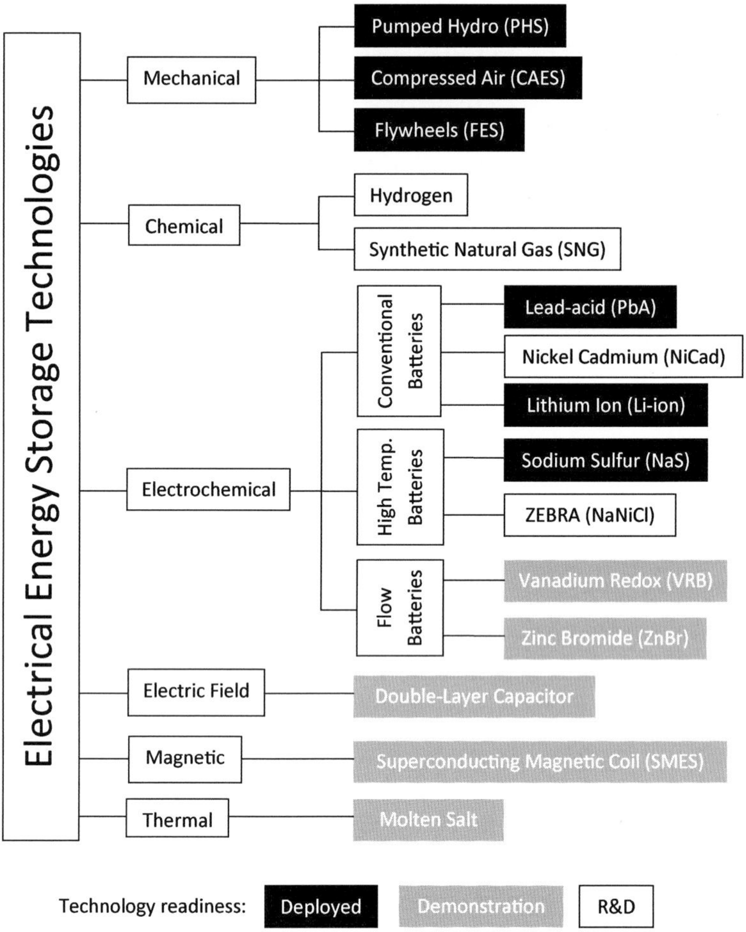

Fig. 11.13. Electrical energy storage technologies and readiness stages. Information in this figure is from (Carnegie, 2013; US-DOE, 2013a).

There are several energy storage technologies deployed in the State of Illinois, as shown in Table 11.14. Although these systems are relatively small, their application indicates that the potential of such technologies exists in the state.

Job creation would be one of the economic impacts of development of large-scale energy storage systems in Illinois. According to the US-EIA (2014b), 88,000 GWh of the total 203,000 GWh

Table 11.14. Energy storage systems in Illinois [US-DOE, 2014e]

Project Name	Storage Type/Storage Catagory	Rated Power [kW]	Duration at Rated Power HH:MM	Status	City
IIT RDSI Perfect Power Demonstration	ZBFB/EC	250	2:0.00	UC	Chicago
AllCell Chicago EV Charging Station	LIB/EC	40	N/A	Op	Evanston
Ener Del Mobile Hybrid Power System	LIB/EC	15	5:20.00	Op	Champaign
Invenergy Grand Ridge Wind Proj. BESS	LITB/EC	1500	0:15.00	Op	Marseilles
Jake Energy Storage Center: RES Americas	LIPB/EC	19800	0:24.00	Con	Joliet
Elwood Ener. Stor. Center: RES Americas	LIPB/EC	19800	0:24.00	Con	West Chicago
Grand Ridge Energy Storage 31.5 MW	LIB/EC	31500	0:23.00	UC	Marseilles
S&C CES: Chicago/PJM Frequency Regul.	LIB/EC	150	N/A	Op	Chicago
NextEra Lee/DeKalb 20 MW Illinois	LIB/EC	20000	N/A	Op	Lee/DeKalb counties
Follett VYCON Flywheel	Flywheel/flywheel	300	0:0.40	Op	Aurora
Underwriters Laboratories	Ice/Thermal	1250	14:0.00	Op	Northbrook
Ice Energy project at 1028 Central St.	Ice / Thermal	10	6:0.00	Op	Evanston
Ice Energy project at 1200 W Ogden Ave	Ice / Thermal	10	6:0.00	Op	Naperville
Ice Energy project at 200 Devereaux Way	Ice / Thermal	10	6:0.00	Op	St Charles

EC = Electrochemical; UC = Under Construction; Op = Operational; Con = Contracted; LIB = Lithium Ion Battery; LITB = Lithium Ion Titanate Battery; LIPB = Lithium Iron Phosphate Battery; ZBFB = Zinc Bromide Flow Battery electricity generation in Illinois is by coal-fired power plants. Utilization of energy storage technologies in the state can reduce the burden of coal-fired power plants and therefore, reduce GHG emissions. Furthermore, energy storage will help lower consumer costs by saving low-cost power for peak times and making renewable energy available when it is most needed (US-DOE, 2013a).

According to the DOE, the four challenges of energy storage implementation include development of cost competitive energy storage technologies, validating reliability and safety of such systems, existence of impartial regulatory environment and industry acceptance (US-DOE, 2013a).

Efficient designs of energy storage systems will help in the management of energy demand variability and lead to cleaner environment. This is because efficient energy storage systems will make power more efficient. This improved efficiency can reduce the need for less efficient power plants to meet peak load demand, and consequently, minimize the negative environmental impact.

Portfolio of energy storage technologies that were discussed in Secs. 11.2 and 11.3 indicates a range of services that can be provided to the electric grid based on their power requirement. Established technologies such as PHS and CAES can reach a capacity of 1 GW and discharge times in tens of hours. Success of these technologies on commercial scale has been proven. In contrast to PHS and CAES's capabilities in large storage capacity and discharge times, electrochemical batteries and flywheels are designed for low power and short discharge time applications. Unlike PHS and underground CAES, these technologies do not require specific geographical features and hence, are more versatile. There are several electrochemical battery technologies that are available commercially, but they have not received widespread acceptance due to their energy density, power performance, charging abilities and cost.

Flywheels have commercially been deployed for frequency regulation (U.D. DOE, 2013a) and despite the fact that they can provide up to an hour of stored energy, their discharge time is short. The

Table 11.15. Summary of energy storage technologies (US-DOE, 2013a; IEA, 2014b).

Technology	Primary Application	Advantages	Challenges	Ini. Inv.*‡ (USD/kW)
PHS	• Energy management • Backup and seasonal reserves • Regulation service also available through variable speed pumps	• Developed and mature technology • Very high ramp rate • Currently most cost effective form of storage	• Geographically limited • Plant site • Environmental impacts • High overall project cost	500–4600
CAES	• Energy management • Backup and seasonal reserves • Renewable integration	• Better ramp rates than gas turbine plants • Established technology in operation since the 1970's	• Geographically limited • Lower efficiency due to roundtrip conversion • Slower response time than flywheels or batteries • Environmental impact	500–1500
Fly wheels	• Load leveling • Frequency regulation • Peak shaving and off peak storage • Transient stability	• Modular technology • Proven growth potential to utility scale • Long cycle life • High peak power without overheating concerns • Rapid response • High round trip energy efficiency	• Rotor tensile strength limitations • Limited energy storage time due to high frictional losses	130–500

(Continued)

Table 11.15. *(Continued)*

Technology	Primary Application	Advantages	Challenges	Ini. Inv.*‡ (USD/kW)
Advanced Lead-Acid	• Load leveling and regulation • Grid stabilization	• Mature battery technology • Low cost • High recycled content • Good battery life	• Limited depth of discharge • Low energy density • Large footprint • Electrode corrosion limits useful life	300–3500
NaS	• Power quality • Congestion relief • Renewable source integration	• High energy density • Long discharge cycles • Fast response • Long life • Good scaling potential	• Operating Temperature required between 250° and 300°C • Liquid containment issues (corrosion and brittle glass seals)	300–3500
Li-ion	• Power quality • Frequency regulation	• High energy densities • Good cycle life • High charge/discharge efficiency	• High production cost-scalability • Extremely sensitive to over temperature, overcharge and internal pressure buildup • Intolerance to deep discharges	300–3500

Flow Batteries	• Ramping • Peak Shaving • Time Shifting • Frequency regulation • Power quality	• Ability to perform high number of discharge cycles • Lower charge/discharge efficiencies • Very long life	• Developing technology, not mature for commercial scale development • Complicated design • Lower energy density	300–3500
SMES	• Power quality • Frequency regulation	• Highest round-trip efficiency from discharge	• Low energy density • Material and manufacturing cost prohibitive	130–515
Electrochemical Capacitors	• Power quality • Frequency regulation	• Very long life • Highly reversible and fast discharge	• Currently cost prohibitive	130–515
Thermochemical Energy Storage	• Load leveling and regulation • Grid stabilization	• Extremely high energy densities	• Currently cost prohibitive	1000–3000

*Ini. Inv.: Initial Investment.

‡All cost data were obtained from IEA (2014a).

unique feature of flywheels is their instantaneous response time and therefore, they are a good choice for uninterruptible power supply (Carnegie *et al.*, 2013).

Flow batteries are different from both conventional and high temperature batteries because of their cell construction. Their unique feature is their ability to independently vary energy and power capacity, but cost and complexity of construction are major disadvantages of flow batteries.

SMES systems have a high cycle life, efficiency and reliability. Their disadvantages are refrigeration energy requirements, low energy density and high cost and thus they are used when there is a need for supplying short burst of electricity into the energy system.

Thermochemical energy storage (TES) is an emerging technology that uses reversible chemical reactions to store thermal energy in the form of hot or cold substances for subsequent utilization. They have 5–20 times more energy density than the conventional storage technologies (US-DOE, 2013a), but their commercial deployment is still in its infancy. A summary of energy storage technologies is presented in Table 11.15.

Bibliography

Acuña, J. and Palm, B. (2013). Distributed Thermal Response Tests on Pipe-in-pipe Borehole Heat Exchangers, *Applied Energy*, **109**: 312–320.

Agyarko, L. B. and Mansoori, G. A. (2011). Solar absorption cooling with alkanes as phase change energy storage medium. In Proceedings of AIChE, Spring Meeting & 7th Global Congress on Process Safety. Chicago, 14–17 March, IL.

Agyarko, L. B. and Mansoori, G. A. (2013). Non-renewable Energy Options in Illinois, *Int. J. Oil, Gas and Coal Tech.*, **6**(3): 288–347.

Agyenim, F., Hewitt, N., Eames, P. and Smyth, M. (2010). A Review of Materials, Heat Transfer and Phase Change Problem Formulation for Latent Heat Thermal Energy Storage Systems (LHTESS), *Renewable and Sustainable Energy Reviews*, **14**(2): 615–628.

Akhil, A. A., Huff, G., Currier, A. B., Kaun, B. C., Rastler, D. M., Chen, S. B., Cotter, A. L., Bradshaw, D. T., and Gauntlett, W. D. (2013). DOE/EPRI 2013 Electricity Storage Handbook in Collaboration with NRECA, Report No. SAND2013-5131, Sandia National Lab., Albuquerque, NM.

Al-Khoury, R. and Bundschuh, J. (eds.) (2014). Computational Models for CO2 Geo-sequestration & Compressed Air Energy Storage (Sustainable Energy Developments). CRC Press.

Allen, R. D., Doherty, T. J., Erikson, R. L. and Wiles, L. E. (1983). Factors affecting storage of compressed air in porous rock reservoirs, Tech. Report No. PNL-4707, Pacific-Northwest Lab., Richland, WA.

Andrepont, J. S and Reifsteck, K. V. (2012). Thermal energy storage (TES) at the University of Illinois-Urbana-Champaign: Providing capacity and reliability, while creating millions in NPV, *IDEA Campus Energy Conf.*, 6–9 February, Arlington, VA.

Baer, S. K. (2013). St. Charles, Lake Forest to get electric-car charging stations. *Chicago Tribune*, Chicago, IL.

Bellù, A. (2009). Energy storage systems: An approach involving the concept of exergy, MSc Thesis, Universita Degli Studi Di Padova, Italy.

Black and Veatch Inc. (2012). Cost and performance data for power generation technologies, Tech. report, Prepared for the National Renewable Energy Lab., 11401 Lamar Ave., Overland Park, KS.

Bonte, M., Stuyfzand, P. J., Hulsmann, A. and Van Beelen, P. (2011). Underground thermal energy storage: Environmental risks and policy developments in the Netherlands and European Union, *Ecology and Soc.*, **16**(1): 1–22.

Bridger, D. W. and Allen, D. M. (2005). Designing Aquifer Thermal Energy Storage Systems, *ASHRAE J.*, **47**(9): S32–S37.

Broom, D. P. (2011). *Hydrogen Storage Materials: The Characterization of their Storage Properties*. Springer.

Cabeza, L. F. (ed.) (2014). *Advances in Thermal Energy Storage Systems*, 1st Edn. Woodhead Publishing, Elsevier.

Calmac Corp. (2009). Underwriters Lab Bank on ice: An ice storage system from CALMAC provides cooling, while it conserves energy, CALMAC Manuf. Corp., 3-00 Banta Pl, Fair Lawn, NJ.

Carnegie, R., Gotham, D., Nderitu, D. and Preckel, P. V. (2013). Utility scale energy storage systems, State Utility Forecasting Group, Purdue Univ., West Lafayette, IN.

Caterpillar (2014). Electric Power Generation — UPS-FLYWHEEL, Caterpillar Corp., Peoria, IL. *cat.com.*

Cavallo, A. J. (2001). Energy Storage Technologies for Utility Scale Intermittent Renewable Energy Systems, *J. Solar Energy Eng.* **123**(4): 387–389.

Chandler, K and Walkowicz, K. (2006). King County Metro Transit Hybrid Articulated Buses: Interim Evaluation Results, Tech. Report No. NREL/TP-540-39742, National Renewable Energy Lab., Golden, CO.

Charlton, A. and Haag, T. (2012). Pumped storage: Investigating development of the Elmhurst quarry project. *Hydro Review*. hydroworld.com.

Chen, H., Cong, T. N., Yang, W., Tan, C., Li, Y. and Ding, Y. (2009). Progress in Electrical Energy Storage System: A Critical Review, *Progress in Natural Science*, **19**(3): 291–312.

Cohen, K. (2012). Natural gas cars: A look under the hood, ExxonMobil Corp., Houston, TX. *exxonmobilperspectives.com.*

Curtin, S., Gangi, J. and Skukowski, R. (2012). State of the States: Fuel cell in America 2012 (4th Edn.), Fuel Cell Tech.s Office, Office of Energy Efficiency & Renewable Energy, U.S. Dept. of Energy, Washington, DC.

Das, T. and McCalley, J. D. (2012). Compressed air energy storage, Educational Chapter, College of Eng., Iowa State Univ., Ames, IA.

Denholm, P. and Kulcinski, G. L. (2004). Life Cycle Energy Requirements and Greenhouse Gas Emissions from Large Scale Energy Storage Systems, *Energy Conversion and Management*, **45**: 2153–2172.

Denholm, P., Ela, E., Kirby, B and Milligan, M. (2010). The role of energy storage renewable Electricity generation, Tech. report (NREL/TP-6A2-47187), National Renewable Energy Lab., Golden, CO.

Dobbins, T., Kamineni, V and Lvov, Y. (2007). Nanoarchitecture of protective coatings for air sensitive metal hydrides, *Material Matters* (A publication of the Sigma-Aldrich Corp.), **2**(2): 19–21.

[EPRI] Electric Power Research Inst. (2010). Electricity Energy Storage Technology Options: A White Paper Primer on Applications, Costs and Benefits, Doc. No. 1020676. EPRI, Palo Alto, CA.

Furukawa, H., Miller, M. A. and Yaghi, O. M. (2007). Independent Verification of the Saturation Hydrogen Uptake in MOF-177 and Establishment of a Benchmark for Hydrogen Adsorption in Metal-organic Frameworks, *J. Mater. Chem.*, **17**: 3197–3204.

Gebbia, M., Simma, S. and Alon, E. (2013). Flat supercapacitors evolve to meet needs, Special Report, Illinois Capacitor Inc., 3757 West Touhy Ave., Lincolnwood, IL.

Genovese, J. E., Harg, K., Paster, M. and Turner, J. (2009). Current (2009) State-of-the-art Hydrogen Production Cost Estimate Using Water Electrolysis, Doc. No. NREL/BK-6A1-46676, U.S. Dep't of Energy Hydrogen Program, National Renewable Energy Lab., Golden, CO.

Genta, G. (1985). *Kinetic Energy Storage: Theory and Practice of Advanced Flywheel Systems*. Butterworth-Heinemann.

Government Fleet (2011). Naperville opens facility that converts yard waste to fleet Fuel. *Gov. Fleet Magazine*, 3520 Challenger Street, Torrance, CA.

Gowda, S. R., Reddy, A. L. M., Zhan, X., *et al.* (2012). 3D Nanoporous Nanowire Current Collectors for Thin Film Microbatteries, *Nano Lett.*, **12**(3): 1198–1202.

Guy, S. (2013). IIT, Argonne get $3.4M grant to improve electric-car battery. *Chicago Sun-Times*, Chicago, IL.

Hadjipaschalis, I., Poullikkas, A. and Efthimiou, V. (2009). Overview of Current and Future Energy Storage Technologies for Electric Power Applications, *Renew. Sustainable Energy Rev.*, **13**(6–7): 1513–1522.

Harris, M. and Playford, L. (2012). What lies beneath: Two examples of large scale borehole thermal energy systems in Sweden and Canada, Energi Öresund, The European Regional Devel. Fund, EU.

Hauer, A. (2013). Thermal energy storage (TES), IEA-ETSAP and IRENA Tech. Brief No. E17, Int'l Renewable Energy Agency. *irena.org*.

Haugen, D. (2011). Scrapped Iowa project leaves energy storage lessons. *Midwest Energy News*, RE-AMP. *midwestenergynews.com*.

Himran, S., Suwono, A. and Mansoori, G. A. (1994a). Characterization of Alkanes and Paraffin Waxes for Application as Phase Change Energy Storage Medium, *Energy Sources*, **16**(1): 117–128.

Himran, S., Samudro, H. M., Sularso, Suwono, A. and Mansoori, G. A. (1994b). Numerical study of heat transfer in a multicomponent material between tubes array during melting process. In Proceedings of the Int. Conf. on Fluid and Thermal Energy Conversion '94, Vol. 2: 269–274.

Hu, Y. H. and Ruckenstein, E. (2006). Clathrate hydrogen hydrate — a promising material for hydrogen storage, Angewandte Chemie Int'l Edition, **45**(13): 2011–2013.

Hyman, L. (2011). *Sustainable Thermal Storage Systems Planning Design and Operations*. McGraw-Hill.

Ice Energy. (2013). City of Naperville and Verde Energy USA successfully install IceBear™ Technology at Naperville's Water Service Center, Press Release Document, Ice Energy, Glendale, CA.

[IEA] Int'l Energy Agency (2009). Energy conservation through energy storage programme, ECES Publications. IEA, Paris, France.

[IEA] Int'l Energy Agency (2011a). Technology roadmap energy-efficient buildings: Heating and Cooling Equipment, (OECD/IEA), ISBN: 9789264118492. IEA, Paris, France.

[IEA] Int'l Energy Agency (2014). Energy storage technology roadmap: technology annex, (OECD/IEA). IEA, Paris, France.

[IEA] Int'l Energy Agency (2014a). Technology roadmap: energy storage, (OECD/IEA). IEA, Paris, France.

[IEA] Int'l Energy Agency (2014b). Energy storage technology roadmap: technical annex. IEA, Paris, France.

[IRENA] Int'l Renewable Energy Agency (2012). Electricity storage and renewables for island power: a guide for decision makers, IRENA Innovation and Tech. Center, Robert-Schuman-Platz 3, 53175 Bonn, Germany.

James, B. D and Spisak, A. B. (2012). Mass production cost estimation of direct H_2 PEM fuel cell systems for transportation applications: 2012 update, Revision 4, U.S. DOE, Washington, DC.

Jenco, M. (2014). A vehicle charging station in Naperville. News section, *Chicago Tribune*, Chicago, IL.

Jiang, J., Zhang, L., Wang, X., Holm, N., Rajagopalan, K., Chen, F. and Ma, S. (2013). Highly Ordered Macroporous Woody Biochar with

Ultra-high Carboncontent as Supercapacitor Electrodes, *Electrochimica Acta*, **113**: 481–489.

Kalantzopoulosa, G. N., Vitillob, J. G., Albaneseb, E., *et al.* (2014). Hydrogen Storage of Mg–Zn Mixed Metal Borohydrides, *J. Alloys and Compounds*, **615**(1): S702–S705.

Kevin, H. (2007). Illinois opens first hydrogen fueling station today. News section, *Hydrogen Cars Now*, Redlands, CA. *hydrogencarsnow.com.*

Knebel, D. E. (1986). Thermal storage with ice harvesting system. In *Proceedings of the 3rd Symp. on Improving Bldg. Systems in Hot and Humid Climates.* Arlington, 18–19 November, TX.

Kreider, J., Rabl, A. and Curtiss, P. (2002). *Heating and Cooling of Buildings: Design for Efficiency*, 2nd Edn. McGraw-Hill.

Li, Y. and Yang, R. T. (2008). Hydrogen Storage in Metal-Organic and Covalent-Organic Frameworks by Spillover, *AIChE J.*, **54**(1): 269–279.

Lim, K. L., Kazemian, H., Yaakob, Z. and Daud, W. W. (2010). Solid-state Materials and Methods for Hydrogen Storage: A Critical Review, *Chem. Eng. Technol.* **33**(2): 213–226.

Maeland, A. and Flangan, T. B. (1966). The Hydrogen-palladium System, *Platinum Met. Rev.* **10**(1): 20–23.

Mahajan, D., Taylor, C. E. and Mansoori, G. A. (2007). An Introduction to Natural Gas Hydrate/Clathrate: The Major Organic Carbon Reserve of the Earth, *J. Petrol. Sci. & Eng.*, **56**(1–3): 1–8.

Mansoori, G. A., Barnes, H. L. and Webster, G. M. (2003). Petroleum waxes, Chapter 19. In *Fuels and Lubricants Handbook*, ASTM Int'l, West Conshohocken, PA, pp. 525–558.

Mekhilef, S., Saidur, R., Safari, A. (2012). "Comparative study of different fuel cell technologies", *Renew Sust Energ Rev*, **16**(1): 981–989

Naperville. (2013). Electric vehicle charging stations: Providing energy boosts to electric vehicle owners, City of Naperville, TED Business Group, Naperville, IL.

[NER] Norden Energy Research. (2004). "Pre-design guide for ground source cooling systems with thermal energy storage", 37 pages, Report #. P-57725-A-1 ADM 02, Soil Cool/Rekyl Project, European Commission Save Programme and Nordic Energy Research.

Nordell, B., Grein, M. and Kharseh, M. (2007). Large-scale utilisation of renewable energy requires energy storage. In Proceedings Int'l Conf.

for Renewable Energies and Sustainable Devel.-7 (ICRESD_07). 21–24 May.

Nordell, B. and Ahlström, A. K. (2007). Freezing problems in borehole heat exchangers, Thermal Energy Storage for Sustainable Energy Consumption, NATO Science Series, **234**: 193–203.

[NREL] National Renewable Energy Lab. (2007). Go/No-Go Recommendation for Sodium Borohydride for On-Board Vehicular Hydrogen Storage, Indp. Review Doc. No. NREL/MP-150-42220, U.S. DOE Hydrogen Program. NREL, Golden, CO.

[NYSEG] New York State Electric and Gas (2009). Compressed air energy storage engineering and economic study, NYSERDA 10467, Final report No. 10-09. New York State Energy Research and Devel. Authority, State of New York, Albany, NY.

Ozakia, M., Tomuraa, S., Ohmurab, R. and Morib, Y. H. (2014). Comparative Study of Large-scale Hydrogen Storage Technologies, *Int'l J. Hydrogen Energy*, **39**(7): 3327–3341.

Peltier, R. (2011). Energy storage enables just-in-time generation. *Power Magazine.*

Pikul, J. H., Zhang, H. G., Cho, J., Braun, P. V. and King, W. P. (2013). High-power Lithium Ion Microbatteries from Interdigitated Three-dimensional Bicontinuous Nanoporous Electrodes, *Nature communications*, **4**.

Qing, G., Ming, L., Yan, J., Ming, L. M. Y., Guang, Q. and Spitler, J. D. (2006). Practice and task developing underground thermal energy storage in China. In *Proceedings of the 10th Int'l Conf. on Thermal Energy Storage*. Richard Stockton College, 31 May–3 June, NJ.

Ramani, V., Kunz, H. R. and Fenton, J. M. (2004). The Polymer Electrolyte Fuel Cell, *Interface*, a publication of The Electrochemical Soc., **13**(3): 17–19.

Ribeiro, P. F., Johnson, B. K., Crow, M. L., Arsoy, A and Liu, Y. (2001). Energy storage systems for advanced power applications. In *Proceedings of the IEEE*, **89**(12): 1744–1756.

Riis, T., Sandrock, G., Ulleberg, Ø. and Vie, P. J. S. (2005). Hydrogen storage-gaps and priorities, HIA-HCG Storage paper 11, Hydrogen Implementing Agreement (HIA), Int'l Energy Agency, Paris, France.

Roth, K. and Brodrick, J. (2009). Emerging Technologies: Seasonal Energy Storage, *ASHRAE J.*, **51**(1): 41–43.

Ruth, M and Joseck, F. (2011). Hydrogen threshold cost calculation, Program Record (Offices of Fuel Cell Tech.), Record No. 11007, U.S. Dep't of Energy, Washington, DC.

Safaei, H., Keith, D. W. and Hugo, R. J. (2013). Compressed Air Energy Storage (CAES) with Compressors Distributed at Heat Loads to Enable Waste Heat Utilization, *Applied Energy*, **103**: 165–179.

Silverman, J. (ed.) (1980). Energy Storage: A vital element in mankind's quest for survival and progress. Transactions of the First International Assembly held at Dubrovnik. Yugoslavia, 27 May–1 June. Pergamon Press.

Sohn, C. W. (1991). Field Measurement of an Ice Harvester Storage Cooling System, ASHRAE Transactions, **97**(2): 1187–1193.

Soloveichik, G. L., Gao, Y., Rijssenbeek, J. *et al.* (2009). Magnesium Borohydride as a Hydrogen Storage Material: Properties and Dehydrogenation Pathway of Unsolvated Mg(BH4)2, *Int'l J. of Hydrogen Energy*, **34**(2): 916–928.

Soloveichik, G. (2007). "Metal borohydrides as hydrogen storage materials" *Material Matters*, **2**(2): 11–15.

Specht, M., Baumgart, F., Feigl, B., *et al.* (2009). Storing renewable energy in the natural gas grid, Renewable Energy Research Assoc. (FVEE), German Renewable Org. (AEE), Topics 2009, pp. 69–77. *www.fvee.de*.

Spendelow, J. and Marcinkoski, J. (2013). DOE fuel cell Tech.s office record, Record No. 13012: fuel cell system cost-2013, Fuel Cell Tech.s Office, U.S. Dep't of Energy, Washington, DC.

Steward, D., Saur, G., Penev, M. and Ramsden, T. (2009). Lifecycle cost analysis of hydrogen versus other technologies for electrical energy storage, NREL Tech. Report No. NREL/TP-560-46719, National Renewable Energy Lab., Golden, CO.

Succar, S. and Williams, R. H. (2008). Compressed air energy storage: Theory, operation, and applications for wind power, Energy Systems Analysis Group, Princeton Env. Inst., Princeton Univ., Princeton, NJ.

Syed, M. T., Sherif, S. A., VezirogÞlu, T. N. and Sheffield, J. W. (1998). An Economic Analysis of Three Hydrogen Liquefaction Systems, *Int'l J. of Hydrogen Energy*, **23**(7): 565–576.

Ter-Gazarian, A. (2011). *Energy Storage for Power Systems*, 2nd Edn. Institution of Eng'g and Tech, Stevenage, UK.

[UIUC] Univ. of Illinois Urbana-Champaign (2011). Thermal energy storage center, Facility Services Factsheet. UIUC, Champaign, IL.

[UIUC] Univ. of Illinois Urbana-Champaign (2013). Team uses forest waste to develop cheaper, greener supercapacitors. *News Bureau*, Public Affairs, UIUC, Champaign, IL.

[US-DOE] U.S. Dep't of Energy (2003). Flywheel energy storage: An alternative to batteries for uninterruptible power supply system, Doc. No. DOE/EE-0286, Federal Energy Management Prog. U.S. DOE, Washington, DC.

[US-DOE] U.S. Dep't of Energy (2008). Hydrogen and fuel cell transit bus evaluations: Joint evaluation plan for the U.S. DOE and the Federal Transit Administration, Tech. Report No. NREL/TP-560-42781-1. U.S. DOE, Washington, DC.

[US-DOE] U.S. Dep't of Energy (2010). Final environmental assessment for Honeywell International Inc. Electric Drive Vehicle Battery and Component Manufacturing Initiative Project, Massac County, IL, Doc. No. DOE/EA-1716, National Energy Tech. Lab. U.S. DOE, Morgantown, WV.

[US-DOE] U.S. Dep't of Energy (2012a). Team led by Argonne National Lab selected as DOE's batteries and energy storage hub. *News*, U.S. DOE, Washington, DC. *energy.gov.*

[US-DOE] U.S. Department of Energy (2012b). "DOE Global Energy Storage Database", U.S. DOE, Washington, DC.

[US-DOE] U.S. Dep't of Energy (2012c). Fuel cell technologies program multi-year research, development and demonstration plan — Section 3.3 hydrogen storage, U.S. DOE, Washington, DC.

[US-DOE] U.S. Dep't of Energy (2013a). Grid energy storage. U.S. DOE, Washington, DC.

[US-DOE] U.S. Dep't of Energy (2013b). Final Environmental Assessment for the Beacon Power Corporation Flywheel Frequency Regulation Plant, Chicago Heights, Illinois (Site 1), and Hazle Township, Pennsylvania (Site 2), Doc. No. DOE/EA-1753, CreateSpace Indep. Publ. Platform.

[US-DOE] U.S. Dep't of Energy (2013c). Batteries for hybrid and plug-In electric vehicles, Alternative Fuels Data Center. U.S. DOE, Washington, DC.

[US-DOE] U.S. Dep't of Energy (2014a). DOE hydrogen and fuel cells program: Hydrogen storage, Office of Energy Efficiency & Renewable Energy, U.S. DOE, Washington, DC.

[US-DOE] U.S. Dep't of Energy (2014b). Market transformation: Accomplishments and progress, Office of Energy Efficiency & Renewable Energy, DOE, Washington, DC.

[US-DOE] U.S. Dep't of Energy (2014c). Hydrogen storage-current technology, Office of Energy Efficiency & Renewable Energy. U.S. DOE, Washington, DC.

[US-DOE] U.S. Dep't of Energy (2014d). Hydrogen storage basics, Office of Energy Efficiency & Renewable Energy. U.S. DOE, Washington, DC.

[US-DOE] U.S. Dep't of Energy (2014e). DOE Global Energy Storage Database, Sandia National Lab. U.S. DOE, Albuquerque, NM and Livermore, CA. *energystorageexchange.org*.

[US-DOE/EPRI] Dep't of Energy and Electric Power Research Inst. (2013). Electricity storage handbook in collaboration with NRECA, Report No. SAND2013-5131, Energy Storage Tech. and Systems, Sandia National Lab. U.S. DOE, Albuquerque, NM; Electric Power Research Inst., Palo Alto, CA.

[US-DRIVE] (2013). Hydrogen storage technologies roadmap fuel cell technical team roadmap, U.S. DRIVE Partnership. *uscar.org*.

[US-EIA] U.S. Energy Info. Admin. (2014a) "Underground natural gas storage", U.S. DOE, Washington, DC.

[US-EIA] U.S. Energy Info. Admin. (2014b) "State profile and energy estimates" , U.S. DOE, Washington, DC.

[US-FHWA] U.S. Federal Highway Admin. (2014). 2013 status of the nation's highways, bridges, and transit: Conditions & Performance, U.S. FHWA,U.S. Dep't of Transportation, Washington, DC.

Verde Energy. (2014). Verde Energy USA announces results Illinois Ice Bear™ projects: Innovative energy efficiency programs yield substantial reduction in electricity use during peak demand hours in three Illinois communities, Press release, Verde Energy USA, Norwalk, CT.

Village of Skokie (2014). Village of Skokie offers electric vehicle charging stations at no charge. News section, *Chicago Sun-Times*, Chicago, IL.

Viswanathan, V., Kintner-Meyer, M., Balducci, P. and Jin, C. (2013). National assessment of energy storage for grid balancing and arbitrage phase II, Vol. 2, Pacific Northwest National Lab., U.S. DOE, Richland, WA.

Wee, J. H. (2007). Applications of Proton Exchange Membrane Fuel Cell Systems, *Renew. Sustainable Energy Rev.*, **11**(8): 1720–1738.

Wernau, J. (2011). 1st fast charging station for electric vehicles is on the grid at Midway Airport: Airport station is 1 of 73 such stations scheduled for installation in Chicago and the suburbs by year's end. *Chicago Tribune*, Chicago, IL.

Wernau, J. (2012). U.S. state to fund battery research at Argonne: Coordinated effort will include universities, private companies. *Chicago Tribune*, Chicago, IL.

Wernau, J. (2014). Electric vehicle chargers unplugged in Chicago court battles, FBI probe crimp ambitious project for area; drivers left to wonder about their working stations' availability, safety. *Chicago Tribune*, Chicago, IL.

Whelpley, R., Houlihan, K. R and Garhan, L. (2013). St. Charles, Naperville, & Evanston Chosen as Sites for Energy Storage Demonstration Projects, Illinois Municipal Review Magazine, Springfield, IL.

Worthington, M. A. (2014). In depth feasibility studies aquifer thermal energy storage (ATES) VAMC Chillicothe and Columbus, OH. *Presented at IDEA's 27th Ann. Campus Energy Conf.*

Wu, Y. (2003, Apr. 14–15). "Hydrogen Storage via sodium borohydride-current status, barriers, and R&D roadmap" Presentation at GCEP — Stanford University.

Xtreme Power (2012). Invenergy and Xtreme Power announce joint energy storage project for PJM frequency regulation market, Press release. *xtremepower.com.*

Yang, X.-F., Yang, J.-H., Zaghib, K.,Trudeau, M. L. and Ying, J. Y. (2015). Synthesis of Phase-pure Li_2MnSiO_4@C Porous Nanoboxes for High-capacity Li-ion Battery Cathodes M *Nano Energy*, **12**: 305–313.

Yu, A., Chabot, V. and Zhang, J. (2013). *Electrochemical Supercapacitors for Energy Storage and Delivery: Fundamentals and Applications*. CRC Press, New York, NY.

Zhao, D., Yuan, D. and Zhou, H. C. (2008). The Current Status of Hydrogen Storage in Metal–organic frameworks, *Energ Environ Sci.*, **1**(2): 222–235.

Zuberbühler, U. and Rieke, S. (2010). Storing green electricity as natural gas, Fraunhofer-Gesellschaft, Press release. *fraunhofer.de/en/press*.

Glossary

Acid rain

An environmental problem caused by deposition of acidic materials from the atmosphere that can harm the nature, human health and buildings.

Actinides

These are radioactive metallic elements starting from actinium through lawrencium on the periodic table.

ADM = Archer-Daniels-Midland Company

Advanced gas-cooled reactor

An English-made nuclear reactor which uses graphite as moderator and carbon dioxide gas as a heat transfer medium.

Algal bio-fixation

A process that uses mass culturing of microalgae to consume carbon dioxide (from power plants flue gases) to produce biomass for conversion into liquid biofuels.

Alkene or **Olefine**

These are unsaturated hydrocarbons with at least one carbon-carbon double bond.

American Recovery and Reinvestment Act (ARRA) of 2009

A form of economic stimulus package also referred to as "The Stimulus" or "The Recovery Act" that was enacted by the U.S. Congress in February 2009. The bill was signed into law on 17 February 2009 by President Barack Obama.

Anaerobic digestion

A biological process where microorganisms breakdown chemical structure of biodegradable materials in absence of oxygen.

ANL = Argonne National Laboratory.

Anoxic conditions

An environment that has no (or very low concentration of) oxygen.

ANR = Alpha Natural Resources Incorporation.

Anthracite = hard coal

This is the hardest coal with the highest percentage of carbon content (86–97%) and calorific value.

API gravity = American Petroleum Institute gravity

A method used in the petroleum industry to determine the heaviness or lightness of petroleum fuels. The API Gravity is measured with a hydrometer in degrees which is corrected and reported at a basis of 60°F (15.6°C). The equation relating API Gravity to Specific Gravity (SG) is defined as

$$\text{API Gravity} = \frac{141.5}{SG_{(@60°\text{F})}} - 131.5$$

ARRA = American Recovery and Reinvestment Act.

a-Si = amorphous Silicon

A Silicon material that has irregular atomic arrangement.

ASP = Alkaline Surfactant Polymer.

ASTM = American Society for Testing and Materials.

Auto ignition temperature

Lowest temperature at which a substance or a liquid fuel under normal atmospheric pressure ignites spontaneously, without external ignition source such as a flame or spark.

Aviation gasoline

An aviation fuel in spark-ignition internal combustion engines used for aircrafts propulsion.

AWEA = American Wind Energy Association.

AWEM = American Wind Energy Management.

BACT = Best Available Control Technology

Air pollution control standards mandated by the U.S. Clean Air Act. The U.S. EPA determines what air pollution control technology must be used to meet a specific air pollution limit. The individual states can choose to have more stringent rules, however, not lower than the EPA standards.

Base load power plant

Electric power plants designed to operate continuously to meet the minimum power demand by consumers, based on reasonable expectations of their needs.

Bathymetric data

Data obtained from the study of underwater depth of lake or ocean floors.

BESS = Battery energy storage systems

BGEU = Borehole geothermal energy utilization

Binary cycle power plant

A power plant having two stages, one at high temperature and one at lower temperature and possibly with two different working fluids.

Biocatalyst

Substance, synthetic or natural, that can speed up a biochemical reaction.

Biocides

Chemical compound(s) capable of terminating living organisms.

Biodegradation processes

A process in which the chemical structures of materials are broken down by living organism(s) or enzymes.

Biodiesel

A type of liquid biofuel produced from base-catalyzed transesterification of natural fats and oils using light alcohol such as methanol. They are chemically known as fatty acid methyl esters (FAME).

Biofuel

A type of fuel (gas or liquid) that is directly or indirectly obtained from biological materials such as living organisms or plants.

Biohydrogen

Hydrogen obtained from biologically-induced process.

Biomass

Renewable organic matter originally from living organisms.

Bio-oil

An organic liquid fuel produced from pyrolysis of lignocellulosic materials.

Biorefinery

A facility or a group of facilities that integrate bio-based materials conversion processes to fuels, chemicals and power.

Bituminous coal (black coal)

A relatively soft coal with a tarlike substance and the broadest range of carbon content (45–86%). The quality of this type of coal is higher than lignite coal, but poorer than anthracite.

Boiling Water Reactor (BWR)

A type of nuclear reactor without a secondary circuit, using enriched uranium fuel and ordinary water as both coolant and moderator.

BOS = Balance of System

Botryococcenes

Oily hydrocarbons obtained from green algae (*Botryococcus braunii*).

Breeder reactor

A nuclear fission reactor designed to generate fissile material as it consumes such material.

Bronchitis

A disease characterized by inflammation of the mucous membrane in the bronchial tubes (airways carrying air from the trachea to the lungs).

Brownfield site

A polluted land previously used for industrial or commercial activities that need to be cleaned up/remediated for re-use.

Butanol

Also known as butyl alcohol and with the chemical formula CH_3-$(CH_2)_3$-OH. It may be used as a fuel in internal combustion engines like ethanol.

BWR = Boiling Water Reactor

CAES = Compressed Air Energy Storage

A technology based on conventional gas turbine technology that uses elastic potential energy of compressed air for energy storage.

CAIR = Clean Air Interstate Rule

A rule issued by U.S. EPA on 10 March 2005, purported to prevent the pollution from power plants drifting from one state to another. CAIR covers 27 Eastern states (including Illinois) and the District of Columbia. The rule implements a cap and trade system to reduce the target pollutants such as sulfur dioxide (SO_2) and nitrogen oxides (NOx) by 70%.

CANDU reactor = CANada Deuterium Uranium reactor

A nuclear reactor made in Canada using natural uranium as the fuel with heavy water (deuterium dioxide) as moderator.

Capacity factor (CF)

Actual output of a power plant recorded over a specific time period, divided by its potential output or rated power.

Capital cost

Total one-time cost of funds to finance an investment/project. This includes the cost of purchasing assets such as building and equipment and cost of going into debt.

Carbon dioxide capture and sequestration (CCS)

A process in which carbon dioxide emissions from industrial facilities and power plants are captured and stored or converted into other materials for reuse.

Carbon dioxide footprint

Total amount of carbon dioxide gas emitted directly or indirectly by human activities.

Carbon number

The number of carbon atoms in an organic molecule which approximately is a measure of the size of molecule.

Carbonatite

A rock that is formed by cooling and solidification of magma consisting of over 50% of carbonate minerals.

Catalytic hydroprocessing

A process that uses chemical catalyst to convert large molecules (like heavy fractions of crude oil) into smaller molecules.

CATUP = Competitive American Tails Upgrade Partnership

A joint partnership between two companies (ConverDyn/ Metropolis Works, in Illinois and Urenco USA, in New Mexico). The partnership aims to meet any potential demand from the U.S. Department of Energy in managing and upgrading stocks of depleted uranium.

CCS = Carbon Capture and Storage; Carbon dioxide Capture and Sequestration.

CCT = Clean Coal Technology

A technology aimed to improve the environmental performance and efficiency of coal-fired power plants.

CdTe = Cadmium Telluride

A chemical compound produced from cadmium and tellurium used in photovoltaics cells.

Cetane number = CN

An indicator used to determine the ignition delay of fuels in diesel engines (i.e., higher cetane number represents shorter ignition delay and better fuel). Diesel engines operate well with a CN from 40–55.

Chemical decontamination

These are measures taken by environmental safety officers to reduce chemical contamination by removing chemical and radiological materials from humans, equipment or a site.

Chronic Obstructive Pulmonary Diseases = COPD

A group of lung diseases including emphysema, chronic bronchitis and bronchiectasis that can cause obstruction of the airways in humans.

CIGS = Copper-Indium-Gallium-Selenide

A compound made of elements copper, indium, gallium and selenide used in thin-film photovoltaic cells.

Clathrate hydrate

A crystalline water-based solid similar to ice in physical appearance, forming a cage of frozen water molecules with hydrogen bonds, inside which a small non-polar molecules made up of gases or polar molecules with large hydrophobic moieties are encapsulated.

Clean Air Act

A U.S. federal law enacted by the U.S. Congress and signed by President Richard Nixon on 31 December 1970 for air pollution control at the national level.

Clean Air Interstate Rule = CAIR

Cloud point

The temperature at which the first solid crystals of paraffin hydrocarbons in liquid fuels are observed, when cooled under controlled conditions.

Coal-bed gas recovery

The recovery of methane gas from coal beds using various methods such as pressurized carbon dioxide.

Coefficient of performance = COP

The ratio of cooling output to power input in refrigeration and air condition systems.

Combined gasification and biodegradation process

A technology for conversion of syngas into pure ethanol (or pure butanol) using microorganisms for fermentation of the gas.

Competitive American Tails Upgrade Partnership = CATUP

Compressed Air Energy Storage = CAES

Concentrating Solar Photothermal = CSP

Solar technology that converts radiation energy of the sun into useful thermal energy, which can be used directly or converted into electrical power.

Concrete curing

A technique used for strengthening and hardening of freshly poured concrete over a period of time.

Control rod

Used in nuclear reactors to control the fission rate of uranium. It absorbs neutrons and thus, prevents chemical chain reaction.

COP = Coefficient of performance

COPD = Chronic Obstructive Pulmonary Diseases

Co-product credit

Monetary value assigned to high value by-products obtained from a plant besides the main product.

Corn stover

Leftover on the field after harvesting corn. It includes stalks, husk, and leaves of corn plants.

Cross-State Air Pollution Rule = CSAPR

A rule finalized by the U.S. EPA on 6 July 2011, for protection of the health of millions of Americans by helping states to reduce air pollution and attain clean air standards. The CSAPR requires states to significantly improve air quality through the reduction of power plant emissions which contribute to ozone and/or fine particle pollution in other states.

CSARP = Cross-State Air Pollution Rule

c-Si = Crystalline silicon

A Silicon material that has orderly atomic arrangement.

CSP = Concentrating Solar Photothermal

CTA = Chicago Transit Authority

CTL = Coal-to-liquid

Also known as coal liquefaction, it is a process of obtaining liquid fuel from coal.

Cushion gas = base gas

The minimum volume of gas, which must be permanently maintained in gas storage reservoirs to provide adequate pressure and deliverability rates during the withdrawal season.

Darcy

A unit of permeability. A porous substance with permeability of one darcy allows flow of a fluid (gas or liquid) with viscosity of 1centipose to pass through a 1 cm^2 area at the rate of 1 cm^3/sec, when the pressure gradient is 1 atm/cm.

DDGS = Dried Distillers Grains with Solubles

The dry-grind ethanol plant by-product used as a lower-cost feed ingredient for beef, dairy, poultry and swine industry.

Dedicated energy crops

Agricultural crops specifically cultivated and grown for biofuels production.

Demand response program

This is an energy efficiency program which incentivizes electricity consumers to reduce their usage when the reliability of the electric grid is in jeopardy during peak load periods.

Depleted uranium

Uranium ore or compound containing a lower concentration of the fissile isotope (uranium-235) than natural uranium.

Diamondoids

Cage saturated hydrocarbons that are major molecular building blocks for nanotechnology.

Dienes = Diolefins

Hydrocarbons containing two carbon double bonds.

DOE = Department of Energy (of the U.S.)

Distributed generation system

Small-scale power generation technology that generates power on-site or close to the end-users of power, as oppose to central power generation, thus eliminating the cost, complexity and inefficiencies associated with transmission.

Dried Distillers Grains with Solubles = DDGS

DSAR = Defueled Safety Analysis Report

DSSC = Dye-sensitized solar cell

Dual flash cycle

A geothermal power generating plant using high and low pressure flash tank to produce high and low pressure steam to run various kinds of turbines.

ECBM = Enhanced Coal-Bed Methane

A method to increase methane production from unmineable coal seams by injecting nitrogen or carbon dioxide gas to a coal mine to extract methane from the surface of coal.

ECRA = Energy Conservation Reauthorization Act

EDV = Electric Drive Vehicle

EER = Energy Efficiency Ratio

The ratio of energy output to energy input in an energy conversion process.

EGS = Enhanced geothermal system

A geothermal system with enhanced steam production by injecting water into a well and contacting the water with hot rocks to generate steam to produce power.

EISA = Energy Independence and Security Act

Electromagnetic uranium enrichment

A separation technology using magnetic force to separate ions of the two isotopes. This is based on the principle that ions with the same energy and electrical charge, but different masses have different trajectories when placed in an electromagnetic field.

Eminent Domain

The power of a state to acquire private property without prior consent of the owner, but with due compensation.

Emission banking and trading

A cap-and-trade system used to set the maximum allowable limit or cap on total emissions by environmental polluters. The cap represents the total number of allowances or permits which is distributed among a group of polluters. The number of allowances or permits held by a polluter determines the limit of pollution allowed by the polluter.

Emphysema

A chronic obstructive pulmonary disease (COPD) characterized by damage to the air sacs (alveoli) in the lungs, making it difficult for the body to get needed oxygen.

Energy efficiency program

A statewide program operated by the electric utility companies which allows judicious use of energy for equipment, buildings, or products without altering the quality of the services provided by these equipment.

Energy Efficiency Ratio = EER

Energy Improvement and Extension Act of 2008 (H.R. 1424)

Energy bill passed by the U.S. Congress in 2008 to provide tax credit of 30% (caped at $2000 prior to 2009) to homeowners who choose to utilize renewable energy technologies.

Enhanced geothermal system = EGS

Enhanced oil recovery = EOR

A technique which employs detergents or high pressure fluids, such as natural gas, nitrogen or carbon dioxide to obtain improved recovery rates from depleted and/or low pressure oil wells.

Enriched uranium

Uranium that has higher proportion (concentration) of ^{235}U than natural uranium ore.

EOR = Enhanced oil recovery.

EPA = Environmental Protection Agency (of the U.S.).

Epigenetic veins

A distinct sheetlike body of crystallized minerals found within a rock that were formed much later than the rocks that enclose them.

ERPI = Electric Power Research Institute

Ethanol

CH_3-CH_2-OH, ethyl alcohol, grain alcohol.

EVCS = Electric vehicle charging station.

EVSE = Electric Vehicle Supply Equipment

Equipment necessary to deliver electricity to an electric vehicle and include conductors, the electric vehicle connectors, attachment plugs and fittings, devices, power outlets, etc.

FAME = Fatty Acid Methyl Esters = biodiesel.

FASOM = Forestry and Agriculture Sector Optimization Model

A dynamic, nonlinear optimization model of the forest and agricultural sectors in the U.S. developed for the U.S. EPA.

Fast neutron reactor

Nuclear reactors designed to use fast neutrons for sustained fission reactions. These reactors do not need moderators and will allow nuclear fuels to be bred from almost all the actinides.

Federal Energy Regulatory Commission (FERC)

The U.S. federal agency with jurisdiction over hydroelectric licensing, wholesale electric rates, interstate electricity sales, oil pipeline rates, and natural gas pricing. Additionally, FERC reviews and authorizes liquefied natural gas (LNG) terminals projects, non-federal hydropower projects and interstate natural gas pipelines projects.

FERC = Federal Energy Regulatory Commission.

Fertile nuclear material

A material that cannot sustain a nuclear fission chain reaction by itself, but can be converted into a fissile material by neutron absorption.

Fibrosis disease

A disease characterized by the formation of excess fibrous connective tissue in an organ or tissue (in a reparative or reactive process), which can be benign or in a pathological state.

First generation biofuels

Biofuels derived from feedstock that are unsustainable in long-term. They include food crops, such as soybeans and corn grains.

Fischer-Tropsch Synthesis = FTS

A chemical process that converts the mixture of hydrogen and carbon monoxide into liquid hydrocarbons.

Fissile Material

Material that can sustain a nuclear fission chain reaction and can be used as a nuclear fuel.

Fission reaction

A type of nuclear reaction in which the nucleus of a larger atom is bombarded by a neutron and splits into two or more smaller nuclei to release energy and more neutrons.

Flash point

The lowest temperature at which the vapor generated above a fuel will ignite on exposure to open flame.

Fluidized bed combustion technology

A combustion technology where solid fuel particles are suspended and circulating due to an upward gas flow in the reactor.

Flywheels

A rotating mechanical device used for energy storage. The energy from intermittent sources can be transferred to a flywheel to rotate the spinning mass over time. When needed, a flywheel can deliver continuous supply of power.

Fossil fuel

Refers to all underground hydrocarbon resources which include natural gas, gas condensate (NGL), petroleum, heavy oil, tar sand, oil shale and coal.

FTS = Fischer-Tropsch Synthesis.

Fuel cell

An electrochemical device that converts chemical energy of a fuel directly into electricity.

Fusion reaction

A nuclear reaction where two or more nuclei of small atoms, like hydrogen, collide and combine to form a new nucleus and release large amount of energy.

FutureGen 2.0

Clean coal technology project supported by the U.S. DOE and industrial partners. The project was based on oxy-combustion technology, which was supposed to upgrade an existing power plant in Meredosia, Illinois to capture approximately 1.3 million metric tons of carbon dioxide each year.

Gas-condensate = natural gas condensate = condensate = NGL

A mixture of low molecular weight hydrocarbons (starting with propane), which separates (in liquid state) from natural gas during well-head gas production.

Gaseous diffusion enrichment

A uranium enrichment technology where gaseous uranium hexafluoride (UF_6) is forced through semi-permeable membranes to separate ^{235}U isotope from ^{238}U.

Gasification Combined Cycle technology = GCC

Clean coal technology which combines the conversion of coal into pipeline quality-natural-gas with electricity generation using some of the gas produced.

GCC = Gasification Combined Cycle technology.

gCO_2-eq = Gram Carbon Dioxide Equivalent

A measure of the effect of a specific greenhouse gas on global warming, using the same amount of carbon dioxide as a reference.

GDP = Gross Domestic Product

The market value of all the final goods and services produced within a country over a given time period.

Geothermal Heat Pumps = GHP = ground source heat pump

A heat transfer system that utilizes the difference in ground and atmospheric temperatures for heating and cooling of buildings.

gge = gallon of gasoline equivalent

GHG = Greenhouse Gas

Tri-atomic and larger gas molecules in the atmosphere (CO_2, CH_4, etc.) able to trap (store) heat in the atmosphere and as a result, contribute to global warming.

GHP = Geothermal Heat Pump.

Green diesel = hydrogenation-derived renewable diesel (HDRD)

A type of biofuel produced by hydroprocessing of fats or vegetable oils through a catalytic reaction. It is used either directly or blended with petroleum diesel.

Greenhouse Gas = GHG.

GSHP = Ground Source Heat Pump.

Half-life

The time required for a radioactive material to decay to half of its original radioactivity.

Hard coal = Anthracite.

HAWT = Horizontal-Axis Wind Turbine

A type of wind turbine designed with the axis of rotation parallel to the direction of wind flow.

HDR = Hot Dry Rock (energy resource)

A geothermal energy source that is devoid of fluids.

HDRD = Hydrogenation-Derived Renewable Diesel = green diesel = 2nd generation biodiesel

The difference between HDRD and biodiesel is addition of hydrogenation step to remove oxygen and nitrogen from the fuel.

Heating value

Total amount of heat released from combustion of specific amount of a fuel, which is equivalent to the enthalpy of combustion. If water is present in the fuel and after combustion is recovered as liquid, then the value is referred to as higher heating value (HHV). If water is present in the fuel and after combustion is recovered as vapor, then the value is referred to as lower heating value (LHV). Heating values are typically reported at room temperature and pressure (25°C, 1 atm or 77°F, 1 bar).

Heavy water reactor

A type of nuclear fission reactor that uses heavy water (the one that has larger amount of hydrogen isotope deuterium) as coolant.

HHV = Higher Heating Value (see Heating Value).

Horizontal closed loop geothermal heat pump (GHP) system

The most commonly used geothermal heat pump system suitable for small residential buildings which consists of pipes laid horizontally to the ground.

HVAC = heating, ventilation and air-conditioning

Hydraulic fracturing = "Fracking"

A technology which allows the production of natural gas and oil from subterranean natural reservoirs, generally at a depth of 5,000–20,000 feet below the surface of the earth. The hydraulic fracture is created by pressurizing fracturing fluid through the wellbore into the geologic formation.

Hydroprocessing

Chemical processes that include hydrogenation, hydrocracking, or hydrotreating.

Hydrothermal energy resources

These are subterranean geothermal reservoirs where steam or hot water are formed by water seeping into the earth and being heated by fractured or porous hot rock.

Hydrothermal liquefaction process

A thermochemical process for the synthesis of biofuels based on application of heat (250–350°C) and pressure to the biomass.

I.C. = Internal combustion

ICC = Illinois Commerce Commission

IDCEO = Illinois Department of Commerce and Economic Opportunity

IDNR = Illinois Department of Natural Resources

IDPH = Illinois Department of Public Health

IEEE = Institute of Electrical and Electronics Engineers

IEPA = Illinois Environmental Protection Agency

IGCC= Integrated Gasification Combined Cycle

Clean coal technology for conversion of coal into synthesis gases (syngas) at high temperature and pressure, while combusting the gas to produce electricity with steam turbine.

Illinois Clean Coal Portfolio Standard

A law under the Public Act 095-1027, requiring emissions from coal-fired power generating plants not to be higher than those from combined-cycle, natural gas-fired plants. The Bill also mandates the coal power plants to capture at least 50% of total CO_2 emissions.

Illinois Power Agency = IPA

An independent agency established in 2007 by the Illinois Public Act 95-0481, subject to the oversight of an Executive Ethics Commission. Among its other duties, the IPA ensures that the process of power procurement in the state of Illinois is conducted in an ethical and transparent manner.

Illinois RPS = Illinois Renewable (energy) Portfolio Standards

A statewide law enacted in 2007, requiring Illinois utility companies to have a certain percentage of the electricity they supply to customers from renewable sources, such as wind energy, solar energy and biofuels.

Inertial confinement fusion reactor

An experimental nuclear fusion reactor, which uses lasers to compress plasma to obtain high temperatures and densities for fusion to take place.

International Thermonuclear Experimental Reactor = ITER

An international project for design and construction of the first nuclear fusion reactor.

Investor-Owned Utility company = IOU

Private business organization which provides services or products to consumers.

IOGA = Illinois Oil and Gas Association.

Ion transport membrane = ITM

A gas separation technology using ceramic membranes to separate an ionized gas.

IOU = Investor-Owned Utility company = public utility company

IPA = Illinois Power Agency.

IPCC = Intergovernmental Panel on Climate Change

ISFSI = Independent Spent Fuel Storage Installations

Facilities constructed for temporary storage of nuclear spent fuel and other radioactive materials.

Isotopes

Atoms of the same element having the same number of electrons and protons, but with different number of neutrons.

ITER = International Thermonuclear Experimental Reactor

ITM = Ion transport membrane

JCESR = Joint Center for Energy Storage Research

A consortium of five DOE national laboratories, five universities and four private companies and led by Argonne National Laboratory, focusing on efforts to develop advanced battery and energy storage technologies.

Joint Center for Energy Storage Research = JCESR.

Joule

The basic energy unit in the metric system which is equal to the work done by a force of one Newton [N], when its point of application moves one meter in the direction of action of the force.

Laser uranium enrichment

A type of uranium enrichment technology based on isotopic separation of uranium, using photoexcitation principle (i.e., exciting the molecules of uranium hexafluoride using laser light).

LCA = Life Cycle Analysis

A method to examine environmental impacts of a product throughout its life cycle; from acquisition of raw material to management of waste at the end of product's life.

LCOE = Levelized cost of energy

The total cost per unit of energy over a predetermined financial life that makes the investment to break even.

Leadership in Energy and Environmental Design = LEED

A set of rating standards to which the design, construction, operation and maintenance of green buildings, homes and neighborhoods are compared.

LEED = Leadership in Energy and Environmental Design.

Levelized cost of energy = LCOE.

LGTE = **LFGE** = Landfill-Gas-To Energy

The gas produced in a landfill as a result of decomposition of organic materials, which can be used as a source of energy.

LHV = Lower Heating Value (see Heating Value).

Life Cycle Analysis = LCA.

Lignite

A type of coal with the lowest energy content among the four types of coal (bituminous, Sub bituminous, anthracite and lignite) and with carbon content ranging from 25–35%.

Li-ion = Lithium ion

A type of rechargeable battery that uses lithium compound as a positive electrode.

Liquefied Natural Gas = LNG

Methane gas that is supercooled and liquefied. Condensation temperature of LNG depends on composition of natural gas and varies between (−120) to (−170°C) [−184 to −274°F].

Liquefied Petroleum Gas = LPG

Hydrocarbon gases produced as by-products of natural gas or petroleum refinery that are used as fuel and can be liquefied at relatively low pressures. Propane, butane and propylene are examples of LPGs.

LMOP = Landfill Methane Outreach Program.

Load leveling

The smoothing of transient fluctuations in power lines due to variability in usage, differences in power transmission efficiency and other small factors.

Low impact crops

Agronomic crops that can be produced sustainably over a wide range of geographical area and at an economically advantageous price over their counterparts.

Low-volatile organic compounds

A term used to describe a product with volatile organic compounds (VOC) content at or below 150 g/L.

LPG = Liquefied Petroleum Gas.

MADRI = Mid-Atlantic Demand Resource Initiative.

Magnetic confinement fusion reactor

A type of experimental nuclear fusion reactor technology that uses magnetic fields (i.e., magnetic materials and a magnetic influence of electric currents) to confine hot fusion fuel in the form of plasma.

Magnetohydrodynamics (MHD)

An electrical power generation process where movement of conducting plasma through a magnetic field creates a voltage.

Matrix permeability

This is defined as the ability of fluids to diffuse through rocks mainly through the interstices between the mineral grains forming the rock, instead of passing through the natural fractures inside the rock.

MATS = Mercury and Air Toxics Standards.

MBB = Molecular Building Block

Small molecules used in the design and assembly of larger molecules (macromolecules).

Metal Organometallic frameworks (MOF)

Metal/oxide groups linked together by organic struts that are synthetic, crystalline and microporous in nature.

Metathesis

A bimolecular process which involves an exchange of bonds between two reacting chemical species.

MGSC = Midwest Geological Sequestration Consortium

Microdarcy = 10^{-6} Darcy.

Midwest Geological Sequestration Consortium = MGSC

Launched in 2003 as Phase I of the U.S. DOE Regional Carbon Sequestration Partnership and was continued in 2005 as Phase II effort. Its objective is to study the geological carbon sequestration potential in the Illinois Basin shared by Illinois, Kentucky and Indiana.

Mineral carbonation

Carbon dioxide sequestration technology for converting CO_2 into stable carbonate minerals (including calcite $CaCO_3$, dolomite $CaMg(CO_3)_2$, magnesite $MgCO_3$ and siderite $FeCO_3$) for storage.

Miscanthus giganteus

A tall hybrid perennial grass native to Japan, which is currently used in the European Union as a commercial energy crop.

MISO = Midwest Independent System Operator.

Mixed oxide fuel = MOX

A type of nuclear fuel containing more than one oxide of fissile material, such as plutonium blended with natural uranium, reprocessed uranium, or depleted uranium.

Moderator

Material used in the core of nuclear reactors to slow down the neutrons released from the fission reaction and thus, control the rate of nuclear reaction.

Monazite

A type of phosphate mineral with reddish-brown color containing thorium and rare earth elements such as cerium (Ce), lanthanum (La) and neodymium (Nd).

MOX = Mixed Oxide Fuel.

Nanofabrication

A technology used for manufacturing of devices and systems using material components with sizes not exceeding 100 nm.

National Institute for Occupational Safety and Health = NIOSH

A U.S. Federal Government agency and a part of the Centers for Disease Control and Prevention (CDC), conducting research and making recommendations to ensure safety and health for people in workplace.

Natural Gas Combined Cycle = NGCC

Combining multiple thermodynamics cycles in order to increase efficiency in a natural gas-fired power generating facility.

Natural gas liquids = NGL = gas-condensate = natural gas condensate = condensate.

NCG = Non-Condensable Gases.

Net metering

A statewide policy purported to boast renewable energy production by providing retail credit to small electricity consumers who own renewable energy systems (solar, wind, etc.,) and generate surplus electric energy.

NETL = National Energy Technology Laboratory.

NFCBP = National Fuel Cell Bus Program.

NGCC = Natural Gas Combined Cycle.

NGL = Natural gas liquids = gas-condensate = natural gas condensate = condensate.

NIOSH = National Institute for Occupational Safety and Health.

NOAA = National Oceanic and Atmospheric Administration.

Non-renewable energy sources

Energy sources that cannot be replenished once depleted. Fossil fuels are considered non-renewable.

NREL = National Renewable Energy Laboratory (of DOE).

NSPS = New Source Performance Standards

Air pollution control standards issued by the U.S. EPA.

Nuclear chain reaction

A type of reaction in which atoms of a fissile material can disintegrate after capturing a neutron, followed by release of more neutrons and tremendous amount of thermal energy. The produced neutrons are captured by other atoms of the fissile material, causing further disintegration of atoms and the release of more neutrons and thermal energy.

O&M = operation and maintenance.

Occlusion pattern

The disruption of porosity of rocks due to capillary blocking or shale infiltration caused by disintegration of carbonates into rubbles.

OOIP = Original-oil-in-place

An estimated volume of oil in a natural underground oil reservoir when it was first discovered.

Open-loop GHP system

Geothermal heat pump system, which utilizes wells or surface body water as heat exchange fluid that circulates directly through the GHP system.

Organic richness

Availability of organic matter in sedimentary rocks.

Original-oil-in-place = OOIP.

ORNL = Oak Ridge National Laboratory.

Oxy-combustion technology

Clean coal technology, which separates oxygen from air and combines the oxygen with coal for combustion to occur in a mixture of carbon dioxide and oxygen medium. The resulting flue-gas has about 70–90% carbon dioxide (on dry basis) and negligible nitrogen oxides.

p.u. = porosity unit.

PACE = Property Assessed Clean Energy.

Particulate matter = PM

A mixture of very small and hazardous solid particles and liquid droplets in the air.

Passivhaus

Voluntary standards for construction of energy-efficient buildings in order to reduce energy requirement for heating/cooling by 90%.

PCM = phase change materials.

Peak load power plant

Smaller-size electric power plants that are quickly brought online to respond to electricity demand during peak load periods.

Pennycress

A member of the mustard family of plants and a potential biofuel plant capable of producing seeds containing about 20–36% of oil, with 38.1% of erucic acid as the major fatty acid.

Permanent Magnet generator = PM generator

Generator that uses permanent magnet instead of coil to convert mechanical energy to electrical energy.

Phase Change Material = PCM

Thermal energy storage material taking advantage mostly of its latent heat and partly of its sensible heat properties.

PHEV = Plug-in Hybrid and Electric Vehicles.

Photovoltaic (PV) energy density

Electrical output of a solar photovoltaic cell per its unit area.

PHS = Pumped Hydro Storage

Energy storage technology that produces electricity to meet high peak demands by using low price off-peak electricity from the grid for pumping water between two water reservoirs at different elevations.

PJM Interconnection

A regional power transmission company in charge of the electric grid for ~54 million customers in 13 states and the District of Columbia in the U.S, including the ComEd region of Illinois.

Plasma state

The state of matter at very high temperatures, when molecules of the substance are converted into a collection of positive ions and electrons.

PM = Particulate Matter, Permanent Magnet.

PMG = Permanent Magnet Generator.

Pneumoconiosis

Occupational and restrictive lung diseases usually caused by the inhalation of dust in mining areas.

Pond and lake loop GHP system

Geothermal heat pump system utilizing a body of surface water such as a pond or a lake as heat transfer fluid for heating and air-conditioning.

porosity unit = p.u.

Percentage of pore space in a unit volume of a porous material.

Post combustion capture technologies

Technologies used for the removal of contaminants from fossil fuels after burning the fuel.

Pour point

The temperature at which a liquid mixture (such as diesel oil) behaves like a gel and can no longer be poured from a container, or be pumped through a tube.

Prairie State Energy Campus = PSEC

Located in Washington County, Illinois, the commercial power generating facility includes a coal-fired power plant and a coal mine.

Pre-combustion capture technology

A technology for the removal of contaminants from fossil fuels (sulfur compounds, etc.) before the fuel is burnt.

Proppant

Small solid materials used in hydraulic fracturing (fracking) and pumped into the ground to keep the pores of a rock open for flow of natural gas or oil.

PSEC = Prairie State Energy Campus.

PT = Photothermal.

Pulse-Width Modulation = PWM

A method to generate analog signals, using a digital source to control power applications.

PUREX = Plutonium and Uranium Recovery by Extraction Technology

An aqueous nuclear reprocessing method based on liquid-liquid extraction ion exchange for the recovery of uranium and plutonium from spent / used nuclear fuel.

PV = Photovoltaic.

PWM = Pulse-Width Modulation.

PWR = Pressurized water reactor

A nuclear power plant with a primary coolant (water) pumped under high pressure to the reactor core, where the water is heated by the fission reaction heat generation.

Pyrolysis

A thermochemical process for decomposition of organic compounds in absence of oxygen.

Radioisotope = Radionuclide.

Radionuclide = Radioisotope

A product of nuclear fission reaction. It refers to an atom, which nucleus has excess energy to make it unstable. This energy can be transported to an electron in the atom or to a radiation particle created within the nucleus.

Renewable energy

Alternative forms of energy derived from sustainable energy sources such as biomass, geothermal, solar or wind.

Residual fuel oil = heavy fuel oil

Viscous fuel that is extracted from the bottom of a refinery's fractionation unit with possibility of high-sulfur content.

RTO = Regional Transmission Organization.

SCWR = Supercritical Water-cooled Reactor

A high pressure and temperature water-cooled reactor operating above the thermodynamic critical point of water (374°C, 22 MPa), in order to achieve a thermal efficiency of approximately 44% or one third higher than the current light water reactors on which the design is based.

Second generation biofuel = Advanced biofuel

Biofuel derived from lignocellulosic feedstock, such as agriculture residue and dedicated energy crops.

Seismic acoustic fracture mapping

A technique used for accurate measurement of the hypocenter of acoustic emissions, which is the result of stress changes in the rock matrix from injection of fluids, gas, proppant or other materials, during hydraulic fracturing.

Sensible heat thermal energy storage

Energy storage based on the specific heat capacity of the storage medium, including the mass and temperature difference of the medium.

Shale gas

Natural gas found in earth's shale formation.

Simultaneous fracturing

A technology for hydraulic fracturing of parallel sections of a reservoir with two horizontal wells, typically about 1,000 feet apart at the same depth.

Solar electric footprint

The land area needed to supply all end-use electricity derived from solar photovoltaics.

Solar Photothermal (PT) collector

A device used for direct conversion of solar radiation into heat.

Solar Photovoltaic (PV) cell

A device used for direct conversion of solar radiation into electricity.

Spent fuel reprocessing

A technology for separation and recovery of fissionable plutonium and other radioisotopes from irradiated nuclear fuel.

Subbituminous coal

Coal rank that has the second lowest energy content after lignite and containing 35–45% carbon.

Subcritical

Refers to a fluid condition (like water) below its thermodynamic critical point (water critical point temperature is 373.95^{o}C (705°F) and its critical pressure is 220.5 bar (3200 psia).

Substitute natural gas = SNG = Synthetic Natural Gas = Syngas

An artificially produced natural gas that can be obtained from gasification of coal, biomass, petroleum coke, or organic solid waste.

Sulfur, Organic

Sulfur content in coal that is in the structure of organic matter in coal.

Sulfur, Pyritic

Pyritic sulfur or pyrite (FeS_2) is an inorganic type and, usually, the dominant sulfur compound found in coal.

Supercapacitor

A high-capacity energy storage that works by using static charge instead of an electrochemical reaction.

Supercritical

Refers to a fluid condition (like water) above its thermodynamic critical point (water critical point temperature is 373.95^{o}C (705°F) and its critical pressure is 220.5 bar (3200 psia).

Sustainable energy

An energy source that is undepletable and environmentally friendly.

Switchgrass

Large prairie grass native to the U.S. Midwestern region, which can be transformed into cellulosic biofuels.

Syngas = Synthetic Natural Gas = SNG = Substitute natural gas

An abbreviation of Synthetic Gas, and is obtained from gasification or pyrolysis of carbonaceous materials. It is a gas mixture comprising mainly of carbon dioxide, carbon monoxide and hydrogen.

TES = Thermal Energy Storage.

TFS = Thin film semiconductor.

Thermal maturation

Time-temperature dependency transformation of sedimentary organic matter to mobile hydrocarbons.

Third generation biofuel = Advanced biofuel

Biofuels obtained from algae.

Tillage

A method of soil cultivation.

Tiltmeter

An instrument for the measurement of very small changes from the vertical level.

Tip speed ratio

The ratio of the speed of the tips of the wind turbine blades to wind speed.

TOC= Total organic carbon.

Tokamak

A system that uses magnetic confinement of plasma to increase the temperature of plasma necessary for self-sustaining fusion reaction.

Transesterification

A class of chemical reactions in which ester-to-ester transformation is done via interchange of alkoxy functional groups.

Transmutation

In the context of nuclear chemistry, this refers to the conversion of one chemical element (or isotope) into another.

Transuranic isotopes

These are isotopes with atomic numbers greater than that of uranium. These isotopes are unstable and decay radioactively into other isotopes.

Trombé wall

Also referred to as solar wall, this is a dense and heat absorbing wall built to store thermal energy of the sun during the day and slowly release the heat during the night.

UCG = Underground Coal Gasification

A process that uses the same chemical reactions as conventional gasification plants for conversion of coal, buried deep underground, into synthesis gas (syngas).

UF_6 = Uranium hexafluoride

A white compound that is solid at room temperature and pressure and consists of one atom of uranium and six atoms of fluorine, used in the enrichment of uranium.

ULSD = Ultra-Low Sulfur Diesel

Diesel fuel with substantially lowered sulfur content (15 ppm in U.S.).

Uranium enrichment

A process that increases the percentage of uranium 235 (^{235}U) in natural uranium through isotope separation.

USC = Ultra-supercritical

Power plants designed to operate at supercritical pressures (above 240 bars) and supercritical temperatures (above 600°C) of water.

USEC= United States Enrichment Corporation

Currently under the name Centrus Energy Corp.: this is a corporation that produces enriched uranium for the U.S. Government.

USGS=Unites States Geological Survey

UTES = Underground thermal energy storage

A concept of energy storage that stores excess heat during summer and extract the heat during winter.

Vertical closed loop GHP systems

Geothermal heat pump systems, requiring vertical boreholes of about 10 cm in diameter and are recommended for large commercial buildings and schools, where the land area requirement for horizontal loops would be expensive.

Vertical-axis wind turbine

A type of wind turbine designed with the axis of rotation of rotor perpendicular to wind current.

Vitrinite

One of the main organic constituents of coal and sedimentary rocks.

Vitrinite reflectance

A method to determine coal rank and shale thermal maturity by measuring the percentage of incident light reflected from the surface of vitrinite particles.

VOC = Volatile Organic Compound

Organic compound that have high vapor pressure at room temperature.

VRB = Vanadium Redox Flow Battery

A type of rechargeable battery that uses two different liquid vanadium solutions to store electricity.

Water-gas shift reaction

A reversible exothermic chemical reaction in which carbon monoxide reacts with steam to produce a mixture of hydrogen and carbon dioxide.

Wound-field synchronous generator

A device used for conversion of electromagnetic energy to mechanical energy (or vice versa), where the frequency is related to mechanical speed and the number of poles.

Yellowcake

Uranium concentrate powder derived from leach solutions, which serves as an intermediate step in the processing of uranium ores.

Notations

A	surface area
Å	Angstrom (10^{-10} meters)
bbl	barrel
bpm	barrel per minute
Bscf	billion standard cubic feet
Btu	British thermal unit
C	specific heat = heat capacity
C	speed of light
CNG	compressed natural gas
COP	coefficient of performance
e	electron
E	energy
g	gravitational acceleration
gpm	gallons per minute
Gt	giga tons (10^9 ton)
GW	gigawatts (10^9 watt)
GWh	gigawatt hour
h	specific enthalpy
ha	hectare
hph	horsepower hour
I	solar irradiance [W/m^2]
I.E.	intrinsic energy

km^2	kilometer square
kW	kilowatts (10^3 watt)
kWh	kilowatt hour
K.E.	kinetic energy
L	latent heat of phase transition (i.e., melting, etc.)
LDV	light duty vehicle
LFG	landfill gas
LL	lower limit
m	mass
Mha	million hectare
MJ/kgDM	megajoule per kilogram dry mass
MMbpd	million barrels per day
MMBtu	million Btu
MMscf	million standard cubic feet
MMT	million metric tons
MOF	metal-organic frameworks
MP	microporous
MTDM	metric tons of dry mass
MW	megawatts (10^6 watt)
MWe	megawatts electrical
MWh	megawatt hour
MWt	megawatts thermal
nmi	nautical mile (1.15 mile)
NCG	non-condensable gases
P	pressure
PCM	phase change materials
P.E.	potential energy
Q	heat
Quad	quadrillion (10^{15})
s	specific entropy
SD	standard deviation
tDM	total dry matter
tU	ton of Uranium

T	absolute temperature
TWe	terawatt electricity
scf	standard cubic feet
Tscf	trillion standard cubic feet
U	total (overall) heat loss
UL	upper limit
USD	U.S. Dollar
x	distance, length
X	mass fraction
w	work
z	elevation/height of the system with respect to the frame of reference

Subscripts

1	initial state
2	final state
a	aperture
A	absorber
C	cold
col	collector (solar)
DC	direct current (electricity)
f	final
G	generator
H	hot
i	initial state
in	inlet
I	initial
l	liquid
m	melting
opt	optical
p	pump

P pressure
R refrigerant, refrigeration
s solid
t throttle, total

Superscripts

- rate = d/dt (like $\dot{x}$ = dx/dt = velocity; $\dot{m}$ = dm/dt = mass flow rate)

Greek Letters

ρ density
η efficiency
Σ summation

Index